THE MOLECULAR DESIGNING
of
MATERIALS AND DEVICES

THE MOLECULAR DESIGNING
of
MATERIALS AND DEVICES

ARTHUR R. VON HIPPEL, EDITOR

Contributors:

ALI S. ARGON — DAVID J. EPSTEIN — JOHN B. GOODENOUGH — SHINYA INOUÉ — THOMAS A. KAPLAN — HENDRIK A. KLASENS — HELEN D. MEGAW — E. W. MÜLLER — HEINZ RAETHER — GERT W. RATHENAU — ALBERT ROSE — MURRAY D. ROSENBERG — WALTER A. ROSENBLITH — HEINZ M. SCHLICKE — FRANCIS O. SCHMITT — JAMES H. SCHULMAN — J. C. SLATER — A. SMAKULA — EDWARD V. SOMERS — S. DONALD STOOKEY — MORRIS TANENBAUM — ALFRED VON ENGEL — ARTHUR VON HIPPEL — CARL WAGNER — HERBERT J. ZEIGER — CLARENCE ZENER

THE M.I.T. PRESS
Massachusetts Institute of Technology
Cambridge, Massachusetts

COPYRIGHT © 1965
BY
THE MASSACHUSETTS INSTITUTE OF TECHNOLOGY

*All rights reserved. This book may not be reproduced,
in whole or in part, in any form (except by reviewers for the public press),
without written permission from the publishers.*

*Some of the material contained in this book was copyrighted by
The Massachusetts Institute of Technology in 1963 under the same title.*

Library of Congress Catalog Card Number 65-22436

PRINTED IN THE UNITED STATES OF AMERICA

PREFACE

Any new approach, once accepted, passes the danger zone of exaggerated expectations. This is the situation for Molecular Science and Molecular Engineering today. The isolation of specialists is ending; all science and engineering begins to be drawn into mutual alliance on the basis of molecular understanding; interdepartmental research centers transform the universities. Since nature designs everything from atoms, we should be able to create with foresight any feasible kind of material and device if we understood the Periodic System in all its implications. Yet—like weather forecasters—we still find ourselves members of the gambling profession. There is no philosopher's stone, only laborious advance; and the number of variables is endless.

What then is the present status of our insight and capabilities? In search for an honest appraisal, the Laboratory for Insulation Research sponsored in 1963 a summer-session course at the Massachusetts Institute of Technology: "The Molecular Designing of Materials and Devices." Its program ranged from the electronic structure of atoms and molecules to the design patterns and operating mechanisms of living systems. Eighty-two registered participants and thirty-seven lecturers from universities, government, and industry laboratories gathered for two weeks of learning and debating—a strenuous experience, but so full of interest and enjoyment that even classical methods assumed a new luster of adventure. In the hope of capturing some of this spirit, the majority of the lecturers voted for a joint publication of their contributions. Thus this "Volume IV" came to be added to the three preceding ones on modern materials research (Volume I, *Dielectrics and Waves*, 1954; Vol. II, *Dielectric Materials and Applications*, 1954; Vol. III, *Molecular Science and Molecular Engineering*, 1959).

Since the participants came from a gamut of professions, course 6.64S was threatened by a confusion of tongues. But chaos was avoided by insisting that every lecturer express his concepts in graphic language before mathematical formulation sharpened or beclouded the issue. This approach the book has tried to preserve; most papers have been rewritten, and the authors have allowed the editor freedom to condense and link. A comprehensive Survey at the outset and a Subject Index at the end should help the cause of unity and long-range usefulness —goals so hard to achieve in such a diversified enterprise.

Course 6.64S was an attempt to draw outstanding specialists as lecturers and listeners into an atmosphere of synthesizing. The book is dedicated to the participants in this effort and the community of scholars for which they stand.

Grateful acknowledgment is due to a number of co-workers: Mr. J. Stein with Miss Aina Sils readied the manuscripts for the printer. Our artist, Mr. J. J. Mara, drew the major part of the illustrations with his usual perception of the essential.

Messrs. V. Ern, J. M. Knudsen, C. W. Nelson, and R. Nevald were of great help in discussions and the laying-out of figures for Chaps. 2 and 20.

Special thanks go to the sponsors of the Laboratory for Insulation Research: the Office of Naval Research, the Army Signal Corps, and the Air Force, for continuous understanding support over many years; and to Dr. J. A. Stratton, President of M.I.T., for his inspiring interest in this summer session.

Arthur von Hippel

Cambridge, Massachusetts
February, 1965

CONTENTS

Survey . . . . Arthur von Hippel   1
*Institute Professor, Emeritus,
Massachusetts Institute of Technology*

1 · **Predictions by *a priori* Theory** . . . . . . J. C. Slater   7
*Institute Professor,
Massachusetts Institute of Technology*

2 · **Building from Atoms** . . . . Arthur von Hippel   9
Introduction — Atomic orbitals and molecular bonds — Atomic orbitals and electron states in crystals — Atomic structure and the boiling points of the elements — The melting points of the elements — Dense-packed structures — Compound formation and polarity — Insulators and conductors

3 · **Competitive Structures and Phase Transitions** . Helen D. Megaw   29
*Associate Director of the
Crystallography Department,
Cavendish Laboratories,
Cambridge University*

Introduction — Packing structures and framework structures — Feldspars and perovskites as framework structures — Structure families, aristotypes, and heterotypes — Character of transitions — Structure changes and properties of perovskites — Free energy and displacive transitions — Displacive changes in feldspars and their interaction with ordering effects — Twin domains and antiphase domains — Antiphase domains in a feldspar (anorthite) — Summary

4 · **Properties Imparted by *d* Electrons** . . . . John B. Goodenough   42
*Group Leader, Lincoln Laboratory,
Massachusetts Institute of Technology*

Introduction — Some consequences of electron localization — Description of localized d electrons — Hydrogenlike wave functions and cubic field splitting — Tetragonal field splitting and spin-orbit splitting — Applications — Exchange coupling and magnetic order — Collective d electrons and crystallographic distortion — Exchange coupling via collective electrons — Transition metals and their alloys

Contents

5 · Magnetic Ordering in Heisenberg Magnets . . THOMAS A. KAPLAN 54
*Staff Member, Lincoln Laboratory,
Massachusetts Institute of Technology*

The Heisenberg Hamiltonian — The classical ground-state problem — Spiral spin configurations — Variations of spirals in simple lattices — Spirals in spinel lattices — Discussion — Appendix

6 · Phase Transitions in Condensed Systems . . S. DONALD STOOKEY 61
*Manager of Fundamental Chemical Research,
Corning Glass Works*

Introduction — The concept of metastability — Metastable states in one-component systems — Metastable states in polycomponent systems — The Li_2O-Al_2O_3-SiO_2 system — Controlled phase transformations in a photosensitive glass-ceramic — Latent silver image formation — Growth of silver particles — Heterogeneous nucleation of Li_2SiO_3 and change to the equilibrium phase ($Li_2Si_2O_5$ + quartz) — Photochromic action — Biological phase changes — Conclusion

7 · Defects in Ionic Crystals . . . . A. SMAKULA 69
*Director, Crystal Physics Laboratory,
The Center for Materials Science and
Materials Engineering,
Massachusetts Institute of Technology*

Vacancies and interstitials — Color centers and their response to purity and composition — Conductivity and crystal defects — Dislocations — X-ray density versus macrodensity

8 · Defects in Metal Crystals . . . . E. W. MÜLLER 77
*Research Professor of Physics,
Pennsylvania State University*

Introduction — Technique of field-ion microscopy — Observation of point defects — Complex defect structures — Conclusion

9 · Deformation and Fracture of Solids . . . . ALI S. ARGON 84
*Associate Professor of Mechanical Engineering,
Massachusetts Institute of Technology*

Introduction — Elastic deformation — Viscous deformation — Plastic deformation by dislocations — Dislocation mills — Observation of dislocations — Deformation of polymers — Hardening of crystals — Solution hardening — Precipitation hardening — Work hardening — Yield strength of single crystals — Fracture — Conclusions

10 · Structure in Living Systems . . . . . . FRANCIS O. SCHMITT 97
*Institute Professor,
Massachusetts Institute of Technology*

Introduction — Self-replicating polymers: a monumental step in evolutionary advance — The proteins — Molecular assemblies: an essential structural hierarchy — Cell membranes — The molecule↔cell, component↔system reciprocal control — The molecule↔neuron↔brain↔mind problem

Contents

11 · Magnetization Process . . . . DAVID J. EPSTEIN . . . 110
*Professor of Electrical Engineering,
Massachusetts Institute of Technology*

Introduction — The fundamental equation — Magnetization of the infinite crystal — Domains and domain walls — Fine particles — Dynamic excitation — Domain-wall dynamics — Magnetic losses — Diffusion damping

12 · Conduction and Diffusion in Ionic Crystals . . CARL WAGNER . . . 122
*Director, Max-Planck-Institut für
Physikalische Chemie,
Göttingen*

Introduction — Conduction in ionic crystals — Defects in stoichiometric binary crystals — Defects due to nonstoichiometry — Change of conductivity by doping — Conduction in ternary ionic crystals — Self-diffusion — Diffusion in substitutional solid solutions — Diffusion in binary crystals involving a gradient of metal-to-nonmetal ratio — Diffusion in ternary compounds — Transport phenomena with diffusion in different phases — Nonisothermal transport — The use of ionic conductors for thermodynamic measurements — The use of ionic conductors in fuel cells

13 · Luminescence in Solids . . . . JAMES H. SCHULMAN . . . 130
*Head, Dielectrics Branch,
Solid State Division,
U.S. Naval Research Laboratory*

General aspects of luminescence — Luminescence efficiency — Types of activators — Configuration-coordinate diagrams — Temperature dependence of absorption and emission — Derivation of configuration-coordinate diagrams — Polarized luminescence of centers — Systems involving energy transfer — Resonance transfer — Forbidden transitions — Other aspects of energy transfer

14 · Photoconducting Phosphors . . . . . . HENDRIK A. KLASENS . . . 139
*Staff Member,
Philips Research Laboratories,
Eindhoven*

Introduction — Intrinsic and extrinsic luminescence — Activators and capture cross sections — Equilibrium states — Effects of excitation intensity — Quenching of fluorescence — Luminescent centers in ZnS — n-Type conductivity and automatic compensation — Vacancies and center structure — Paramagnetic studies

15 · Electrons and Holes in Semiconductors . . . HERBERT J. ZEIGER . . . 149
*Staff Member, Lincoln Laboratory,
Massachusetts Institute of Technology*

Introduction — Electrons and holes in bands — Landau levels and cyclotron resonance — Excitons — Donor and acceptor impurities — Impurities and excitons in a magnetic field — Interband magneto-optical phenomena — Impurity states and stimulated emission

16 · **Electrons in Metallic Systems** . . . . . . GERT W. RATHENAU 153
Staff Member,
Philips Research Laboratories,
Eindhoven

Metals and energy bands — Properties caused by high Fermi energy of typical metals — The Hume-Rothery rule — Band calculations with one-particle wave functions — Specific heat and magnetoresistance — The de Haas–van Alphen effect and cyclotron resonance — Cohesion — Transition metals — Magnetic interaction through conduction electrons — Transition metal solute atoms in alloys — Exchange fields and Knight shift — Electron correlation in plasma oscillations — Superconductivity — Model for soft superconductors — Occupation function and gap parameter

17 · **Avalanches and Gas Breakdown** . . . . . HEINZ RAETHER 168
Institute for Applied Physics,
Hamburg University

The formation of avalanches — Experimental study of single avalanches — Breakdown conditions — Many-generation and streamer mechanisms — The spark chamber — Verifications of the preceding concepts — Extension to large gap distances and to low or high pressures

18 · **Electric Discharges and Excited Species** . . . ALFRED VON ENGEL 173
Fellow of Keble College,
Clarendon Laboratory,
Oxford University

Ionization in gases — Excited species in discharges — Excitation cross sections and coefficients — Application to discharges — Outlook

19 · **Space-Charge-Limited Currents** . . . . . ALBERT ROSE 178
Physicist, RCA Laboratories

Vacuum capacitor versus solid-state capacitor — Shallow traps and trap densities — Transient currents — Space-charge-limited currents in semiconductors — Performance of photoconductors and solid-state triodes — Double injection

20 · **Conduction and Breakdown** . . . . . . ARTHUR VON HIPPEL 183

Electric strength and external electron supply — Dynamic field distortion and plasma formation — Failure of Paschen's law; emergence of new breakdown conditions — Feature changes in condensed systems — Prebreakdown currents in liquid hexane — Breakdown phenomena in liquids and solids

21 · **The Living Cell** . . . . MURRAY D. ROSENBERG 197
Career Scientist, Rockefeller Institute

Introduction — Cytology — Dimensions of cells — Molecular constituents and their reproduction — Nucleus and cytoplasm — The cytoplasmic matrix — Dynamic activities — The cell surface — Packing and cell dynamics — Cells on mono- and multilayer substrates — Experiments on liquid-liquid interfaces — Possible causes of cellular motion and surface movements — Interfacial dynamics — Outlook

Contents

22 · **Studies of Cell Division with an Improved Polarizing Microscope** . . . . SHINYA INOUÉ 211
*Professor of Cytology,
Dartmouth College*

Spindle fibers and mitosis — Improvement of the polarizing microscope — Verification of spindle-fiber formation

23 · **Mutual Substitution of Networks and Materials** HEINZ M. SCHLICKE 215
*Manager, H-F Laboratories,
Allen-Bradley Company*

Simulation of material properties by network approaches — Specific targets — Distributed parameter effects in high-dielectric-constant ceramics — Suppression of resonances in ceramic interference filters — Transformation properties of simple elements — Quasi materials created by series-parallel feedback — Generalized verters — Design procedures — Parametric transmutation — Conclusions

24 · **New Combinations of Physical Phenomena in Device Design** . . . . MORRIS TANENBAUM 227
*Director, Engineering Research Center,
Western Electric Co.*

Introduction — Light beams as information carriers — Modulation of light beams — The magneto-optical modulator — The electro-optical modulator — The zigzag modulator — Piezoelectric semiconductor devices — Magneto-elastic devices — Conclusion

25 · **Principles of Energy Conversion** . . . . . CLARENCE ZENER 238
*Director of Research Laboratories,
Westinghouse Electric Corporation*

Introduction — Free energy of mixing: sea water versus fresh water — Chemical energy: oxygen-concentration cell — Radiant energy: solar cells — Magnetic sieves — Electronic heat engines — Material limitations on thermionic converters — Material limitations on thermoelectric converters — Nernst-Ettingshausen generator

26 · **Energy-Conversion Devices** . . . . EDWARD V. SOMERS 248
*Manager, Energy Conversion Department,
Westinghouse Electric Corporation*

Introduction — Thermoelectric generators — Thermionic generators — Fuel-cell generators — Magnetohydrodynamic generators

27 · **Sensory Coding in the Nervous System** . . . WALTER A. ROSENBLITH 255
*Professor of Biophysics,
Massachusetts Institute of Technology*

Introduction — Some vital statistics and a rudimentary map of a sensory system — Changes in patterns of neuroelectric activity that may lend themselves to stimulus coding — Conclusion

Subject Index 265

Index of Chemicals 271

SURVEY

Arthur von Hippel

Our book advances its theme in three stages: from the architecture of materials to the properties produced by structures and compositions and to the devices employing such properties with increasing sophistication. Since this vast panorama is conjured by twenty-six contributors with highly individual outlook and emphasis, the reader—like Mr. Escher's Thinker*—might

get lost in puzzled confusion. This survey hastens to his rescue by tracing problems and thoughts as they connect chapter to chapter and by summing up the insight that emerges.

At the outset, the theoretical physicist speaks *in abstracto* (Chap. 1): In principle, the fundamental laws for calculating the structures and properties of matter are known; rapid strides are being made to carry out such calculations; our *a priori* knowledge will steeply increase with the perfection of computers. This sounds discouragingly as if the adventure were finished before it really began, but actually such foresight applies only to simple equilibrium constellations, e.g., the electronic structure of the free atoms.

Thus Chap. 2 inquires: Starting with the atoms of the periodic system as well-known building stones, how do we design? The atomic orbitals in their energy position and stereostructure provide a satisfactory basis for deriving the electron bonds of molecules and the electron bands of crystals. The trends of the melting and boiling points of the elements become understandable. Simple concepts lead from dense-packed elemental structures to polar crystals, to framework structures and molecules, and from insulators to metals. For any specific discussion of aggregation, however, the experimentalist has to supply structure parameters and charge distributions.

At this point the structure analyst takes over (Chap. 3): Frequently the interatomic distances and number of nearest neighbors can be predicted geometrically by assigning experimentally determined radii to atoms and ions and packing them closely, with due regard that all cation bonds directed to one anion (and vice versa) produce electroneutrality. Such "packing structures" of Goldschmidt have to be contrasted with "framework structures," where polyhedra (SiO_4 tetrahedra, TiO_6 octahedra, etc.) are fitted together with specified bond angles. By filling the cavities in frameworks with counterions, one can return to packing concepts. These ideas, applied with imagination to the feldspars

---

* We are greatly indebted to Mr. C. Escher—Dutch artist and friend of the Laboratory for Insulation Research—for this woodcut.

and perovskites as prototypes, lead to a deeper understanding of displacive structure changes and abrupt phase transitions, domain formation and ordering effects, and to a new identification of structure families.

"To explain phase transitions by architectonics alone is oversimplification," the physicist is bound to counter. The ordering of electric dipoles or magnetic spins can be a fortuitous adjunct to crystal design but frequently will prove the trigger of phase transitions. Herewith the battle is joined about the priority of the hen (crystal structure) or the egg (near-order exchange energies). Simultaneously the discussion expands to correlations between structure and properties.

Clear evidence for the importance of short-range-order terms is provided by the actions of localized d electrons (Chap. 4). While the outer s and p electrons shape the main band structure of a crystal, partially filled d shells produce localized magnetic moments, electronic conductivity by trapping and release processes, and a wide variety of electron-ordering transitions. Such phenomena can be understood in detail by considering the electron-energy states of lattice polyhedra, the term splitting caused by various interaction effects, the lowering of energy levels by lattice distortion (Jahn-Teller effect), etc. Only when the cation-cation separation reaches a lower critical limit is the localized d-electron-bond approach supplanted by collective d-electron-band action.

Obviously, priority must be conceded to either hen or egg, depending on circumstances. Still, tacit agreement is maintained that the electronic structure finally emerging will also conform to crystal symmetry. This conviction is rudely shattered by the next speaker (Chap. 5): The magnetic ordering of spins is not required to reflect the periodicity of the lattice. Spins can form all kinds of spiral configurations—in the extreme, of no periodicity. Various examples, verified by neutron diffraction, are given of this insight based on theoretical calculations. In retrospect, the outcome is not completely surprising: The existence of magnetic resonance and of spin waves testifies to the frequently loose coupling between spin orientation and crystal lattice.

Herewith the discussion turns to phase transitions in glaseous systems, where long-range order is initially absent (Chap. 6). The high viscosity of glasses provides unusual experimental opportunities: to keep metastable phases in suspended animation; to trigger nucleation and crystal growth by photoeffect, temperature, or catalysis; to start and stop reactions at will and to direct them along devious paths not foreseen by equilibrium phase diagrams. The outcome is not only new insight but materials, devices, and techniques that surprise and delight: (nearly) unbreakable glass-ceramics, glasses that preserve photographic images and allow their transformation into three-dimensional cutouts by chemical machining, filters adjusting automatically to light intensity, etc. A new era begins in the understanding and control of the amorphous state, and investigations of glaseous systems promise fresh knowledge about molecular organization, possibly even for living systems.

The road between glass and crystal can be traveled in either direction; thus the discussion now takes the opposite tack: the upset of perfect order and its consequences. The theme opens (Chap. 7) with color centers, close relatives to the reversible color centers just discussed in glasses. Books have been written for many years about such electron and hole traps, but the subject of even simple point defects continues to pose riddles. The response of the F-center band to purity and temperature and the formation of mixed crystals are examples in kind. Ionic conductivity, optical absorption, and the comparison of macroscopic with X-ray density (to six significant decimal places) provide additional information on vacancies and interstitials and explain, for example, the raison d'être of diamond Types I and II.

While these techniques disclose defects by indirection, the field-ion microscope achieves true atomic resolution (Chap. 8); suddenly we can actually observe any disturbance of crystalline order from generation to demise. Vacancies, quenched-in or created by bombardment in clusters up to ten; interstitials, revealed by surface deformation; impurities protruding from the atomic rows; dislocations and other complex defect structures; all these are shown by this instrument of ingenious simplicity and can be followed into depth by controlled field evaporation. Precipitation processes, alloy organization, adsorption, and reaction phenomena, can be studied *in situ*, and the structures of domain boundaries and dislocation cores lie open for inspection. As of now, the application of the microscope is limited to high-melting metals and to situations where the enormous electric field can be tolerated.

A perfect crystal is a Platonic ideal. This is made clear once more by the mechanical strength of materials (Chap. 9), which in general lags several orders of magnitude behind molecular expectation. Dislocations, i.e., line defects in the crystal structure, are responsible for this shortcoming yet cause properties of extreme usefulness. Without them the mechanical response of solids would be confined to the extremes: reversible elastic deformation versus irreversible destruction. It is the generation and migration of dislocations that allow plastic deformation, and it is the obstruction of dislocation movement that hardens materials. Pileup of dislocations may lead to local overstress and crack formation; annealing can come to the rescue, reduce the dislocation density, and restore their mobility. In short, dislocation dynamics is the key to the molecular designing of mechanical properties.

Successful practice of this dynamics demands the space perception and skill of handling molecular lassos. Still, these difficulties shrink to insignificance before the complexities unfolded by the biologist (Chap. 10). Previously we could discern relatively simple structure elements (atoms, molecules, unit cells, etc.) and understandable laws of architecture. Now we are confronted with biological components and systems, mysteriously designed to provide the interacting mechanical, chemical, and electrical responses that life demands. The question of how and why these marvelous devices have been built and function must be asked simultaneously. It is as if the man in the moon had stumbled on a miniaturized guidance system and tried to understand its composition and action.

Slightly awed by the width of the gulf that must be bridged from crystals to living systems, we return to simpler correlations between macroscopic characteristics and molecular events, to the properties: magnetization and conduction.

A study of the magnetization process (Chap. 11) leads from the electronic spins of atoms via mesoscopic spin arrays to macroscopic response characteristics. Geometrical considerations enter through domain-wall configurations and boundaries with their closing fields; crystal parameters influence through microstructure and composition, anisotropy and magnetostriction. To foresee the macroscopic magnetic response displayed by hysteresis loops and in magnetic resonance and relaxation spectra, one has to analyze the structure of spin systems and the torques to which they are subjected. Procedures are outlined and results presented for typical situations.

While for spin systems a crystal acts in essence as a deformable container, conduction and diffusion in ionic crystals provide the key to solid-state reactions (Chap. 12). The thermodynamic equilibrium between phases is ruled by the law of mass action and by the condition of electroneutrality. To establish equilibria, charge carriers and neutral particles must be mobilized by thermal activation, and chemical reactions must proceed at boundaries with gaseous and condensed phases. Atomic particles in crystals can move effectively only by taking advantage of interstitial space and of lattice defects. Electronic carriers in pure crystals, not hampered by space demands, are normally buried in deep-lying valence bands and only accessible if a step-ladder of localized intermediate energy states is provided for their mobilization. It is the task of the molecular scientist to foresee the possible reactions and to set the desired ones in motion by judicious substitution and doping, proper adjustment of temperatures and gaseous phases, and even by invoking electron and hole injection. Thus a chemistry of controlled diffusion and conduction emerges as prerequisite for the effective designing of solid-state devices. Simultaneously new insight is gained for the metallurgist on processes of corrosion, alloy formation, and precipitation, because ionic systems allow discernment of molecular actions that in metals are concealed by their electron screen.

The scope of the preceding discussion was limited to thermal activation processes and to the drift of charge carriers according to Ohm's law. In the subsequent chapters these restrictions are dropped one by one. As a first step, photoexcitation to higher energy states is admitted, and the complex ways are explored by which a solid system rids itself of such localized quantum energy. "From the Lightning Bug to the Laser" should be our all-encompassing theme, as Chap. 13 suggests, but a much more modest inquiry into processes of luminescence already leads to formidable complications. The experimental procedure is to trigger a solid by a well-defined light input and to analyze the subsequent light emission as to intensity, spectral distribution, polarization, time and temperature dependence, and even localization. By correlating that response to the structure and composition of the solid, one hopes to gain understanding and control of the pathways of energy dissipation.

In the simplest case, the absorber is also emitter. This luminescence center can be characterized by configuration diagrams presenting the ground state and various excited states before and after light absorption with due regard to the Franck Condon principle. Complete diagrams would also show the activation energies needed to switch to radiationless transitions, the effect of doping agents in lowering these activation barriers thereby becoming "killers" of the luminescence, the stereostructure of the center, etc. Much has been learned about the control of luminescence, but only the onset of such detailed information exists.

In more complicated situations, the absorber transfers its energy to some emitter center. This may be done by quantum-mechanical resonance—there the coupling distance between the centers enters decisively —or by transport through excitons or uncoupled electrons and/or holes. The last-mentioned method of shipping excitation energy by electronic carriers and releasing it by recombination processes is realized in photoconducting phosphors, vehicles of entertainment and witchcraft since ancient times (Chap. 14). Until recently their preparation was an art because of the extreme sensitivity of such phosphors to impurities and lattice defects. A large part of these difficulties became understandable when insight was gained on electron and hole mobilization and motion in semiconductors. But puzzling problems specific to phosphors remained, such as steering the recombination act away from unwanted killers to effective luminescence centers. These processes are best understood today for phosphors of the ZnS type. Activators like copper or silver, introduced by fluxing or coactivators, are instrumental, and by judicious manipulation of reaction equations the desired emission can be obtained. Also the structure of specific luminescence centers has now been clarified.

Luminescence and phosphorescence emphasize the terminal stages of electron and hole transfer, the mobilization and recombination processes. Semiconductor physics turns its primary attention to the electron band structure of solids, to the motion of electrons and holes in such bands and to their interband transitions (Chap. 15). The complex band structure in k space is now relatively well known for Si, Ge, and the III-V compounds important in semiconductor devices. Thus it has become clear where the electrons and holes congregate, what their effective masses are, how they are excited in direct and indirect transitions, and what kind of changes occur in band structure and electron motion under the influence of magnetic fields. Donors and acceptors can be represented by hydrogenlike models and excitons as analogue of positronium. Interacting impurities may coalesce into impurity bands; injection of electrons and holes can be driven to population inversion and laser action.

In insulators, localized bonds tend to dominate, and the motion of charge carriers is greatly affected by the halo of countercharges. In semiconductors, such bonds still interfere with the transfer of electronic carriers in conduction and valence bands. In metal single crystals, the electron-band theory comes fully into its own and provides a wealth of information on the distribution of electrons and holes and the propagation of electron waves in periodic structures (Chap. 16). However, in advancing theoretical predictions and interpretations one must remember that calculations are normally restricted to one-electron wave functions; hence, collective interactions may not be properly represented.

After the first triumph of explaining the low specific heat and the temperature-independent paramagnetism of electrons in metals of high Fermi energy, advances in understanding demanded more and more detailed information on the actual shape of the Fermi surface and on the band structure in k space. Such information was supplied experimentally by the ingenious application of electric measurement techniques in magnetic fields, which can exert directed forces of much greater intensity than electric ones in metallic systems. Magnetic forces furthermore do not change the kinetic energy of the electronic carriers—band-splitting effects exempted—but provide direction of motion. Hence, electrons can feel out the contour of Fermi surfaces by describing closed paths on them or coming to grief on junctions and terminations. Topographic studies by cyclotron resonance, de Haas-van Alphen effect, etc., led not only to a colorful description of such paths through the k-space landscape as dog's-bone orbits, bellies, and goosenecks, but also measured the effective masses and relaxation times of electrons and holes in their orbits. Detailed information resulted on electron spin interactions, including pair formation in superconductors. Photoexcitation helps to measure band gaps and transition probabilities. And finally nuclear spins have been enlisted as detectors, e.g., in the Mössbauer effect, to measure s-electron distribution and spin ordering, isotope effects and impurity action, coupling of s electrons with d electrons, etc. The day approaches when biographies of individual metals can be written with a wealth of intimate detail.

The preceding chapters clarified dissipation processes of localized excitation energy and outlined the intricate pathways in k space traversed by electronic carriers in weak electric fields. Now we turn to strong electric fields, which cause destruction of the insulating properties of dielectric systems. Transitions from insulator to conductor presuppose a vast increase in the number of available charge carriers. One way of accomplishing this augmentation in gases is electron-impact ionization and avalanche formation, first suggested by Townsend more than fifty years ago (Chap. 17). Avalanches *per se* do not imply instability but may trigger it. How this is accomplished, has been the subject of hot debate for many years. The issue remained clouded until modern electronics could pursue causes of instability into the nanosecond range. Methods are now available to study the build-up of individual avalanches in space (cloud chamber, image intensifier) and in time (oscillographic records of current rise and photon emission). Two main types of avalanche breakdown can clearly be distinguished: a relatively leisurely "multiavalanche" build-up, with generation producing successor generation through electron release at the cathode (primarily by photoeffect); and a "single-avalanche" breakdown converting the gas by plasma formation. The inherent cause of instability in either case is field distortion by space-charge action.

Gas breakdown by impact ionization would occur at lower voltages if the electronic excitation states of the particle would not syphon off kinetic energy from the electrons and squander it in secondary processes like light emission. In special cases, however, this protective method backfires, and the excitation itself becomes the primer of gas discharges (Chap. 18). Excited atoms and molecules can transform into long-living metastable species, ready to create ionization in step reactions or by resonance transfer of their energy to ionizable collision partners, as in the Penning effect (e.g., $Ne^* \rightarrow Ar^+ + e^-$). Electrons may also be ejected from the walls by excited particles or photoeffect, as well known from the restriking of gas discharges. More surprisingly, discharges like the mercury arc seem actually to be sustained by the interplay between light emission of excited mercury atoms and photoelectrons released by that emission from the cathode. Detailed studies of excited species presently gain new impetus through the development of gas lasers and plasma physics. They promise not only new special insight and devices but ultimately a mastery of the complex chemistry of gas discharges.

Current flow through dielectrics, when measured externally, does not normally obey Ohm's law, because space charges distort the field distribution (Chap. 19). Such space-charge-limited currents are well known from vacuum diodes, where the Childs-Langmuir $V^{3/2}$ law describes the flow of electrons delivered by a hot cathode in abundance and moving in free fall from cathode to anode. In that case a cloud of electron space charge forms in front of the cathode. For solid dielectrics the situation is modified: Free acceleration gives way to ohmic friction, and traps and recombination centers remove electrons and holes from the free carrier stream. Studies of such space-charge-limited current flow with d-c and transient voltages, optical-absorption and photoresponse measurements give detailed information on average trap distribution, life and transit times of carriers. The high quantum yield of the indirect photoeffect becomes understandable by space-charge-compensation effects. Electron-migration studies provide a quantitative tool to judge crystal perfection and to give optimum design criteria for solid-state photoelectric devices.

The stage is set for a comprehensive inquiry, how phenomena of electric imbalance arise as one progresses from gases to liquids and solids (Chap. 20). In gas breakdown one can distinguish between a "statistical time lag" caused by paucity of the electron supply and a "formative time lag" characterizing the avalanche build-up. A Townsend-type breakdown criterion in the multiavalanche range links the regeneration probability of primary electrons to a critical avalanche height. The field strength across the free path and the number of free paths across the gap are the decisive yardsticks; hence, Paschen's similarity law results. After breakdown sets in, dynamic field distortion by positive space charge contracts the effective gap length into a cathode fall.

At a critical space-charge density, plasma formation begins and positive and negative spark paths develop. This occurrence can be promoted in a variety of ways: by lengthening the holding time of a voltage pulse or raising the overvoltage, by increasing gap length or gas pressure, by geometrical field distortion or augmentation of the number of simultaneously available starting electrons through field emission or radioactive ionizers. If plasma can set in before the first avalanche traverses the full gap length, Paschen's law loses its validity and new breakdown conditions emerge. The breakdown field strength becomes independent of the thickness of the dielectric, rises linearly with gas density, and depends strongly on material and surface condition of the cathode.

No discontinuity marks the breakdown from gas to liquid near the critical point. Gradually the concept of free path gives way to phenomena of permanent coupling; the friction barrier of electronic excitation states loses its importance as the phonon modes of the coupled system develop; charge carriers begin to shield themselves against recombination by electronic, ionic, and dipolar counteratmospheres. Ionic prebreakdown currents trigger field emission, and offhand contradictory observations may actually not be in conflict because of the startling variety of molecular phenomena that hide behind the macroscopic concept "breakdown." We begin to understand the electric strength of materials.

Any premature optimism about our intelligence, however, is squashed when the spotlight again turns to biology, with an inquiry into the elementary units of life (Chap. 21). A short survey on structure and properties of cells is followed by a description of experiments to clarify the causes of cell motion in contact with artificial boundaries. It is very revealing for the nonbiologist to learn how his colleagues must thread their way through a maze of variables until relations of significance can be isolated. Even the discerning of structure elements poses formidable problems, not to mention the continuous interlocking of their chemical, mechanical, and electrical actions.

How a specific cell-structure problem was isolated and solved ingeniously is described in Chap. 22. The existence of spindle fibers, the pulley system for the separation of chromosomes in cell division, had been placed in doubt because electron microscopy failed to reveal any such structure. A most sensitive, wide-aperture polarization microscope was designed, and now the formation of the spindle system in the cytoplasm could actually be detected by its weak birefringence. Molecular units aligned a few minutes before division started and dissipated again after the work was performed. Temperature changes or illumination in sensitive spectral ranges thwarted the process. Obviously, the electron microscope must fail to discern phenomena that proceed only when life remains undisturbed.

Devices and the "making of materials to order" received only sporadic attention in the foregoing account; the last five chapters make this subject their central issue.

The engineer cannot wait until basic insight offers the most elegant solutions. Thus, he had to invent, for instance, a compromise transition between classical network theory and molecular engineering by combining available materials and circuits so that "they act like materials we wish to have but do not have" (Chap. 23). Examples of specific targets are given which can be realized by the imaginative use of two simple "building block units": a cylindrical body of high-dielectric-constant ceramic and an amplifier with series and parallel feedback impedance. Resonances of the dielectric body can be utilized to create VHF cavities that produce modes and magnetic field patterns not found in standard microwave technology. Suppression of these resonances leads to low-pass filters. The discussion expands to the manipulation of whole networks and their transformation properties. This transposing of networks, including the switching of their function by light beams, opens possibilities for microminiaturization, which add an exciting new chapter to modern electrical engineering.

The solid-state physicist has become an acknowledged electronics engineer since the advent of the transistor. The span between new insight and application, previously measured in decades, has shrunk to a few months; strikingly new electronic devices are on the horizon (Chap. 24). Lasers and masers have made coherent beams of unprecedented monochromacy and parallelism available. The use of such beams as information carriers becomes practical by electro-optical and magneto-optical modulation devices. To achieve the critical matching between the phase velocities of carrier and modulator waves, schemes like the zigzag modulator are introduced. Perfection of crystals and intelligent molecular design of their electric and optical properties are prerequisites.

Piezoelectric devices are the basis of transmitters and receivers for sonar and ultrasound applications. Recently new devices have become feasible by the interaction between ultrasound waves and electron streams in piezoelectric semiconductors. Since the velocity of the electrons can be adjusted by the applied field, ultrasound waves may be amplified or electron beams modulated. Proper doping of such crystals (e.g., CdS) changes the semiconductor into an insulator; hence, all functions of an amplifier may be combined in one monolithic piece of crystal. New magnetoelastic devices of adjustable Faraday rotation are another example of the creative phantasy pervading the field of molecular engineering.

Turning from device design to the fundamental problem of energy conversion, the molecular physicist finds the world full of intriguing and unexploited possibilities (Chap. 25). The power of a Niagara Falls 840 feet high and carrying all the waters of the Mississippi is wasted by discharging that river into the Gulf of Mexico and freely letting it mix its "sweet" water with the salt water of the ocean. Unfortunately no economical conversion of that power is in sight, but the reverse problem of desalting sea water by distillation or electrodialysis has led to ingenious plant designs and will soon alleviate our water shortage.

Petroleum, gas, and coal are our greatest stored energy resources outside of nuclear energy. At present they are exploited in heat by combustion with oxygen. Fuel cells, designed with the proper solid-state electrolytes, promise an elegant way of direct conversion into electricity. Solar cells also may come into their own with high efficiency; magnetic sieves operating with gas plasmas might compete with fuel cells; electronic heat engines and thermionic converters are under study. The complacency of the power engineer is rudely challenged, and the old monopoly of rotating machinery may be broken in the not too distant future. Chapter 26 surveys how far the generators based on thermoelectricity, fuel cells, or magnetohydrodynamics have by now become a competitive reality.

In closing we turn once more to biology and its devices. Understanding begins to dawn about the action of neurons in sensory perception (Chap. 27). Brain research joins with electronics, physiology, and psychology to make us slightly comprehensible to ourselves; but how these intricate devices operate on the molecular level is still a mystery.

1 · PREDICTIONS BY *a priori* THEORY

J. C. Slater

My remarks, as a theorist, will mainly be devoted to the question of how far purely *a priori* theoretical calculations can predict the properties of matter. If we make a theory of a whole range of phenomena, the theory being less than perfect, and check some of the predictions by experiment, extrapolation methods on the discrepancies between theory and experiment can correct our forecasts and probably do quite well. This semiempirical procedure, very valuable in practice, is however not straight theory.

The theoretical basis for the study of materials is, as far as we know, complete and accurate, with trivial exceptions. Quantum theory, or wave mechanics, plus electromagnetic theory, thermodynamics, and so on, are capable of answering our questions provided we can work out the mathematics. Unfortunately the mathematics is more difficult than that in any other branch of science. We are making great progress using digital computers, but we are still far from perfection.

The theory has been checked in many simple cases. The hydrogen atom, the helium atom, the hydrogen molecule, are worked out with great mathematical precision, and the results agree with experiment to spectroscopic accuracy. The study of the heavier atoms has gone far enough so that we can predict the properties of any atom, no matter how heavy. Recent unpublished studies of the whole periodic system of the atoms have gone far beyond anything in the literature in giving details of the structure of all the atoms. These calculations, however, are far less precise than those for hydrogen and helium. Errors of several parts in a thousand are common for the heavier atoms, compared to better than parts per million for hydrogen and helium. For some of the atomic properties, existing calculations cannot do much better than an error of a percent. The trouble lies in the approximation used, not in the theory. Do not forget that an atom of atomic number 100 involves the mechanics of a 100-body problem, whereas the astronomers have for centuries pointed out the difficulties even in the 3-body problem.

The hardest problem for the theorists is unfortunately the one of most direct interest for the experimenter: How are the atoms arranged in a given type of matter? In principle, we know exactly how to solve this problem. We work out the energy of the system as a function of the positions of the atoms (or free energy, if we are not working at the absolute zero of temperature). The position leading to the lowest energy is the stable one. Of two structures, that of the lower energy will exist. And as the pressure or temperature is changed, modifying the free energy of the various forms of the substances, the free energies of two phases may cross, leading to a polymorphic transition. All this depends on a calculation of free energy, accurate enough to decide which of two free energies is lower. The differences with which we are concerned are measured in kilogram-calories per mole. But the total energies required to strip all the electrons off for moderately heavy atoms may come to many thousands of Rydberg units (a Rydberg unit is 13.6 electron volts, and an electron volt about 23 kilogram-calories per mole). We are often dealing with energy differences of a millionth of the total energy.

Fortunately, the greater part of this total energy arises from the inner electrons of the atoms, which are not appreciably affected by chemical change. The modified parts of the energy are only a few tens or hundreds of times the magnitude of the change, so that we are within sight of enough accuracy in the calculations to give an adequate account of chemical stability. Let me give a few examples. KCl and KBr are known to undergo a polymorphic transition under pressure. Calculations of the free energy for both phases, on a completely *a priori* basis, are good enough to indicate the approximate pressure at which the change should occur. Many calculations have been made on simple molecules; the interatomic distances corresponding to the lowest energy agree pretty well with the distances observed. Angles between bonds can be computed; a recent calculation of the angle between the two OH bonds in water gives a minimum energy fairly close to the observed angle. These are only a few cases compared with the enormous amount of data we should like to explain, but they show that we are on the right track.

If, however, instead of demanding that theory predict the stable atomic arrangement in the material we take that information as given, then theory can give properties of the substance with much better success. The best example is the calculation of the electronic energy levels of the system. These are of vital importance. If we are dealing with a solid, these levels form the energy bands, and from their study we can tell at once whether the substance should be a metal, a semimetal, a semiconductor, or an insulator. The calculations of energy bands are now good enough so that this information can be got correctly from *a priori* calculation.

Many experiments can be understood once we know the energy levels, and we are rapidly getting to the point where we can predict these. Examples are optical properties, such as absorption and index of refraction; electric and magnetic properties as cyclotron resonance, the de Haas-van Alphen effect; magnetic resonance; the optical transitions concerned in laser action; and so on. We are still far from complete quantitative forecasts, but theory is a great help even when indicating the general form of behavior, so that experimental results can be tentatively fitted with slight adjustments of constants and parameters.

Another field of relatively good *a priori* calculations is that of finding the electronic charge density in a molecule or solid. We can calculate this charge density well enough to agree almost perfectly with that measured by X-ray and neutron diffraction. In molecules, charge density leads to such observable properties as the electrical dipole moment; the best calculations of dipole moment are in perfect accord with experiment. At last, theory has had the satisfaction of computing one such quantity—the dipole moment of the LiH molecule—before it was measured, and of finding that the experiment verified the calculation within the limits of error.

In conclusion then: Theory is getting to the point where really valuable checks with experiment may be expected. The situation is rapidly improving; five years ago one could not have been nearly as hopeful. Improvements in methods now being worked out give much encouragement for anticipating increased success for *a priori* calculations in the years immediately ahead.

SOME RECENT REFERENCES ON THEORY

H. Brooks, "Quantum Theory of Cohesion," *Nuovo cimento*, Ser. 10, *Suppl.* to Vol. 7, 165 (1958).

J. Callaway, "Electronic Energy Bands in Solids," *Solid State Physics*, Vol. 7, F. Seitz and D. Turnbull, Eds., Academic Press, New York, 1958, p. 99.

C. A. Coulson, *Valence*, Clarendon Press, Oxford, 1957.

The Fermi Surface, W. A. Harrison and M. B. Webb, Eds., John Wiley and Sons, New York, 1960.

F. Herman, "Theoretical Investigation of the Electronic Energy Band Structure of Solids," *Revs. Modern Phys.* 30, 102 (1958).

H. Jones, *The Theory of Brillouin Zones and Electronic States in Crystals*, Interscience Publishers, New York, 1960.

P.-O. Löwdin, "Quantum Theory of Cohesive Properties of Solids," *Advances in Physics* 5, 1 (1956).

P.-O. Löwdin, "Correlation Problems in Many-Electron Quantum Mechanics," *Advances in Chem. Phys.* 2, 207 (1959).

N. F. Mott, "The Cohesive Forces in Metals and Alloys," *Repts. Progr. in Phys.* 25, 218 (1962).

S. Raimes, *The Wave Mechanics of Electrons in Metals*, Interscience Publishers, New York, 1961.

J. R. Reitz, "Methods of the One-Electron Theory of Solids," *Solid State Physics*, Vol. 1, F. Seitz and D. Turnbull, Eds., Academic Press, New York, 1955, p. 1.

J. C. Slater, "The Electronic Structure of Solids," *Handbuch der Physik*, Vol. 19, 3rd ed., Springer-Verlag, Berlin-Göttingen-Heidelberg, 1956, p. 1.

J. C. Slater, *Quantum Theory of Atomic Structure*, Vols. 1 and 2, McGraw-Hill Book Co., New York, 1960.

J. C. Slater, *Quantum Theory of Molecules and Solids*, Vol. 1, *Electronic Structure of Molecules*, McGraw-Hill Book Co., New York, 1963 (Vol. 2, *Electronic Structure of Solids*, and further volumes, to appear).

R. A. Smith, *Wave Mechanics of Crystalline Solids*, John Wiley and Sons, New York, 1961.

2 · BUILDING FROM ATOMS

Arthur von Hippel

Introduction — Atomic orbitals and molecular bonds — Atomic orbitals and electron states in crystals — Atomic structure and the boiling points of the elements — The melting points of the elements — Dense-packed structures — Compound formation and polarity — Insulators and conductors

Introduction

The electronic systems of atoms are well known and conveniently characterized by completed inner shells, which act mainly as shields of the nuclear charge, and by incompleted outer shells, whose valence electrons participate in chemical reactions. Obviously, this description is not strict: The shell structure becomes blurred for many-electron systems, complete outer subshells may shield incomplete inner subshells in precarious stability, and electrons of just completed subshells may be forced back into chemical bonding.

If the atoms are relatively well understood, one should expect that they can be used as building stones for design purposes with reasonable foresight. How far this is possible today, and where some of the difficulties enter, is the topic of this chapter. Simultaneously the discussion introduces background information for an easier understanding of subsequent chapters.

Atomic orbitals and molecular bonds

The hydrogenlike wave functions of the atoms can be written as the product of a radial part dependent on the principal quantum number n and of an angular part, prescribed by the azimuthal quantum number and imparting stereoproperties to the atoms. The radial functions for $n = 1, 2, 3, 4$ (designating the K, L, M, and N shells) and the angular shapes for $l = 0, 1, 2$ (characterizing the s, p, and d orbitals) are shown in Fig. 2.1. The $+$ and $-$ signs in the amplitude modes of standing electron waves signify phase relations in space as in standing electromagnetic waves. Overlap of wave functions for electrons from different atoms with areas of like sign ($++$ or $--$) indicates addition of the amplitudes, hence attraction and bonding; overlap of areas of opposite sign ($+ -$ or $- +$) means subtraction, repulsion, and antibonding. In the s-wave functions ($l = 0$), the $+$ and $-$ areas follow each other in spherical symmetry; the electronic motion is restricted to radial pulsations. In the other types, consecutive wings of the wave function alternate in sign, indicating a rotatory orbital motion as it expresses itself in orbital angular momenta.

By combining two atoms into a diatomic molecule, one replaces spherical with cylindrical symmetry and has to refer to the resultant orbital angular momenta around the cylinder axis. The $s, p, d, f, \cdots$ designation of the atomic orbitals is replaced by the corresponding Greek letter designation $\sigma, \pi, \delta, \phi, \cdots$, with a subscript indicating the symmetry of the wave function, if the molecule has a center of symmetry (Fig. 2.2).

The complications increase when one advances to the stereostructures of polyatomic molecules. The designations σ and π refer now to specific bond axes, e.g., the threefold rotation axes of the tetrahedron or the four-

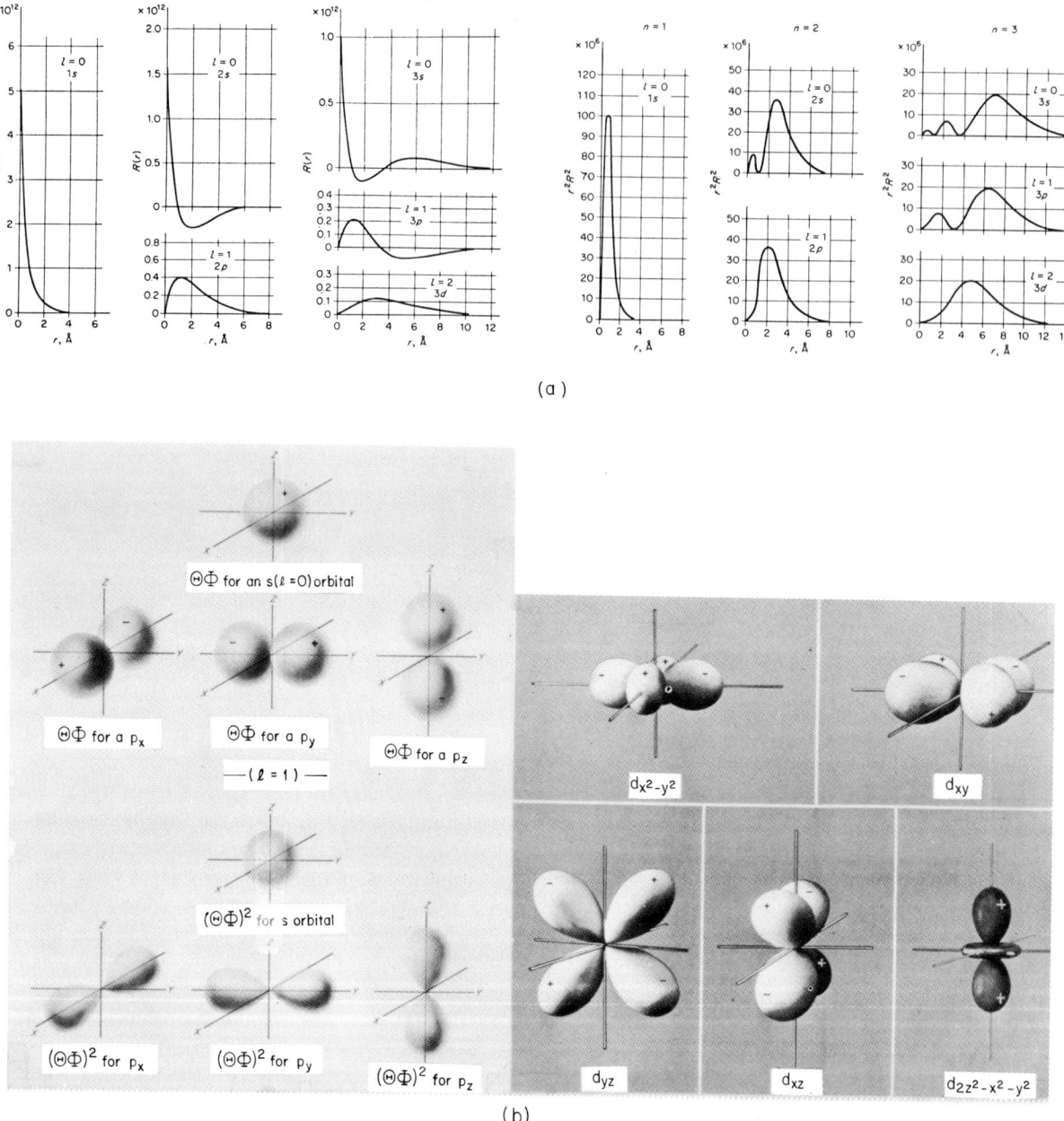

Fig. 2.1. (a) Radial and (b) angular shapes of atomic wave functions.

fold ones of the octahedron. Simultaneously one begins to distinguish in such configurations between the "primary" atom located in the interior of the constellation and its "ligand" atoms, the surrounding near-neighbors (cf. Fig. 2.2).

To obtain a qualitative picture of the possible molecular bonds and their sequence in the energy scale, it normally suffices to consider the $(n-1)d$, ns, and np valence orbitals of the atoms. One first visualizes the atoms in their stereoconstellation at infinite separation. It then becomes clear which σ, π, and δ bonds might be formed between the valence orbitals of the primary atom and the molecular orbitals of the ligand atoms (Fig. 2.3).[1] The actual coupling at molecular distances splits each orbital into two, because the overlap of the wave functions may occur with either like or unlike signs: Bonding lowers, antibonding raises the energy levels. Since each orbital can take two electrons with antiparallel spins and since the number of available valence electrons is given by the atoms, the electronic

[1] Cf. C. W. Nelson, *Tech. Rep. 179*, Lab. Ins. Res., Mass. Inst. Tech., May, 1963.

Table 2.1. Possible combinations of atomic orbitals producing σ and π bonds
for some molecular symmetry constellations

Representation	Metal Orbitals	Ligand σ Orbitals	Ligand π Orbitals
A_{1g}	$4s$	$\frac{1}{\sqrt{6}}(\sigma_1 + \sigma_2 + \sigma_3 + \sigma_4 + \sigma_5 + \sigma_6)$	
E_g	$3d_{z^2}$	$\frac{1}{2\sqrt{3}}(2\sigma_5 + 2\sigma_6 - \sigma_1 - \sigma_2 - \sigma_3 - \sigma_4)$	
	$3d_{x^2-y^2}$	$\frac{1}{2}(\sigma_1 + \sigma_2 - \sigma_3 - \sigma_4)$	
T_{1u}	$4p_x$	$\frac{1}{\sqrt{2}}(\sigma_1 - \sigma_2)$	$\frac{1}{2}(\pi_{3x} + \pi_{4x} + \pi_{5x} + \pi_{6x})$
	$4p_y$	$\frac{1}{\sqrt{2}}(\sigma_3 - \sigma_4)$	$\frac{1}{2}(\pi_{1y} + \pi_{2y} + \pi_{5y} + \pi_{6y})$
	$4p_z$	$\frac{1}{\sqrt{2}}(\sigma_5 - \sigma_6)$	$\frac{1}{2}(\pi_{1z} + \pi_{2z} + \pi_{3z} + \pi_{4z})$
T_{2g}	$3d_{xy}$		$\frac{1}{2}(\pi_{1y} - \pi_{2y} + \pi_{3x} - \pi_{4x})$
	$3d_{xz}$		$\frac{1}{2}(\pi_{1z} - \pi_{2z} + \pi_{5y} - \pi_{6x})$
	$3d_{yz}$		$\frac{1}{2}(\pi_{3z} - \pi_{4z} + \pi_{5y} - \pi_{6y})$
T_{2u}			$\frac{1}{2}(\pi_{3x} + \pi_{4x} - \pi_{5x} - \pi_{6x})$
			$\frac{1}{2}(\pi_{1y} + \pi_{2y} - \pi_{5y} - \pi_{6y})$
			$\frac{1}{2}(\pi_{1z} + \pi_{2z} - \pi_{3z} - \pi_{4z})$
T_{1g}			$\frac{1}{2}(\pi_{1y} - \pi_{2y} - \pi_{3x} + \pi_{4x})$
			$\frac{1}{2}(\pi_{1z} - \pi_{2z} - \pi_{5x} + \pi_{6x})$
			$\frac{1}{2}(\pi_{3z} - \pi_{4z} - \pi_{5y} + \pi_{6y})$

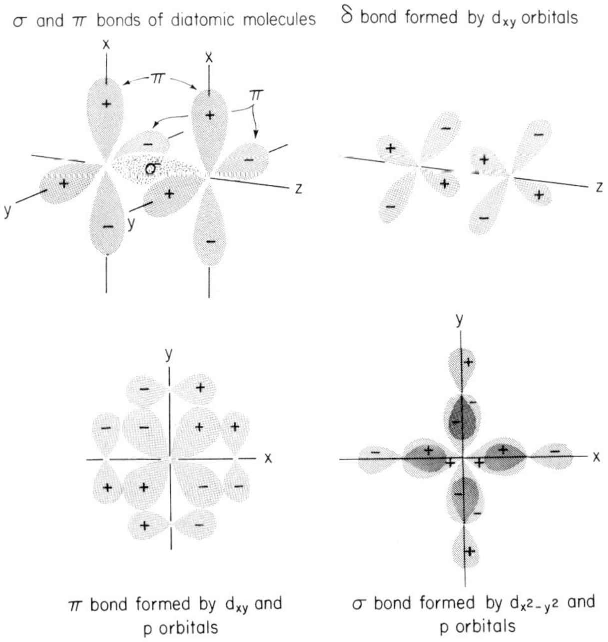

σ and π bonds of diatomic molecules δ bond formed by d_{xy} orbitals

π bond formed by d_{xy} and p orbitals σ bond formed by $d_{x^2-y^2}$ and p orbitals

Fig. 2.2. Designation of molecular orbitals.

state of the molecular group is defined. At the same time the diagram makes clear which electronic states are available for excitation and which orbitals cannot be used for a prescribed stereoconstellation.[2]

The last remark needs amplification. Any stereostructure is characterized by certain symmetry elements (axes, inversion axes, planes of reflection), which can be represented by a matrix. Only those molecular orbitals can serve that conform to this group symmetry. Tables have been worked out[3] that show the possible combinations of atomic orbitals producing the various

[2] Cf., e.g., F. A. Cotton, *Chemical Applications of Group Theory*, Interscience Publishers (John Wiley & Sons), New York, 1963; C. J. Ballhausen, *Introduction to Ligand Field Theory*, McGraw-Hill Book Co., New York, 1962; C. K. Jørgensen, *Absorption Spectra and Chemical Bonding in Complexes*, Pergamon Press, London, 1962; H. B. Gray, *J. Chem. Educ.* 41, 2 (1964).

[3] G. E. Kimball, *J. Chem. Phys.* 8, 188 (1940); J. C. Eisenstein, *ibid.* 25, 142 (1956); E. B. Wilson, Jr., D. C. Decius, and P. C. Cross, *Molecular Vibrations*, McGraw-Hill Book Co., New York, 1955.

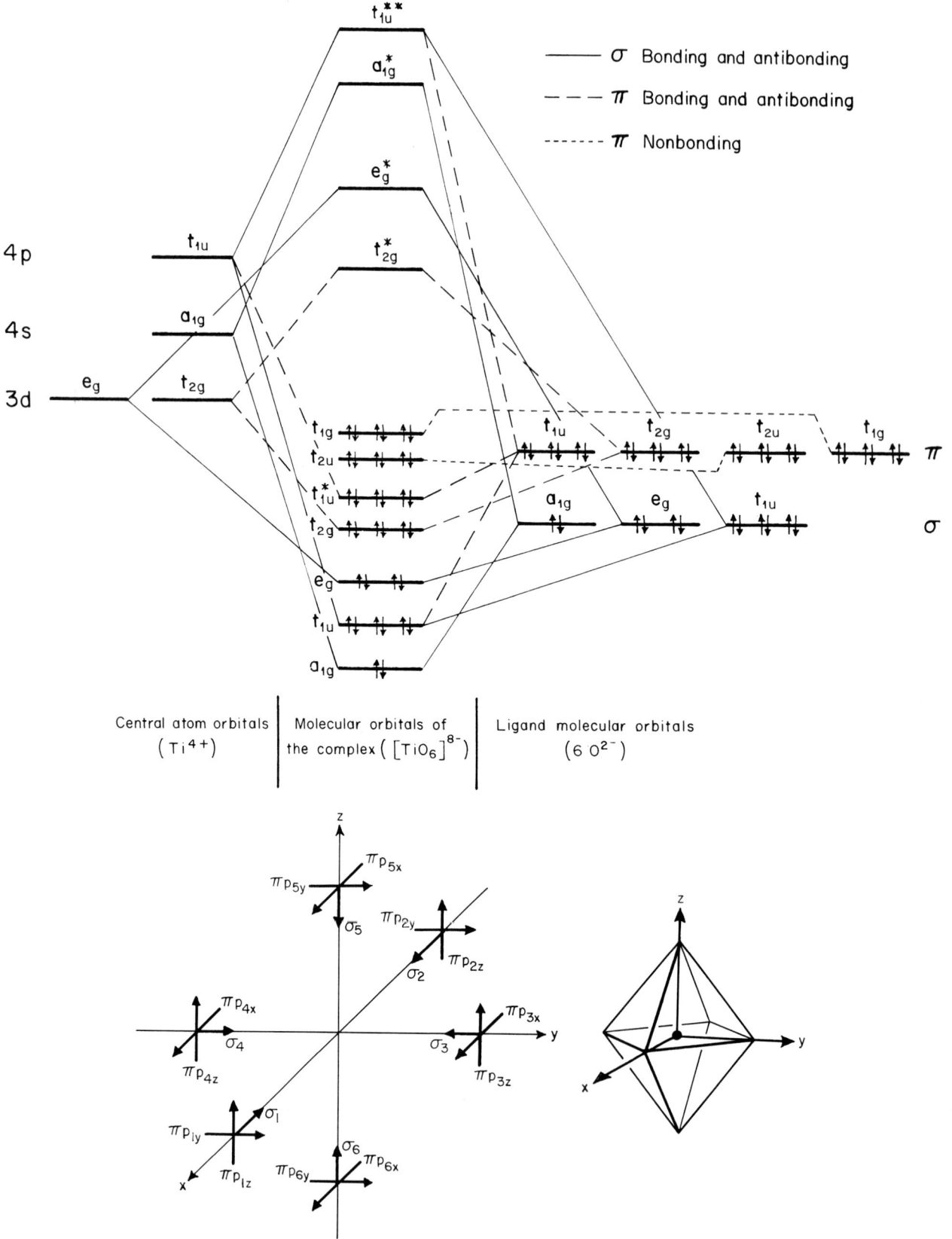

Fig. 2.3. Molecular orbitals for the octahedral TiO$_6$ configuration.

σ and π bonds for each molecular symmetry constellation. Table 2.1 gives a survey of this kind for the octahedral coordination (O_h).[4]

Atomic orbitals and electron states in crystals

The electronic charge of atoms theoretically extends to infinity. If atoms are combined into crystal structures, wave functions must exist which spread the electron charge over the whole crystal. Since the Schrödinger equation contains the potential energy, these wave functions are modulated by the periodic potential of the crystal as first written by Bloch,[5]

$$\psi_k = u_k(r)e^{jk \cdot r}, \qquad (2.1)$$

where $u_k(r)$ is the amplitude modulated with the period of the lattice. Any electron state spread as a band through a periodic crystal lattice can be built up by Fourier series of such modulated plane waves.

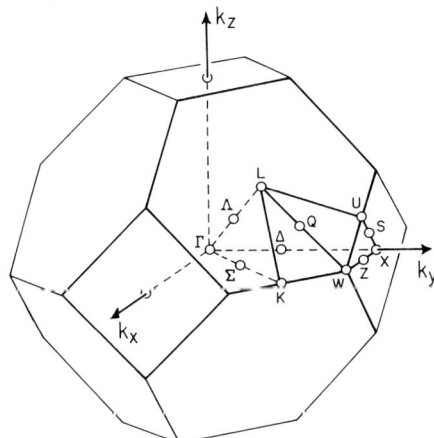

Fig. 2.4. First Brillouin zone of face-centered cubic crystal.[7]

When a traveling plane wave (wave vector $k = 2\pi/\lambda$) strikes in normal incidence a principal set of lattice planes of spacing d that fits the Bragg condition for total reflection

$$n\lambda = 2d; \text{ i.e., } k = n\frac{\pi}{d}, \text{ with } n = 1, 2, 3, \cdots, \qquad (2.2)$$

the three-dimensional crystal waveguide cuts off in that direction. By drawing the cutoff planes for the principal crystallographic orientation to intersection, k space can be divided into polyhedra, the Brillouin zones of allowed propagation,[6] bounded by planes of total reflection (Fig. 2.4).[7] The consecutive zones of increasing k (corresponding to the fundamental wavelengths and harmonics) are identified by integers $n = 1, 2, 3$, etc.

An electron represented by a plane wave in free space has only kinetic energy, correlated by the de Broglie equation,

$$\mathbf{p} = \frac{h}{\lambda} \equiv \hbar \mathbf{k}, \qquad (2.3)$$

to the wave vector as

$$\mathcal{E}(k) = \tfrac{1}{2}mv^2 = \frac{\hbar^2}{2m}k^2. \qquad (2.4)$$

The energy is represented by a parabola; its tangent at any point determines the group velocity v of the electron and the curvature (constant for a parabola) of the

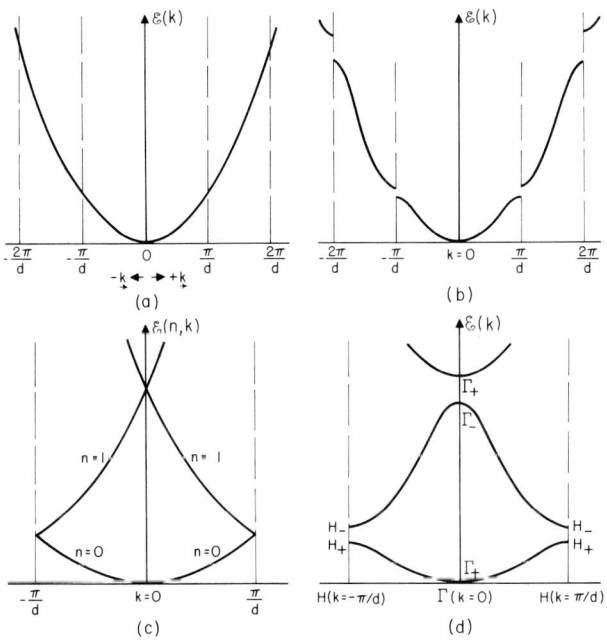

Fig. 2.5. Energy as function of wave vector for free electron and electron in crystal lattice.

electron mass m (Fig. 2.5a). For a crystal, these $\mathcal{E}(k)$ characteristics—now representing the total energy as

---

[4] Calculated by C. W. Nelson, Lab. Ins. Res., Mass. Inst. Tech.

[5] F. Bloch, *Z. Physik* **52**, 555 (1928); S. Raimes, *The Wavemechanics of Electrons in Metals*, North-Holland Publishing Co., Amsterdam, 1961.

[6] L. Brillouin, *Wave Propagation in Periodic Structures*, McGraw-Hill Book Co., New York, 1946; H. Jones, *The Theory of Brillouin Zones and Electron States of Crystals*, North-Holland Publishing Co., Amsterdam, 1960.

[7] The zone depicted in Fig. 2.4 corresponds in k space to a body-centered cuboctahedron with x, y, and z the cube-edge directions, $\Gamma \to K$ a face diagonal, and $\Gamma \to L$ a space diagonal; the edges of the square and hexagonal faces are face diagonals; Σ, Λ, and Δ are general points on the three high-symmetry directions.

function of the wave vector—must be interrupted at the edge of the Brillouin zones by a gap of nonpropagation due to the cutoff feature of the crystal waveguide (Fig. 2.5b).

In a mathematical point lattice the electrons would encounter sharp Bragg reflection planes. A change of electron energy would produce a corresponding change in electron wavelength; the energy gaps of total reflection would be infinitely narrow. In an actual crystal the planes do not correspond to sudden discontinuities but are composed of broad potential humps formed by the electron clouds of the atoms. The wavelength of the traversing electrons responds to this potential energy variation by dispersion according to the Schrödinger equation; it remains clamped to the value for total reflection over some energy range; the forbidden energy gaps are widened.

The $\mathcal{E}(k)$ characteristics of such plane electron waves can be compressed (by folding over) into the repeat interval $k = 0 \to \pm \pi/d$ with the Brillouin zone number n (harmonic number) and the wave vector $\mathbf{k}$ (for a certain k-space direction) as parameters (Figs. 2.5c, d). The positive and negative curvature regions of the electron waves in the crystal lattice, caused by interference effects, show now the band shapes typical for the electron and hole description. Each curve in such a "reduced zone" $\mathcal{E}(n, k)$ diagram represents a cut in a certain crystallographic orientation through an allowed energy surface in k space. For a crystal of N primitive cells this surface contains N discrete energy states accommodating two electrons, each with antiparallel spins.

Frequently a "combined representation diagram" is used that shows in one figure the shape of the bands for various principal crystallographic directions, mapped out by paths across and along the Brillouin zone poly-

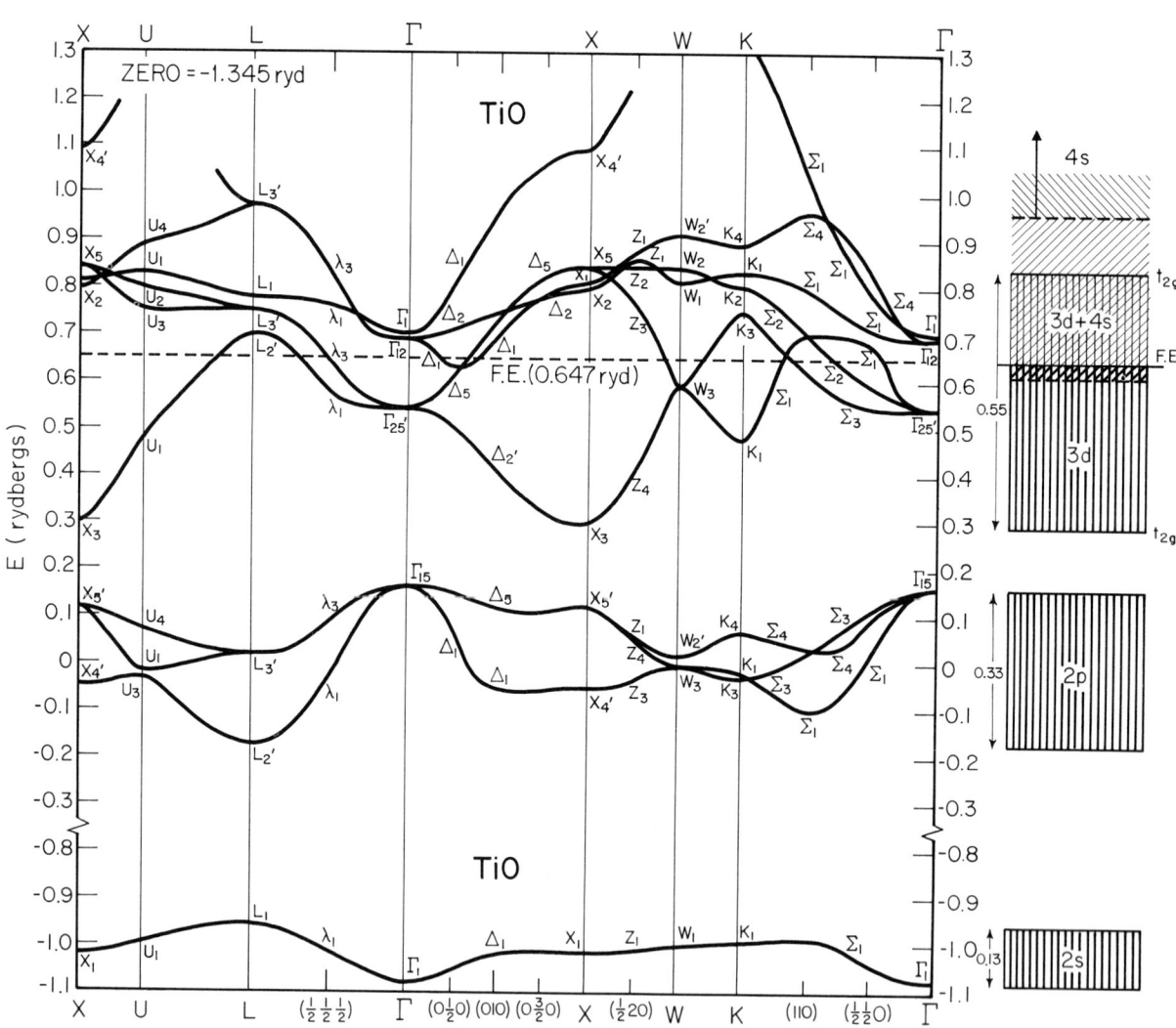

Fig. 2.6. Some $\mathcal{E}(k)$ curves for metallic TiO.

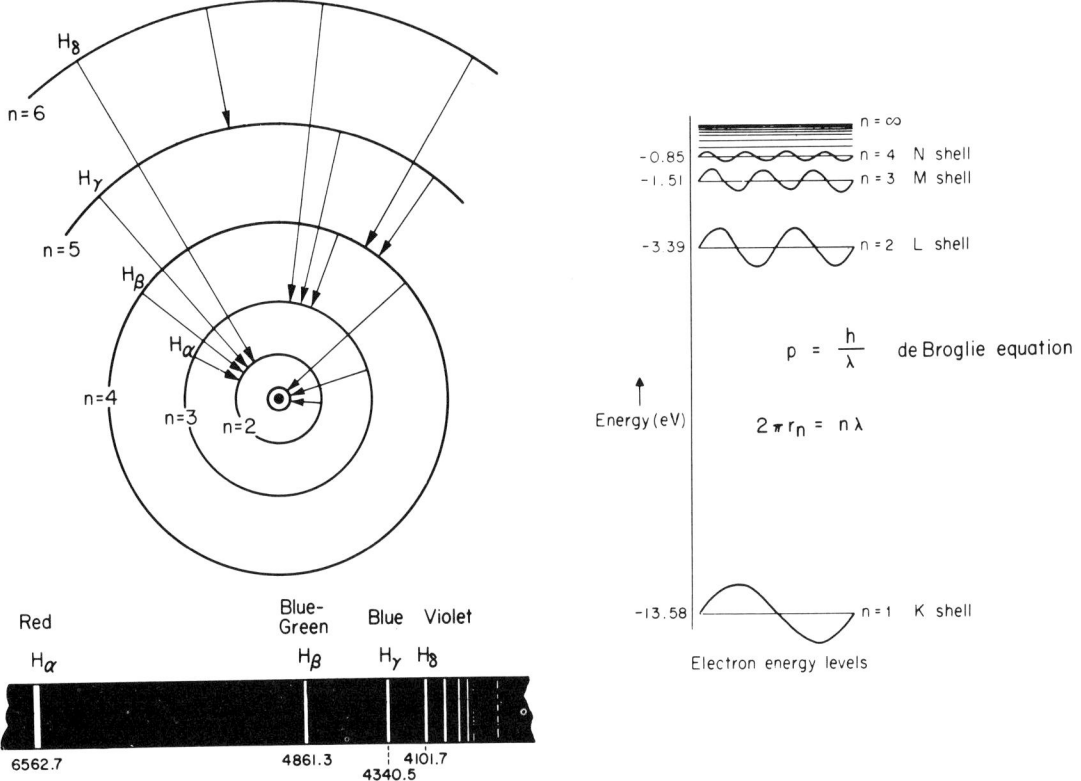

Fig. 2.7. Bohr's circular orbits as standing-wave patterns.

hedron (Fig. 2.6).[8] The capital Greek letters shown here have become customary for direction and thus band designations.[9]

In this way one tries to measure and visualize the surfaces of constant energy in k space; the gradient of this $\mathcal{E}(k)$ field, which determines the group velocity of the electron waves; and its curvature, which signifies the effective mass m^* of electron or hole for the outside observer:

$$v_g = \frac{1}{\hbar} \mathrm{grad}_k \, \mathcal{E}(k) \qquad (2.5a)$$

$$\left(\frac{1}{m^*}\right)_{ij} = \frac{1}{\hbar^2} \frac{\partial^2 \mathcal{E}(k)}{\partial k_i \partial k_j}. \qquad (2.5b)$$

This "one-electron model" for the crystal—electron interaction effects beyond those contained in the potential energy are neglected—is an analogue of the one-electron model of the atom. The hydrogenlike orbitals (cf. Fig. 2.1) are characterized by discrete energy values $\mathcal{E}(n, l, m)$. The principal quantum number n selects a sequence of shells, characterized by circular Bohr radii r_n, which correspond to standing-wave modes

$$n\lambda = 2\pi r_n \quad \text{or} \quad \mathbf{k} = n\frac{1}{r_n}, \text{ with } n = 1, 2, 3, \cdots \qquad (2.6)$$

(Fig. 2.7). A sequence of quantized angular momenta is compatible with each integer n, prescribed by the azimuthal quantum number l as

$$p' = \sqrt{l(l+1)}\hbar, \quad \text{with } l = 0, 1, 2, \cdots, (n-1). \qquad (2.7)$$

An electron in any of these orbitals (n, l) spreads over all space, bound to a central field. To localize its orbital with reference to a space-coordinate system requires some physical reference axis, e.g., a magnetic field. The magnetic quantum number m then specifies the quantized component of the angular momentum in field direction. Thus three quantum numbers prescribe an atomic orbital in space. In crystal space, the

---

[8] Diagram of metallic TiO (nondefective NaCl structure, $a = 4.181$ Å, calculated by the APW method), derived by V. Ern, *Tech. Rep. 192*, Lab. Ins. Res., Mass. Inst. Tech., October, 1964. Bands along several principal symmetry directions of the Brillouin zone of Fig. 2.4. Starting from right to left, the diagram shows the behavior of the eigenvalues with k along the paths Γ-K-W-X-Γ [in the (001) plane] and Γ-L-U-X. The bands are drawn in the reduced zone scheme, and the symmetries of the wave functions labeled according to the standard notation of Ref. 9. On the right a schematic representation of the different bandwidths and overlaps. Labels show the predominant character of each band. The electronic conduction is predominantly in the $3d$ band of t_{2g} symmetry. F.E. = Fermi level.

[9] Cf. L. P. Bouckaert, R. Smoluchowski, and E. P. Wigner, *Phys. Rev.* **56**, 58 (1936).

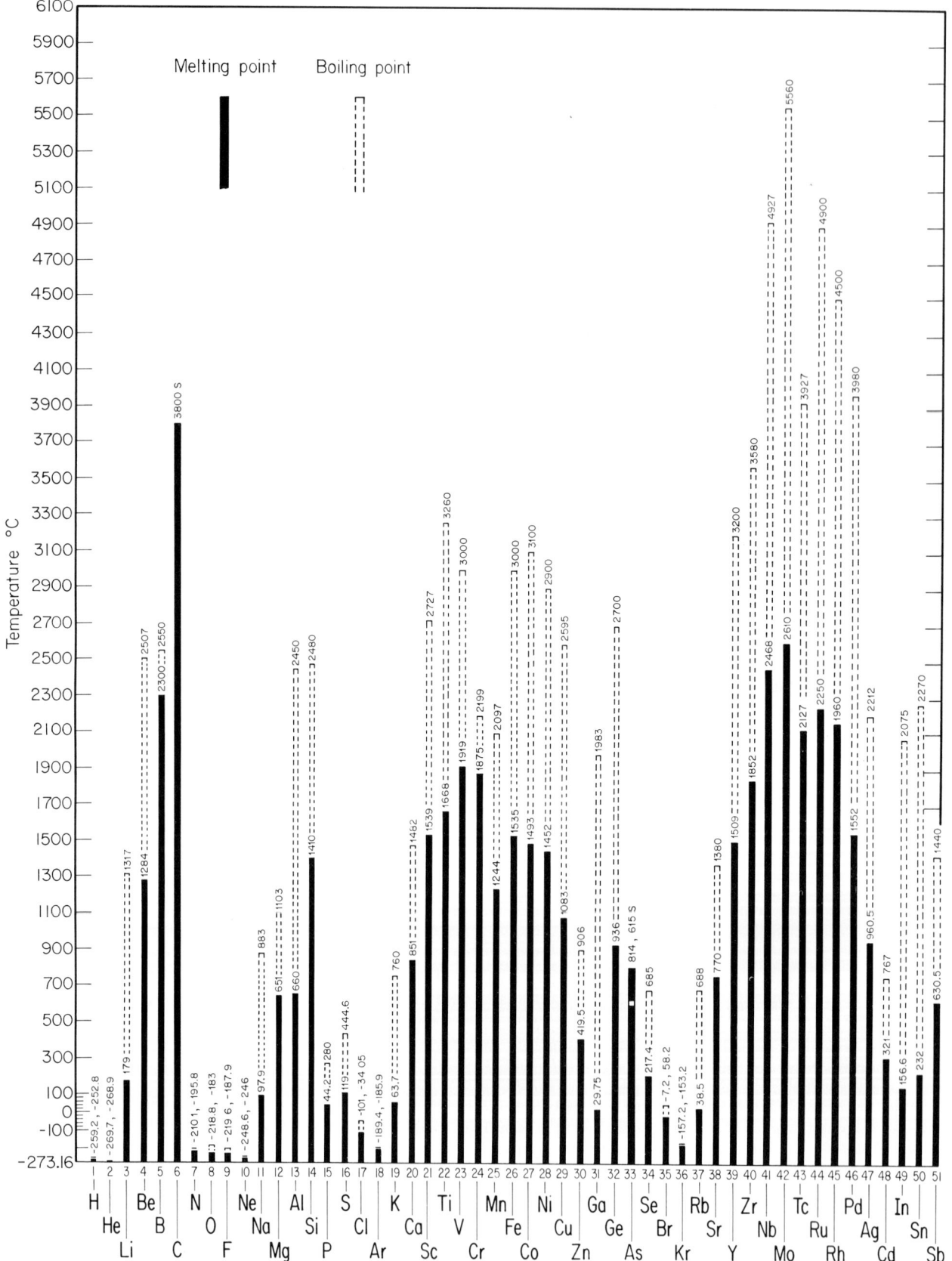

Fig. 2.8. Melting and boiling

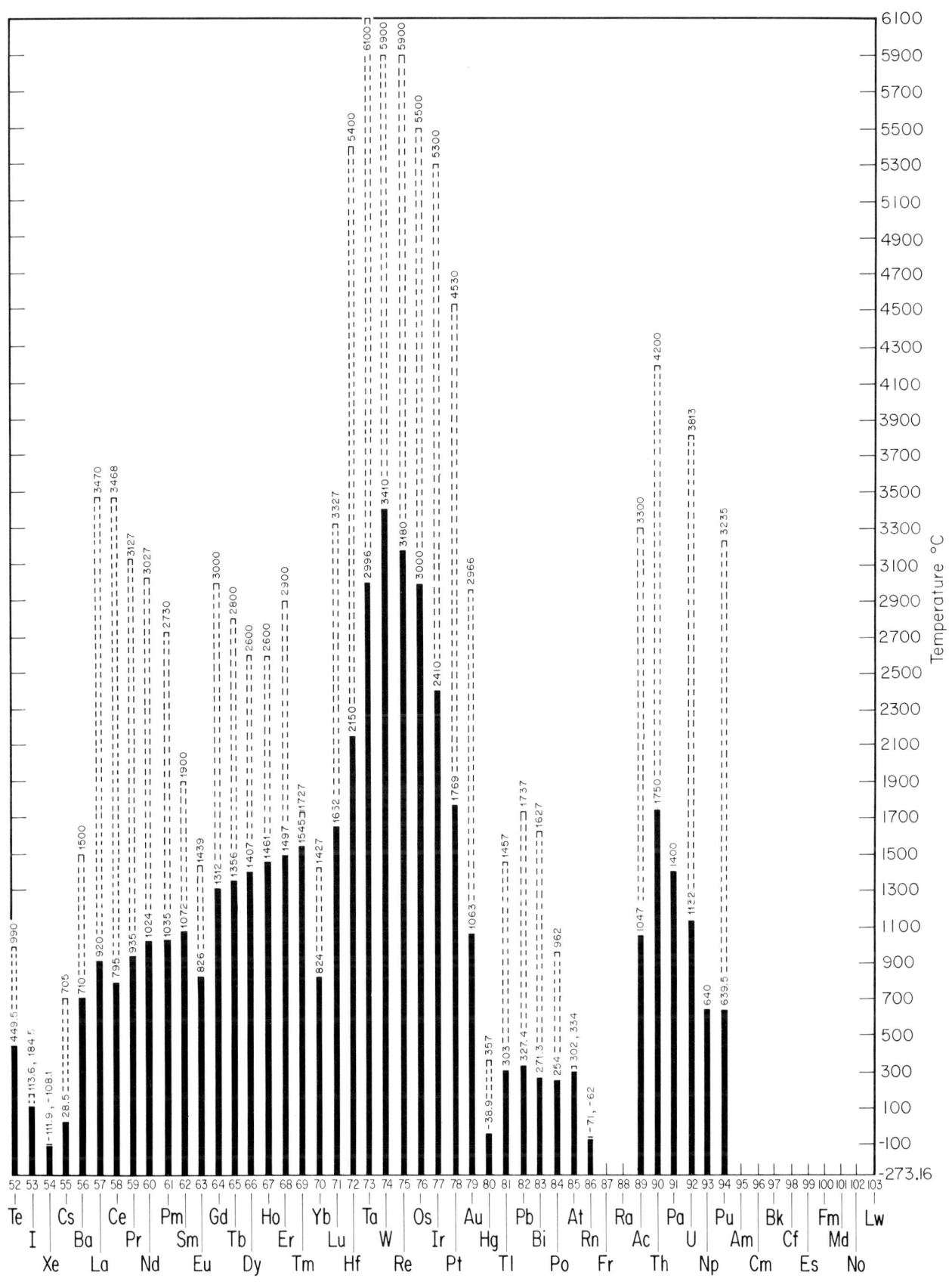

temperatures of the elements.

role of these quantum numbers is taken over by the three momentum components in k space,

$$\mathcal{E} \equiv \mathcal{E}(k_i, k_j, k_k). \qquad (2.8)$$

In making actual calculations of these "one-electron states" of crystals, the atom types alone do not suffice; one must introduce at present the actual crystal structure and the nuclear distances as experimental information. When the atoms are inserted into their lattice positions, the potential-energy curves—obtained from Hartree-Fock calculations[10]—intersect. At the closest approach distance between two neighboring atoms that point is marked where the intersection potential curves are of equal height. This distance from the two nuclei is used to define their atomic radii. The atomic shells thus described cut off a certain amount of electronic charge from the free atoms; hence, these lattice atoms have a positive net charge. Their residual electronic charge is spread through the interstitial space as Bloch waves, properly joined to the truncated atomic wave functions of the atoms (APW method).[11] The joining procedure adds localized electronic charge to the atom spheres. This pileup of charge between atoms corresponds to the molecular bonds discussed before.

The $\mathcal{E}(n, k)$ diagrams of the crystal (where n is not a quantum number) are thus correlated to the wave functions of the composing atoms. The energy bands, in addition, conform to the group symmetry of the crystal lattice, because the Bloch wave functions contain this symmetry.

Atomic structure and the boiling points of the elements

When atoms coagulate, the main energy release takes place in condensation to the liquid state (condensation energy $\mathcal{E}_c$); any subsequent crystalline ordering (heat of fusion $\mathcal{E}_f$) represents in general a minor adjustment. The trend in boiling temperatures T_b should therefore reveal much about the electronic interaction of the partners.

A first glance at the boiling and melting temperatures of the elements (Fig. 2.8) conveys a silhouette as complex as the skyline of Manhattan. Only the minima are easily explained: in New York as parks or blighted areas, in the periodic system as condensed phases of saturated atoms or molecules, with H_2 and the rare

---

[10] Cf. F. Herman and S. Skillman, *Atomic Structure Calculations*, Prentice-Hall, Englewood Cliffs, N. J., 1963.

[11] APW = Augmented Plane Wave (method), cf. J. C. Slater, *Phys. Rev.* **92**, 603 (1953); M. M. Saffren and J. C. Slater, *ibid.* **92**, 1126 (1953); J. H. Wood, *ibid.* **126**, 517 (1962); G. A. Burdick, *ibid.* **129**, 138 (1963); there are also various other methods in use for band calculations.

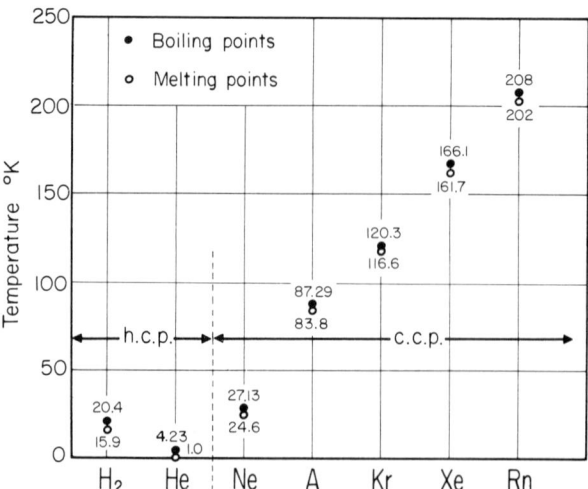

Fig. 2.9. Melting and boiling points of H_2 and the rare gases.

gases as prime examples (Fig. 2.9). He and Ne are shielded by a completed K shell $(1s^2)$ and L shell $(2s^2 2p^6)$, respectively, and the other rare gases by completed subshells $s^2 p^6$. The rising trend in melting and boiling points for the heavier rare gases reflects growing van der Waals–London attraction by phase relations between increasing numbers of electrons.

Lewis, Kossel, and others emphasized at an early date the importance of such octet shells and the tendency of atoms to surround themselves with this kind of saturated electron environment by pair-bond formation. The "8-N rule" was formulated, where N, the column number of an atom in the periodic system, ranges from I to VIII (Fig. 2.10), and $8 - N$ signifies

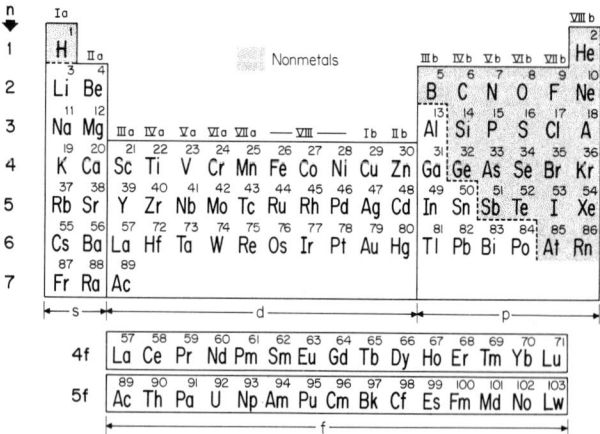

Fig. 2.10. Periodic table of the elements, emphasizing 8-N grouping and distribution of metals.

the number of bonds needed to form a simile of an inert rare-gas shell $(s^2 p^6)$ around a reference atom. When this rule is applied to ascertain the simplest

Atomic Structure and Boiling Points of Elements

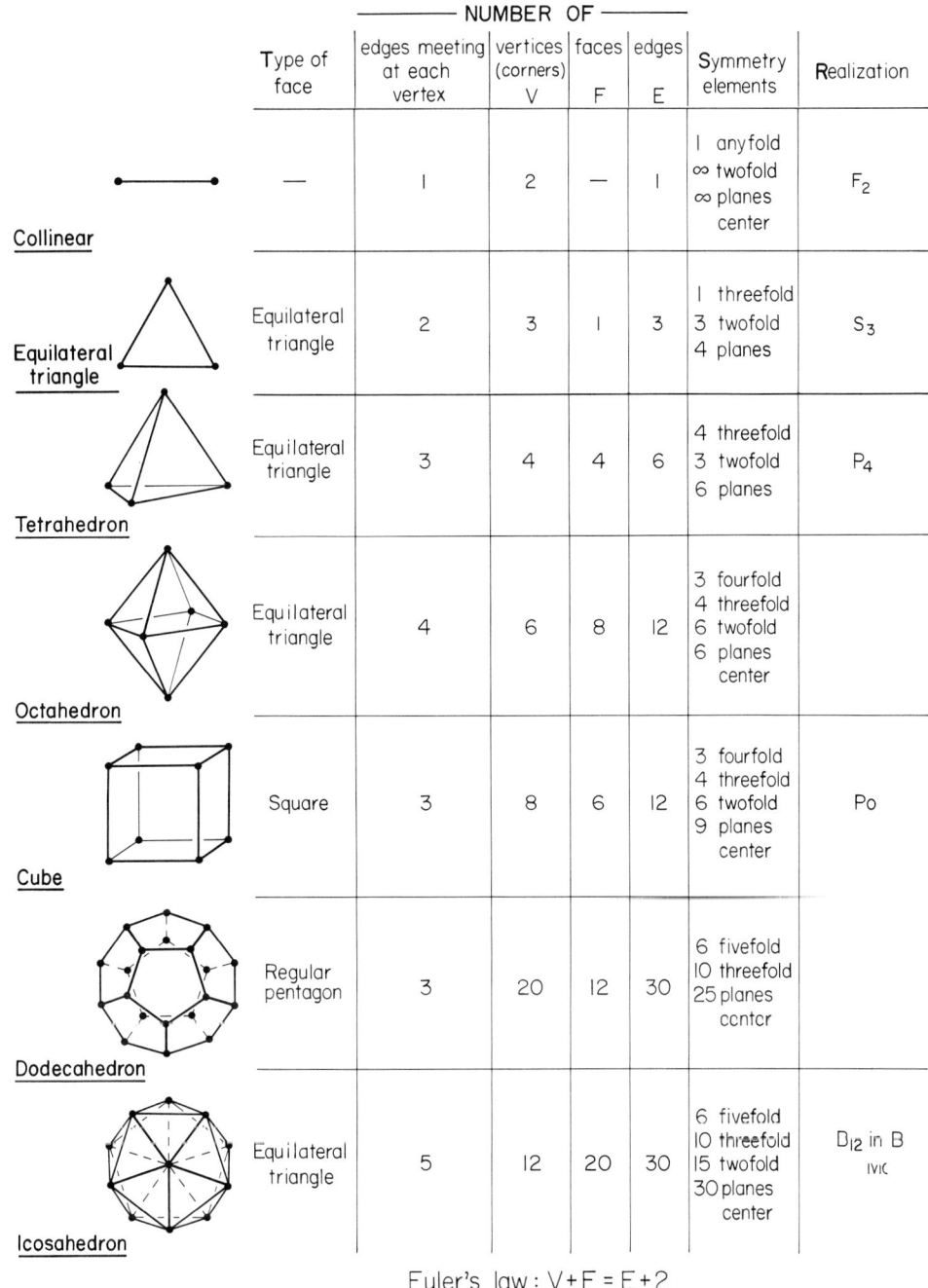

Fig. 2.11. Combination of alike atoms in geometrical constellations according to 8-N rule.

possible molecule structure for alike atoms on identical sites, Fig. 2.11 results. For $N = 7$, diatomic molecules are formed; for $N = 6$, triangular ring structures; for $N = 5$, the atoms are placed at the corners of a *tetrahedron* (the *cube* with 8 and the pentagonal *dodecahedron* with 20 atoms would also qualify); for $N = 4$, the corners of an *octahedron*; and for $N = 5$, those of an *icosahedron* might serve. At this stage the possibility of building molecules as *regular* polyhedra ends. There exist no more than the five regular polyhedra (italicized), as is well known since early Greek times. If more than five bonds (edges) are to meet at one corner atom, unequal sites have to be admitted, or structures must be spun out into crystals, e.g., into the simple cubic lattice of Po with six bonds per atom (cf. Fig. 2.14).

A closer look at the 8-N rule shows that its application cannot simply consist in counting electrons missing from an octet shell and furnishing such electrons from a corresponding number of nearest neighbors. The octet constellation is subdivided into four orbitals (one s and three p orbitals or hybrids formed therefrom), each able to hold two electrons with antiparallel spins. From $N = 8$ to $N = 4$, the individual atom by itself can still supply at least one electron per orbital and thus establish a framework linking its electrons in fixed stereoconstellations to the surroundings. The atoms from $N = 3$ down can provide only fractional electron charges per orbital from their own supply. If not able to steal electrons outright from electropositive neighbors, they must "resonate" electrons between various bond positions (as in Pauling's structure concepts) or pool them as in metals and the electron-band approach.

The maximum number of electrons that can be made available for bonding per atom in this s, p situation is four; hence, $N = 4$ (the group IV atoms) should provide the high-water mark of cohesion and a primary divide between metals (resonating bonds) and nonmetals (fixed bonds). The boiling-point characteristics for the rows of elements completing $s^2 p^6$ shells confirm this (Fig. 2.12): The main maxima lie at $N = 4$. Secondary trends toward both ends of the 8-N curves add finer details to the understanding of condensation.

At the metal side (left) the condensation energy stems mainly from the difference in binding energy of the valence electrons, held by the free atom versus the atom in liquid surroundings. The rigidity of the bonds connecting the valence electrons to their own core is reflected in the energy and spectral type of the first atomic excitation $\mathcal{E}_e$. Frequently the condensation energy is smaller than this excitation energy, but as the balance $\mathcal{E}_c - \mathcal{E}_e$ tends to become more positive, the boiling temperatures should rise, because condensation can proceed in face of a larger thermal randomizing energy. Table 2.2 bears this out and accounts for a variety of trends: the surprising shift in T_b for $8 - N = 7$ and 6; the secondary maxima for Cu, Ag, and Au reflecting the added bonding action of just completed but not yet firmed-up $3d$, $4d$, and $5d$ subshells; the effect for

Table 2.2. **Predicted sequence of boiling points***

Sequence of Elements		$\mathcal{E}_e$	$\mathcal{E}_c$	$\Delta \mathcal{E}$	Boiling-Point Sequence Predicted by $\Delta \mathcal{E}$ (from high to low)
8 − N = 7		$s \to p$			
	Li	1.8	1.4	−0.4	Li
	Na	2.1	0.95	−1.15	Cs
	K	1.6	0.8	−0.8	right — K
	Rb	1.54	0.71	−0.83	Rb — wrong
	Cs	1.37	0.68	−0.68	Na
		$d^{10}s \to d^9 s^2$			
	Cu	1.4	3.2	+1.8	Au
	Ag	3.7	2.6	−1.1	right — Cu
	Au	1.1	3.3	+2.2	Ag (pulled down by incipient $4f$ shell)
8 − N = 6		$s^2 \to sp$			
	Be	2.7	3.1	0.4	Be
	Mg	2.7	1.3	−1.4	Ba
	Ca	1.85	1.56	−0.29	right — Ca
	Sr	1.75	1.4	−0.35	Sr
	Ba	1.5	1.53	+0.03	Mg
		$s^2 \to sp$			
	Zn	3.95	1.2	−2.75	Zn
	Cd	3.68	1.04	−2.64	? Cd (pulled down by incipient $4f$ shell)
	Hg	4.6	0.6	−4.0	Hg

* $\mathcal{E}_e$ = lowest electronic excitation energy of atom; $\mathcal{E}_c$ = condensation energy of atom; $\Delta \mathcal{E} = \mathcal{E}_c - \mathcal{E}_e$ (energies given in eV).

Atomic Structure and Boiling Points of Elements

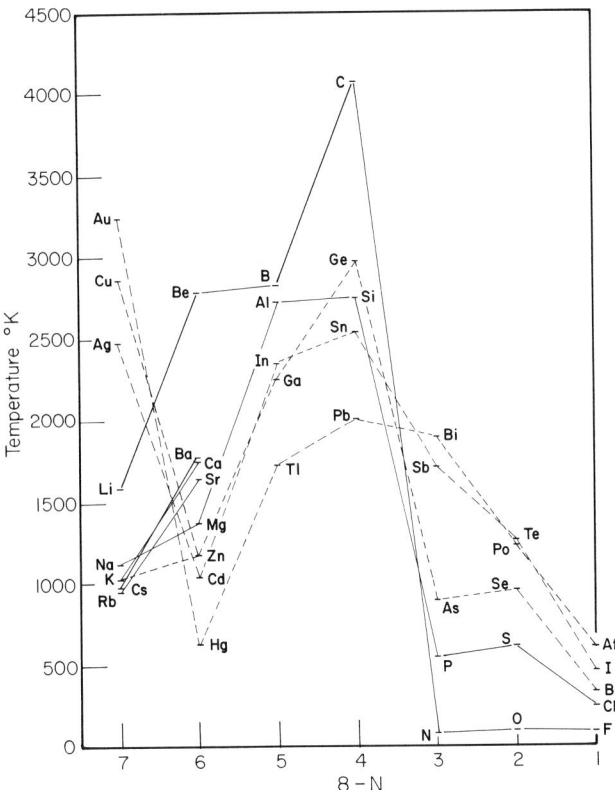

Fig. 2.12. Boiling-point characteristics of elements completing s^2p^6 shells.

silver as being smallest, because the incipient filling up of the $4f$ shell exercises its inward pull; the scrambled sequence of the alkaline-earth metals, etc. Only the sequence Na → K does not fit such simple analysis; it requires a more sophisticated quantum-mechanical appraisal of the electron energies in the Brillouin zones.

Toward the nonmetal region (right) the crossing over of the boiling-point curves is caused by systematic bond-type changes. The dominance of the saturated triple bond (N≡N) and double bond (O=O) in the second row enforces molecule formation and low melting and boiling points (Fig. 2.13). It is broken in the third

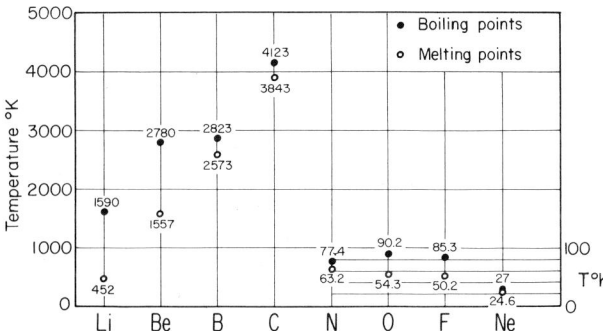

Fig. 2.13. Melting and boiling points of the elements of the second row.

row of the elements: three single bonds for P or two single bonds for S become stronger than a triple or double bond, respectively. In the still higher rows the van der Waals–London binding energy of the many-electron systems becomes increasingly pronounced and pulls the structures into metallic binding (cf. Fig. 2.10). A prototype example of the systematic transition from single bonding to metallic structure is provided by the sequence Se → Te → Po with its change from chain lattice to simple cubic structure (Fig. 2.14).

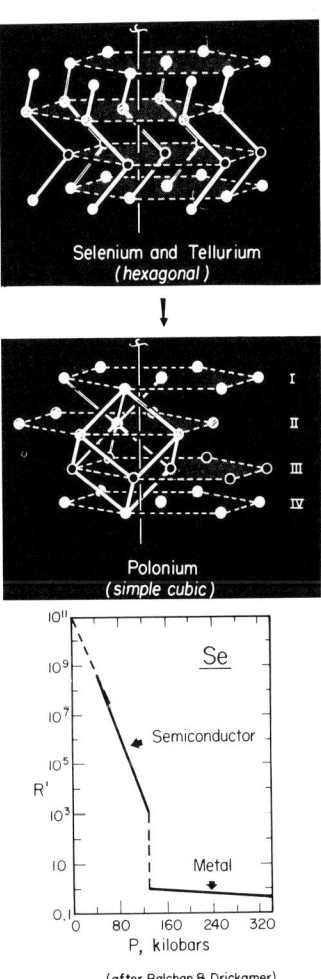

Fig. 2.14. Transition from single bonds in chain lattice to cubic metal structure.

The boiling-point curves for these s^2p^6 elements show a general decrease from L to P shell: the better the nuclei are shielded by firmed-up filled shells or sub-shells, the lower the Coulomb binding. An opposite trend dominates the d-shell transition elements $(10 - N)$ (Fig. 2.15). The half-filled $3d$ subshell is relatively stable and tends to hold on to its unpaired electrons, to the detriment of cohesive binding: T_b

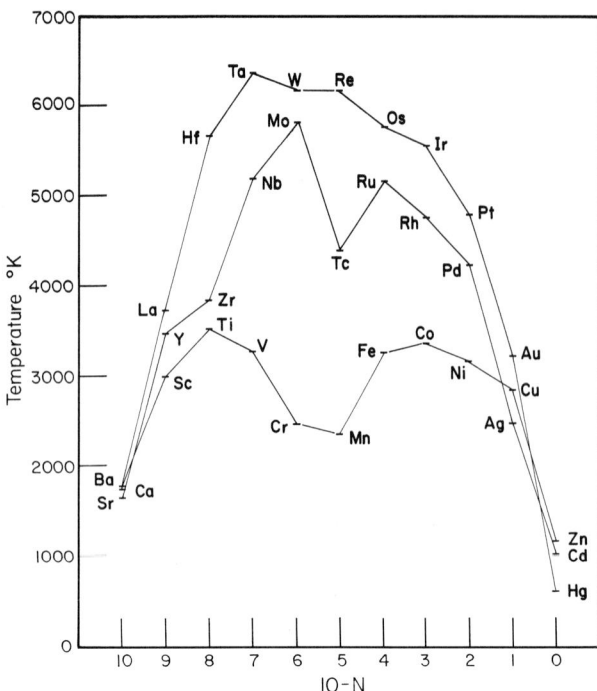

Fig. 2.15. Boiling points of elements completing d shells.

$(14 - N)$ enters the energy balance only through inward pull on the valence electrons; the boiling points are depressed (Fig. 2.16). The stability of the half-filled subshell ($N = 7$) reduces the central force field of the atoms; hence, a maximum of T_b is observed for gadolinium.

The melting points of the elements

The qualitative trend of the boiling-point characteristics of the elements becomes understandable when one considers the number of binding electrons provided per atom and the attraction of the core in competition with the surroundings. The melting-point curves (Figs. 2.17 to 2.19) image in a general way the habits of the T_b

has a minimum at $N = 5$. The $4d$ and $5d$ subshells overlap successively more with their surroundings, thus increasing the mutual bonding. A maximum for T_b near $N = 5$ results, in analogy to that at $N = 4$ for the octet shells.

The well-shielded $4f$ shell of the rare-earth elements

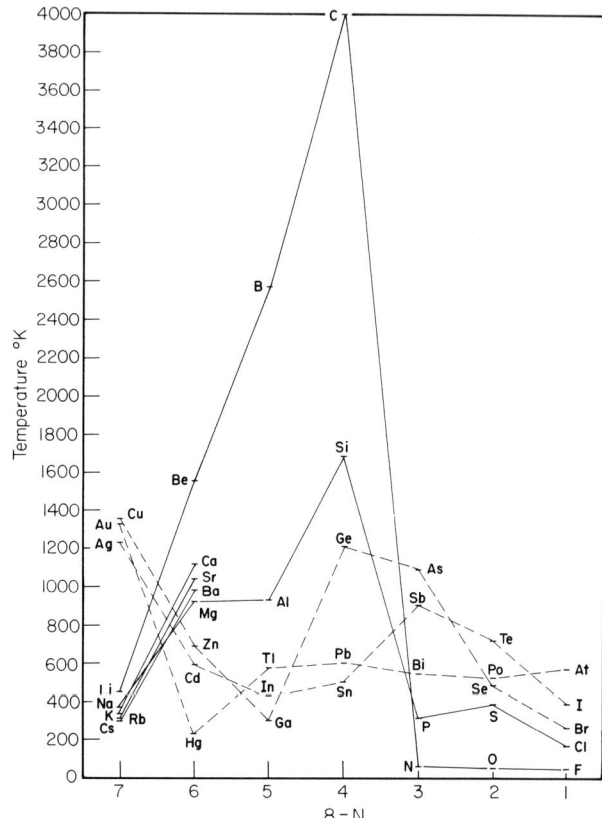

Fig. 2.17. Melting points of elements completing s^2p^6 shells.

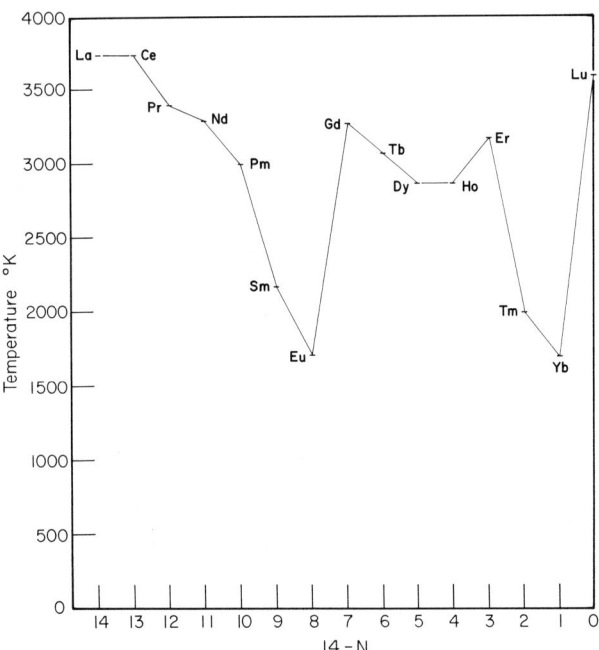

Fig. 2.16. Boiling points of rare-earth elements.

curves: $N = 4$ still shows maxima in the $8 - N$ diagrams; the value for Ag remains depressed in reference to those of Cu and Au, and Mg stays at the bottom of the alkaline-earth metals. Such aspects tempt to coin the slogan "good coupling condenses and orders." However, a look at the very low melting points of Al, Ga, and In warns that the second half of this pronouncement needs closer scrutiny.

The crystallizing tendency of condensed matter

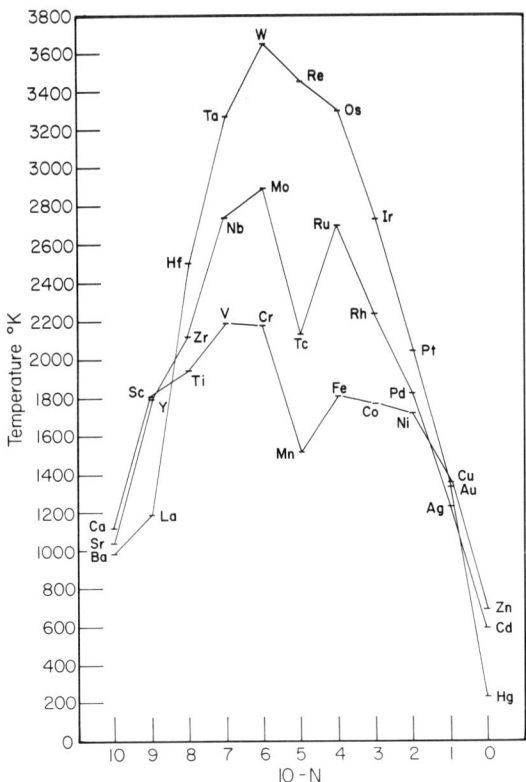

Fig. 2.18. Melting points of elements completing d shells.

stems from the improvement in the overlap of orbitals that can be realized for an organized versus a randomized array. When one operates with well-defined eccentric wave functions, e.g., the sp^3 tetrahedral bonds of C, Si, Ge, the establishment of long-range order obviously improves the average overlap; hence, the melting points are high. For Sn, however, the tetrahedral bonds begin to be submerged in multielectron

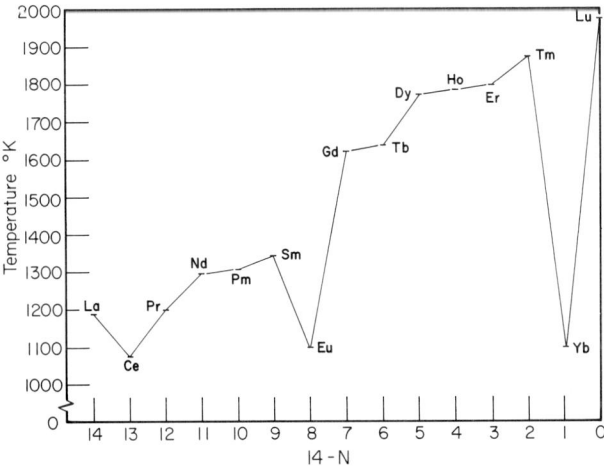

Fig. 2.19. Melting points of rare-earth elements filling the $4f$ shell.

attraction, as the phase transition from the diamond to the h.c.p. structure testifies. The energy gain attainable by ordering is lowered and the melting point depressed.

More spectacular is the sudden drop in melting temperature in the sequence Al, Ga, and In. A reasonable explanation can be offered by considering the first atomic excitation state. For boron $(2s^2, 2p)$ this excitation increases the bonding electrons by two $(2s, 2p^2)$, and the eccentric p orbitals favor crystallization; hence, the melting point of B is high. For Al, Ga, and In the first excited state promotes the p electron into the next higher s orbital $(3s^2 3p \to 3s^2 4s,$ etc.); the s^2 pair is not broken up, and the spherical symmetry of the lone s electron is not conducive to crystallization. Only when the metal is well cooled down will the unexcited p state come into its own and make long-range ordering attractive.

If this type of argument is correct, the change of average electronic structure with temperature may play an important role in crystallizing certain phases.

Dense-packed structures

About three quarters of all elements condense as metals (cf. Fig. 2.10). Their structure is dominated by three arrangements providing the greatest number of near neighbors: the hexagonal close-packed (h.c.p.) and cubic close-packed (c.c.p.) arrays with 12 equally distant atoms, and the body-centered cubic (b.c.c.) structure of 14 neighbors, 8 at distance $a\sqrt{3}/2$, 6 at distance a (Fig. 2.20).

Packing concepts conjure the picture of definite atomic and ionic radii, used so successfully in structure analysis by Goldschmidt, Pauling, and others.[12] Identical atoms, packed like oranges in a crate, would produce two types of close-packed arrays with layer repeat sequence $ABAB\cdots$ (h.c.p.) or $ABCABC\cdots$ (c.c.p.). Which of the three metal structures is actually realized depends on the electron sharing in the condensed phases.

A simple correlation between electronic configuration of the atom and metal structure has been proposed by Engel[13] and applied with apparently spectacular success by Brewer[14] to the prediction of high-temperature metallic phase diagrams. The "Engel rule" states that only the s and p valence electrons decide the

---

[12] Cf., e.g., the summarizing table in A. von Hippel, *Molecular Science and Molecular Engineering*, The M.I.T. Press and John Wiley and Sons, New York, 1959, p. 148.

[13] N. Engel, *Kem. Maanedsblad*, Nos. 5, 6, 8, 9, 10 (1949).

[14] L. Brewer, *Report UCRL-10701*, University of California, Radiation Laboratory, July, 1963.

arrangement and that the b.c.c.(I), h.c.p.(II), and c.c.p.(III) structures correspond to the presence of 1, 2, and 3 valence electrons, respectively.

At first glance, this seems oversimplification. Why should the d orbitals play no part in enforcing a specific long-range order? The answer for most metals seems to be that the strong overlap of d orbitals does not extend beyond nearest neighbors. Thus the influence of the s and p electrons on the next-nearest neighbors decides the long-range ordering of such closely related structures.

In contrast to this atomic-orbital approach, the same structure competition in dense-packed metals has been explained previously by invoking the electron-band

Compound formation and polarity

When unlike atoms are admitted into structures, a new design element enters through the polarity of the partners. Rare-gas-type electron constellations can now be built in two ways: The electronegative atoms accept electrons to build up toward saturated configurations, while the electropositive ones donate electrons and tend to regress to subshells previously completed. Striving thus toward two types of balanced atmospheres by oxidation and reduction, the anions and cations organize their electronic surroundings so that the electron spins are in accord with the formal ionic charges Cl^-, O^{2-}, Na^+, Ca^{2+}, etc.

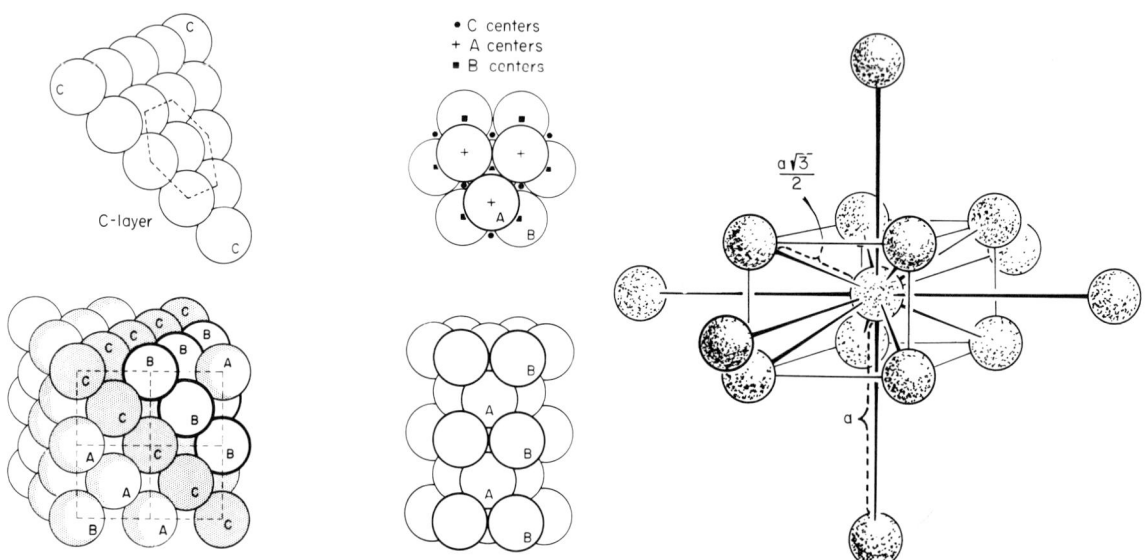

Fig. 2.20. The close-packed structures of identical spheres and the body-centered cubic structure.

model and the electron density in these bands. Hume-Rothery[15] observed that the appearance of structures I to III in alloys depends on certain ratios of valence electrons to atoms. Since the electron distribution in the bands is ruled by Fermi-Dirac statistics, the "Hume-Rothery rules" imply that it is energetically more profitable to switch to a new crystal structure when the filling of energy states approaches the edge of the Brillouin zone (cf. Chap. 16).

It seems that both "rules" actually explain the same situation with different words, as long as d electrons can be treated as localized.

[15] W. Hume-Rothery, *Structures of Metals and Alloys*, Institute of Metals, London, 1936; *Elements of Structural Metallurgy*, Institute of Metals, London, 1961.

It is meaningful to assign to ions average radii for packing considerations but misleading to endow these spherical ions with their formal charges in the expectation that they will behave electrostatically like spheres thus charged. Large potential differences cannot be maintained in the presence of bonding electrons (a statement akin to Pauling's electroneutrality rule[16]). The electrostatic effect of Fe^{3+}, for example, may be smaller than that of Fe^{2+}, because the compensating electrons of the anion counteratmosphere penetrate closer to the nucleus.

A regular array of polar interpenetrating electron atmospheres, as compared with a similar nonpolar

[16] Cf. L. Pauling, *The Nature of the Chemical Bond*, 3rd ed., Cornell University Press, Ithaca, N. Y., 1960.

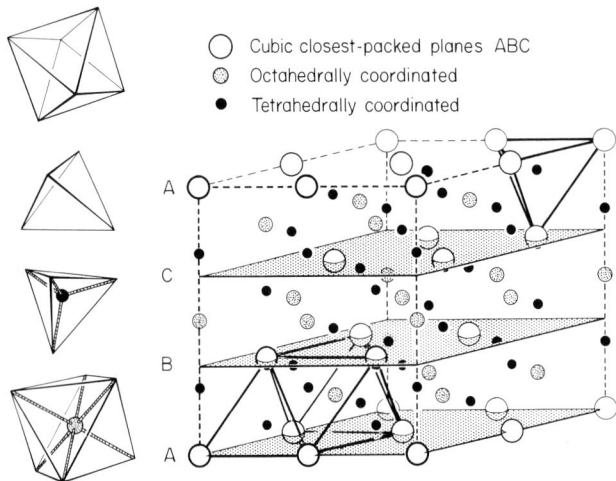

Fig. 2.21. Close-packed structures built from tetrahedral and octahedral modular units.

built up from octahedral and tetrahedral modular units (cf. the model approach of Loeb[17]) (Fig. 2.21). The centers of the octahedra lie halfway between two lattice planes, the centers of the tetrahedra at $\frac{1}{4}$ and $\frac{3}{4}$ of the separation distance. Since the centers of the octahedral and of the two tetrahedral positions form close-packed arrays by themselves, polar crystals can be constructed by proper use of interpenetrating lattices as cation and anion sites.

Occupation of the octahedral sites of a c.c.p. anion lattice with cations leads to the rock salt structure; similarly, the calcium fluoride structure is obtained from a c.c.p. alkaline-earth cation lattice by filling all tetrahedral sites with fluorine anions (Fig. 2.22). These three interpenetrating polar sublattices provide still stronger bonding. The melting points for such compounds of saturated polar design are relatively high and tend (in general) to decrease, as larger ions of lower polarizing power and electron affinity are substituted (Fig. 2.23).

Systematic structure alteration can be made by filling only a fraction of the cation sites through insertion of positive ions of higher valency, as for example in the corundum or spinel lattice (Fig. 2.24). Also some anions may be replaced by cations or anion sites left

array, will strengthen cohesion, i.e., raise the boiling points. Also, the melting points should increase because crystallization can produce higher energy gain than in the case of identical particles, by maximizing in long-range order the polar attraction and minimizing the polar repulsion terms. However, if the ionicity

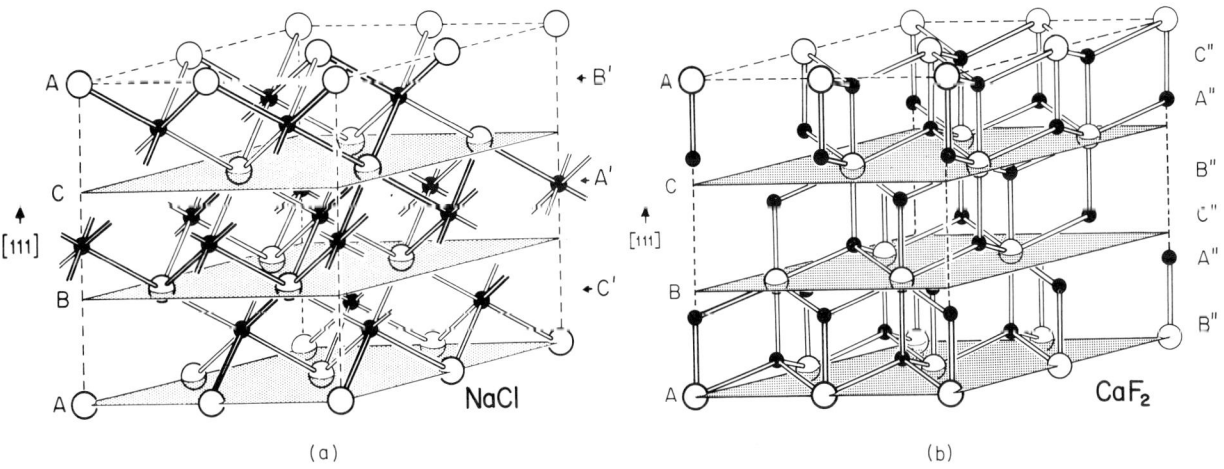

Fig. 2.22. Derivation of (a) NaCl and (b) CaF_2 structure from interpenetrating dense-packed lattices.

leads to clustering in saturated groups, the structure is weakened and its breakup into molecules facilitated. Both trends are clearly discernible in the melting and boiling points of binary halides and oxides.

For a convenient description of the situation we return to the concepts of dense packing. A close-packed structure of N lattice points contains N octahedral and $2N$ tetrahedral interstices, hence can be

empty, as in the perovskite and WO_3 structures (Fig. 2.25). How far such partial substitutions weaken the cohesion depends on the balance between increased bond strength caused by higher cation valency versus reduced bridging due to partial site occupation. Clearly, when one increases the clustering too far, the

[17] A. L. Loeb, Acta Cryst. 11, 469 (1958); A. L. Loeb and G. W. Pearsall, Am. J. Phys. 31, 190 (1963).

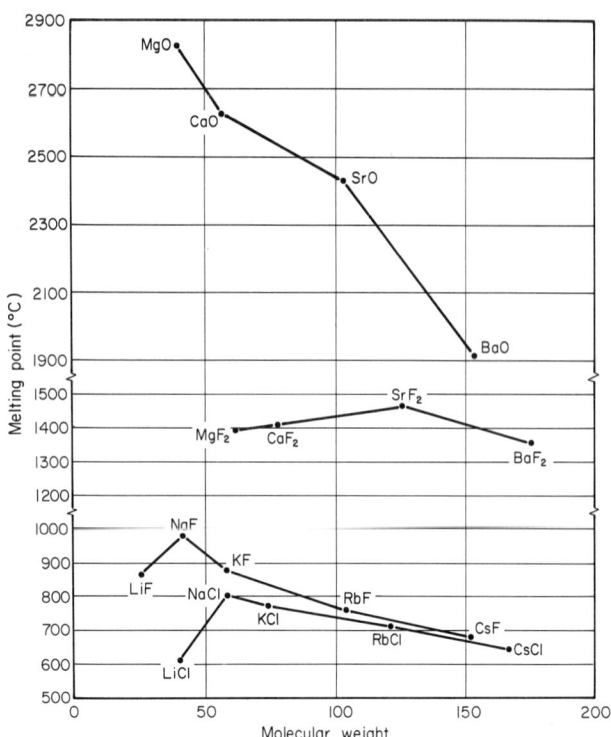

Fig. 2.23. Melting points of some binary halides and oxides.

structure tends to fall apart and saturated molecules form, like SiF_4, SF_6, or Mn_2O_7, condensing at low temperatures (Fig. 2.26).

Cations of the transition metals offer new design possibilities and complications due to multivalency caused by incompleted d or f shells. Here we find the materials producing magnetic spin lattices and components leading in systematic transition from insulator to metal (e.g., $TiO_2 \rightarrow Ti_2O_3 \rightarrow TiO \rightarrow Ti$) (cf. Chap. 4).

The partial filling of interstitial sites, furthermore, allows a subtle shift in architecture from dense-packed to framework structures emphasizing individual bonds. Thus the vista opens to the endless variety of designs dominant in carbon and silicate chemistry (cf. Chap. 3) and the building concepts of the organic world.

Insulators and conductors

The "one-electron model for crystals" discussed earlier distinguishes (in its most simple application) between insulator, semiconductor, and metal by comparing the optical gap width ε_g between filled and empty electron states with the average thermal energy kT. One judges by the number n_c of conducting electrons per unit volume intrinsically made available

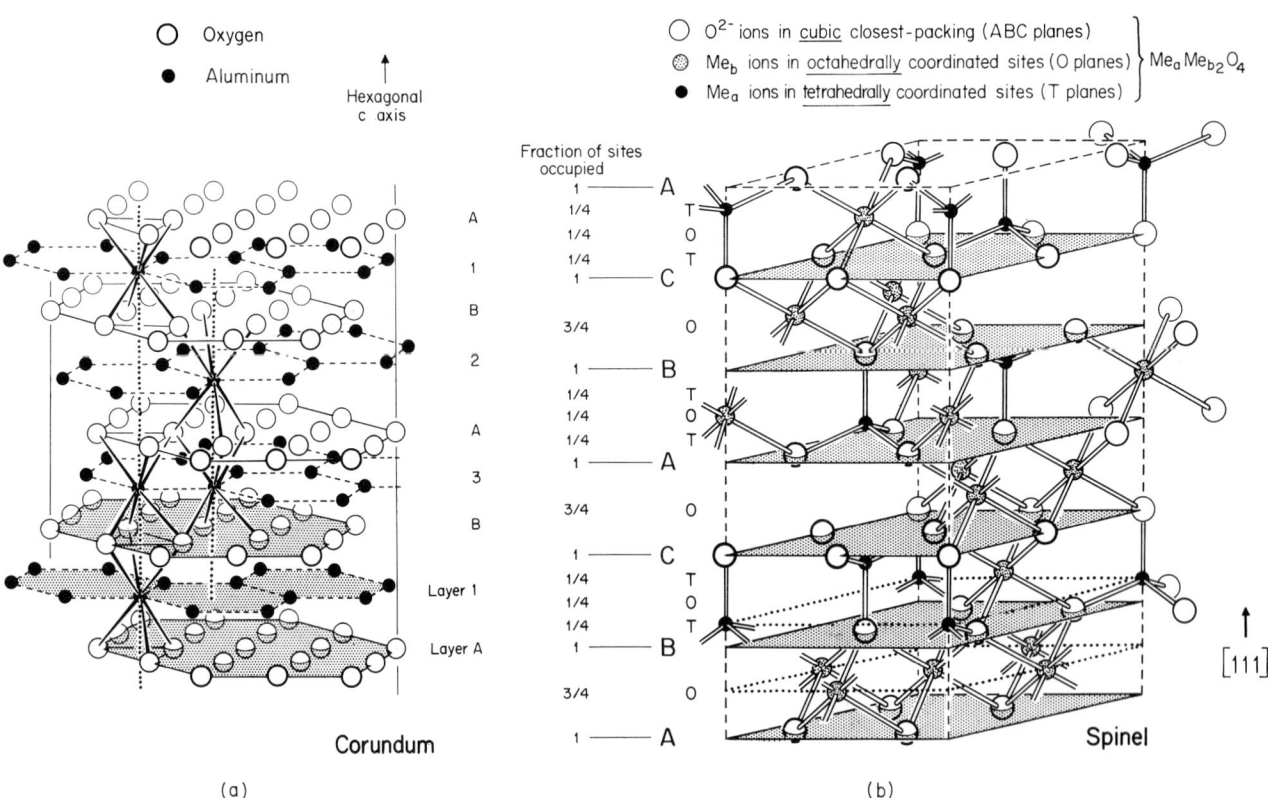

Fig. 2.24. The structure of (a) corundum and (b) spinel.

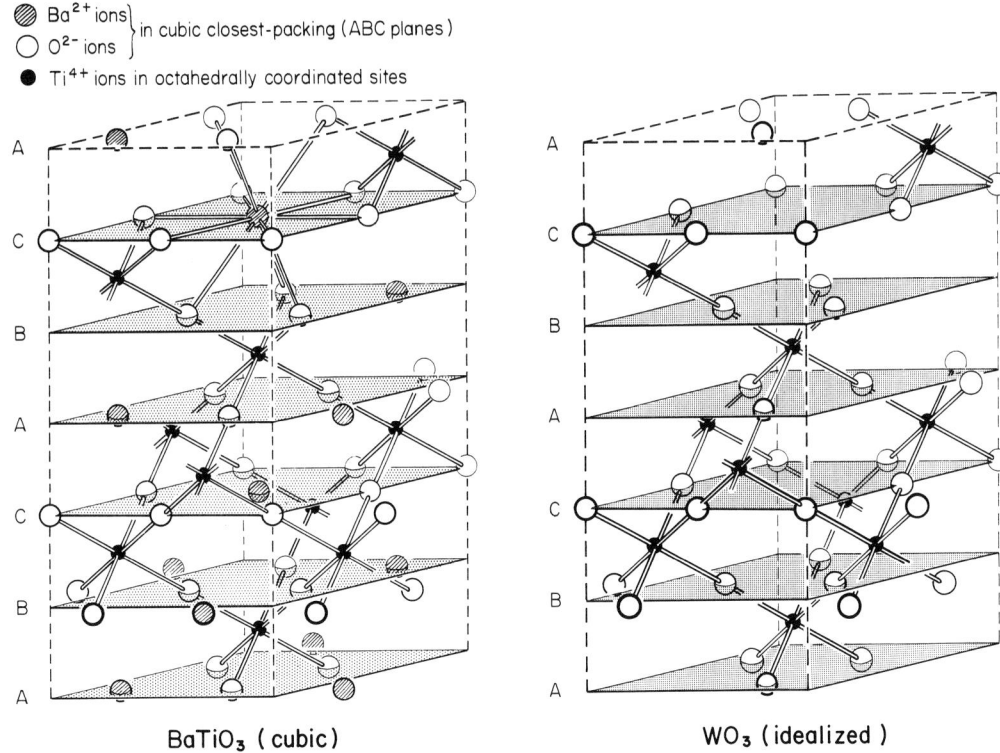

Fig. 2.25. The perovskite and WO₃ structures.

through thermal excitation according to Fermi-Dirac statistics:[18]

$$n_c = 2 \left(\frac{2\pi m^* kT}{h^2}\right)^{3/2} e^{-\varepsilon_g/2kT}. \qquad (2.9)$$

At room temperature the density of conducting electrons is about

$$n_c \simeq 10^{19} e^{-\varepsilon_g/2kT} \quad [\text{cm}^{-3}]. \qquad (2.10)$$

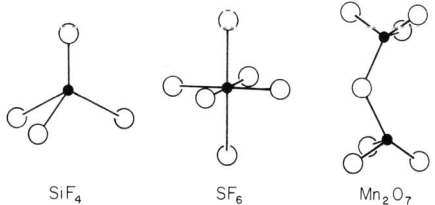

Fig. 2.26. Saturated polar molecules.

The conductivity caused by these electrons of charge $e \simeq 1.6 \times 10^{-19}$ [coul] and mobility b is

$$\sigma = ebn_c \quad [\text{ohm}^{-1}\,\text{cm}^{-1}]. \qquad (2.11)$$

The electrical engineer likes to draw some defining boundary lines between insulators, semiconductors, and metals on the basis of such conductivity,

Insulator	$\sigma < 10^{-9}$
Semiconductor	$10^{-9} < \sigma < 10^2$
Metal	$\sigma > 10^2$

For a representative mobility $b \approx 1$ [cm²/volt sec] the boundary between insulator and conductor falls at $n_c \approx 10^{10}$ (Fig. 2.27).

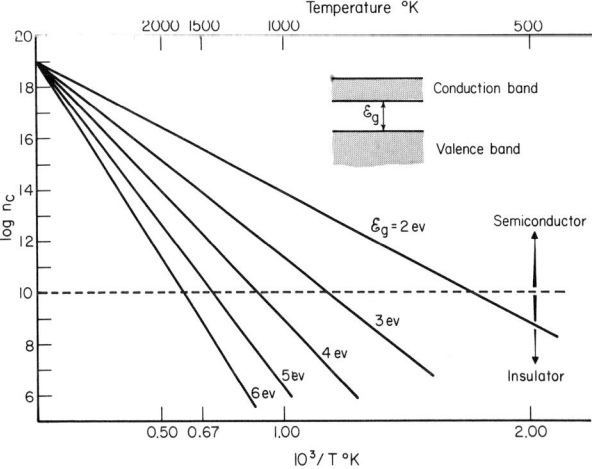

Fig. 2.27. Thermal excitation of electrons into conduction band as function of temperature and gap width.

[18] Cf., e.g., N. B. Hannay, *Semiconductors*, Reinhold Publishing Corp., New York, 1959.

While such yardsticks give a useful order-of-magnitude feeling, any detailed discussion has to go back to $\mathcal{E}(\mathbf{k})$ diagrams (cf. Fig. 2.6) and to visualize the location and shape of the Fermi energy surface $\mathcal{E}_F$[19] defined by the Fermi-Dirac probability function for thermal occupation

$$f(\mathcal{E}) = \frac{1}{1 + e^{(\mathcal{E} - \mathcal{E}_F)/kT}}. \tag{2.12}$$

It has to determine how doping and defects, surface states, carrier injection, and space charge affect the Fermi level and to what extent the electrons and holes really act collectively, as the band model implies, or must be treated as localized. In the latter case we return to molecular-orbital considerations and ligand-field theory.[1,2]

The more polar or imperfect a crystal, the more its conducting electrons tend to advance by trapping and release processes. This trend can be carried to the extreme by obliteration of the long-range order. Now electrons in low fields travel practically with the mobility of ions, ionic conduction begins to dominate the scene, and the localizing activation energies depend on the compensating countercharges—the ionic and electronic atmospheres—and the statistically available interstitial space.

As one thus proceeds from the simple order of ideal to the complexities of real systems, the predicting ends and imaginative experimentation takes over. This theme will dominate all subsequent chapters. But whatever one learns turns finally into an enlarged capability of "building from atoms."

---

[19] *The Fermi Surface*, W. A. Harrison and M. B. Webb, Eds., John Wiley and Sons, New York, 1960.

3 · COMPETITIVE STRUCTURES AND PHASE TRANSITIONS

Helen D. Megaw

Introduction — Packing structures and framework structures — Feldspars and perovskites as framework structures — Structure families, aristotypes, and heterotypes — Character of transitions — Structure changes and properties of perovskites — Free energy and displacive transitions — Displacive changes in feldspars and their interaction with ordering effects — Twin domains and antiphase domains — Antiphase domains in a feldspar (anorthite) — Summary

Introduction

Historically, our present understanding of the structure of solids is largely derived from the work of Goldschmidt in the 1920's,[1] followed by that of Bragg and his Manchester school.[2] The underlying ideas are (a) that atoms (or ions) have a definite radius, approximately constant from one compound to another, so that one can always predict interatomic distances (to within a few percent) by adding radii; (b) that the number of anion neighbors of any cation does not depend directly on valency but tends to be the largest number that can fit around the cation. These ideas are not self-evident; they are empirical generalizations drawn from the study of actual structures. Yet they have contributed more to our understanding of the nature of nonmetallic solids than any theoretical work to date.

It is hard now to realize how revolutionary was the concept of coordination number (CN) as a feature—more important in some respects than valency. Nowadays we accept, for example, that Al^{3+} may have either 4 or 6 oxygen neighbors (very rarely 5) and that the difference in structural role between 4-coordinated Al^{3+} and 4-coordinated Si^{4+} is very much less than that between 4-coordinated Al^{3+} and 6-coordinated Al^{3+}. For the chemist in the 1920's it was difficult to become used to the idea that valency did not play a primary role in determining structure.

Goldschmidt's rules are usually discussed in terms of ions rather than of atoms. This is encouraged by a generalization drawn from structural work, that if all the atoms are treated as fully ionized, their charges are locally balanced. The electrostatic valence of a bond is defined as the valency of the cation divided by its coordination number; Pauling's "Electrostatic Valence Rule" states that the sum of the electrostatic valences of all bonds to an ion tends to equal the charge on the anion. Though this rule has been of great value in predicting structures, it does not imply that the bonds concerned are fully ionic.

I shall use the words "atom" or "ion" indifferently

[1] V. M. Goldschmidt, "Geochemische Verteilungsgesetze der Elemente," *Skrifter det Norske Videnskaps-Akad*, Oslo I. Matem.-Naturvid. Klasse, 1926.

[2] W. L. Bragg, *Atomic Structure of Minerals*, Cornell University Press, Ithaca, N. Y., 1937.

without trying to decide the degree of ionization. This is not to blur the distinction between chemically different states of ionization: For example, Ti^{3+} is chemically a different species from Ti^{4+}, whatever the position of the latter between the extremes of complete ionization and complete homopolar bonding.

Packing structures and framework structures

The early work was largely concerned with "packing structures." Coordination numbers 4, 6, and 8, which are very common, give rise to symmetrical cation-anion polyhedra: the tetrahedron, octahedron, and cube, respectively (Fig. 3.1). Simple periodic structures

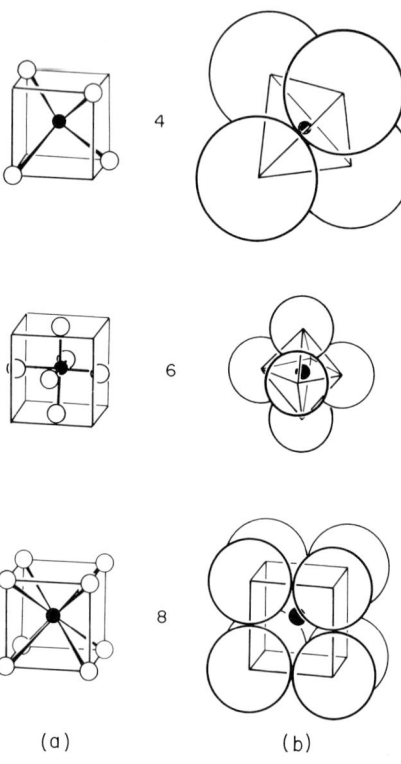

Fig. 3.1. Tetrahedral, octahedral, and cubic coordinations. (a) Positions of atomic centers in relation to cubic axes (cations ●, anions ○); (b) packing diagrams for the anions (cations ●).

result by building these polyhedra together in space-filling ways. Alternatively, such structures can be regarded as a space-filling array of large anions into whose interstices the small cations have been inserted in a perfect periodic pattern; the interstices give the cations their correct coordination. Examples are CsCl, with a simple cubic array of chlorine atoms and cesium atoms in the 8-coordinated or cubic interstices; MgO, with a face-centered cubic array of oxygen atoms and magnesium atoms in the 6-coordinated or octahedral

interstices; spinel ($AlMg_2O_4$) with a similar anion array and the different types of cation in appropriate octahedral or tetrahedral interstices. Nothing in the geometry of these structures contradicts the hypothesis that the bonds are semipolar. Thus the environment of the silicon atom will be the same, regardless of whether the neighboring oxygens are simply packed around it or joined to it by bonds directed toward the corners of a regular tetrahedron.

Sodium and calcium ions are always difficult to place in a simple packing structure where the anions are oxygen. They are too large to fit into 6-coordinated interstices without pushing the oxygens far apart, and rather too small for 8-coordination, which corresponds to a slightly less closely packed anion array. In consequence, they tend to give rise to more complicated patterns (Fig. 3.2).

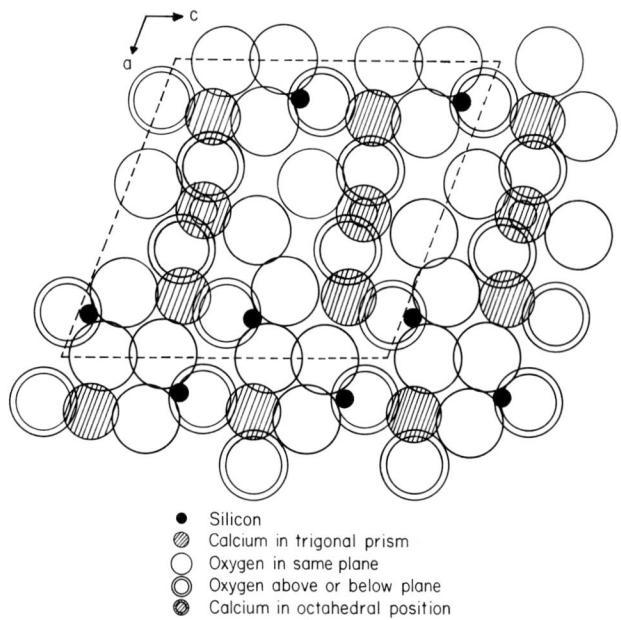

Fig. 3.2. Coordination of Ca in afwillite, $Ca_3Si_2(O_3OH)_2 2H_2O$.

By contrast with the packing structures, there are others that cannot be explained in this way, e.g., the three well-known forms of silica: quartz, tridymite, and cristobalite (Fig. 3.3). All three consist of SiO_4 tetrahedra linked to other tetrahedra by their corner oxygens, so that every oxygen has two silicon neighbors. The linkage patterns are different in each of the three structures, but in none of them are the oxygens packed to fill space. In fact, the oxygens are not even midway between their two silicon neighbors, but are so placed as to form an angle of about 130° between the two Si—O bonds. Obviously such a structure cannot be explained in electrostatic terms; the picture of the SiO_4

tetrahedron as a tetravalent silicon ion surrounded by four divalent oxygen ions was always a very unrealistic one and is now superseded.

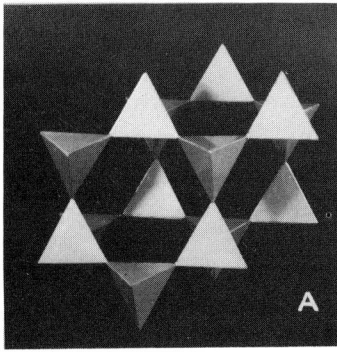
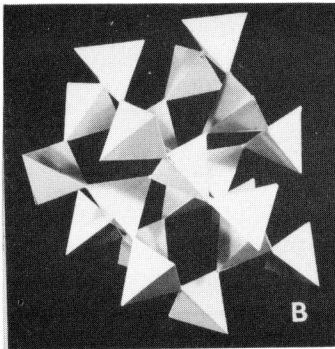
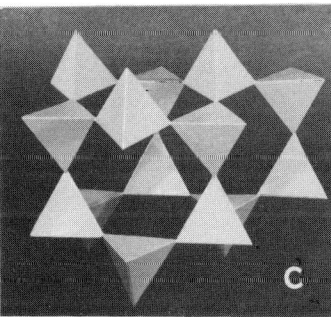

Fig. 3.3. Three forms of silica, showing different linkage schemes of SiO_4: (A) cristobalite, (B) quartz, (C) tridymite. (From A. F. Wells, *Third Dimension in Chemistry*, Clarendon Press, Oxford, 1956.)

We may perhaps distinguish these "linkage structures" or "framework structures" from the packing structures as follows: In the packing structures the polyhedra are put together to give as nearly uniform as possible a set of distances between anions; in the framework structures they give as nearly uniform as possible a set of bond angles at the anions.

The distinction between the two types of structure is not an absolute one; there can be intermediate cases. Consider a close-packed array of anions associated with two kinds of cation of very different electrostatic valence. For some such structures the polyhedra of high electrostatic valence, taken by themselves, form a three-dimensional framework, accommodating the cations of low electrostatic valence in the cavities. Whether we regard such a structure as a framework structure or a packing structure is a matter of convenience; it is, however, necessary to be aware of both points of view.

Starting at the other end, with obvious framework structures, it is not essential for all cations to be incorporated in the framework. There are cavities that can accommodate other cations. The over-all charge must of course be neutral, but how much of the positive charge is contributed by the framework cation and how much by the cavity cation is not fixed. One might envisage a series of compounds in which the framework cation would decrease in valency from one member to the next and the cavity cation increase correspondingly. Eventually the stage would be reached where the bonds holding the cavity cation in position are nearly as strong as those of the framework itself, and the concept of framework ceases to have much physical meaning. We shall consider only cases where cavity cations have a very much lower electrostatic valence than the framework cations.

The size of the cation which will fit in the cavity is important. Goldschmidt's rule suggests that cations which would be bad misfits for the cavities of a particular framework will simply not enter into compounds having this framework, however suitable their valency and chemical character might appear to be; this is verified in practice. But the fit does not have to be perfect to be tolerable: Cations which are only slightly too small allow the framework to crumple around them and those which are only slightly too large tend to push the atoms of the framework a little apart. In neither case do they affect its topology; their general role is that of spacers.

I shall illustrate these generalizations with two different types of framework structure: the feldspars and the perovskites.

Feldspars and perovskites as framework structures

The feldspar framework is built from tetrahedra containing Si^{4+} or Al^{3+}, with a characteristic pattern of linkages (Fig. 3.4). The larger interstices hold monovalent or divalent cations, Na^+, K^+, Ca^{2+}, Ba^{2+}. We shall here consider only pure end members, though solid solutions between them are also of great interest and importance. The cavity cations K^+ and Ba^{2+} are

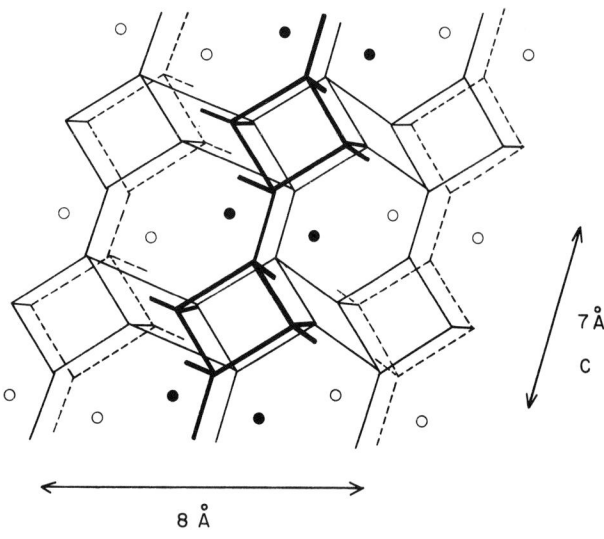

Fig. 3.4. Linkage scheme of feldspar. Si or Al in corners; lines represent (Si, Al)—O—(Si, Al) bonds.

large, Na^+ and Ca^{2+} of only moderate size. Suppose we start with an idealized version of a framework as symmetrical as the linkage scheme will allow. The cavity in which the cation must be inserted is lined with oxygen atoms: Three of these form a triangle in one plane, and four form a rectangle, very nearly a square, in a plane at right angles to it; there are two others, rather more remotely linked, that we need not consider. The large cations K^+ and Ba^{2+} are large enough to touch all four oxygen atoms of the square as well as all three of the triangle. They fill out the framework and keep it symmetrical. But the moderate-sized cations Na^+ and Ca^{2+}, clamped in position by the triangle, cannot simultaneously touch all four atoms of the square. One corner is pushed away, and the whole framework is sheared (with distortion of bond angles) to a configuration of lower symmetry (Fig. 3.5). Still smaller cations, such as Mg^{2+}, simply do not give feldspar structures, nor do very large ones.

By contrast, the perovskite framework is built from octahedra linked by all their corners to other octahedra, and the linkage scheme is a particularly simple one—the octahedra are in parallel orientation with their centers on a simple cubic lattice (Figs. 3.6 and 3.7). The array of anions can be thought of as a face-centered cubic array from which one atom in four has been systematically removed; the octahedral interstices between the others contain one kind of cation, B, while the site where the anion X was taken out is filled with the other cation A, giving a chemical formula ABX_3—in our examples, ABO_3. The dimensions of the framework are determined by the B—O distance, and if A is sufficiently large, but not too large, the framework remains in its most symmetrical form, simple cubic. If A is not large enough to touch all twelve anions simultaneously, the octahedra tilt in relation to one another till they grip it firmly with a smaller coordination number; the symmetry is thus lowered. Examples are $CaTiO_3$ and $NaNbO_3$ (Fig. 3.8). (Distortion of the framework may occur in the absence of a stuffing A cation, cf. WO_3; but for small A cations, such as Mg, the perovskite structure is not formed.) If A is rather too large, so that it forces all the anions apart and makes the B—O distances abnormally large, the B cation rearranges itself

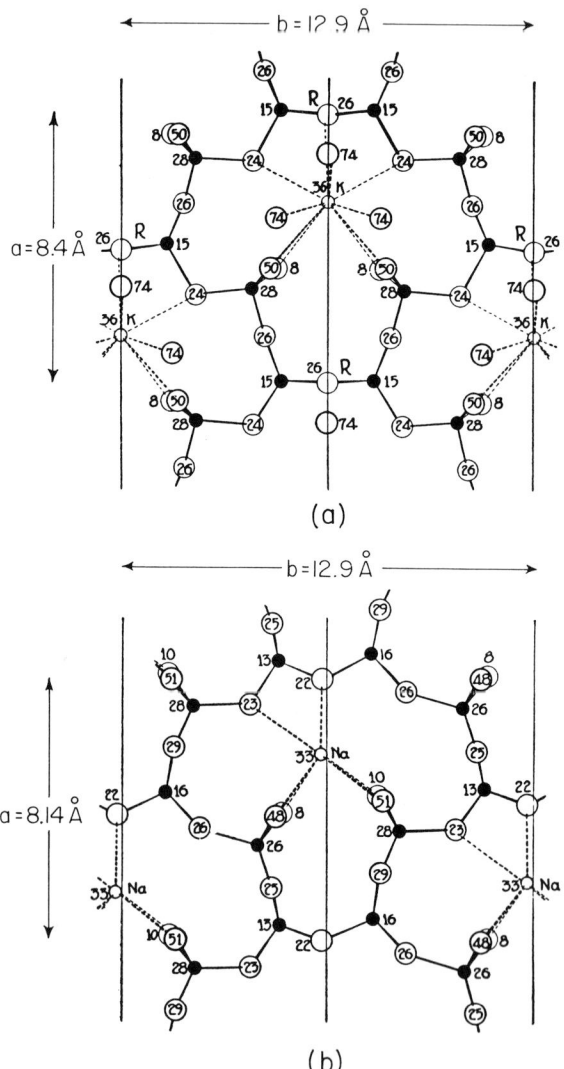

Fig. 3.5. Relation of monoclinic sanidine (a) and triclinic albite (b). (After Bragg.[2]) Projection of part of the structure along z axis.

Structure Families, Aristotypes, and Heterotypes

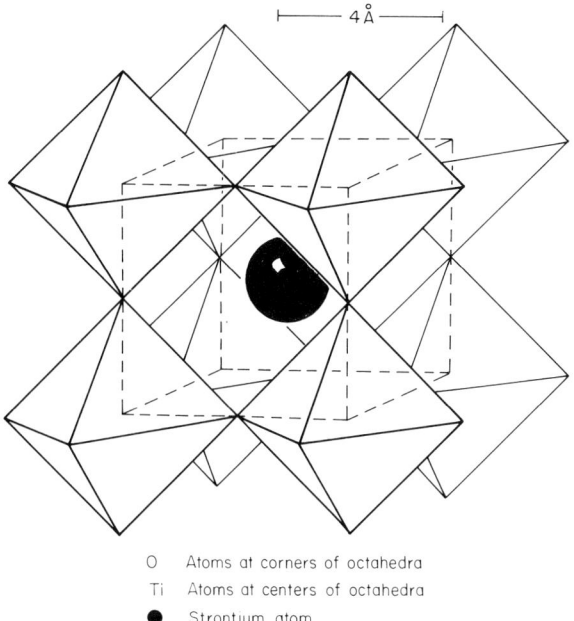

Fig. 3.6. Idealized perovskite structure, ABO₃ (structure of SrTiO₃).

off-center in the octahedron, in a way which cannot be explained by simple packing considerations but indicates some other change in the force system acting on it. This is what happens in BaTiO₃ (Fig. 3.9a, b) and KNbO₃.

The perovskites are less obviously framework structures than the feldspars; they might even be thought of as packing structures as long as they keep the ideal symmetry. But for dealing with effects when the ideal symmetry is lost, the framework description is the more helpful one. The nature of the distortions can be understood only as some sort of compromise between a packing structure in which the spacing of the large

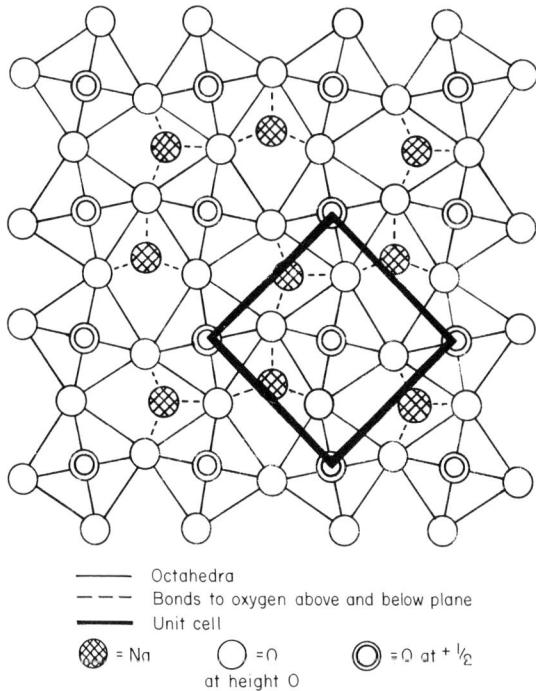

Fig. 3.8. Projection of NaNbO₃ structure.

atoms contributes most to the energy and a framework structure in which the bond angles at oxygen contribute most.

Structure families, aristotypes, and heterotypes

In the description of these framework structures it becomes clear that the over-all pattern of linkages—the topology of the structure—is much more important than the detailed symmetry or even the actual atoms present. The topology characterizes the *family* of structures. It is convenient to call the simplest and most symmetrical structure of the family the *aristotype* and all other variants the *heterotypes*. The perovskite aristotype (idealized perovskite structure) has a simple cubic lattice, with one formula unit per cell. One heterotype is the room-temperature BaTiO₃ structure, in which the unit cell is tetragonal and all the atoms have arbitary z parameters. Some heterotypes may involve a doubling (or other multiplying) of the unit cell. For example, in room-temperature NaNbO₃, the

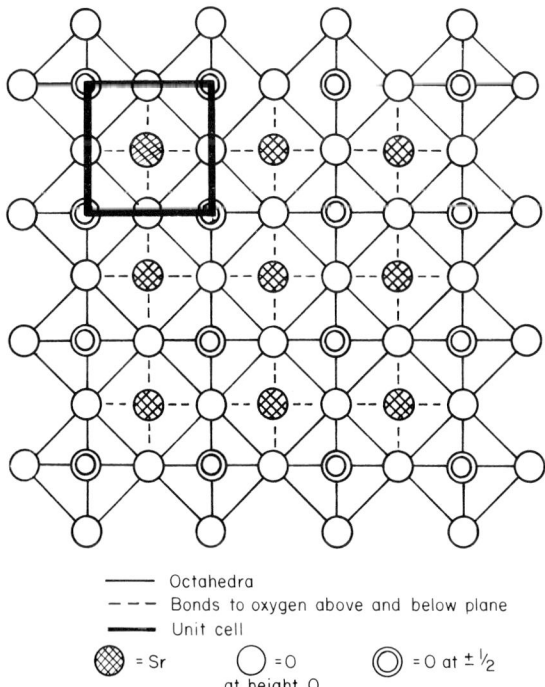

Fig. 3.7. Projection of idealized perovskite structure.

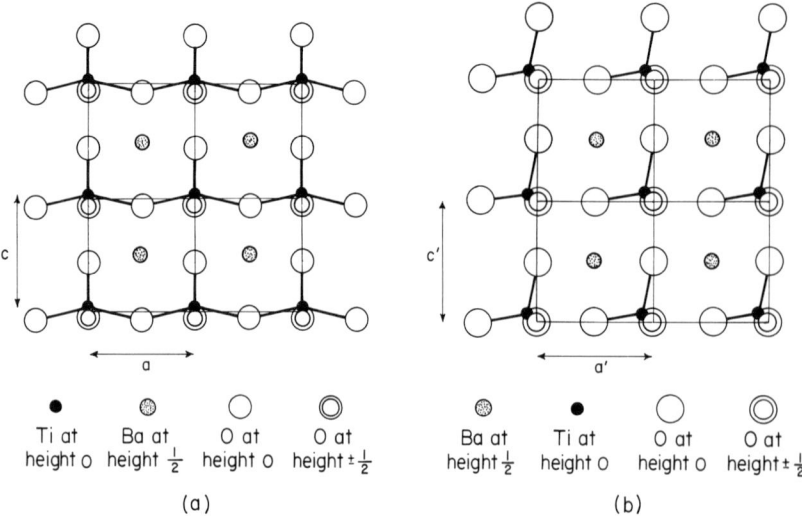

Fig. 3.9. Tetragonal (a) and orthorhombic (b) heterotypes of BaTiO₃.

base of the unit cell has for edges the face-diagonals of the aristotype, and its height is four times that of the aristotype. Such doubling, or multiplying, represents a reduction of symmetry, but this is not made obvious in merely naming the crystal class or system. For example, room-temperature KNbO₃ and room-temperature NaNbO₃ are both orthorhombic, but the structure of the former is very much simpler, as indicated by its smaller unit cell, which may be regarded as possessing a higher density of symmetry elements. The volume per lattice point for KNbO₃ is equal to that of the aristotype, while for NaNbO₃ it is eight times as great; this implies that in KNbO₃, as in BaTiO₃, the atoms must all remain in special positions on symmetry elements, while in NaNbO₃ they may be in general positions. It is clear that doubling of the unit cell is needed to allow the tilting of octahedra relative to one another; if there is only one octahedron per primitive unit cell, all octahedra in the crystal must be oriented parallel to one another.

Three points, often forgotten, should be noted in dealing with families of structures. First, what is physically and chemically interesting depends on the different distortions of the polyhedra, their relative orientations, and the kind of cavity they leave for the larger cation; facts about cell dimensions are of interest (except for pure description) only insofar as they allow these features to be deduced. Second, the aristotype may not display all the physically important features of the family: For example, the idealized perovskite structure cannot show the departures from 180° bond angles at oxygen which are vital to the understanding of the heterotypes. Third, while it is easy to deduce the aristotype from any given heterotype, the converse process is not automatically possible; one can never enumerate all the ways in which a particular aristotype may be distorted geometrically without losing its original topology, because there is no geometrical limit to the factors by which the sides of the original cell may be multiplied.

Character of transitions

The transition from one heterotype to another or to the aristotype can be accomplished without breaking any bonds and without any interchange of atomic positions. It is therefore likely to be associated with small differences of energy. By contrast, the more familiar type of transition involves a complete breakup and reorganization of the structure, as pointed out by Buerger.[3] He called the latter type reconstructive, the former displacive.

Examples of both reconstructive and displacive transitions are found among the forms of silica. Conversion of any one of the forms quartz, tridymite, or cristobalite (cf. Fig. 3.3) into any of the others involves a reconstructive transition, giving a wholly different pattern of linkages. But each of the three has also a displacive transition which is reversible; the best known is the high-low quartz transition at 550° C.

I shall not be further concerned with reconstructive transitions, because properties cannot be traced through from one phase to the other. In displacive transitions one can hope to correlate changes of properties with differences of structural features. The trouble is that

[3] M. J. Buerger, "Crystallographic Aspects of Phase Transformations," *Phase Transformations in Solids*, R. Smoluchowski, Ed., John Wiley and Sons, New York, 1951, p. 183.

we need to know the details of the structural differences in the two forms and not merely the difference in some feature (such as symmetry) arbitrarily picked out because of its ease of observation. Studies of this detailed kind are difficult and time-consuming, but we have already a number of examples. With them as guide, we can use the additional fragmentary evidence of more easily observed features with less danger of assuming that they are always the physically important ones.

Two points about displacive transitions are very important: Since they involve very small over-all distortions, they can take place reversibly in single crystals; and since they involve no diffusion of atoms, they cannot easily be quenched in. (Both statements need some qualification: There are effects of twinning and domain formation on the one hand and thermal hysteresis on the other, dealt with later, but they modify the picture only slightly.)

The character of displacive transitions can be illustrated in the feldspar and the perovskite families.

Structure changes and properties of perovskites

Of the perovskites, $NaNbO_3$ at room temperature was already mentioned as an example of the crumpling of the framework around a moderate-sized cation, giving a complicated structure with tilted octahedra (cf. Fig. 3.8). With increasing temperature it undergoes several successive transitions and finally, above 640° C, it is ideal cubic. This implies that the Na atom, which manages only 6 to 7 oxygen neighbors at room temperature, can at high temperatures tolerate 12; either the effective radius or the thermal-vibration amplitude of Na has increased greatly. At room temperature the Nb atom was not central in its octahedron; in the cubic structure it is central. Which of the intervening transitions correspond to displacement of Nb within its octahedron and which to differing tilts of the octahedra we do not yet know.

$KNbO_3$ is simpler; at 210° C it changes the direction of displacement of Nb within the octahedron, and at 410° C, when the structure becomes cubic, the Nb is central and all octahedra are in strictly parallel orientation throughout.

$CaTiO_3$, which at room temperature has tilted octahedra with Ti centrally placed in them, has a transition at about 1260° C, where it is believed to become cubic.

All these transitions are accompanied by changes in physical properties, often very small—for example, the latent heats of the two $KNbO_3$ transitions are 85 and 190 cal/mole, respectively. However, where the octahedron has a dipole moment that changes direction or disappears at the transition, there may be very high dielectric constants immediately above the transition. Such peaks in the dielectric constant often have been the first indication that a transition occurred.

If the octahedra—or, more generally, any tightly bonded structural units—have dipole moments and the units are so arranged that their net dipole moment does not cancel out by symmetry, the material has a spontaneous polarization. Such a material is a true pyroelectric. If, as often happens, the direction of the resultant moment can be reversed by a field, the material is ferroelectric; by definition, therefore, a ferroelectric is a pyroelectric whose spontaneous polarization is reversible without a change of structure.[4,5] This last point is important, and sometimes not understood. The reversibility is not of a dipole changing orientation relative to the framework, but of the framework as a whole turning itself inside out, so that it finishes up as the same structure pointing in the opposite direction (Fig. 3.10). Obviously the reversing of the structure is

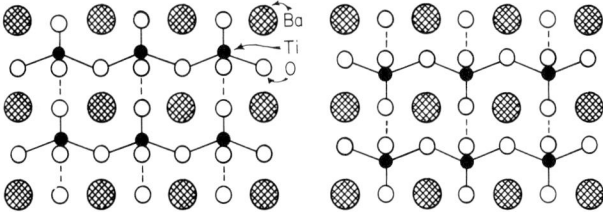

Fig. 3.10. Reversal of dipole moment in the ferroelectric structure of $BaTiO_3$.

likely to happen only if there is a symmetrical form of only slightly greater energy through which it can pass; such a form is likely to be stable at some higher temperature. Hence, ferroelectrics generally undergo displacive transitions. Their net dipole moment, macroscopically measurable, makes them very interesting cases to study but does not mean that they have to be treated fundamentally differently from other materials with displacive transitions.

Free energy and displacive transitions

In many cases the exact temperature of a displacive transition cannot be determined unambiguously. There is a thermal hysteresis effect, the change in the property studied taking place at a higher temperature when the crystal is heated than in the reverse direction. This is always an indication of a first-order transition, one in which the first derivative of the free energy has

---

[4] A. von Hippel, *Revs. Modern Phys.* **22**, 221 (1950); *Z. Physik* **133**, 158 (1952).

[5] H. D. Megaw, *Acta Cryst.* **5**, 739 (1952).

a discontinuity. It is easy to describe the transition qualitatively, in terms of free-energy curves. Suppose the free energy is plotted against some convenient parameter x describing the geometry of the structure. A stable structure will be represented by a minimum. Since there are two structures very nearly alike, there will, at the transition temperature, be two equal minima differing slightly in their x coordinate and separated by an energy hill (Fig. 3.11). Above and below the tran-

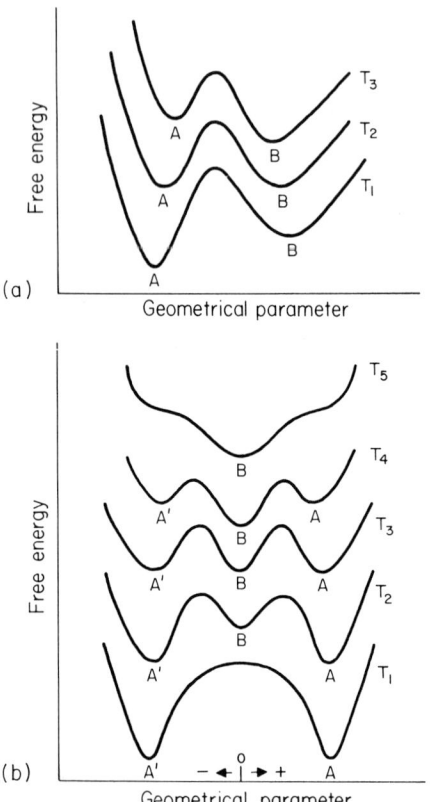

Fig. 3.11. Free-energy curves for a displacive transition; T_1 to T_5 represent curves for successively higher temperatures. (a) Transition without change of symmetry, (b) with change of symmetry.

sition, the free-energy curve changes smoothly with temperature; the existence of the transition means that below it one minimum (A) becomes deeper and the other (B) shallower, while above it the reverse is true. On heating the crystal, it stays in state A until this minimum is not only shallower than B but sufficiently far above it for the height of the hill bounding it to be comparable with the thermal energy; on cooling, it stays in state B until the hill can be crossed in the reverse direction. The difference of the two temperatures, the thermal hysteresis, is thus a measure of the height of the energy hill.

Free energy is, of course, a function of other variables besides temperature, e.g., pressure and electric field. Usually these have only a subordinate effect on a displacive transition, modifying the temperature at which minima become of equal depth rather than changing the whole shape of the curve. Such effects of pressure or field on the "Curie point"—the transition temperature of a ferroelectric—are well known. It is also theoretically possible that pressure or electric field may deepen minima which are otherwise unimportant and hence allow a transition to a completely new heterotype. This seems to occur in $NaNbO_3$. The normal room-temperature form is not ferroelectric; in a strong field there is a displacive transition to a ferroelectric form[6] (Fig. 3.12). There is some evidence[7] that this field-

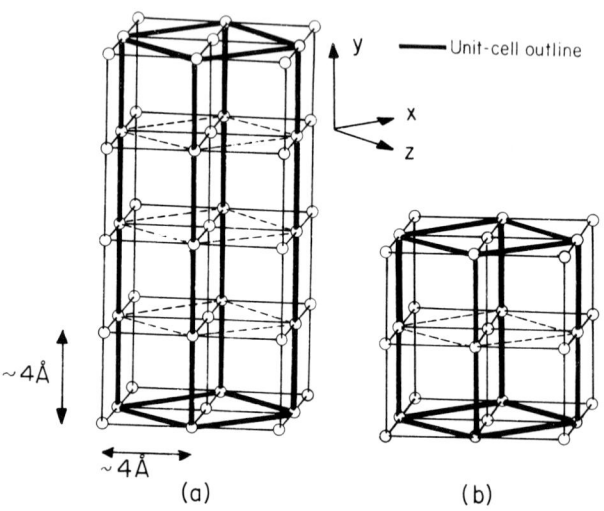

Fig. 3.12. Unit cells of two forms of $NaNbO_3$. ○ = Nb atoms in the aristotype. Other atoms not shown. (a) Normal room-temperature form, antiferroelectric; (b) ferroelectric form.

induced form is one of the phases found when $NaNbO_3$ is heated in the absence of a field and that it can exist metastably without field at room temperature; its structure is probably the same as that of room-temperature, field-free $(K_{0.025}Na_{0.975})NbO_3$.[8]

Displacive changes in feldspars and their interaction with ordering effects

In the feldspar family it is more difficult to recognize displacive transitions because of the complicating

---

[6] Observed in electrical studies by L. E. Cross and B. Nicholson, *British Research Suppl.* 7, s 36 (1954), and investigated with X rays by E. A. Wood, R. C. Miller, and J. P. Remeika, *Acta Cryst.* 15, 1273 (1962).

[7] I. Lefkowitz and H. E. Megaw, unpublished.

[8] Studied electrically by L. E. Cross, *Nature* 181, 178 (1958), and with X rays by M. Wells and H. D. Megaw, *Proc. Phys. Soc. (London)* 78, 1258 (1961).

effects of Si/Al disorder. However, recent work shows that displacive transitions do occur and suggests that they are affected only to a minor extent by the state of Si/Al order or disorder. The most obvious example is anorthite, $CaAl_2Si_2O_8$.[9] With a suitable (though un-

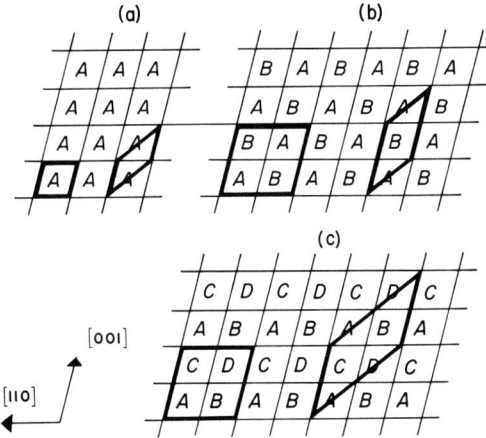

Fig. 3.13. Relations between true cells and subcells for (a) albite (or sanidine), (b) celsian (or anorthite above transition), and (c) anorthite (low).

usual) choice of axes, it can be shown that the primitive unit cell is double that of celsian, $BaAl_2Si_2O_8$ (Fig. 3.13).[10] The Si/Al order is exactly the same in both structures, and so is the topology of the framework, but

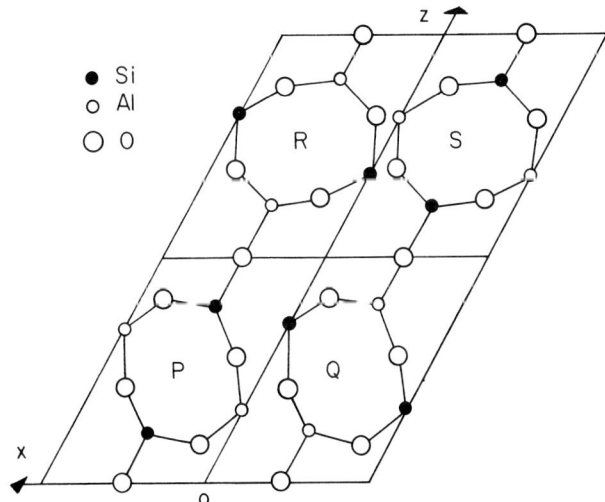

Fig. 3.14. Parts of anorthite structure, showing projections of atomic positions (at 300° C transition, P becomes identical with S, and Q with R).

in $CaAl_2Si_2O_8$ the framework is crumpled to give the doubled cell (Fig. 3.14). On heating anorthite, the

X-ray pattern changes at about 300° C to one with the same systematic absences as that of the celsian heterotype; on cooling, the anorthite pattern reappears.[11] This is exactly comparable to the changes in the perovskites: In both structures, at high temperature the moderate-sized cation in the cavity suddenly becomes able to tolerate a larger number of neighbors, and the high-temperature heterotype is like that containing the larger cation at room temperature.

The unit cell of celsian—and that of anorthite above 300° C—is still double the primitive unit cell of the feldspar aristotype (found in sanidine, $KAlSi_3O_8$, which has complete Si/Al disorder). This doubling results from the ordered Si/Al arrangement: Both halves contain Si and Al in equal amounts but in exactly interchanged positions. It is believed that no change of Si/Al order (or at least no substantial change) occurs up to the melting point, and hence there cannot be any further transition to a heterotype of smaller volume or to the aristotype.

For high albite ($NaAlSi_3O_8$) quenched from a high temperature, there is substantial (perhaps complete) Si/Al disorder, allowing further possibilities. It has been suggested[12] that the actual room-temperature structure is like that of anorthite, and that it undergoes either one or two displacive transitions with increasing temperature.

When a displacive transition occurs in a system with variable states of order, there may be interaction between the two effects. The configuration of the framework in the high-temperature phase cannot be retained below the transition point; crumpling will necessarily occur. In this sense the transition cannot be quenched in. But the equilibrium state of order associated with the high-temperature configuration may be markedly different from that of the low, and this can be quenched in. In all but the simplest structures the specification of "state of order" must include the pattern as well as the degree of order; the equilibrium pattern of order may be determined by the configuration and change discontinuously at the transition, while the degree of order may change continuously with temperature. The latter can be handled by conventional thermodynamic arguments, the former cannot.

Ordering processes are slow, depending on diffusion rates, while displacive processes are likely to be propagated at velocities comparable with that of sound. Moreover, if the patterns of order in the two phases are very different, the conversion from one to the other may need, as an intermediate state, a degree of disorder

---

[9] Structure determined H. D. Megaw, C. J. E. Kempster, and E. W. Radoslovich, *Acta Cryst.* 15, 1017 (1952).

[10] Structure determined by R. E. Newnham and H. D. Megaw, *Acta Cryst.* 13, 303 (1960).

[11] W. L. Brown, W. Hoffmann, and F. Laves, *Naturwissenschaften* 50, 221 (1963).

[12] H. D. Megaw, *Norsk Geol. Tidsskr.* 42.2, 105 (1962).

greater than the equilibrium degree for either. At any temperature, therefore, the state of order is not necessarily an equilibrium one but reflects the thermal history of the crystal. To some extent, the actual state of order must of course influence the detail of the configuration, but this effect seems subordinate and does not mask (though it may sometimes blur) the difference between the phases on either side of a displacive transition, whatever their state of order. One might say that the ordering or disordering of Si and Al is analogous to the effect of a frictional or damping force on the mechanism of the displacive transition, and not its driving force. These ideas are tentative but open up possibilities for further study.

Twin domains and antiphase domains

We have taken it as one of the indications of a displacive transition that it can occur reversibly in a single crystal. This needs qualification. It is generally true when the material is heated through the transition, but when it is cooled there may be formation, either twin or antiphase domains. There is always some distortion in going from one heterotype to another or to the aristotype. For a low-symmetry form growing from a high-symmetry form, at least two equally good orientations exist, related by a twin axis or plane normal that coincides with the symmetry direction lost at the transition. If growth nuclei are numerous and randomly distributed, both orientations will occur in equal amounts. Moreover, the lowest over-all volume strain will be achieved if the individual extensions and contractions of separate domains neutralize one another

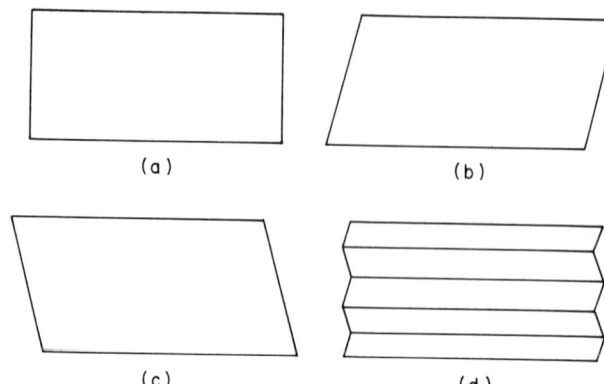

Fig. 3.15. Effect of twinning in reducing volume strain. (a) High-temperature form, (b) and (c) two orientations of low-temperature form, (d) combination of (b) and (a) to match shape of (a).

(Fig. 3.15). This effect is particularly important if the crystal is clamped so as to prevent volume changes; for a large crystal, its own weight may clamp.

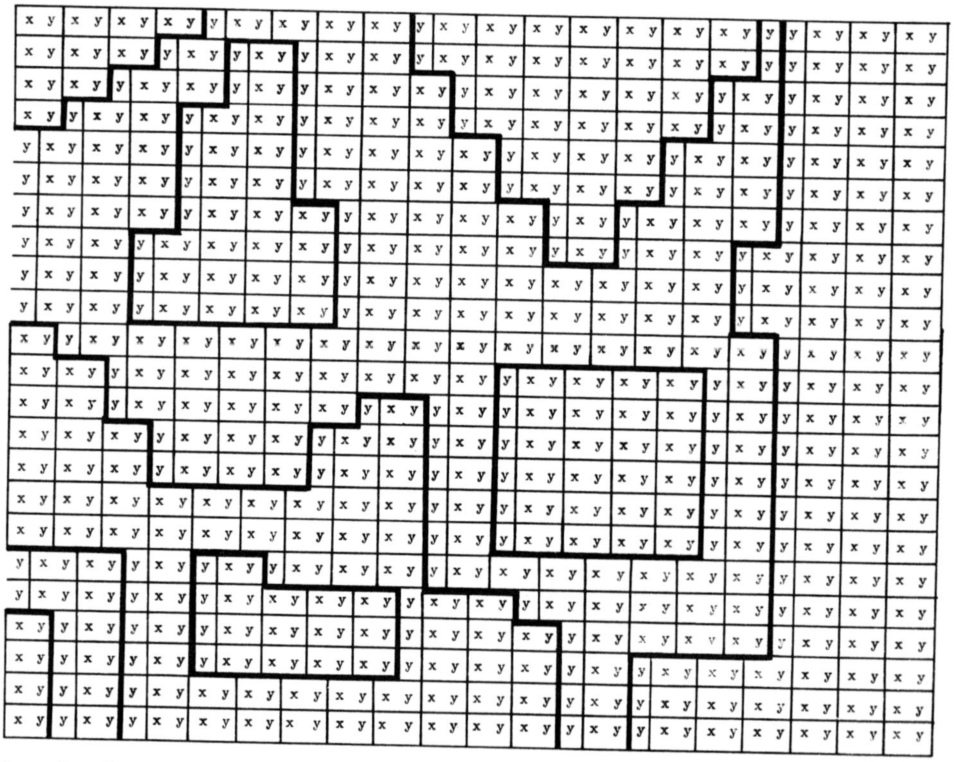

Fig. 3.16. Antiphase domain texture. (True cell composed of two slightly different subcells x and y, showing boundaries between rows, between columns, or along diagonal lines.)

Whereas twin domains can form when a symmetry axis or plane is lost, antiphase domains can form when a translation-repeat is lost (i.e., when the length of the cell edge of the low-symmetry phase is a multiple of that of the high-symmetry phase).[13] Just as the unit cell of the low-symmetry can have two (or more) orientations relative to that of the high-symmetry phase, so the unit cell of n-fold length can have n difference origins. If growth nuclei at the transition are numerous and randomly distributed, a low-temperature phase with a doubled cell will develop as a mosaic of domains in parallel orientation but related to one another by a translation equal to the length of the original single cell (Fig. 3.16).

If the volume considered is that of the original single crystal, domain formation appears as a process of disorder. The domain texture is a compromise between the reduction of over-all strain energy and the increased energy stored in the domain walls. Often it is not an equilibrium state but represents the frozen-in response to locally fluctuating conditions during the transition from the high-temperature phase. If, however, we consider volumes small enough to fit into the interior of domains without trespassing across boundaries, we deal with a structure that is still perfect, though of a heterotype different from that above the transition. Effectively, then, the material has two phases: one (the domain interior) with a perfect three-dimensional structure capable of independent existence, the other (the domain walls) with a two-dimensional structure that can exist only on a substrate of the first. This description is likely to be useful only if the domain size is not extreme. If the domains are very large, the effect of the walls is negligible. If they are very small, the surface forces in the walls may distort the structure of the interior; and if their extension is less than, say, two unit cells in any direction, the attempts to distinguish their interiors from their walls becomes artificial. Indeed, if the domains appear very small and very uniform in size, it may be better to regard the whole as a new and perfect, though complicated, heterotype and abandon the description in terms of domains. But, between the extremes of the very small and the very large, "domain disorder" below a transition must be treated differently from random disorder of atomic substitution.

So far, the discussion has included twin domains as well as antiphase and out-of-step domains; from this point on, it will be restricted to the latter. Suppose we are dealing with a material of perfect stoichiometric composition and with no disorder of any kind except the existence of out-of-step domains. There is some degree of misfit in the domain walls (Fig. 3.17); this implies

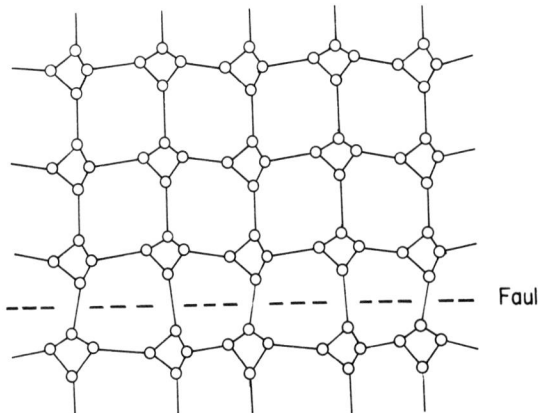

Fig. 3.17. Effect of antiphase mistake. (True repeat unit consists of a pair of nonidentical 4-rings, alternating along the rows.)

strain and increased energy. Since there is no other compensating reduction of energy from the presence of the walls, it is not the equilibrium state. On annealing (if no other kind of disorder arises below the annealing temperature), the walls will move out and the whole crystal becomes a single domain. If, however, there are foreign atoms present—as, for example, the "guest atoms" of a solid solution—they are likely to increase the energy least, or perhaps even to decrease it if they are located in boundary regions. Then, on heating, either the misfit atoms will migrate to the boundaries, the boundaries will move till they lock into position on the misfit atoms, or perhaps both processes will occur together. It can easily be seen that the boundaries will not anneal out, and a domain texture will remain as the equilibrium state.

It must be emphasized that the atoms in the boundary are not interstitial; they play a normal part in the topology of the framework and the occupation of its cavities, which continue unchanged across the boundaries. All that happens is that the atoms in this region are slightly displaced, so that their detailed coordinates are not exactly those of either half of the unit cell possessed by the structure of the domain interior, and that the proportions of different kinds of atoms occupying topologically similar sites in the domain wall and domain interior may be different. In an extreme case, the domain boundary may itself be perfectly ordered and may be regarded as a two-dimensionally extended

---

[13] Strictly speaking, the term "antiphase domains" refers only to the very common case when n is even and the slip between domains is half the cell length of the low-symmetry heterotype; the term "out-of-step domains" covers all cases when the slip is m/n.

heterotype representing a chemical phase different from that of the domain interior.

Antiphase domains in a feldspar (anorthite)

"High anorthite" is anorthite heated for some time above 1300° C and returned to room temperature. The domain texture is shown by the fact that certain X-ray reflections have become systematically diffuse—those reflections which would be absent for the celsian heterotype. The most obvious explanation, and a physically reasonable one, is that a transition to the celsian structure occurs at a high temperature, and on cooling the anorthite structure reappears, but with an antiphase domain texture. If the material is re-annealed at about 1100° C and returned to room temperature, the diffuse reflections have all become sharp again; either the domain boundaries have been annealed out altogether or at least have moved too far apart to affect the sharpness of reflections. This suggested a transition at about 1200° C. It was therefore very unexpected when it was discovered[11] that examined at a temperature not much above 300° C the reflections disappeared altogether, but they reappeared with unchanged sharpness on cooling. The transition itself must happen near 300° C; and whatever happens above 1200° C must introduce imperfections which can be quenched in and which, harmless as long as the structure is the high-symmetry form, somehow determine the position of domain boundaries as soon as the transition to the low-symmetry form makes antiphase domains possible.

The obvious hypothesis about the nature of the imperfections is that they are sites of local Si/Al disorder in an otherwise perfectly ordered framework. In a complicated structure there is no reason to suppose that disorder is equally difficult, or equally easy, for all chemically similar atoms if they are crystallographically different. In anorthite, with 8 different Si's and 8 different Al's, or in celsian, with 4 of each, there is clearly a wide range of choice. Suppose that the activation energy needed to allow the interchange of one particular Si/Al pair is very much less than that for any other. Then the interchange of this pair will take place with a probability increasing with temperature in the usual way, while the other sites remain perfectly ordered and maintain the over-all pattern. On quenching, the number of nuclei for domain boundaries depends on the number of Si/Al interchanges. The hypothesis thus explains the observations of Laves and Goldsmith[14] that the diffuseness of the reflections of "high anorthite," and hence the smallness of the domains in the quenched material,[15] changes progressively with the temperature of heat treatment above 1200° C. There is a little direct evidence from the structure analysis[16] of a natural "high anorthite" which indicates a small amount of Si/Al disorder at certain sites, but the magnitude of the effect is too near the limit of error to be conclusive.

The unusual feature in anorthite is the larger interval between the temperature of the displacive transition and that where interchange disorder sets in. This allows the separate effects of the two processes to be distinguished. Moreover, the disorder is confined to isolated groups of atoms. If we think of the set of Si/Al interchanges in a group as an excitation process, the excitation of a group in one unit cell does not affect the probability of excitation of similar groups in neighboring unit cells. (It is not essential that only one Si/Al pair per unit cell be excited, provided the whole excited group is a small and isolated part of the unit cell.)

Summary

1. Framework structures very often possess a series of heterotypes, variants of an idealized structure or aristotype, all possessing the same topology but with different detailed symmetry and cell sizes.

2. Different heterotypes may be illustrated by compounds of different compositions, by the same compound at different temperatures, or (in at least one case) by the same compound with and without an applied electric field.

3. Transitions between heterotypes are displacive; they take place reversibly in a single crystal. They may show thermal hysteresis and domain formation.

4. For crystals with no other disorder, and unclamped, the equilibrium state is single-domain.

5. For a heterotype with a unit cell double that of the aristotype, antiphase domains may occur but are stabilized only if there is substitution disorder (perhaps due to nonstoichiometric composition); in this case the domain walls are anchored to the sites of disorder.

6. The presence of a large amount of substitution disorder does not prevent the occurrence of displacive transitions but makes the domains so small that they are hard to study.

7. In any consideration of order, the pattern of order as well as the degree of order is significant.

---

[14] F. Laves and J. R. Goldsmith, *Acta Cryst.* 7, 465 (1954).

[15] Some authors call these materials "transitional anorthites," referring to the transition from complete sharpness to complete diffuseness; for them, "high anorthite" would be the hypothetical end member, with the reflections so diffuse as to have become invisible.

[16] P. H. Ribbe, Ph.D. Thesis, Cambridge University, 1963.

Summary

In all cases, the "crystal structure" is the three-dimensionally periodic structure within a domain, effectively independent of domain size. Comparison of the properties and structures of different heterotypes is likely to give more useful understanding of the relation of properties to atomic configuration and interatomic forces than can be gained from comparison of totally different structures; once we know what must be attributed to volume effects, we shall be in a better position to sort out the separate effects of surface layers and domain walls. The ideas put forward have been illustrated almost entirely from two families of structures, the feldspars and the perovskites, because they have been studied in more detail than most others, but the same sort of approach should be valid for all framework structures.

SOME GENERAL REFERENCES

General Background Reading (fairly elementary)

W. L. Bragg, *Atomic Structure of Minerals*, Cornell University Press, Ithaca, N. Y., 1937.

R. C. Evans, *An Introduction to Crystal Chemistry*, Cambridge University Press, Cambridge, 1939.

A. F. Wells, *Structural Inorganic Chemistry*, 3rd ed., Oxford University Press, London, 1962.

L. V. Azaroff, *Introduction to Solids*, McGraw-Hill Book Co., New York, 1960.

Specialist Books and Review Articles

M. J. Buerger, "Crystallographic Aspects of Phase Transformations," *Phase Transformations in Solids*, R. Smoluchowski, Ed., John Wiley and Sons, New York, 1951, p. 183.

H. D. Megaw, *Ferroelectricity in Crystals*, Methuen and Co., Ltd., London, 1957.

H. D. Megaw, *Order and Disorder in the Feldspars*, Min. Mag. 32, 226 (1959) (an introductory paper).

4 · PROPERTIES IMPARTED BY *d* ELECTRONS

John B. Goodenough

Introduction — Some consequences of electron localization — Description of localized d *electrons — Hydrogenlike wave functions and cubic field splitting — Tetragonal field splitting and spin-orbit splitting — Applications — Exchange coupling and magnetic order — Collective* d *electrons and crystallographic distortion — Exchange coupling via collective electrons — Transition metals and their alloys*

Introduction

Transition-metal atoms in solids give rise to a rich variety of physical properties that have been extensively investigated and exploited in recent years. Pursuit of these studies with imagination requires a relevant description of the outer *d* electrons of such atoms. This chapter outlines the fundamental physical ideas on which our present concepts are based.

In dielectrics the electronic orbitals built from the outer *s* and *p* atomic orbitals are split into a *filled* valence band and an *empty* conduction band; the Fermi level lies in the large energy gap (ca. 1 to 10 eV) between these bands. The outer *s* and *p* electrons thus make only indirect contributions to the magnetic and electric properties. The most important of these are their contributions to indirect exchange coupling, to the magnitude of the ligand-field splittings, and to the dielectric constant.

The *d* orbitals in a crystal mix with the crystalline *s-p* orbitals of corresponding symmetry. Nevertheless, for semiempirical arguments it is possible to use simple atomic orbitals rather than the *d*-like crystalline orbitals because the essential physical arguments follow from pure symmetry considerations. Mixing is treated as a perturbation. For ionic crystals this approach leads to qualitative rules for the signs and relative magnitudes of interactions between localized magnetic moments on neighboring cations. These rules have been empirically verified in oxides, halides, and many sulfides. (In nitrides and carbides, on the other hand, the approximation breaks down, because the mixing of anion *s-p* orbitals and cation *d* orbitals is large.)

In halides and oxides, the cation-cation separation is usually—but not always—larger than the metal-metal distances in the transition metals. Therefore 3*d* electrons are usually localized. Even if the cation-cation separations are so small that the *d* electrons must be treated as belonging to the crystal as a whole, the width of the band of these collective energy states ($\Delta\epsilon$) is smaller than Δ_c, the splitting of the atomic *d* levels by the cubic component of the crystalline field.

While most of our discussion will be concerned with localized *d* electrons, at the end we will consider briefly what happens if cations are so close that the outer *d* electrons belong to the collective cation sublattice. We do not discuss the interesting metallic oxides in which each oxygen has three or fewer transition-ion neighbors, since these are not, in general, magnetic. However, ferromagnetic, metallic CrO_2 belongs to this class of compound which is discussed elsewhere.[1]

Some consequences of electron localization

Localized electrons[2] in filled shells contribute little to the electronic and magnetic properties of crystals; partially filled d shells may give rise to localized magnetic moments, electrical conductivity, and a wide variety of electron-ordering transitions.

Atomic moment. A localized atomic moment is expressed as

$$\mathbf{m}_A = gJ\mathbf{m}_B, \quad (4.1)$$

with g the spectroscopic splitting factor, $J = L + S$ the total angular momentum quantum number, and $\mathbf{m}_B$ the Bohr magneton. For isolated atoms, Hund's rule of highest multiplicity prescribes as lowest energy state the one that maximizes J. We shall see that in solids this rule may be broken in the presence of large crystalline fields.

Conductivity. Conductivity takes place by a hopping of electrons from one cation to the next. Therefore the charge-carrier mobility (b_h for holes, b_e for electrons) obeys Einstein's diffusion equation:

$$b_h = -b_e = \frac{eD}{kT} = \frac{b_0}{T} e^{-\varepsilon_a/kT}, \quad (4.2)$$

where $-e$ is the electron charge and ε_a the activation energy required by an electronic carrier to transfer to a neighboring cation site (e.g., from $Fe^{2+} \rightarrow Fe^{3+}$). The anions relax about the cations to form a "trap" for the electron, thus localizing it; the electron has to acquire an energy ε_a to free itself.

In a doped or nonstoichiometric material the number of available charge carriers may remain roughly constant with temperature. In a stoichiometric compound, separated hole-electron pairs have to be created before conduction can occur. If ε_g is the energy required to do this, the number of electrons and holes becomes

$$n_e = n_h = n_0 e^{-\varepsilon_g/2kT} \quad (4.3)$$

and the conductivity

$$\sigma = -n_e e b_e + n_h e b_h = \left(\frac{\sigma_0}{T}\right) e^{-U/kT}, \quad (4.4)$$

where U (stoichiometric) $= \varepsilon_a + (\varepsilon_g/2)$ and U (nonstoichiometric) $= \varepsilon_a$. Usually the energy ε_g is large.

[1] J. B. Goodenough, *Bull. soc. chim. France*, Fasc. 4, 1200 (1965).

[2] J. S. Griffith, *The Theory of Transition-Metal Ions* (Cambridge University Press, London, 1961) provides an excellent theoretical treatment of localized d electrons in dielectrics.

It is for this reason that stoichiometric compounds with an integral number of electrons per cation (e.g., MnO or NiO) tend to be insulators. However, ε_g decreases sensitively with decreasing cation-cation separation R and is sharply reduced by any screening of the electron from its hole by neighboring hole-electron pairs. Mott[3] pointed out that this screening effect introduces a feedback into the reduction of ε_g that should cause it to drop abruptly to zero at some critical separation distance R_c. Therefore, a sharp operational definition for such a critical R_c may exist: R_c is that distance at which $\varepsilon_g \rightarrow 0$. This is an important parameter to establish experimentally, because localized-electron theory can be valid only if $R > R_c$. If $R < R_c$, a collective-electron theory must be used.

Electron-ordering transitions. Since the melting point is primarily determined by bonding via outer s and p electrons (cf. Chap. 2), ordering of the localized (or narrow-band) d electrons may take place below the melting temperature. This gives rise to a variety of phase transitions caused by electron ordering. Such fast, diffusionless transitions cannot be prevented by quenching and may take place at low temperatures. Typical examples are (a) magnetic ordering, i.e., ordering of the relative directions of the atomic moments (cf. Chap. 5); (b) ionic ordering, as in ordering of Fe^{2+} and Fe^{3+} ions on octahedral sites of the spinel Fe_3O_4; (c) Jahn-Teller ordering (to be discussed); (d) spin-orbit coupling plus magnetic ordering (to be discussed); and (e) high-spin-state $\rightleftharpoons$ low-spin-state transitions (to be discussed).

In the case of collective electrons, ordering of electron-electron correlations leads to (f) homopolar-bond formation (to be discussed), (g) bonding-band formation (to be discussed); and (h) superconductivity (cf. Chap. 16).

Description of localized d electrons

A description of the outer d electrons begins with the Schrödinger wave equation

$$\mathcal{H}\psi = \varepsilon\psi, \quad (4.5)$$

where $\mathcal{H} = \mathcal{H}_0 + V_{el} + V_{cf} + V_{LS} + \mathcal{H}_I + V_\lambda + \mathcal{H}_{ex}$. The first five terms of the Hamiltonian $\mathcal{H}$ are one-ion terms, the last two introduce interactions between cations.

The Hamiltonian for an isolated ion in the one-electron approximation is $\mathcal{H}_0$; all the other electrons are lumped in a spherical potential distribution. This leads to hydrogenic wave functions and is the correct

[3] N. F. Mott, *Can. J. Phys.* 34, 1356 (1956); *Nuovo cimento* 7 [10], Suppl., 312 (1958).

approximation for a single electron outside closed electron shells. For more than one outer d electron, a correction is introduced by the term V_{el}, which gives rise to Hund's rule of highest multiplicity. V_{el} causes intra-atomic exchange splitting between states with parallel versus antiparallel electron spins of the order ~ 1 eV (or 10^4 cm^{-1}).

Since the cations are not isolated but embedded in an ionic crystal, perturbation of the ion energies by ligand fields must be considered. The crystalline field $V_{cf} = V_0 + V_c + V_t$ introduces a constant V_0, which merely shifts all the d states uniformly, and two terms that split the degenerate d states of the isolated ion. The cubic component V_c is usually much larger than the noncubic component V_t; it can cause splittings as large as 1 eV and may interfere with Hund's rule.

If the ligand fields do not quench the orbital angular momentum (represented by the quantum number L), there will also be a spin-orbit coupling energy $V_{LS} = \lambda \mathbf{L} \cdot \mathbf{S}$, which may introduce splittings as large as 0.1 eV (10^3 cm^{-1}).

The perturbation H_I represents the interaction of the d electrons with the nuclear magnetic moment. It is important for nuclear magnetic resonance but causes splittings of only $\sim 10^{-2}$ cm^{-1}, negligible compared with the other one-ion terms.

The parameters V_λ and H_{ex} represent many-body elastic and magnetic energies due to long-range interactions between cations. V_λ enters wherever electron ordering at the cations induces a distortion of the interstices from higher to lower symmetry: The crystalline elastic energies are minimized if the individual distortions act cooperatively. H_{ex} is an interaction energy between cationic magnetic moments of the form

$$H_{ex} = -\sum_{ij} \{ J_{ij} \mathbf{S}_i \cdot \mathbf{S}_j + \gamma_{ij}(\mathbf{S}_i \cdot \mathbf{S}_j)^2 \\ + D_{ij}[\mathbf{S}_i \times \mathbf{S}_j] + \mathbf{S}_i \cdot \mathbf{\Gamma}_{ij} \cdot \mathbf{S}_j \}. \quad (4.6)$$

The scalar J_{ij} is the conventional, isotropic Heisenberg exchange parameter. The other terms, though much smaller, may give rise to observable effects. For example, $\gamma_{ij}/J_{ij} \lesssim 0.05$ may cause the M_s-vs.-T curve to deviate measurably from a Brillouin function, and $D_{ij}/J_{ij} \sim \Delta g/g$ can induce parasitic ferromagnetism by canting each sublattice moment of a collinear antiferromagnet toward a common perpendicular vector. The tensor quantity $\Gamma_{ij}/J_{ij} \sim (\Delta g/g)^2$ is even smaller: Δg represents the deviation of the spectroscopic splitting factor from its spin-only value $g = 2$.

For most problems in magnetism, the simple Heisenberg expression is sufficient. For example, from a knowledge of the J_{ij} and the individual atomic moments ($\mathbf{m}_A = g\mathbf{S}_i\mathbf{m}_B$), one can usually derive what type of long-range magnetic order minimizes H_{ex} (cf. Chap. 5). It permits us to understand phenomenologically not only ferromagnetism but also antiferromagnetism, ferrimagnetism, and metamagnetism.[4]

Hydrogenlike wave functions and cubic field splitting

The angular part of the hydrogenic wave functions consists of spherical harmonics. Broken down for the d orbitals into real components,

$$\psi_{\pm 2} = R(r) \sin^2 \theta e^{\pm 2i\phi} = R(r)(d_{x^2-y^2} \pm id_{xy});$$
$$\psi_{\pm 1} = R(r) \sin^2 \theta e^{\pm i\phi} = R(r)(d_{zx} \pm id_{yz}); \quad (4.7)$$
$$\psi_0 = R(r)(3\cos^2 \theta - 1) = R(r)d_{2z^2-x^2-y^2}.$$

They are shown graphically in Fig. 4.1. The two states $d_{x^2-y^2}$, $d_{2z^2-x^2-y^2}$ are directed along Cartesian axes, the other three along the diagonals. The former are said to have E_g, the latter T_{2g} symmetry. We will refer to these orbitals by that notation.

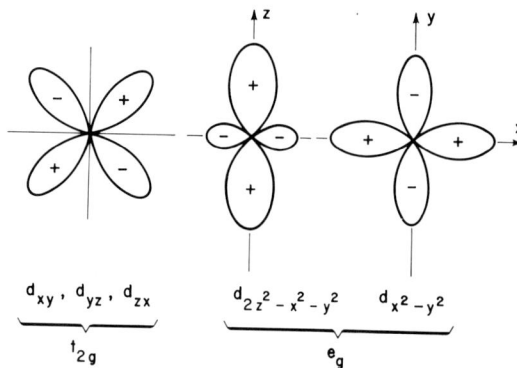

Fig. 4.1. Angular part of hydrogenic d-wave functions (real components).

What happens to the energies of these fivefold degenerate states if the transition metal cations find themselves in octahedral or in tetrahedral interstices? In octahedral position (Fig. 4.2) an electron in e_g orbitals (directed toward neighboring anions) suffers greater electrostatic repulsion than in the t_{2g} orbitals, pointing between the anions. This causes splitting of the pre-

---

[4] Like a ferromagnet, a ferrimagnet contains a net saturation moment below the Curie temperature. However, the ferrimagnetic crystal contains two or more cation subarrays having unequal magnetic moments that are not aligned parallel to one another. Usually there are two subarrays that are aligned antiparallel, so that the net saturation moment is the difference of the two subarray moments. A metamagnetic crystal is antiferromagnetic at lowest temperatures but exhibits an antiferromagnetic $\rightleftharpoons$ ferromagnetic phase change that is a function of both temperature and low (relative to exchange fields) magnetic field.

viously fivefold degenerate levels into two groups: three degenerate, more stable t_{2g} states and two less stable e_g states. Since both e_g states are real, e_g electrons carry no angular momentum. For a purely ionic model and first-order perturbation, the total energy is conserved,[5] and the splitting Δ_c, usually designated as $10Dq$, is distributed as shown in Fig. 4.2.

states of different spin. States with parallel spin are stabilized relative to states with antiparallel spin by an amount Δ_{ex} (Hund's rule). These fivefold degenerate atomic states are split by the octahedral fields as in the one-electron case. The ground state has two electrons (designated XX in Fig. 4.4a) with parallel spin in the lowest t_{2g} state.

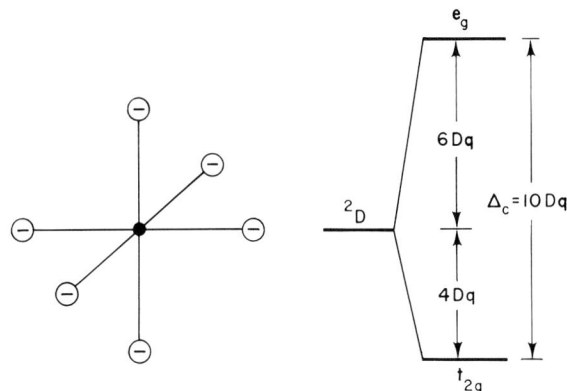

Fig. 4.2. Splitting of single d-electron orbitals in octahedral site.

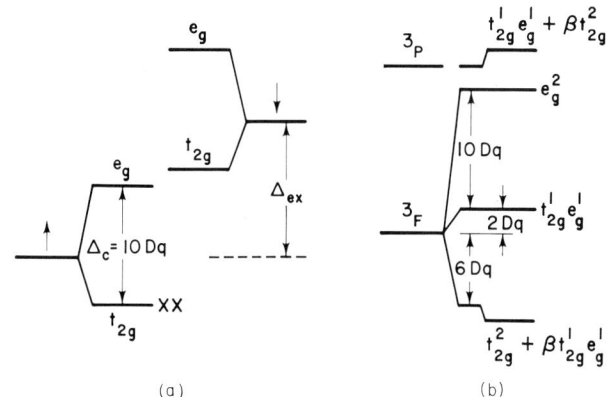

Fig. 4.4. Splitting for two d electrons in octahedral site: (a) one-electron levels, (b) two-electron levels.

In tetrahedral interstices (Fig. 4.3) the e_g orbitals point between the anions, and the t_{2g} orbitals are directed toward the neighboring anions; hence, the e_g states are stabilized relative to the t_{2g} states. Further, with four rather than six neighboring anions, the total splitting $10(Dq)'$ is considerably smaller than the splitting $10Dq$ for the octahedral case. From purely electrostatic arguments one estimates $(Dq)' \approx 4Dq/9$.

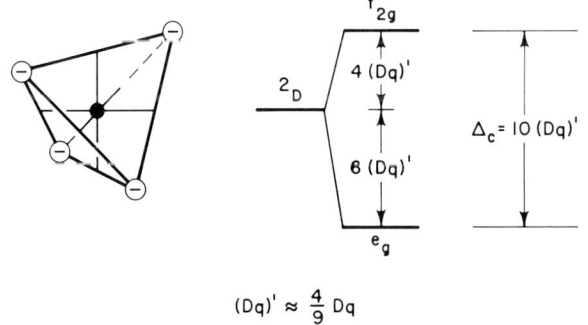

Fig. 4.3. Splitting of single d-electron orbitals in tetrahedral site.

Figure 4.4 shows what happens when a cation at an octahedral site has two outer d electrons. From a one-electron point of view (Fig. 4.4a) one has to distinguish

This one-electron solution omits important electron-correlation energies. If the two-electron problem is solved rigorously, the two lowest levels for an isolated ion are ^3F and ^3P (Fig. 4.4b). (The superscript 3 signifies parallel electron spins; F corresponds to $L = 3$, P to $L = 1$.) The sevenfold-degenerate ^3F state splits in an octahedral crystalline field into two threefold-degenerate states and an unstable nondegenerate state e_g^2 with two e_g electrons. The stable state corresponds to two t_{2g} electrons (t_{2g}^2) as in the one-electron scheme, plus a small admixture of $t_{2g}^1 e_g^1$ from the ^3P level. Obviously, significant errors are introduced into any quantitative argument based on the one-electron model, but the essential qualitative features are contained in it. Since the one-electron model is easier to visualize, it will be used in the remainder of this discussion.

Tetragonal field splitting and spin-orbit splitting

If an octahedral interstice is distorted to tetragonal ($c/a > 1$) symmetry (Fig. 4.5), the distortion destabilizes orbitals in the x-y plane relative to those outside this plane, and the e_g and t_{2g} levels are split as shown. Such a distortion leaves an orbitally degenerate one-electron ground state that contains an imaginary component. Therefore, the orbital momentum is not completely quenched, and spin-orbit coupling splits this level by $\Delta_{LS} = 2\lambda LS$. A tetragonal ($c/a < 1$) dis-

---

[5] Unequal mixing of s-p with e_g-vs.-t_{2g} states (i.e., mixing with σ-vs.-π orbitals) perturbs this energy conservation. This effect is important in estimating which lattice sites are preferred. It also contributes significantly to the magnitude of $10Dq$. At present this magnitude must be determined empirically.

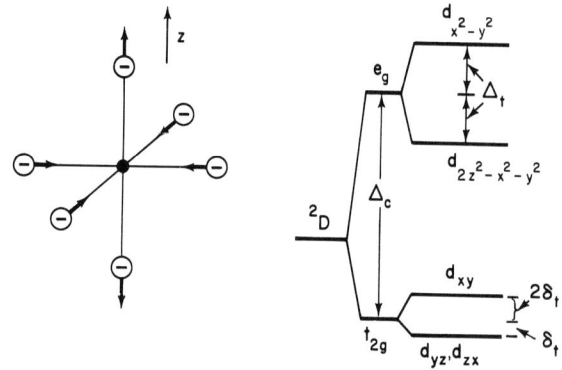

Fig. 4.5. Splitting of one-electron states in octahedral site for tetragonal $(c/a > 1)$ distortion.

tortion would have stabilized d_{xy} relative to $\psi_{\pm 1} = R(r)(d_{yz} \pm id_{zx})$, quenching the spin-orbit splitting.

The analogous situation for a tetragonally distorted $(c/a > 1)$ tetrahedral site is shown in Fig. 4.6. Here

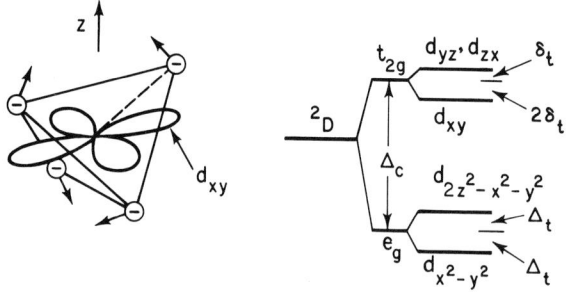

Fig. 4.6. Splitting of one-electron states in tetrahedral site for tetragonal $(c/a > 1)$ distortion.

the orbital momentum has been quenched by Δ_c; there is no spin-orbit splitting of the one-electron ground state with or without the presence of a distortion.

Applications

To demonstrate the significance of these ideas, we apply them to four practical situations: site preference, spin quenching, Jahn-Teller versus spin-orbit distortion, and magnetic coupling.

Site preference. Several factors contribute to site preference, but frequently the dominant energy contribution stems from ligand-field splitting.

A purely ionic model for Fe^{3+} or Mn^{2+} (each with five outer d electrons) gives no ligand-field stabilization; the total energy of the five states is conserved and each state occupied. Here tetrahedral-versus-octahedral-site s-p bonding and unequal mixing of these bonding states with e_g versus t_{2g} states control the site-preference energy.

In the case of Cr^{3+}, in contrast, the three outer d electrons are stabilized by a total of $12Dq$ in an octahedral site, but only by $8(Dq)' \approx (32/9)Dq$ in a tetrahedral site (Fig. 4.7).[6] Therefore Cr^{3+} is much more

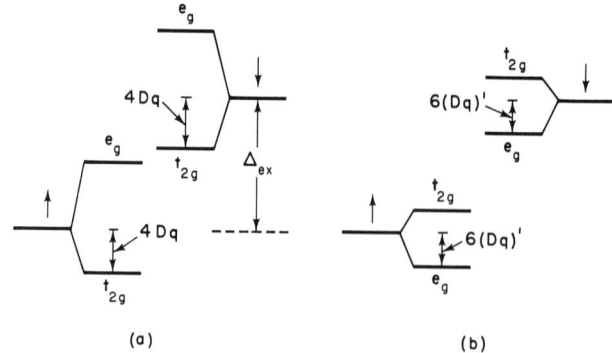

Fig. 4.7. Site-preference energies for Cr^{3+}; contribution for (a) octahedral sites versus (b) tetrahedral sites.

stable in an octahedral than a tetrahedral interstice, and presumably for this reason Cr^{3+} is never found in a tetrahedral site. (With a covalent rather than a purely ionic model for the splitting Δ_c, the stabilization is even larger.)

Spin quenching. The magnitude of the atomic moment $\mathbf{m}_A$ in Eq. 4.1 contains two components: a spin component $\mathbf{m}_S = gS\mathbf{m}_B$ and an orbital component $\mathbf{m}_L = gL\mathbf{m}_B$. If the orbital component is quenched by the ligand-field splitting (Δ_{lf}), $\mathbf{m}_A = \mathbf{m}_S$ and $g \approx 2$ in good approximation. The magnitude of $\mathbf{m}_S$ depends upon the relative magnitudes of the ligand-field splitting Δ_{lf} and the exchange splitting Δ_{ex}. If $\Delta_{ex} > \Delta_{lf}$, $\mathbf{m}_S$ corresponds to the highest possible S. However, if $\Delta_{ex} < \Delta_{lf}$ so that Hund's rule is violated, the spin is quenched and the cation is said to be in a low-spin state.

Consider, for example, Co^{3+} (with six outer d electrons) in an octahedral site. For $\Delta_{lf} \equiv \Delta_c < \Delta_{ex}$, five spins are parallel and the sixth spin is antiparallel (Fig. 4.8a); hence, $\mathbf{m}_S = 4\mathbf{m}_B$. This moment is approached in the high-temperature ($650° < T < 1210° K$) perovskite $LaCoO_3$. In the spinel Co_3O_4, on the other hand, the trivalent cobalt ions are diamagnetic because $\Delta_c > \Delta_{ex}$. Here the six outer electrons are all t_{2g} electrons, three with spin up and three with spin down (Fig. 4.8b). Since Δ_{lf} decreases with increasing temperature because of thermal expansion while Δ_{ex} stays constant, a low-spin-state $\rightleftharpoons$ high-spin-state transition

[6] From Fig. 4.7, the three electrons of octahedral site Cr^{3+} are in the threefold-degenerate t_{2g} state, so that each electron is stabilized $4Dq$ by the ligand fields. For tetrahedral site Cr^{3+}, two electrons are in the twofold-degenerate e_g state and one electron is in the t_{2g} state, giving a total stabilization of $(2 \times 6 - 4)(Dq)'$.

can occur at a temperature T_t where $\Delta_{lf} \approx \Delta_{ex}$. In LaCoO$_3$, this apparently occurs for a large fraction of the cations[7] in the interval 400° K < T < 650° K (see Fig. 4.8c).[8]

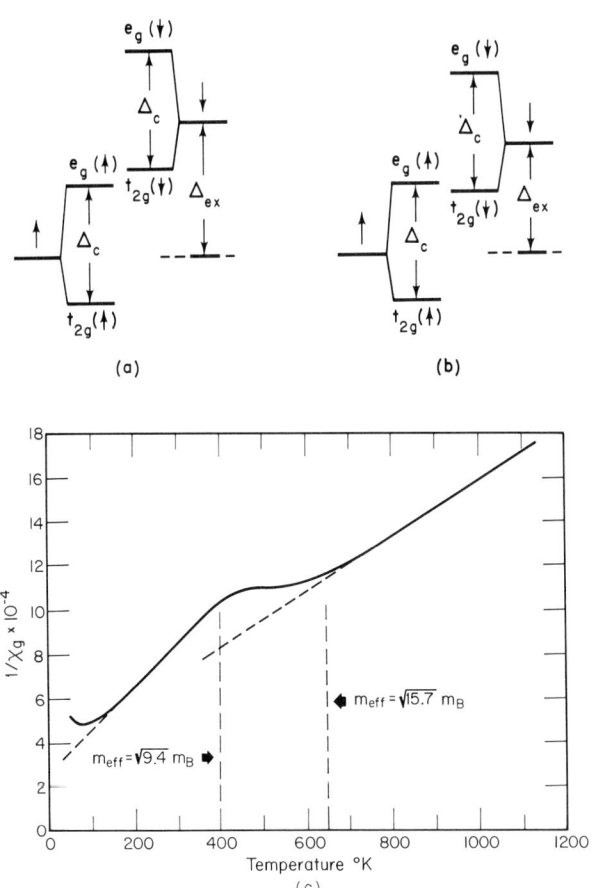

Fig. 4.8. Spin quenching by strong crystal fields. (a) High-spin state $\Delta_c < \Delta_{ex}$; (b) low-spin states: $\Delta_c > \Delta_{ex}$; (c) magnetic susceptibility for LaCoO$_3$. (After Heikes et al.[8])

Usually cations of the first long period are in the high-spin state; cations of the second and third long periods tend to be in a low-spin state. This is one reason that $3d$ elements dominate in materials of interest in magnetism.

Jahn-Teller versus spin-orbit distortions. Three different cations may illustrate the Jahn-Teller effect: octahedral-site Mn^{3+}, tetrahedral-site Fe^{2+}, and tetrahedral-site Ni^{2+}.

From Fig. 4.5 it is clear that a high-spin state Mn^{3+} ion (four outer $3d$ electrons) $t_{2g}^3 e_g^1$ in an octahedral interstice would be stabilized by an amount Δ_t by distortion from cubic to tetragonal symmetry. The

[7] J. B. Goodenough, *J. Phys. Chem. Solids* **6**, 287 (1958).
[8] R. R. Heikes, R. C. Miller, and R. Mazelsky, *Physica*, **30**, 1600 (1964).

Jahn-Teller theorem is based on this concept. It states that if the ground state of an ion contains an orbital degeneracy and no other perturbation is present, there will be a spontaneous distortion of the crystal to some lower symmetry to remove the degeneracy. The ground state of high-spin Mn^{3+} has a twofold degeneracy associated with the choice of one out of two e_g orbitals. Since there is no orbital momentum associated with the degenerate e_g states, spin-orbit coupling is not a competitive perturbation for the removal of the degeneracy, and a Jahn-Teller electron-ordering transition can be anticipated. However, it should be noted that a distortion of ratio $c/a < 1$ rather than $c/a > 1$ would stabilize the Mn^{3+} ion equally well; hence the *sign* of the Jahn-Teller distortion cannot be predicted from these considerations alone.

The situation is further complicated by the fact that distortions to *orthorhombic* symmetry would also stabilize the Mn^{3+} ion equally.[9] The two degenerate modes of vibration (orthorhombic and tetragonal) that stabilize octahedral Mn^{3+} are shown in Fig. 4.9.

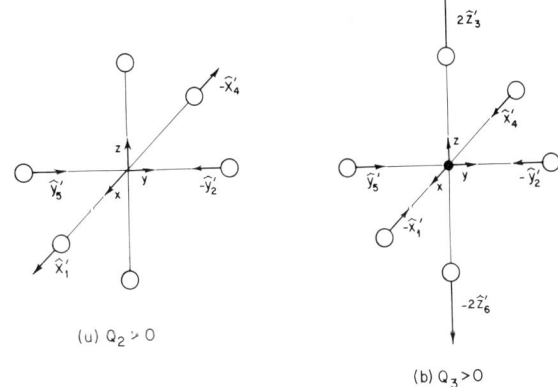

Fig. 4.9. Degenerate vibrational modes with E_g symmetry for octahedral complex.

Because of the degeneracy of these two vibrational modes, there would be no static Jahn-Teller distortion if there were no second-order effects; there would only be an interaction between the electron configuration and the vibrational modes. In the strong-coupling limit the electrons order in the state minimized by the vibrations at any given moment; the constellation vibrates freely among various linear combinations of the degenerate normal modes (dynamic Jahn-Teller stabilization). Presumably the dynamic stabilization, which has greater entropy, occurs at higher temperatures. At low temperature there is usually enough

[9] J. H. Van Vleck, *J. Chem. Phys.* **7**, 472 (1939); J. Kanamori, *J. Appl. Phys.*, Suppl., **31**, 14S (1960).

anisotropy from second-order elastic energies to stabilize a static distortion.

An important second-order energy comes from the anharmonic terms in the vibrational energy: More energy is required to reduce the lattice parameter to $a_o - \Delta a$ than to increase it to $a_o + \Delta a$. Since Δa is greatest along the c axis in the tetragonal Q_3 mode, tetragonal ($c/a > 1$) distortions are the most stable for an isolated octahedral-site complex. In a crystal, the elastic coupling energy V_λ between neighboring interstices introduces a *cooperative* static distortion. The particular cooperative distortion that minimizes the elastic energy depends on crystal structure as well as on anharmonic vibrational energy. Three examples illustrate various situations:

1. In spinels containing sufficient Mn^{3+} in octahedral sites, all these sites become tetragonal ($c/a > 1$) with parallel c axes; a macroscopic tetragonal ($c/a > 1$) distortion results. The spin system $Co_{3-x}Mn_xO_4$ (Fig. 4.10)[10] illustrates this. It remains cubic[11] until sufficient Mn^{3+} concentration ($x > 1.2$) induces a cooperative, macroscopic distortion. The magnitude of the tetragonal c/a ratio increases with Mn^{3+} concentration, reaching a maximum at $x \approx 2.0$. For higher concentrations of manganese, Mn^{2+} substitutes for Co^{2+} in tetrahedral sites. (That the initial manganese substitutes as Mn^{3+} for octahedral-site Co^{3+} is clear from the magnetic data shown in Fig. 4.10.)

2. In perovskites or the ReO_3 structure, the distortion of the individual interstices is more nearly orthorhombic;[12] the long and short cation-anion distances alternate within an a-b plane (Fig. 4.11). This distortion is stabilized because it induces a much smaller elastic energy. Orthorhombic perovskites with $a < c/\sqrt{2} < b$ remain orthorhombic below the distortion, but change to $c/\sqrt{2} < a \leq b$.[13]

3. Although $CrS(Cr^{2+}, 3d^4)$ exhibits a Jahn-Teller distortion, Mn^{3+} compounds of NiAs structure do not. In MnAs, MnSb, and MnBi, the absence of a Jahn-Teller distortion may also indicate sufficiently strong mixing of e_g and anion s-p orbitals so that the e_g elec-

[10] D. G. Wickham and W. J. Croft, *J. Phys. Chem. Solids* 7, 351 (1958).

[11] With slow cooling, compositions of smaller x appear to form small, inhomogeneous, Mn^{3+}-rich regions of tetragonal symmetry. The c axes of these isolated regions are distributed randomly over the cubic [100] axes; hence, the over-all symmetry appears cubic. However, there is evidence that such regions produce the spontaneous squareness of the hysteresis loops of memory-core ferrites used in computers.

[12] M. A. Hepworth and K. H. Jack, *Acta Cryst.* 10, 345 (1957).

[13] J. B. Goodenough, "The Magnetic Properties of Perovskites," *Landolt-Börnstein Tabellen*, 6th ed. Vol. 2, Pt. 9, Springer-Verlag, Berlin-Göttingen-Heidelberg, 1962, p. 2—187.

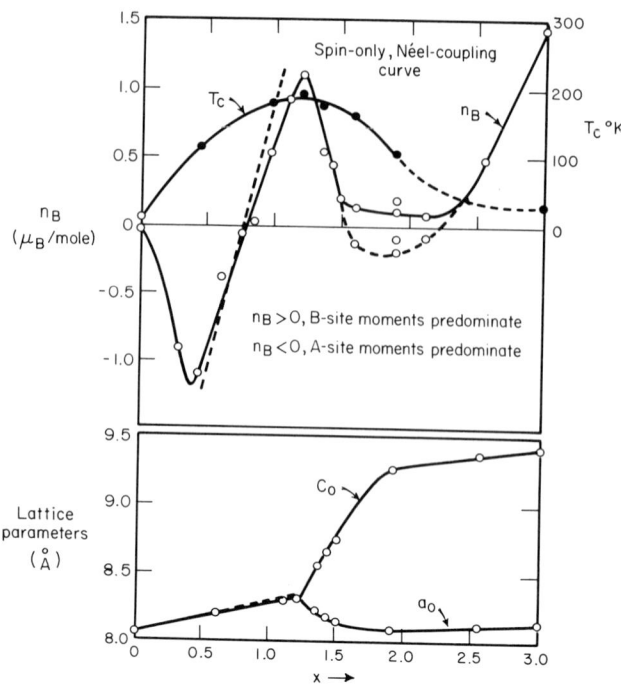

Fig. 4.10. Room-temperature lattice parameters, Curie temperature T_c, and spontaneous molecular moment for the system $Co_{3-x}Mn_xO_4$.[10] (Co_3O_4 has antiferromagnetically coupled tetrahedral-site Co^{2+}, diamagnetic octahedral-site Co^{3+}. Néel coupling curve assumes ferromagnetic tetrahedral-site Co^{2+} coupled antiferromagnetically with octahedral-site Mn^{3+}. Magnetic order in tetragonal phase is complex.)

trons are no longer localized but collective. Cation-anion-cation coupling via electrons in these band orbitals would be ferromagnetic.

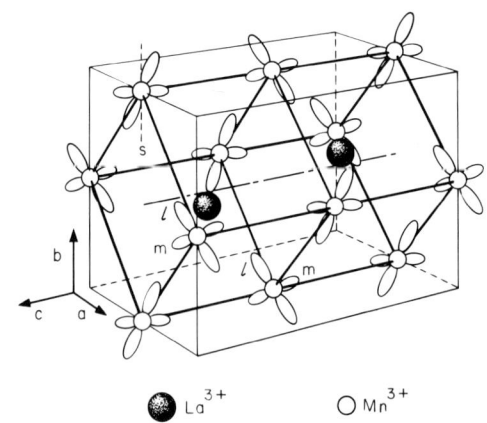

Fig. 4.11. Electronic ordering and site distortion at Mn^{3+} ions in perovskite.[12]

Tetrahedral-site Fe^{2+}, with six outer d electrons and therefore one e_g hole, is the analogue[14] of octahedral-site

[14] J. B. Goodenough, *J. Phys. Chem. Solids* 25, 151 (1964).

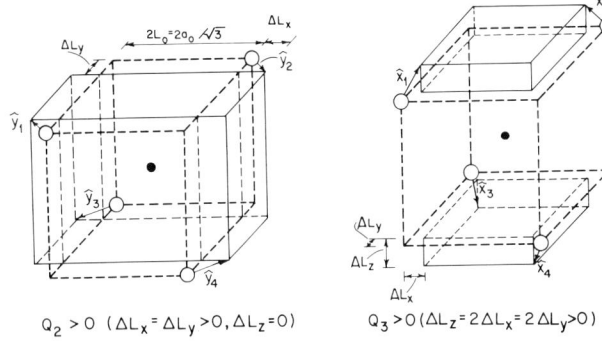

Fig. 4.12. Degenerate vibrational modes with E_g symmetry for a tetrahedral complex.

Mn^{3+}. The pair of degenerate vibrational modes of E_g symmetry (Fig. 4.12) does not change the cation-anion distances. This reduces the second-order energies, since anharmonic terms do not contribute to the vibrational energy, and the sign and character of the distortion is determined by the long-range forces in the crystal. This is illustrated for the spinel system $Fe^{2+}Fe_{2-x}^{3+}Cr_x^{3+}O_4$ with lattice parameters at $-183°$ C (Fig. 4.13).[15] The system exhibits not only orthorhom-

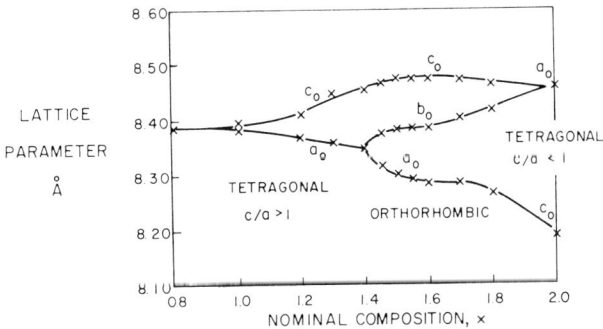

Fig. 4.13. Lattice parameters at $-183°$ C for the spinel systems $Fe_{2-x}Cr_xO_4$. (After Francombe.[15])

bic symmetry at low temperature but also tetragonal symmetry of both signs. In this case the character of the distortion at the tetrahedral-site Fe^{2+} ions is determined by the type of cations on the octahedral sites.

Tetrahedral-site Ni^{2+} represents a case where spin-orbit and Jahn-Teller stabilization are competitive.[16] Ni^{2+} has eight outer d electrons, i.e., two holes among the spin-down orbitals. From Fig. 4.6 it is clear that a tetragonal ($c/a > 1$) distortion stabilizes the Ni^{2+} energy by $2\delta_t$ but quenches any spin-orbit stabilization. A tetragonal ($c/a < 1$) distortion stabilizes the Ni^{2+} energy by δ_t without quenching the spin-orbit energy.

[15] M. H. Francombe, *J. Phys. Chem. Solids* 3, 37 (1957).

[16] J. B. Goodenough, *J. Phys. Soc. Japan* 17, Suppl. B-I, 185 (1962).

Therefore the *sign* of the anticipated distortion depends upon the relative magnitudes of $2\delta_t$ vs. $\delta_t + \lambda \mathbf{L} \cdot \mathbf{S}$, or of δ_t vs. $\lambda \mathbf{L} \cdot \mathbf{S}$. Since $\lambda \approx 0.1$ eV and $\delta_t \sim 0.01$ eV, it follows that for spins aligned collinearly below a magnetic ordering temperature, $\lambda \mathbf{L} \cdot \mathbf{S} = \lambda LS > \delta_t$; hence a tetragonal distortion ($c/a < 1$) occurs.[17] If the spins are not aligned collinearly, a great deal of elastic energy must be associated with noncooperative spin-orbit distortions of the individual interstices. In this case either elastic energies will force collinear spins on the distorting ions or the distortion will carry the sign of the Jahn-Teller effect. The former situation presupposes a magnetic ordering temperature.

In the spinel system $NiCr_tFe_{2-t}O_4$, tetrahedral-site Ni^{2+} induces both types of distortion (Fig. 4.14).[18]

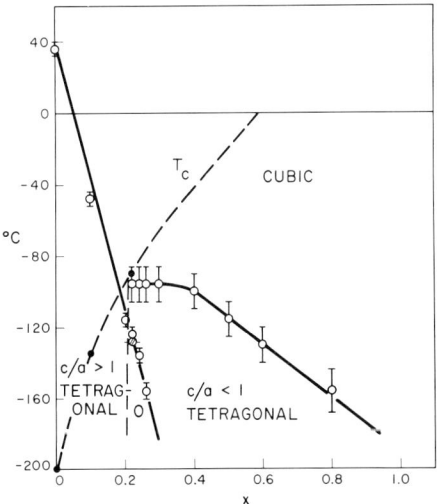

Fig. 4.14. Transition temperatures for the system $NiCr_tFe_{2-t}O_4$. (After Arnott et al.[18])

Magnetization data show that the spins are not collinear for $t > 1.0$, but distortions below the Curie temperature T_c induce collinear spins on tetrahedral sites and tetragonality ($c/a < 1$). As $t \to 2.0$, the distortions occur below $T_t > T_c$. They are also tetragonal, but $c/a > 1$, and no collinearity is forced on the spins.

Exchange coupling and magnetic order

The signs and relative magnitudes of the J_{ij} in the exchange Hamiltonian $\mathcal{H}_{ex}$ (Eq. 4.5) depend upon the electron ordering induced by the crystalline fields.

[17] Distortions determined by spin-orbit coupling for octahedral sites are found below the Néel temperature in CoO and FeO_{1+x}; octahedral-site Co^{2+} and Fe^{2+} have partially filled t_{2g} states with spin down.

[18] R. J. Arnott, A. Wold, and D. B. Rogers, *J. Phys. Chem. Solids* 25, 161 (1964).

Table 4.1. Cation-anion-cation and cation-cation superexchange interactions

CASE	OUTER-ELECTRON CONFIGURATION	DELOCALIZATION SUPEREXCHANGE		CORRELATION SUPEREXCHANGE		SUM
		$e_g - e_g$	$t_{2g} - t_{2g}$	$p\sigma$	$p\pi$	
1	d^5 ... d^5	STRONG ↑↓	WEAK ↑↓	STRONG ↑↓	WEAK ↑↓	STRONG ↑↓
2	d^3 ... d^3	—	WEAK ↑↓	WEAK TO MODERATE ↑↓	WEAK ↑↓	WEAK TO MODERATE ↑↓
3	d^5 ... d^3	MODERATE ↑↑	WEAK ↑↓	MODERATE ↑↑	WEAK ↑↓	MODERATE ↑↑
4	d^3 ... d^8	WEAK ↑↑	WEAK TO MODERATE ↑↑	WEAK ↑↑	—	WEAK TO MODERATE ↑↑

Table 4.1 gives four situations: In the first three, an anion separates two cations; in the last, the orbitals of neighboring cations overlap directly. Exchange interactions via an anion intermediary are called cation-anion-cation superexchange; those without an anion intermediary are called cation-cation superexchange.

Case 1. Each cation contains half-filled orbitals directed toward the anion (e.g., interaction between two Fe^{3+} ions). Cation-anion covalent bonding causes mixing of the cation d states and bonding p states; the d-like wave functions of the σ bonds spread out over the anion to overlap one another. Bonding between these overlapping orbitals can occur only if the electrons are antiparallel (Pauli exclusion principle). Relatively strong antiferromagnetic coupling results.

If the intermediate anion were not present, the argument would be identical so long as the cations were close enough for significant overlap, but not so close that collective rather than localized-electron d orbitals result.[19] This contribution is known as "delocalization superexchange."

Another important contribution is present in the case of an intermediate anion: "correlation superexchange" (originally called semicovalent exchange)[20] involving an excited state in which a double electron "hop" occurs from the anion to each side. This contribution always adds to the delocalization superexchange with the same sign.

Case 2. In the second situation (e.g., two Cr^{3+} ions in octahedral sites) no e_g electrons are directed toward the anion: The interaction takes place via π-electron overlap. Again antiferromagnetic coupling is predicted, but with reduced magnitude, because it is due to the overlap of π rather than σ electrons.

Case 3. In the interaction between an Fe^{3+} and a Cr^{3+} ion, the σ orbital electron on the Fe^{3+} ion may hop into an empty Cr^{3+} σ orbital without spin restriction by the Pauli principle. However, by Hund's rule the intra-atomic energy of the excited chromium ions is less for spin-parallel (ferromagnetic) coupling. This coupling is of intermediate strength because, although it involves σ orbitals, intra-atomic exchange energies also enter the problem.

Case 4. The fourth case corresponds to cation-cation superexchange between a Cr^{3+} ion and a Ni^{2+} ion. The excited state represents electron transfer from a nickel t_{2g} orbital to a chromium t_{2g} orbital. By Hund's rule, it requires less energy to hop an electron of antiparallel spin, and by the Pauli exclusion principle this electron can go only to the chromium if the nickel and chromium net spins are parallel. The 90° cation-anion-cation contribution is like Case 3 and is also ferromagnetic.

These rules have been found generally valid in insulators and semiconductors.[21] Perovskites containing Mn^{3+} ions represent a specially clear example.

In low-temperature $LaMnO_3$, for example, the orthorhombic symmetry is $c/\sqrt{2} < a < b$, characteristic of Jahn-Teller ordering (cf. Fig. 4.12). The magnetic order exhibits ferromagnetic a-b planes coupled antiferromagnetically along the c axis. This anisotropic coupling between like cations is compatible with Table 4.1: The coupling within the a-b planes corresponds to Case 3, that along the c axis to a weakened Case 1.

Evidence for a dynamic Jahn-Teller stabilization in the strong-coupling limit comes from the appearance of ferromagnetic Mn^{3+}-O^{2-}-Mn^{3+} coupling if the static Jahn-Teller distortions are destroyed either by raising the temperature or by diluting the Mn^{3+} array with nonmagnetic trivalent ions.[22] A dynamic Jahn-Teller stabilization, in the strong-coupling limit, would correlate the electron configurations on opposite sides of an anion (Case 3) for most of the time (Fig. 4.15).

Collective d electrons and crystallographic distortion

It is important to know how sharp a transition in physical properties takes place as the cation-cation separation R is reduced from $R > R_c$ (localized elec-

---

[19] P. W. Anderson, "Theory of Magnetic Exchange Interactions: Exchange in Insulators and Semiconductors," *Solid State Physics*, Vol. *14*, F. Seitz and D. Turnbull, Eds., Academic Press, New York, 1963, p. 99.

[20] J. B. Goodenough and A. L. Loeb, *Phys. Rev.* *98*, 391 (1955).

[21] J. B. Goodenough, *Magnetism and the Chemical Bond*, Interscience Publishers (John Wiley and Sons), New York, 1963.

[22] J. B. Goodenough, A. Wold, R. J. Arnott, and N. Menyuk, *Phys. Rev.* *124*, 373 (1961).

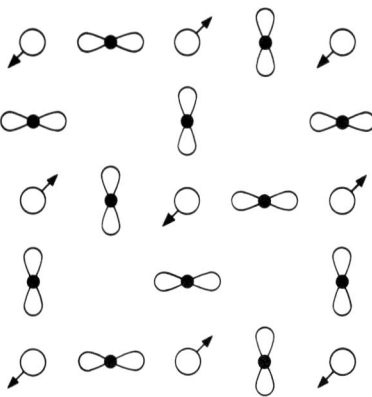

Fig. 4.15. Projection on a (100) plane of pseudocubic ReO$_3$-type structure, showing coupling of anion displacements and cationic single-e_g-electron configuration that induces ferromagnetic interactions via Case 3, Table 4.1.

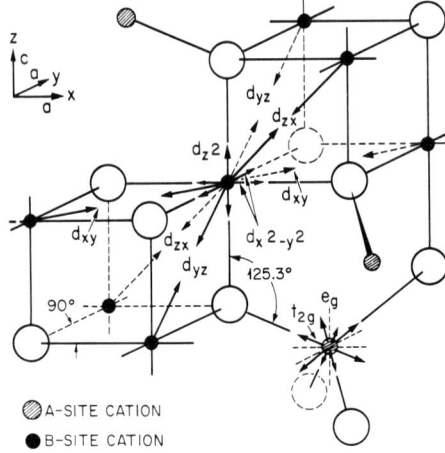

Fig. 4.16. Orientation of cation d orbitals with respect to the spinel structure. (After Wickham and Goodenough.[24])

trons) to $R < R_c$ (collective electrons). The essential feature of a collective-electron description is addition of *translational symmetry* to the problem (Bloch wave functions of conventional band theory). The initial basis functions may be either plane waves, as in broadband theory, or atomic orbitals, as in the tight-binding approximation for narrow-band theory. Electrons are localized at a given cationic site because $U = \varepsilon_a + (\varepsilon_g/2)$ is large (cf. Eq. 4.4). If $\varepsilon_g \to 0$ and ε_a is simultaneously small, translational symmetry must be considered; hence, $\varepsilon_g \to 0$ can serve as a definition of R_c.

To determine R_c requires a series of stoichiometric compounds in which R can be changed through R_c without altering the cation array whose d electrons are being studied.

One obvious experimental technique would be high pressure, but it is difficult to find a suitable highly compressible compound. Instead, a series of spinels $M^{2+}[V_2^{3+}]O_4$ was studied with M = Mn, Fe, Mg, Zn, and Co.[23] The V^{3+} ions occupy octahedral (B) sites that share common edges with six neighboring octahedral V^{3+} sites (Fig. 4.16).[24] Partially occupied orbitals of t_{2g} symmetry (V^{3+} has two outer d electrons) are directed toward neighboring V^{3+} ions between the anions. For $R < R_c$, the translational symmetry for the V^{3+}-ion array is important; there is a narrow t_{2g} band of collective-electron states. If $R > R_c$, translational symmetry is not important; the t_{2g} electrons are localized at individual V^{3+} sites. Measurements of the energy $U = \varepsilon_a + (\varepsilon_g/2)$ (cf. Eq. 4.4) as a function of R should reveal whether the transition $R \to R_c$ is abrupt or gradual. The separation R can be varied by changing the size of the tetrahedral-site M^{2+} ion.

Table 4.2. Activation energies and V^{3+}—V^{3+} separation for several vanadium spinels

Formula	V/M^{2+}	R(V—V) Å	q eV
MnV$_2$O$_4$	2.015 ± 0.005	3.014	0.37 ± 0.01
FeV$_2$O$_4$	1.990 ± 0.006	2.990	0.25 ± 0.01
MgV$_2$O$_4$	1.986 ± 0.004	2.974	0.18 ± 0.01
ZnV$_2$O$_4$	2.003 ± 0.006	2.973+	0.16 ± 0.01
CoV$_2$O$_4$	2.009 ± 0.004	2.973−	0.070 ± 0.005

The outcome of the experiment is shown in Table 4.2. The fact that U decreases from 0.37 to 0.07 eV as R changes by only 0.041 Å indicates that R_c is sharply defined as ≈ 2.97 Å for V^{3+} ions in this oxide structure. Attempts to prepare NiV$_2$O$_4$, which would presumably have $R < R_c$, have been unsuccessful. This fact and similar negative results for other systems which we have attempted to prepare with $R \approx R_c$ suggest that this critical separation is intrinsically unstable.

Apparent illustrations of this conjecture are given by spontaneous crystallographic distortions in which two or more subarrays of the cation sublattice move toward one another, each cation being displaced from the center of symmetry of its anion interstice. Spontaneous distortions induced by electrons localized at particular cations leave the cations in the centers of symmetry of their anion interstices. These are illustrated by the Jahn-Teller and spin-orbit-coupling distortions discussed earlier. Spontaneous distortions induced by narrow-band electrons, on the other hand, reflect the collective character of the electrons. The character-

---

[23] D. B. Rogers, R. J. Arnott, A. Wold, and J. B. Goodenough, *J. Phys. Chem. Solids* **24**, 34 (1963).

[24] D. G. Wickham and J. B. Goodenough, *Phys. Rev.* **115**, 1156 (1959).

istic feature of collective electrons in a partially filled band is the existence of a *Fermi surface* in momentum space. A change in the translational symmetry that introduces a Brillouin-zone surface at the Fermi surface stabilizes the energies of the occupied states and destabilizes the energies of the unoccupied states. Since the stabilization energies of the occupied orbitals vary linearly with the displacements, whereas the elastic restoring energies vary with the square of the displacements, finite distortions are anticipated below a critical temperature T_t.[25] This temperature increases sharply as R increases toward R_c. For a half-filled, cation-sublattice d band having $R \approx R_c$, spontaneous distortions create cationic clusters within which $R < R_c$ and between which $R > R_c$. In this case the band electrons of the high-temperature phase become localized within the cationic clusters.

There are numerous examples of this type of crystallographic distortion. In cases like VO_2 (distorted rutile) and $CoSb_2$ (arsenopyrite or distorted marcasite),[26] isolated homopolar bonds form between cation pairs. In others,[27] such as MnP (distorted NiAs structure), bonding bands are formed; the cation sublattice of MnP is composed of two subarrays shifted toward one another as a result of metal-metal bonding.[28] In the "compressed marcasite" structures,[26] such as FeP_2, and the Al5 structures like Nb_3Sn, metal-metal bonding takes place via both σ and π d electrons; the metal ions remain evenly spaced along linear chains, but the metal-metal spacing is abnormally short. These cooperative transitions are of first order and often show, besides symmetry changes, a remarkable change in electrical conductivity and magnetic susceptibility. In VO_2, for example, the conducting d electrons become localized at isolated V^{4+}-V^{4+} pairs at low temperatures; this semiconductor $\rightleftarrows$ metal transition produces a conductivity change of six orders of magnitude within $\Delta T < 1°$ K. Spin pairing in homopolar cation-cation bonds ($R < R_c$) simultaneously drops the low-temperature magnetic susceptibility.

Exchange coupling via collective electrons

Some confusion has arisen in the literature because of failure to appreciate that formation of metal-metal bonds, or bonding bands, in the narrow-band limit is essentially identical to the formation of antiferromagnetic order in the localized-electron limit. In the localized-electron problem, antiferromagnetic spin correlations between cation subarrays can be calculated via H_{ec} (Eq. 4.6).[29] Also, since localized electrons give localized atomic moments, their spin correlations can be observed directly with neutron diffraction. In the narrow-band region, where $R \rightarrow R_c$, the spin correlations are probably similar, but their calculation for Bloch electrons has proven difficult. The correlations are also difficult to observe directly. Bonding Bloch electrons are more concentrated where there is spin pairing, so that Bloch electron moments are much smaller. Also, long-range correlations undoubtedly break down unless molecular fields at the cations from localized electrons are simultaneously present. Therefore, Bloch electrons do not contribute an observable spontaneous atomic moment unless localized electrons are simultaneously present. (This is a second reason why $4d$ and $5d$ compounds, with their larger R_c than $3d$ compounds, are in general not magnetic.)

Exchange coupling between localized electrons via collective electrons gives an indirect measure of the spin correlations among collective electrons. Empirical data are compatible with the rules of Table 4.1. This implies that it is valid to extrapolate these rules for the sign of the cation-cation exchange coupling to narrow-band electrons:

Case 1: Electrons of a filled bond, or bonding band, couple the localized atomic moments of the two subarrays antiferromagnetically. The magnitude of the moment is the sum of the localized electron moments and of an induced collective-electron moment of roughly $0.5\mathbf{m}_B$ per collective d electron.

Case 2: Antibonding Bloch electrons couple the localized atomic moments ferromagnetically. The antibonding electrons contribute to the atomic moment $1\mathbf{m}_B$ per antibonding electron or hole.

Transition metals and their alloys

Application of these concepts to transition metals and their alloys presupposes knowledge of the number of d electrons present at any given metal atom. This information can be readily obtained from the formal charges for binary compounds of large electronegativity difference between constituents. For metals and alloys, however, it is necessary to determine the degree of overlap of the broad-band and narrow-band (and/or

---

[25] J. B. Goodenough, "Metallic Oxides: I. Operational and Theoretical Criteria" (to be published).

[26] W. B. Pearson in a private communication first drew my attention to the marcasite structures.

[27] J. B. Goodenough, *Phys. Rev. 117*, 1442 (1960).

[28] J. B. Goodenough, *J. Appl. Phys.*, Suppl., *35*, 1083 (1964); *Technical Report No. 345*, Lincoln Laboratory, Mass. Inst. Tech., 28 January 1964.

[29] Translational symmetry enters perturbation theory [19] in the form of transfer integrals b_{ij}, where $J_{ij} \sim 2b^2_{ij}/\epsilon_g$. The perturbation theory leading to this localized-electron superexchange expression becomes invalid as $R \rightarrow R_c$, because $\epsilon_g \rightarrow 0$.

localized) states. This will depend on structure, bandwidths, and relative energies of the atomic states from which they are formed. The problem is complicated, but experiments on binary compounds of large electronegativity difference begin to clarify the essential physical concepts for an understanding of narrow-band d electrons, whether they occur in such compounds or in metals and alloys. A valid correlation between structure, magnetic coupling, and atomic moments of the transition elements and their alloys[21,30] is being investigated (cf. also Chap. 16).

[30] J. B. Goodenough, *Phys. Rev. 120*, 67 (1960).

5 · MAGNETIC ORDERING IN HEISENBERG MAGNETS

Thomas A. Kaplan

The Heisenberg Hamiltonian — The classical ground-state problem — Spiral spin configurations — Variations of spirals in simple lattices — Spirals in spinel lattices — Discussion — Appendix

The Heisenberg Hamiltonian

The problems of magnetic ordering to be discussed here presuppose that the electron magnets be localized. In a material like MnO, for example, the O^{2-} ions with their closed shells are magnetically inert. Mn^{2+}, however, has 5 outer d electrons that, in the lowest state, have all their spins parallel (Hund's rule). It takes roughly 1 eV to reverse one of these 5 spins; hence, such reversal will not be caused thermally until well above the melting temperature. As long as the Mn^{2+} ion interacts only weakly with its surroundings, the 5 electron spins remain rigidly parallel. The spin angular momentum **S** associated with each electron has a maximum z component $S_z = \frac{1}{2}$; there is a corresponding magnetic dipole moment $\mathbf{m}_B$ which has a maximum z component of one Bohr magneton. The Mn^{2+} ion has therefore a maximum z component of spin $\mathbf{S}_z$ and magnetic moment **m**:

$$\mathbf{S}_z = \tfrac{5}{2}; \quad \mathbf{m} = 5\mathbf{m}_B. \tag{5.1}$$

As long as this ion does not interact with its surroundings, any direction of its spin vector is equally likely. If, however, there is an interaction between a pair of Mn^{2+} ions, a lower energy will be obtained by having the two atomic-spin vectors correlated. Usually the most important type of interaction (cf. Chap. 4) is the "exchange" interaction arising from the Pauli exclusion principle. This is expected to be represented (for small interaction) by an energy dependence of the form

$$\mathcal{E}_{n,m} = -2J_{nm}\mathbf{S}_n \cdot \mathbf{S}_m \tag{5.2}$$

with $\mathbf{S}_n$ and $\mathbf{S}_m$ the spins of the ions n and m, and J_{nm} the exchange parameter (independent of the spins). If J is positive, minimum energy for this ion pair is attained with parallel spins (ferromagnetic interaction); for J negative, antiparallel spins minimize the pair energy (antiferromagnetic interaction). When many ions interact sufficiently weakly, the total exchange energy is simply the sum

$$\mathcal{E} = -\sum_{nm} J_{nm}\mathbf{S}_n \cdot \mathbf{S}_m. \tag{5.3}$$

This is the so-called Heisenberg Hamiltonian.

There are, of course, other spin-dependent forces, such as for example the well-known magnetic dipole-dipole interaction between two point dipoles. Normally these effects are very much smaller and will be neglected here unless stated otherwise.

If Eq. 5.3 is a valid approximation for ionic crystals, the J_{nm} are proportional to overlap integrals between atomic charge clouds and thus expected to decrease rapidly with distance. The same type of expression is also valid for the rare-earth metals, where the localized magnetic electrons are in $4f$ instead of in $3d$ orbitals.

The 4f-4f overlap is probably very small, and the atomic magnets seem to interact through conduction electrons (cf. Chap. 16). In this case the J_{nm} drop off more slowly, in accordance with theoretical prediction.[1] Our problem is to investigate the validity of the Hamiltonian (5.3) and obtain information about the J_{nm}.

The classical ground-state problem

As originally derived, the spins **S** are quantum-mechanical angular momentum operators, and the implied quantum-mechanical problem is essentially insoluble today. One very useful approximation is that of the Weiss self-consistent internal field (local field). It requires a variational calculation completely analogous to the Hartree-Fock approximation for the electronic states of atoms. The ground-state problem in the Weiss approximation turns out to be equivalent to the classical Heisenberg problem: to find the set of classical vectors **S** of fixed length that minimize the energy of Eq. 5.3 for fixed J_{nm}.[2] Without loss of generality, we can take the $\mathbf{S}_n$ as unit vectors, so that our constraints are

$$\mathbf{S}_n^2 = 1. \qquad (5.4)$$

Historically this classical ground-state problem of minimizing Eq. 5.3 subject to the constraints (Eq. 5.4) has been considered quite simple for two reasons: There are cases of physical interest where the solution is obvious, and it had been believed that the spin arrangement would reflect the lattice symmetry (i.e., the symmetry of the J_{nm}).

As an example where the solution is obvious, if all the $J_{nm} > 0$, then parallel $\mathbf{S}_n$ clearly minimize $\mathcal{E}$ (ferromagnetism). If all the $J_{nm} = 0$ in the b.c.c. lattice except for nearest-neighbor n and m, then the $\mathbf{S}_n$ up at the cube corners, down at the cube centers, obviously minimizes $\mathcal{E}$ when the nearest neighbor $J_{nm} < 0$ (antiferromagnetic state) (Fig. 5.1). In these cases the interactions are said to be "noncompeting": We can find a set of spins (corners up, centers down, in the b.c.c. example) such that every interacting pair is in its own minimum energy state. The problem becomes challenging only when competing interactions are present.

In our b.c.c. example, antiferromagnetic next-nearest-neighbor interactions (between the cube corners as well as between the cube centers) are examples of

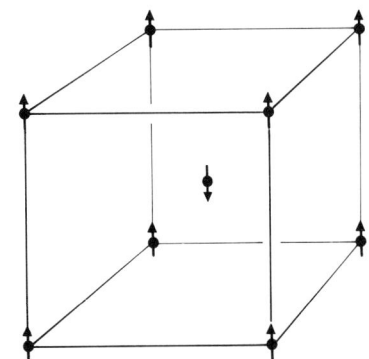

Fig. 5.1. An antiferromagnetic state in b.c.c. lattice.

such competing interactions because they are maximized by the set of spins that minimizes the nearest-neighbor interactions. The spins must now somehow compromise in order to minimize the sum of the competing interactions, the total energy.

To illustrate the assumption that lattice symmetry dominates the spin arrangements, let us recall the history of the theory of spinel structures. In a normal cubic spinel with six magnetic atoms per primitive unit cell (two A's and four B's, Fig. 5.2) the A's lie in octa-

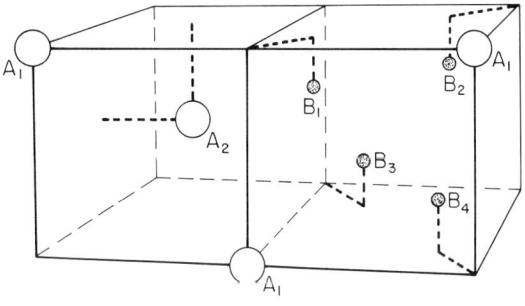

Fig. 5.2. Cations in normal spinel and magnetic ordering proposed by Néel and Yafet-Kittel.

hedral interstitial sites and the B's in tetrahedral ones of a c.c.p. oxygen lattice. Néel[3] made an essential contribution to the physics of ferrites by assuming that all the A spins were parallel and all the B's were parallel; this reduced the ground-state problem to a trivial one, involving only two spin directions. Then, by assuming antiferromagnetic A-B interactions, he was

---

[1] Cf., e.g., T. A. Kaplan and D. H. Lyons, *Phys. Rev.* **129**, 2072 (1963) and references contained there.

[2] The solution to the "classical ground-state problem" is also fundamental in the spin-wave approximation, providing its starting point.

[3] L. Néel, *Ann. phys.* [12] **3**, 137 (1948).

led to the famous Néel state (A spins up, B spins down). In general, the A and B magnetic moments do not cancel, causing a net magnetic moment for the crystal; Néel called this ferrimagnetism.

Yafet and Kittel[4] later generalized this problem: Since neither A_1 and A_2 nor $B_1 \cdots B_4$ are crystallographically equivalent, there might be six different spin directions, one for each of these sublattices. They solved the resulting simple minimization problem and claimed this to be the general solution. In other words, one expected to find a sublattice division that would reduce the total number of spin variables to a very small number.

Yafet and Kittel then found that if the B-B interactions were antiferromagnetic and comparable in magnitude to the A-B interactions, the B spins would cant relative to the A's, as shown in the lower part of Fig. 5.2. This result (1952) was felt to be so reasonable that no suspicion arose that it could be wrong, until seven years later the author attempted a spin-wave calculation assuming that this state (Y-K) gave the lowest energy and found it unstable![5]

Let us state the situation more precisely. When one includes only nearest-neighbor A-B and B-B interactions with cubic symmetry, the Y-K state proved *not* to be the classical ground state for any values of these interactions. Furthermore, this instability is not a result of symmetry: Small changes of the J_{AB} and J_{BB} which destroy the cubic symmetry of the lattice do not remove the instability.

What then is the true ground state? The instability clearly showed that the intuitive assumption that all the A_1 spins were parallel, etc., i.e. that the spins obey the translational symmetry of the lattice, was simply not justified. Obviously the problem had to be considered in full—as a many-body problem. Temporarily dropping this complicated spinel lattice, the author set up an analogous situation in a similar lattice, the b.c.c. structure, and considered the following problem: If the only interactions were between nearest neighbors and they were antiferromagnetic, then we would have the classical Néel state with cube corners up and cube centers down. If we now introduce second- and third-neighbor antiferromagnetic interactions in competition with nearest-neighbor forces, then, if they are large enough, our collinear state will become unstable.

Spiral spin configurations

This study[5] of the small-deviation modes that lowered the energy led to the result that the absolute

[4] Y. Yafet and C. Kittel, *Phys. Rev. 87*, 290 (1952).
[5] T. A. Kaplan, *ibid. 116*, 888 (1959).

minimum energy was attained by a spiral spin configuration. At about the same time such spiral configurations (also referred to as helices or screw arrangements) were independently discovered, also on theoretical grounds, by Yoshimori[6] in Japan and Villain[7] in France. Spirals and related configurations have since been found in many materials.

A fairly general theorem[8] could be proved by us: For any Bravais lattice the absolute minimum of the classical Heisenberg energy is always attained by a simple spiral

$$\mathbf{S}_n = \hat{x} \cos \mathbf{k} \cdot \mathbf{R}_n + \hat{y} \sin \mathbf{k} \cdot \mathbf{R}_n. \quad (5.5)$$

Its k vector is $\mathbf{k}_0$, the one that maximizes

$$\mathcal{J}(\mathbf{k}) = \sum_{\mathbf{R}} J(\mathbf{R}) \exp (i\mathbf{k} \cdot \mathbf{R}). \quad (5.6)$$

By a Bravais lattice is meant the following: The positions $\mathbf{R}_n$ of all the spins are translationally equivalent; i.e., if $\mathbf{a}_1$, $\mathbf{a}_2$, $\mathbf{a}_3$ are three independent vectors, then $\mathbf{R}_n = n_1\mathbf{a}_1 + n_2\mathbf{a}_2 + n_3\mathbf{a}_3$ with the n_i being integers, and the interactions have this translational symmetry:

$$J_{nm} = J(\mathbf{R}_n - \mathbf{R}_m); \quad (5.7)$$

also, periodic boundary conditions are invoked. By virtue of this theorem, the problem of finding the classical ground state has been reduced from one with $\sim 10^{23}$ degrees of freedom to one with only three variables (the components of $\mathbf{k}_0$).

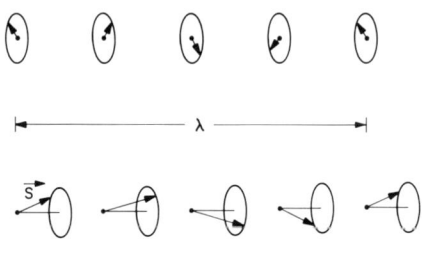

Fig. 5.3. Special spirals on linear chain.

To get a feel for this spiral configuration, we note that $\hat{x}$ and $\hat{y}$ are a pair of orthonormal vectors; hence, the $\mathbf{S}_n$ lie parallel to a given plane fixed in space. The orientation of this plane relative to the crystal is completely arbitrary since the energy of Eq. 5.3 is isotropic, i.e., invariant to any uniform rotation of all spins. Small anisotropic forces will remove this degeneracy,

[6] A. Yoshimori, *J. Phys. Soc. Japan 14*, 807 (1959).
[7] J. Villain, *J. Phys. Chem. Solids 11*, 303 (1959).
[8] D. H. Lyons and T. A. Kaplan, *Phys. Rev. 120*, 1580 (1960).

fixing the x-y plane relative to the crystal axes. We further note that our spiral is a plane wave with propagation vector $\mathbf{k}$; i.e., all ions lying in any one plane perpendicular to $\mathbf{k}$ have parallel spins and the spiral has constant pitch. Its axis $\mathbf{k}$ need not point in a symmetry direction, and the wavelength ($\lambda = 2\pi/|\mathbf{k}|$) need not be some small number of unit cells.

A special case is illustrated in the upper part of Fig. 5.3. Here the wavelength λ is commensurate with the lattice, λ being 4 spacings. However, by a slight change in the parameters J_{nm} (which determine $\mathbf{k}$), λ may be changed ever so slightly from 4, and then the spins never actually repeat. In such a case the magnetic ordering would change the material from crystalline to noncrystalline, if we define a crystal as a periodic array of identical units.

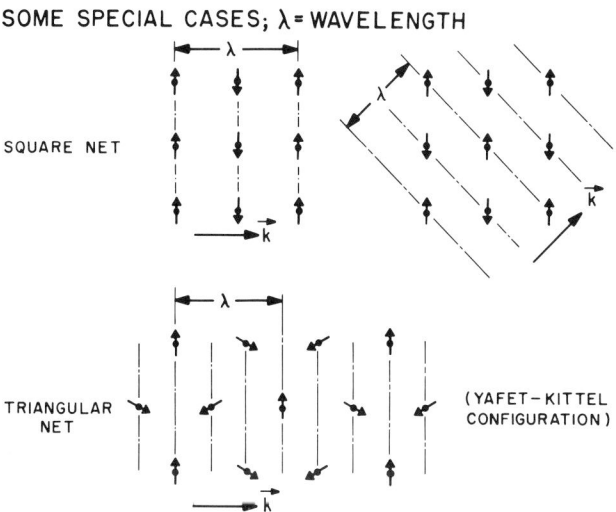

Fig. 5.4. Special spirals in some two-dimensional lattices.

Figure 5.4 gives some special examples of spirals. In the upper sections we have collinear arrangements on a square net: The spiral has plane-wave character, and the $\mathbf{k}$ is not unique; the spiral with $\mathbf{k}$ in the (1, 1) direction is the same as one with $\mathbf{k}$ rotated into (1, −1). The spiral of the lower figure is precisely the configuration proposed by Yafet and Kittel[4] for the triangular net; it can be shown to be the ground state for the triangular net with nearest-neighbor antiferromagnetic interactions, but not for the cubic spinel (a non-Bravais lattice), as already noted.

Variations of spirals in simple lattices

Let us stay temporarily with simple lattices and indicate some variations on these spirals that can result from anisotropy and finite temperature. We refer in this context to some of the heavy rare-earth metals in which many spirals and variations were found experimentally.[9] In Fig. 5.5 the observed ordering as a

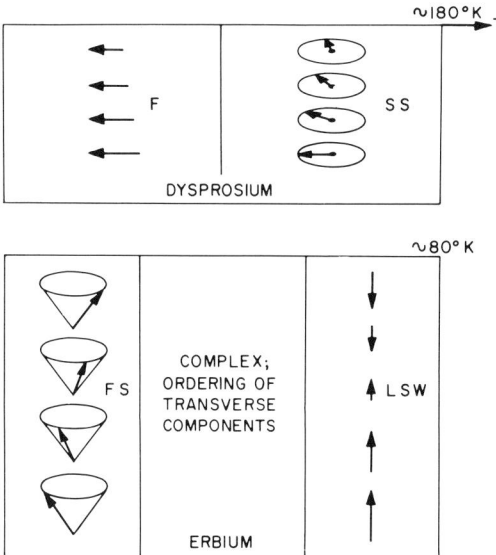

Fig. 5.5. Magnetic ordering in Dy and Er.

function of temperature is indicated for dysprosium and erbium. The $\mathbf{k}$ vector of these hexagonal close-packed crystals is found[9] to be parallel to the c axis; hence, with the vertical axis parallel to c, each arrow in the figure represents the moment direction for all the rare-earth atoms lying in a plane perpendicular to the c axis.

Dy exhibits two ordered magnetic phases, both simple spirals [at low temperatures Dy is a simple ferromagnet, corresponding to a simple spiral (Eq. 5.5) with $\mathbf{k} = 0$]. The wavelength for the high-temperature phase is of the order of 6 atomic layers but changes continuously (by about 40 percent) with temperature. Er, on the other hand, exhibits interesting variations. At low temperatures the configuration is a combination of a simple spiral with nonzero $\mathbf{k}$ for the x-y components lying in the basal plane, and a ferromagnetic z component. In other words, in this "ferromagnetic spiral" the spins lie on the surface of a cone; the material has a net moment. In the ordered high-temperature phase we see again a quite surprising variant of our simple spiral. The spins are collinear, but the magnitude varies as a simple sine wave function along the z direction. (LSW stands for longitudinal spin wave.) In the intermediate phase, the ordering is somewhat more complex.

[9] M. K. Wilkinson, W. C. Koehler, E. O. Wollan, and J. W. Cable, *J. Appl. Phys.*, Suppl., *32*, 48S and 49S (1961); W. C. Koehler, *ibid. 32*, 20S (1961).

Most of the mystery concerning the existence of this large variety of types of magnetic ordering was removed by phenomenological analysis.[10] The essential idea is that addition of anisotropy terms to Eq. 5.3 (e.g., $K\Sigma_n S_{nz}^2$, where K is a constant and S_{nz} the component of $\mathbf{S}_n$ along the c axis), coupled with the assumption of competing J_{nm}, led to a satisfactory explanation of much of the observed complexity.

At first sight, the high-temperature phase of Er seems to violate Eq. 5.4. This however is not true: the arrows represent the *averages* of the spins, whose directions fluctuate thermally; these averages are correlated in the sinusoidal way indicated. The intuitive reason for the occurrence of this LSW form is given in Kaplan, p. 334 of Ref. 10.

Spirals in spinel lattices

Let us return to the spinel ground-state problem, with Heisenberg exchange forces only. The proof of the spiral theorem for Bravais lattices (cf. Eq. 5.5) was made possible by a method due to Luttinger and Tisza,[11] as discussed in the Appendix. Unfortunately, their method does not work for the spinel problem; a generalization was required.[12]

In the remainder of this section some principal results of this (GLT) method and a comparison with recent neutron-diffraction experiments are given. The work was carried out in collaboration with Dwight, Menyuk, and Lyons.[12-16]

If we limit ourselves first to cubic spinels with nearest-neighbor antiferromagnetic A-B and B-B interactions, the classical ground state depends only on the parameter

$$u \equiv \frac{4}{3}\frac{J_{BB}S_B}{J_{AB}S_A}. \qquad (5.8)$$

In Fig. 5.6 the energies for various states are plotted as $f(u)$. The Yafet-Kittel state goes below the Néel one

[10] T. A. Kaplan, *Phys. Rev.* **124**, 329 (1961); R. J. Elliott, *ibid.* **124**, 346 (1961); H. Miwa and K. Yosida, *Progr. Theoret. Phys.* (*Kyoto*) **26**, 693 (1961).

[11] J. M. Luttinger and L. Tisza, *Phys. Rev.* **70**, 954 (1946); J. M. Luttinger, *ibid.* **81**, 1015 (1951). See particularly the second paper.

[12] D. H. Lyons and T. A. Kaplan, *Phys. Rev.* **120**, 1580 (1960).

[13] T. A. Kaplan, K. Dwight, D. H. Lyons, and N. Menyuk, *J. Appl. Phys.*, Suppl., **32**, 13S (1961).

[14] D. H. Lyons, T. A. Kaplan, K. Dwight, and N. Menyuk, *Phys. Rev.* **120**, 1580 (1960).

[15] N. Menyuk, K. Dwight, D. H. Lyons, and T. A. Kaplan, *ibid.* **127**, 1983 (1962).

[16] K. Dwight, N. Menyuk, D. H. Lyons, and T. A. Kaplan, to be published.

for $u > 1$. Stability calculations[5,17] showed not only that the Yafet-Kittel state is unstable but that the stability of the Néel state is limited to $u \leq \frac{8}{9} \equiv u_0$.

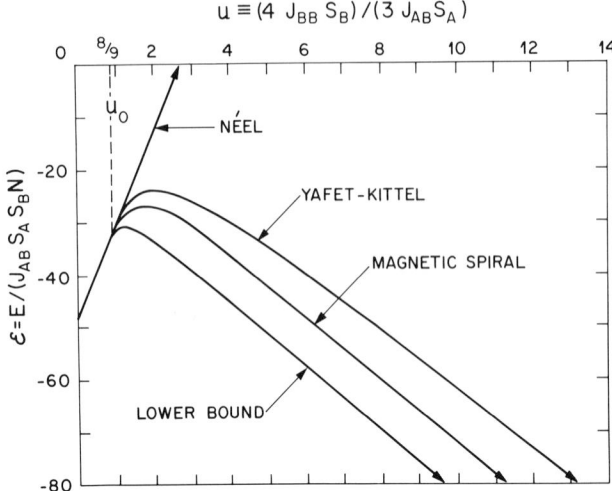

Fig. 5.6. Energies as functions of u (cubic spinel).

The GLT method proved the Néel state the absolute ground state whenever it is locally stable (i.e., it is never metastable). In pushing past u_0 we took a hint from the local stability calculations which showed that the spin deviation causing instability at $u \geq u_0$ is a sine wave with $\mathbf{k}$ vector in the [110] direction. Applied first to "[110] configurations" (i.e., configurations with Fourier components having $\mathbf{k}$'s in the [110] direction only), the GLT method gave the lowest [110] state, a magnetic spiral (Fig. 5.6). It satisfies the rather stringent condition of going below the Néel state precisely at $u = u_0 + 0$. Geometrically it consists of a conical or magnetic spiral type (cf. Fig. 5.3) on each of the sublattices. Their cone axes are mutually collinear, and they all have the same wave vector. However, there are three different cone angles and two phase angles, all single-valued functions of u. Two $\mathbf{k}$ vectors enter into the Fourier analysis: $\mathbf{k} = 0$ for the spin components along the cone axis and $\mathbf{k} \neq 0$ for the perpendicular spin components.

We tried to prove that this magnetic spiral is the ground state but failed (cf. Appendix). We were still able to prove, however, some interesting theorems about the spiral configuration, e.g., that the [110] magnetic spiral is the lowest of all magnetic spirals for u at least up to 1.35. Since these spin configurations in general have 14 independent parameters (k_x, k_y, k_z, 6 cone angles, and 5 phases), the result is not trivial.

Testing for local stability, we found the spiral locally

[17] T. A. Kaplan, *Phys. Rev.* **119**, 1460 (1960).

stable for the range $u < u' \cong 1.298$ and unstable for $u > u'$.

Summarizing: The lowest magnetic spiral for a large range of u is this spiral; it is very likely the true ground state for $u < u'$, but definitely not for $u > u'$.

The first experimental test of these predictions was a neutron-diffraction study[18] on $MnCr_2O_4$ powder. This normal cubic spinel showed a number of magnetic satellite peaks in addition to the fundamentals (the latter can be indexed on the crystallographic cell, hence correspond to the Fourier component $\mathbf{k} = 0$; the former cannot be so indexed). Some of the satellites were comparable in magnitude with the largest fundamentals. Our spiral predicts satellite peaks (arising from the nonzero k component); their intensities are proportional to a structure factor containing the cone angles, phases, and $\mathbf{k}$ vector, all of which are determined theoretically as functions of u. The parameter u can be determined by measuring the total magnetic moment and assuming reasonable moments for the cations. (The $\mathbf{k}$ vector, which determines the locations of all satellite peaks, is very insensitive to the adjustable parameter u.) The theoretically predicted peaks fit the experimental locations precisely!

Recently $CoCr_2O_4$ was studied by neutron diffraction[19] at low temperatures; excellent over-all agreement in peak intensities with predictions was obtained, but the $\mathbf{k}$ vector proved slightly in error (~ 5 percent).

Thus in these two cubic-spinel materials (manganese and cobalt chromite) the calculated magnetic spiral appears to be a good representation of the actual low-temperature magnetic order. Somewhat similar calculations for tetragonally distorted spinels show[5,13,15,16] that various tetragonal distortions of the parameters J_{nm} lead not only to various spirals with $\mathbf{k}$ in [110] but also in other directions, and to configurations more complicated than magnetic or conical spirals. For sufficient distortion of a certain type, the Yafet-Kittel state is the ground state, in agreement with measurements[20] on $CuCr_2O_4$.

Recent diffraction work on MnV_2O_4[21] (apparently a normal cubic spinel) agrees neither with the spiral nor the Néel model; the pattern corresponds more to the old Yafet-Kittel model. This plus our theory suggests strongly that something unusual happens in the vanadite (i.e., something goes wrong with the Heisenberg model). The V-ion configuration $(3d)^2$ is orbitally degenerate in the cubic crystal field; this degeneracy probably remains in the trigonal field (as seems indicated by the fact that the V moment differs appreciably from its spin-only value). The Cr ions probably do not show such orbital degeneracy. The existing derivations of the Heisenberg Hamiltonian have always assumed an orbitally nondegenerate ionic state. This may be the cause of the disagreement; calculations on this question are in progress.

Discussion

The magnetic ordering problem based on the Heisenberg model has been discussed historically. It was originally felt that spins ordered only in collinear configurations. Yafet and Kittel attacked that superstition by suggesting a canted-spin model, but the feeling remained that the symmetry of the lattice must be reflected in the minimum-energy spin configuration in a simple way (which was thought to be understood). Recently it has been found that this feeling is incorrect: Given only the lattice symmetry, i.e., the J_{nm}, any collection of spin directions is *a priori* a "possible" ground state. However, out of such chaos a considerable degree of order has been resurrected by the concept of spirals and their modifications, as well as the development of theoretical methods[22] to predict them. The theorem concerning Bravais lattices, the theoretical and experimental work on spinel lattices, and some of the work on the heavy rare-earth metals have been mentioned. More examples of materials with interesting magnetic order are cited by Bertaut.[23]

Comparison of theory with experiment at very low temperatures on $CoCr_2O_4$ and $MnCr_2O_4$ suggests strongly that these are Heisenberg magnets, i.e., that the Heisenberg Hamiltonian (Eq. 5.3) is a good approximation. However, the neutron-diffraction behavior at higher temperatures (but still below T_c) disagrees rather violently in certain respects with the predictions of the Weiss internal field approximation for the statistical behavior of such magnets.[19] An improved treatment of the statistical mechanics of that model is needed.

Preliminary experiments at Lincoln Laboratory on the microwave resonance absorption of Co and Mn chromites have given unusual results that are not yet fully understood; the general problem of the dynamics of a Heisenberg spin system with magnetic spiral ordering has to be investigated.

[18] J. Hastings and L. Corliss, *Phys. Rev.* **126**, 556 (1962).
[19] N. Menyuk, K. Dwight, and A. Wold, *J. phys.* **25**, 528 (1964).
[20] E. Prince, *Acta Cryst.* **10**, 554 (1957).
[21] R. Plumier, *Compt. rend.* **255**, 2244 (1962).

[22] For a method not mentioned that gives considerable insight, cf. D. H. Lyons and T. A. Kaplan, *J. Phys. Chem. Solids* **25**, 645 (1964).
[23] E. F. Bertaut, "Spin Configurations of Ionic Structure: Theory and Practice," *Magnetism*, Vol. 3, G. T. Rado and H. Suhl, Eds., Academic Press, New York, 1963, p. 150.

Comparison of experiments on several vanadium spinels with the theory of magnetic ordering based on the Heisenberg Hamiltonian suggests that the latter is not a good approximation for materials in which the individual cations have orbitally degenerate ground states. The correct Hamiltonian has yet to be found.[24]

A further important aspect of the theory is application of the Landau-Lifshitz thermodynamic theory of second-order phase transitions to magnetic ordering.[25,26] Briefly described: A density function ρ is expanded[25] in terms of a complete set of functions ϕ_i with coefficients c_i. In part I of the theory the c_i are independent of position; in part II spatial variations of the c_i are considered.

The essential idea of part I is that standard group-theoretical arguments are applicable to the ordering problem in the neighborhood of a second-order phase transition (at temperature T^*). These arguments restrict the possible spin configurations for $T \lesssim T^*$ (lower symmetry phase) in a way that depends on the symmetry at T just above T^*. The general rule is that the higher the symmetry for $T \gtrsim T^*$, the greater the restriction for $T < T^*$. When Curie or Néel points are of second order (as usual), symmetry arguments can give important information[25,27] about the high-temperature magnetic ordering. In other words, the high-temperature ordering problem is much simpler than the ground-state problem, for which the standard symmetry arguments are not directly applicable.[28] Part II of the Landau-Lifshitz theory is questionable (see also Ref. 28).

APPENDIX

The problem is to minimize the energy (Eq. 5.3) subject to the constraints (Eq. 5.4). A straightforward approach requires the energy to be stationary, giving (either by elimination of say S_{iz}^2 via Eq. 5.4 or by the method of Lagrange multipliers):

$$\sum_j J_{ij} \mathbf{S}_j = \lambda_i \mathbf{S}_i, \quad (A5.1)$$

with

$$\lambda_i \equiv \mathbf{S}_i \cdot \sum_j J_{ij} \mathbf{S}_j. \quad (A5.2)$$

Equation A5.2 follows directly from Eqs. A5.1 and 5.4. But this approach is hopeless, as emphasized by Luttinger,[11] because the λ_i as well as the $\mathbf{S}_i$ are unknown; even though J_{ij} possesses high symmetry (that of the lattice without spins), all sets of λ_i consistent with Eqs. 5.4 and A5.1 must be considered, and sets of very low symmetry do exist.[29]

The method of Luttinger and Tisza (L-T) is as follows: From Eq. 5.4 it follows that

$$\sum_i \mathbf{S}_i^2 = N, \text{ total number of spins.} \quad (A5.3)$$

Let us for the moment forget our original "physical" problem, and consider the minimization of the energy (Eq. 5.3) with the single constraint Eq. A5.3. The stationary equations analogous to A5.1 are now

$$\sum_j J_{ij} \mathbf{S}_j = \lambda \mathbf{S}_i, \quad (A5.4)$$

with $\lambda = N^{-1} \Sigma_{ij} J_{ij} \mathbf{S}_i \cdot \mathbf{S}_j = E/N$. Thus the low-symmetry λ_i are replaced by a single number λ, and Eq. A5.4 possesses all the symmetry of the lattice (i.e., of the J_{ij}). This problem is generally tractable. If the solution to this second, so-called "weak-constraint" problem happens to satisfy the original "strong" constraints (Eq. 5.4), a little reflection shows that this solution is the *exact solution of the original physical problem*. It turns out that for Bravais lattices [i.e., for $J_{ij} = J(\mathbf{R}_i - \mathbf{R}_j)$, where $\mathbf{R}_i = l_i \mathbf{a}_1 + m_i \mathbf{a}_2 + n_i \mathbf{a}_3$, l_i, m_i, and n_i running over all integers modulo $N^{1/3}$, and the $\mathbf{a}_u$ are the basic lattice translations] the method works, leading to the theorem stated previously. For our spinel problem, however, the method fails: The solution to the weak-constraint problem does not satisfy the strong constraints. But, as indicated, we are able to generalize the method in a useful way. The basic theoretical concepts, including a discussion of various misconceptions in the literature, recently have been summarized.[29]

---

[24] In this connection, cf. K. Dwight, N. Menyuk, and A. Wold, *Proc. International Conference on Magnetism*, Nottingham, 1964, to be published.

[25] L. D. Landau and E. M. Lifshitz, *Statistical Physics*, translated from the Russian by E. Peierls and R. F. Peierls, Pergamon Press, London, and Addison-Wesley Publishing Co., Reading, Mass., 1958, Chap. 14.

[26] K. P. Belov, *Magnetic Transitions*, translated from the Russian by W. H. Furry, Consultants Bureau, New York, 1961.

[27] J. O. Dimmock, *Proc. International Conference on Magnetism*, Nottingham, 1964, to be published.

[28] The Landau-Lifshitz theory has been extended to finite temperature intervals below T_C by J. O. Dimmock, *Phys. Rev.* 130, 1337 (1963) (limitation $T_0 < T < T_C$, with T_0 a first-order transition temperature).

[29] T. A. Kaplan, *Bull. Acad. Sci. (U.S.S.R.), Phys. Ser.*, 28, No. 3, 328 (1964).

6 · PHASE TRANSITIONS IN CONDENSED SYSTEMS

S. Donald Stookey

Introduction — The concept of metastability — Metastable states in one-component systems — Metastable states in polycomponent systems — The Li_2O-Al_2O_3-SiO_2 system — Controlled phase transformations in a photosensitive glass-ceramic — Latent silver image formation — Growth of silver particles — Heterogeneous nucleation of Li_2SiO_3 and change to the equilibrium phase ($Li_2Si_2O_5$ + quartz) — Photochromic action — Biological phase changes — Conclusion

Introduction

When any thermodynamic system—gas, liquid, or solid—is subjected to a change of environment admitting a new phase of lower energy, that phase tends to form. However, energy barriers may impede the rearrangement of molecules and the formation of nuclei from which the phase can grow. Activation energies may even interfere after stable particles of the new phase are present.

The existence of such barriers may preserve an intermediate metastable state ranging in lifetime from nanoseconds to millennia. Influences impressed on the system in this state are crucial in determining the structure of the new phase (size, shape, number of particles, and the chemical composition in multicomponent systems). Their absence may prevent any transition. The researcher interested in controlling structure and composition of new phases therefore has to study the nature of transition barriers, the metastable state, and the catalytic agents capable of influencing transitions.

The same general principles apply to transitions as diverse as the growth of single crystals of silicon or quartz; crystallization of sugar; synthesis of diamond or of stereospecific polymers; decomposition of calcium carbonate on heating; formation of polycrystalline glass ceramics from glass; and possibly reproduction of identical molecules in living cells and the functions of enzymes in catalyzing biosynthetic processes.

The concept of metastability

In view of the importance of the terms "metastable state" and "energy barrier," their meaning is illustrated in Fig. 6.1. The decreasing potential energy of a system during a phase transition is analogous to that of a ball rolling downhill. The ball may come to rest in one of three types of equilibrium states. Position 2 is metastable: Any small motion must be uphill and increase the potential energy, but increase beyond the energy barrier ΔE permits the ball to descend to the stable equilibrium position 4, the lowest possible energy. In position 3 the ball is in unstable equilibrium.

Metastable states in one-component systems

Figure 6.2 illustrates the meaning of metastable states in a simple one-component system. Solid lines show the equilibrium phase diagram. At point O, three phases are in equilibrium: vapor, liquid, and crystal modification B. Curve OY is the melting-point curve, representing equilibrium between liquid and crystal phase B; PZ, the equilibrium curve for transition between crystal modifications A and B; WP and

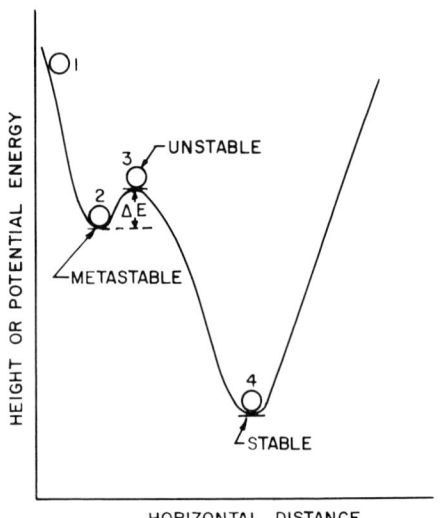

Fig. 6.1. Metastable, unstable, and stable equilibrium.

OP are sublimation curves, representing equilibrium between vapor and crystal, and OX the boiling-point curve between liquid and vapor. The shaded areas are the metastable zones under discussion.

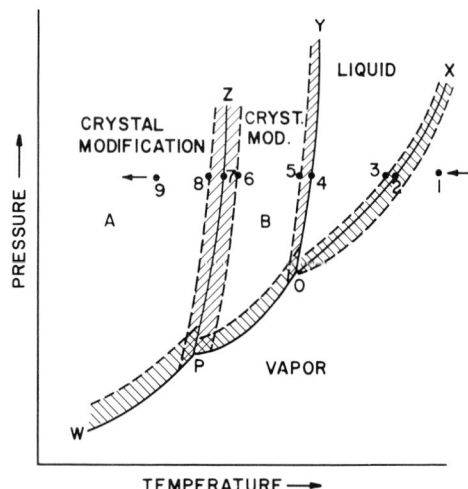

Fig. 6.2. Metastable zones in one-component system.

On starting in the vapor phase at point 1 and progressively cooling at constant pressure to a temperature between 2 and 3 (slightly below the boiling point), the vapor becomes supersaturated with respect to its liquid. In this metastable zone, droplets of liquid do not nucleate spontaneously, for reasons discussed later. However, the vapor may condense to liquid on any foreign surface present (heterogeneous nucleation) or on artificially added droplets of the stable liquid phase (homogeneous nucleation). In the absence of foreign nuclei, it can be cooled to point 3 before spontaneous nucleation and condensation to liquid occur.

Further cooling to a temperature between 4 and 5 (slightly below the freezing point) brings the liquid into another metastable zone. Although crystal modification B is now the stable phase, spontaneous crystallization does not occur; but appropriate foreign nuclei or catalysts can cause its crystallization.

As the temperature is lowered into the range of slight supercooling between points 7 and 8, crystal modification A becomes the stable phase. A solid-state transition may occur in this zone. The structure of the new crystal phase A is often related to that of the old phase B in such a way that a particular crystal plane of B initiates growth of A, or that preferential growth along a particular axis can be catalyzed by dislocations or impurity concentrations. Conversely, the transition may be prevented along one or more axes by local concentrations of certain poisoning impurities.

While nearly every possible phase transition is confronted by energy barriers, the melting and sublimation of crystals seem to be exceptions to the rule. Paradoxically, the boiling of liquids appears subject to superheating and nucleation at surfaces, indicating metastable zones of superheating.

It is sometimes possible to supercool or supersaturate a system so rapidly that the intermediate phase (e.g., crystal B in Fig. 6.2) is passed over. The system would form crystal modification A directly from the liquid, contrary to the phase diagram. Here it must be remembered that the solid lines of Fig. 6.2 represent an equilibrium diagram, i.e., refer to infinitely slow changes of temperature and/or pressure. In actual systems these parameters change relatively rapidly. The more sluggish the reaction, the greater the range of the apparent metastable zone; and consequently, the more effective are nucleating agents and catalysts in directing the course of the transition.

The example makes clear that metastable zones of supersaturation or supercooling exist in even the simplest system. Their extent—in terms of temperature, pressure, or concentration—depends on the height of the energy barrier blocking transition; and the structure of the new phase (size, number, and shapes of particles) is subject to the ways in which this barrier is breached. For example, if a melt of pure silicon metal is supercooled a few degrees below its melting point and a single crystal nucleus of silicon introduced, all of the silicon may form one large crystal. Cooling to a lower temperature would produce spontaneously many nuclei and a solid consisting of small crystals.

Metastable states in polycomponent systems

In polycomponent systems, catalysts can influence not only the physical structure but also the chemical composition of new phases. In some cases, catalysts can induce formation of nonequilibrium phases not even hinted at by the equilibrium diagram. This is especially true in silicate systems. The rapid increase in viscosity as the homogeneous melt cools may allow the liquid to freeze to a solid glass, supersaturated with respect to as many as ten different crystal phases. We will use certain silicate systems as examples to illustrate the general principles of controlled phase transitions, since the writer is most familiar with silicates and these high-viscosity systems are good models for high transition barriers.

Many of the phenomena observed in slow motion in silicates may occur more rapidly in other condensed systems. Metastable phases, even if too transitory to be observed, will frequently influence the rates of transition and the nature of the final product.

The Li_2O-Al_2O_3-SiO_2 system

The lithia-alumina-silica system has recently become technologically important because it is the basis for photosensitive glasses that can be chemically machined and for other glasses that can be crystallized to transparent or opaque polycrystalline "glass-ceramics" of very low expansion coefficients, resistant to breakage by sudden thermal shock. These new materials owe their existence to catalyzed selective crystallization.

By incorporating small concentrations of various catalysts into the batch before melting, a glass that in the absence of catalysts would not crystallize at all is induced to precipitate submicroscopic crystals in high concentrations. It is thereby converted to a polycrystalline ceramic. Moreover, the catalyst determines what crystal phase precipitates first.

Let us consider a composition: 73 wt % SiO_2, 15% Li_2O, and 12% Al_2O_3. It melts to a glass and remains a glass in thermal cycling. However, if 100 parts per million of gold[1] are dissolved in the melt and the gold precipitated as a highly dispersed sol by a photographic process, these gold crystals selectively catalyze the crystallization of lithium metasilicate when the glass is reheated at about 600° C for an hour.

[1] This can be done by incorporating a solution of gold chloride in the batch before melting.

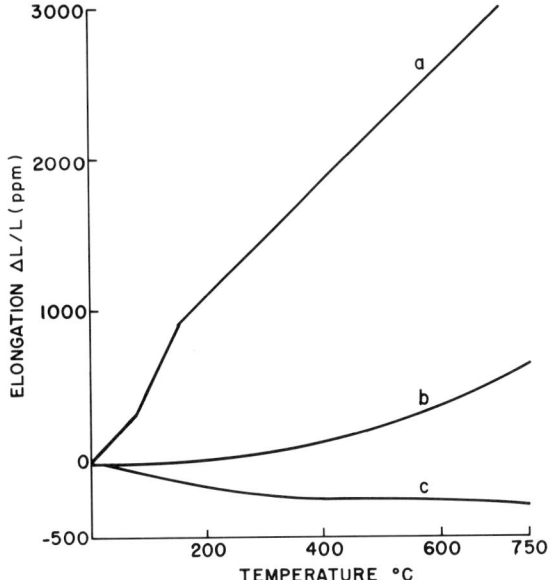

Fig. 6.3. Expansion coefficients of various glasses: (a) alpha-cordierite glass-ceramic, used for radomes for high-speed missiles; (b) beta-spodumene glass-ceramic, used for cooking ware; (c) eucryptite glass-ceramic.

A solution of 5 wt % TiO_2 in the melt results in the formation of myriads of submicroscopic droplets of an immiscible low-silica liquid phase as the glass cools. On reheating at about 850° C, a transparent, polycrystalline, low-thermal-expansion material is formed, in which the only crystal phase is a solid solution of beta eucryptite ($Li_2O \cdot Al_2O_3 \cdot 2SiO_2$) and silica. This beta eucryptite is a strange crystal: When heated it shrinks instead of expanding. Thus articles can be made containing this material that have expansion coefficients near zero or slightly negative (Fig. 6.3) and are immune to breakage by thermal shock. Further heating at about 950° C produces a white, opaque substance, also low in thermal expansion coefficient, in which the crystal phase has altered to a solid solution of beta spodumene ($Li_2O \cdot Al_2O_3 \cdot 4SiO_2$) and silica. According to the equilibrium phase diagram,[2] the beta spodumene-silica solid solution is the primary crystal phase for this composition.

Very probably still other catalysts can be found that would alter the course of crystallization of this glass in other directions.

Controlled phase transformations in a photosensitive glass-ceramic

A related example of a complex nonequilibrium system in which we have learned to control phase

[2] R. Roy and E. F. Osborn, *J. Am. Ceram. Soc.* **71**, [6] 2086 (1949).

transitions for useful purposes, is the photosensitive glass-ceramic, Corning Code 8603.[3] This commercially available material is useful because it can be chemically etched in very complex and precise three-dimensional photographic patterns. The "chemically machined" product may be either a transparent glass or a polycrystalline ceramic containing a high concentration of randomly oriented submicroscopic crystallites of an average diameter below 1 micron.

The manufacturing process involves first melting and forming a homogeneous, transparent glass[3] (composition, Table 6.1).[4] Table 6.2 outlines the subsequent

Table 6.1. Composition of a photosensitive chemically machineable glass-ceramic[4]

Material	Wt %
SiO_2	77.5
Li_2O	12.5
Al_2O_3	10.0
AgCl	0.002
CeO_2	0.02

Table 6.2. Phase transformations in Fotoform (chemically machineable glass)

1. Lithium silicate glass +
    dissolved silver ions and sensitizers
    uv irradiation ↓ + heat, 550° C ½ hr
2. Lithium silicate glass + precipitated silver crystals
    ↓ heat, 600° C ½ hr
3. Lithium silicate glass +
    Li_2SiO_3 crystal nuclei formed on silver crystals > 80 Å
    ↓ heat, 600° C 1 hr
4. Lithium metasilicate crystal growth
    completed (HF-soluble Fotoform)
    ↓ etch in 10% HF solution, 25° C
5. Chemically machined glass object
    ↓ repeat steps 1 through 4,
        plus heat at 850° C 1 hr
6. Lithium disilicate and quartz crystals
    (Fotoceram glass-ceramic)

processing steps. First a three-dimensional photographic latent image is produced by exposure to collimated ultraviolet light through a photographic negative. The latent image, believed to consist of silver nuclei each containing one to four atoms, is undetectable except by small changes in ultraviolet absorption and electron-spin resonance. It is developed by heating at 500° to 600° C (slightly above the annealing temperature) into silver particles growing progressively through the smallest measured size (10 to 12 Å, determined by small-angle X-ray scattering), to a diameter greater than a "critical" size of 80 Å. When the silver crystallites[5] exceed this critical size,[6] each crystal forms a "heterogeneous" nucleus for growth of a lithium metasilicate crystal ($Li_2O \cdot SiO_2$).

After the photographic pattern of lithium metasilicate crystals has been developed (Fig. 6.4), the glass

Fig. 6.4. Test pattern (Army Diamond Laboratories), selectively etched in photosensitive glass by Corning Glass Works to show feasibility of optical fabrication techniques.

is cooled. The crystallized regions are now very soluble in dilute hydrofluoric acid; they can be etched away by immersion of the glass in an acid bath, leaving a "chemically machined" glass structure (Figs. 6.5 and 6.6).

This finished glass article can now be altered to a high-strength polycrystalline glass-ceramic.[7] In this case the glass is given an all-over exposure to ultraviolet light and heated as before, but with an additional treatment at the higher temperature of 850° C for about 1 hour. This results in precipitation of lithium disilicate ($Li_2O \cdot 2SiO_2$) and quartz, the equilibrium phases for this composition. The resultant glass-ceramic is harder and stronger than glass, has much

---

[3] S. D. Stookey, *Ind. Eng. Chem.* 45, 115 (1953).
[4] S. D. Stookey, U. S. Patent 2,684,911, July 27, 1954.
[5] A fundamental question still to be answered: At what stage of their growth do these silver aggregates become crystals?
[6] R. D. Maurer, *J. Appl. Phys.* 29, 1 (1958).
[7] S. D. Stookey and R. D. Maurer, "Catalyzed Crystallization of Glass-Theory and Practice," *Progress in Ceramic Science*, J. E. Burke, Ed., Vol. 2, Pergamon Press, New York, Oxford, London, Paris, 1962, p. 77.

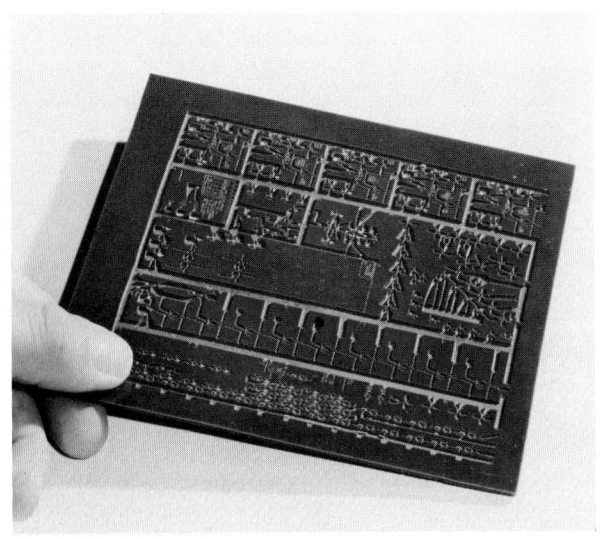

Fig. 6.5. Treated Fotoform test pattern of Fig. 6.4, resulting in Fotoceram glass-ceramic with greater strength and temperature resistance.

higher electrical resistivity, and is no longer easily soluble in HF solution.

The glass in its initial homogeneous transparent state at room temperature was in metastable equilibrium with respect to several silicate crystal species and probably capable of maintaining this state for thousands of years, as the volcanic obsidian glasses have done.

At the same time the glass, as solvent for ions of silver and cerium, holds these species frozen in metastable equilibrium as an oxidation-reduction reaction system in which high-temperature equilibrium is retained at room temperature. The enormous increase in viscosity of the cooling melt provides the barrier against phase transitions.

Latent silver image formation

While the latent image cannot be observed directly, we know that the glass before irradiation contains

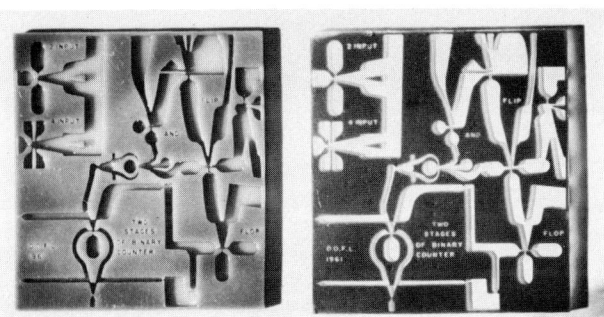

Fig. 6.6. Magnified views of a section from the test pattern of Figs. 6.4 and 6.5.

essentially all of the silver as Ag^+ ions. After irradiation and before heating, electron spin resonance shows the presence of Ag^{2+} ions and trapped electrons, while optical absorption measurements show that Ce^{3+} has been oxidized to Ce^{4+}. After heating, we observe (by optical absorption, small-angle X-ray scattering, and electron microscopy) progressive diffusion-controlled growth of silver particles. We conclude that irradiation produces the reactions:

$$2Ag^+ + h\nu \rightarrow Ag^{2+} + Ag^0, \qquad (6.1)$$

and

$$Ce^{3+} + h\nu \rightarrow Ce^{4+} + e^-, \qquad (6.2a)$$

$$Ag^+ + e^- \rightarrow Ag^0. \qquad (6.2b)$$

Extrapolation of the particle-size characteristic as f (heating time) to zero time seems to indicate that the initial silver nucleus (latent image) contains four atoms

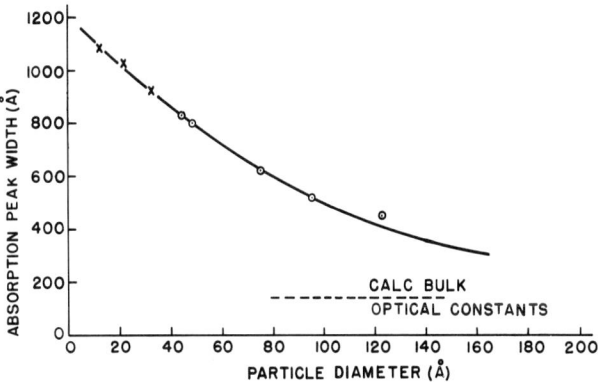

Fig. 6.7. Optical-absorption-peak width of colloidal silver in glass as function of silver-particle diameter. (After Maurer.[6])

or less. Figure 6.7 shows a measurement of absorption-peak width as function of particle diameter.

Growth of silver particles

The growth of silver particles could occur by coalescence of silver atoms or by diffusion of silver ions to silver nuclei and reduction at the metal-particle surface. All evidence supports the latter mechanism operating in one of two possible ways:

$$(Ag^0)_n + Ag^+ + e^- \rightarrow (Ag^0)_{n+1}, \qquad (6.3a)$$

or

$$(Ag^0)_n + m\,Ag^+ + Me^x \rightarrow (Ag^0)_{n+m} + Me^{x+m}, \qquad (6.3b)$$

where Me^x stands for the reduced form of a polyvalent ion, such as Ce^{3+}, capable of reducing silver to the atomic state while being oxidized to higher valency.

The number of silver crystals that can be developed per unit volume increases with irradiation time and intensity. Sufficient heating precipitates all of the silver. Underexposure produces a relatively small number of larger crystals, overexposure a large number of small crystals, after development through heat treatment. Hence, all silver is in a metastable oxidation state, and as soon as the reaction is initiated by the formation of nuclei it can go to completion.

Heterogeneous nucleation of Li_2SiO_3 and change to the equilibrium phase ($Li_2Si_2O_5$ + quartz)

The equilibrium crystal phases to which this glass should devitrify according to its phase diagram are $Li_2O \cdot 2SiO_2$ and quartz. Yet at 600° C the glass crystallizes only in the presence of foreign nuclei; and when these nuclei are photochemically formed silver, it crystallizes to $Li_2O \cdot SiO_2$, a nonequilibrium phase.

The chief barrier to crystallization lies here in the nucleation step. Illumination can control number, size, and location of the silicate crystals. Evidently the structure of the silver is sufficiently similar to that of one of the crystal-lattice planes of Li_2SiO_3, so that it is catalyzed in preference to the stable $Li_2Si_2O_5$.

A critical nucleus size of approximately 80 Å (below which the silver crystal will not catalyze Li_2SiO_3 crystallization) is well established. When the silver particles are purposely kept under this size by overexposure of the glass to ultraviolet light, even long heat treatment fails to crystallize Li_2SiO_3.

At sufficiently high temperature (cf. Table 6.2), the Li_2SiO_3 crystals react with the surrounding higher-silica glass to form $Li_2Si_2O_5$ crystals; simultaneously, quartz crystals grow. Prolonged heating at this higher temperature might eventually produce this crystal composition *a priori* but its microstructure would be completely different. In the absence of silver nuclei, crystallization of the glass starts at surfaces, and oriented growth of large crystals perpendicular to such surfaces ensues.

Photochromic action

The newest example of useful phase transitions in glass is the precipitation of silver halide microcrystals that are reversibly darkened when the transparent glass is exposed to light. These new "photochromic" glasses have a number of potential applications as automatic light valves, in windows, sunglasses, and optical devices.

The crystals are principally silver chloride but may contain bromide and iodide and are sensitized with traces of cuprous ion. They are much smaller (40 to 100 Å) than those in photographic emulsions.

The glass is melted and formed by conventional practices; the desired silver halides are normally added to the batch before melting and go into homogeneous solution in the glass melt. Depending on the concentration of silver halide, the product may either be photochromic after normal annealing and cooling, or (at lower concentrations) may require reheating for minutes or hours in the range between the annealing and softening temperatures. This heating results in nucleation and growth of colloidal droplets of molten silver halide that subsequently crystallize as the glass cools.

Excessive heating or silver halide concentration may result in a translucent or opaque glass because of crystallite growth to a size that scatters light.[8] Such glasses are also photochromic but generally less useful than the transparent ones.

Depending on composition and heat treatment, a wide range of photochromic properties (reaction rates of darkening and fading, spectral sensitivity, temperature coefficients of reaction rates, etc.) is obtainable. Three processes occur simultaneously during illumination: darkening (creation of color centers), optical bleaching, and thermal bleaching. The steady state is the resultant of these competing processes. Darkening and optical fading rates follow curves typical of first-order reactions. Thermal fading requires further studies to clarify its reaction kinetics. It appears that more than one kind of color center is formed.

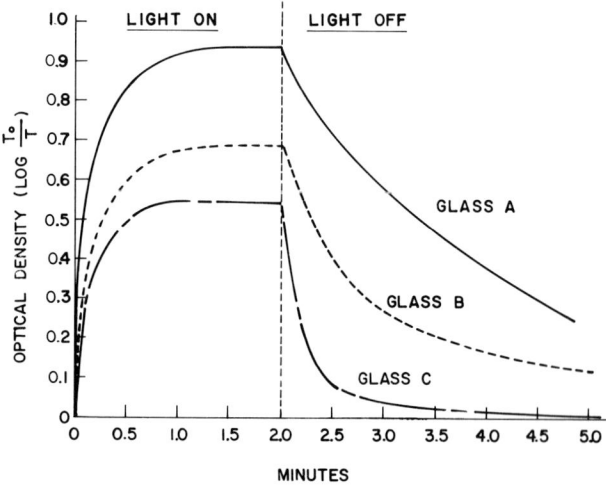

Fig. 6.8. Photochromic darkening and clearing of three glasses.

Typical response curves are shown in Fig. 6.8. The color resulting is generally a neutral gray or gray-brown,

[8] W. H. Armistead and S. D. Stookey, *Science* 144, 150 (1964).

corresponding to a broad absorption extending from the near ultraviolet to the near infrared.

The wavelengths inducing darkening extend from the near ultraviolet into the visible spectrum, depending on silver halides (Fig. 6.9). Silver chloride glasses,

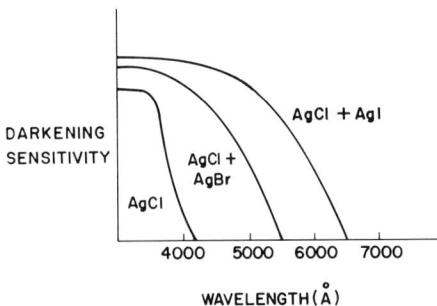

Fig. 6.9. Wavelengths effective in darkening photochromic glass.

sensitive only to ultraviolet, darken readily in outdoor sunlight. Exposure to visible and near infrared in the broad absorption band of the "color centers" produces accelerated fading.

Darkening in sunlight in a typical case approaches its maximum (decrease of transmittance to 25 percent) in a minute; the same effect is produced by a 3-μsec flash of high-intensity light (1000 joules discharged through a xenon flash lamp). The clearing rates in the dark (to 50 percent density decrease) range from seconds to hours, depending on composition, previous heat treatment, and temperature of the glass. At liquid-nitrogen temperature, bleaching does not occur.

The optical transmittance of a transparent photochromic glass before darkening is similar to that of window glass. The steady-state optical transmittance for quarter-inch thick glass plates ranges as low as 1 percent for normal sunlight exposure.

Curves of steady-state optical density ($\log T_0/T$) versus intensity I of incident light for photochromic glasses of two different fading rates are shown in Fig. 6.10. When the color centers are destroyed almost as fast as they are formed, their concentration increases in proportion to the light intensity causing formation.

The photochromic process seems basically that of silver halide photography with the difference that the photographic process is irreversible (formation of stable silver particles), while in glass the silver and halogen can return to their original states in the dark, because only photoelectrons, not atomic diffusion, are involved.

The primary photolytic reaction

$$Ag^+ + Cl^- + h\nu \rightarrow Ag^0 + Cl^0 \qquad (6.4)$$

transfers an electron from the Cl^- to the Ag^+ ion. The silver atoms, or very small aggregates of such atoms, are the color centers. In the case of copper-sensitized photochromism,

$$Ag^+ + Cu^+ + h\nu \rightarrow Ag^0 + Cu^{++}, \qquad (6.5)$$

an electron is freed from the Cu^+ and trapped by the Ag^+ ion. Here the cupric ion as well as the silver con-

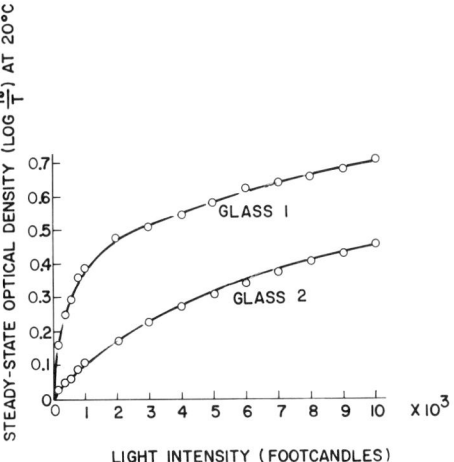

Fig. 6.10. Steady-state optical density versus intensity at 20° C.

tributes to visible light absorption. These reaction products are metastable and can return to the original state, but if permitted to diffuse or react with other species, the reaction becomes irreversible. Extensive aggregation of Ag^0 would form stable silver colloid particles, and if the Cl^0 were permitted to diffuse as in ordinary photographic films, its loss would prevent restoration of the original state.

Irradiation of the glass with ultraviolet light above 400° C, where some diffusion of ions and atoms might be expected, produces in prolonged exposure a permanent colored image containing colloidal silver. The reversible photochromic coloration, on subsequent room-temperature exposure, decreases in intensity with increase of the exposure time at 400° C and finally disappears. Diffusion and precipitation of the photolytic silver produces an irreversible photolysis of silver halide.

All photochromic properties vary in wide limits with thermal history, beginning with the initial cooling of the glass melt. Not only size and number of the microcrystals but also the chemical composition and the atomic arrangement within each crystal are affected. A complex interplay exists between nucleation and growth of the molten salt droplets, relative solubilities of various ions in the glass, in the molten salt, and in the crystal, and interdiffusion rates of ions between glass and salt. Also interfacial phenomena may be important, since the particles are so small that they consist essentially of surface.

Biological phase changes

Some of the most exciting biological discoveries in recent years appear to show that the fundamental factors controlling heredity depend on "nucleation." Without stretching the language of modern biology textbooks[9] too far, it seems an accepted fact that the presence of traces of a chemical substance having a definite molecular structure in a "pool" or "substrate" of amorphous material, containing a disorganized mixture of long chain protein molecules, initiates the organization of the amorphous phase into a partial or complete duplicate of the trace additive. The "pool" may be considered a complex phase, in a metastable state, so that a "nucleating agent"—a gene, or an enzyme—can initiate and determine the structure of the new phase that develops.

Conclusion

Anyone trying to understand and guide the kinetics of chemical reactions and phase transitions needs to study metastable states and the influences at work while nonequilibrium conditions exist.

---

[9] W. D. Elroy, *Cellular Physiology and Biochemistry*, Prentice-Hall, New York, 1961, p. 104.

7 · DEFECTS IN IONIC CRYSTALS

A. Smakula

Vacancies and interstitials — Color centers and their response to purity and composition — Conductivity and crystal defects — Dislocations — X-ray density versus macrodensity

Single crystals are characterized by the long-range order of their constituents (atoms or ions). Many properties can be correlated to the symmetry of this ordered arrangement; others indicate that crystals contain a variety of defects, either built in during growth or generated afterwards.[1-5] Of the large variety of defects and of methods for their study only a few will be discussed here.

Vacancies and interstitials

The simplest type of defects are vacancies and interstitials (point defects). Vacancies can be accommodated in all crystals, but electroneutrality has to be preserved. Interstitials can be formed when atom types are small enough to fit such sites. In ionic crystals positive ions qualify preferentially for these positions. Covalent crystals with small filling factor (e.g., diamond) can easily accommodate interstitials.

Vacancies and interstitials can be produced thermally by foreign ions of different valence.[6,7] The equilibrium concentration of point defects formed by thermal activation is

$$n = \text{const} \exp\left(-\frac{U}{kT}\right). \qquad (7.1)$$

The constant can vary from 10^2 to 10^4 and is smaller for interstitials than for vacancies because of the opposing influence of lattice vibrations. The activation energy U is ca. 2 eV in alkali halides and ca. 1 to 1.5 eV in silver halides.

Thermal point defects are generally of low concentration and random distribution. The majority of them are surrounded by ions in normal position. Purely statistical distribution will bring some of the defects close together, leading to the formation of pairs, triplets, or higher aggregates, depending on temperature and crystal history. The concentration of point defects caused by foreign ions depends on the amount and distribution of such ions.

---

[1] *Imperfections in Nearly Perfect Crystals*, W. Shockley, J. A. Hollomon, R. Maurer, and F. Seitz, Eds., John Wiley and Sons, New York, 1952.

[2] *Report on the Conference on Defects in Crystalline Solids* (Bristol), The Physical Society, London, 1955.

[3] A. Seeger, "Theorie der Gitterfehlstellen," *Encyclopedia of Physics*, Vol. 7, Pt. 1, Springer-Verlag, Berlin-Göttingen-Heidelberg, 1955, p. 383.

[4] H. G. van Bueren, *Imperfections in Crystals*, North-Holland Publishing Co., Amsterdam, 1960.

[5] *Direct Observation of Imperfections in Crystals*, J. B. Newkirk and J. H. Wernick, Eds., Interscience Publishers (John Wiley and Sons), New York, 1962.

[6] N. F. Mott and R. W. Gurney, *Electronic Processes in Ionic Crystals*, Clarendon Press, Oxford, 1946.

[7] O. Stasiw, *Elektronen- und Ionenprozesse in Ionenkristallen*, Springer-Verlag, Berlin, 1959.

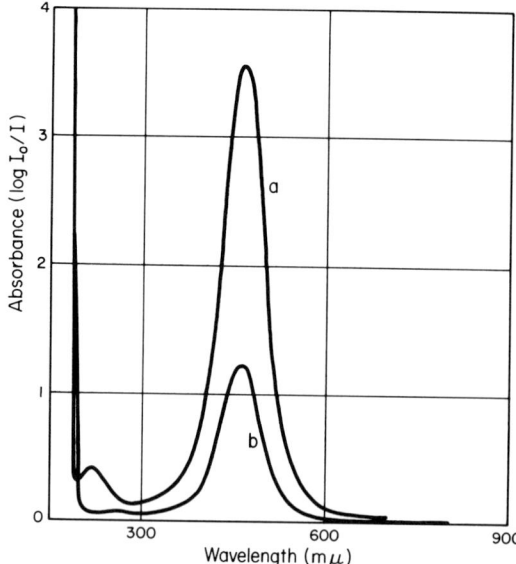

Fig. 7.1. F band in NaCl crystal colored by 3-MeV electrons $(5 \times 10^5 \text{ rads})$ at $20°$ C: (a) c.p. material, (b) material purified by ion exchange and HCl treatment.

The presence of vacancies or interstitials causes a local distortion of the lattice by shifting the neighboring ions (a few percent for vacancies and up to 20 percent for interstitials) outward from their normal positions and changing the polarization.

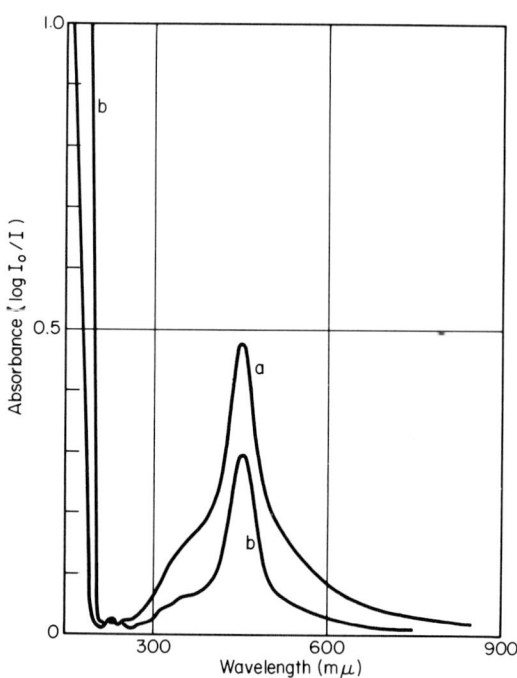

Fig. 7.2. F band in NaCl crystal colored by 3-MeV electrons $(5 \times 10^5 \text{ rads})$ at $-190°$ C: (a) c.p. material, (b) material purified by ion exchange and HCl treatment.

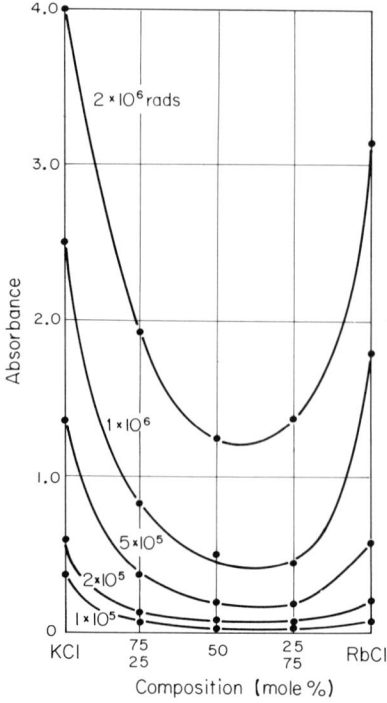

Fig. 7.3. Influence of composition on F-band coloration in KCl-RbCl mixed crystals.

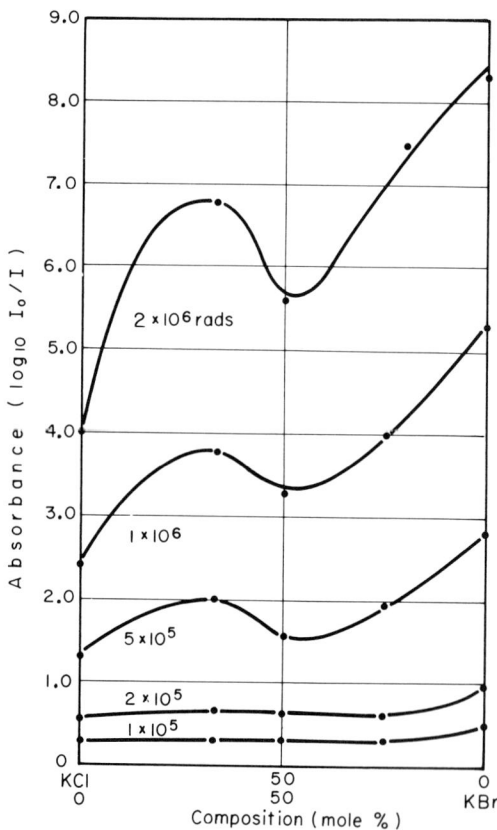

Fig. 7.4. Influence of composition on F-band coloration in KCl-KBr mixed crystals.

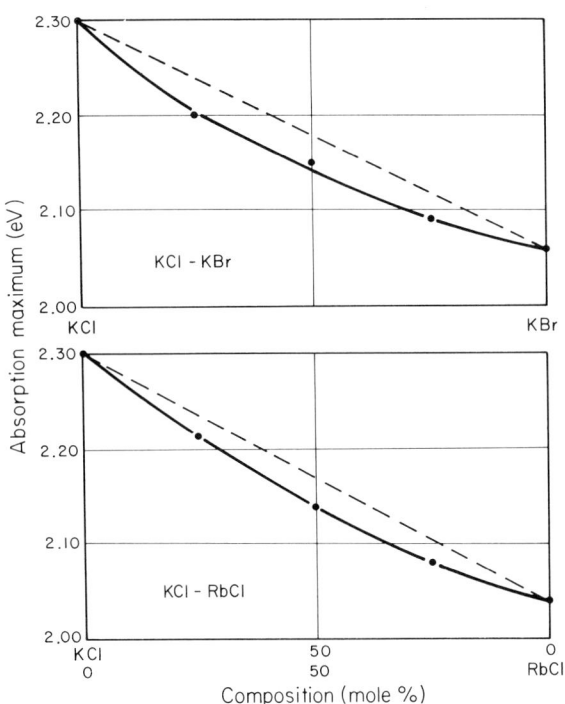

Fig. 7.5. Influence of composition on spectral position of the F band at $-190°$ C.

Point defects can be investigated by a variety of methods; only three will be discussed here: the trapping of charge carriers (color centers), electronic and ionic conductivity.

Color centers and their response to purity and composition

In ionic crystals electrons can be trapped in negative, and holes in positive ion vacancies. Such carriers can

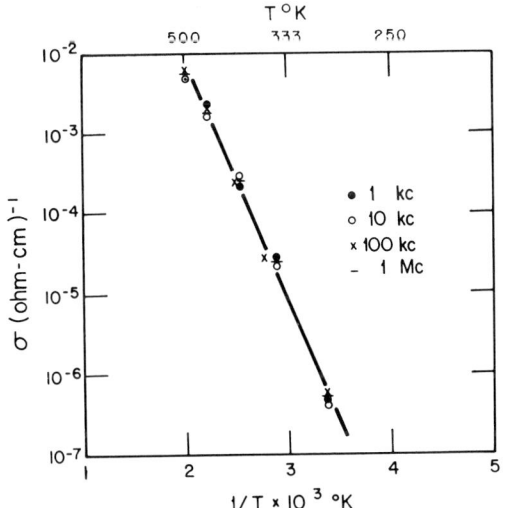

Fig. 7.6. Conductivity of NiO crystal ($U = 0.66$ eV).

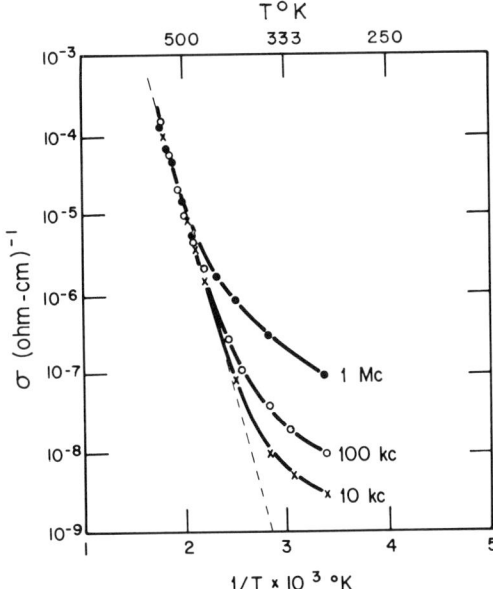

Fig. 7.7. Conductivity of CoO crystal ($U = 0.82$ eV).

best be introduced by internal ionization produced, e.g., by high-energy X-ray irradiation. The trapped charge carriers can be detected by optical-absorption measurements. The absorption bands, when extending into the visible, cause a coloration of the material, hence the name "color centers."[8] The best known of these is the

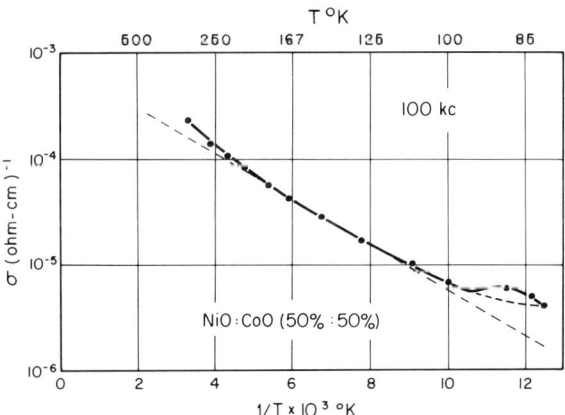

Fig. 7.8. Conductivity of NiO·CoO (1:1) mixed crystal ($U = 0.05$ eV).

F center; it corresponds to an electron trapped in a negative ion vacancy.

The integrated intensity of the F band is directly proportional to the concentration N of the trapped carriers; thus it can be used also to measure the number of vacancies in a crystal. This concentration was

---

[8] Cf., e.g., J. H. Schulman and W. D. Compton, *Color Centers in Solids*, Pergamon Press, New York, 1962.

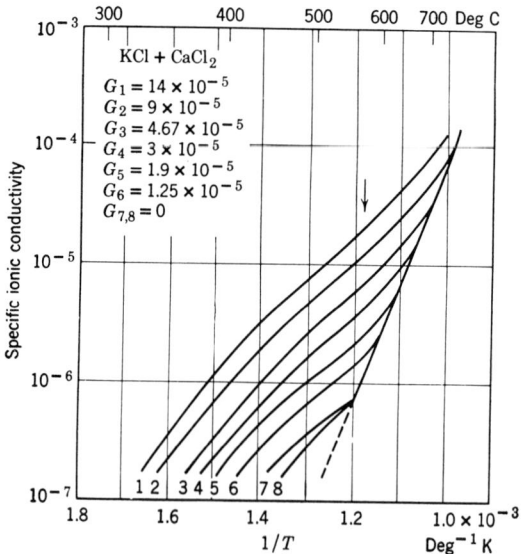

Fig. 7.9. Ionic conductivity of KCl doped with CaCl$_2$. (After Kelting and Witt.[19])

originally computed by assuming a classical oscillator embedded in the dielectric host crystal and subjected to a Lorentz field; the shape of the F band was taken as

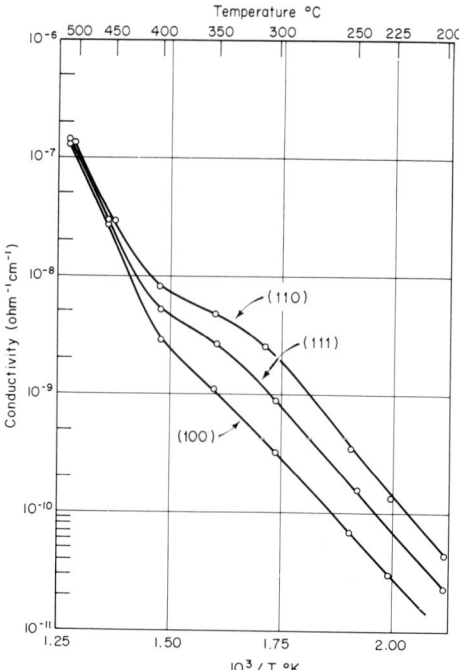

Fig. 7.10. Influence of crystal-growth direction on extrinsic ionic conductivity of KBr crystals.

Lorentzian (i.e., caused by attenuation). The well-known equation resulted:[9]

[9] A. Smakula, *Z. Physik* **59**, 603 (1930).

$$N = \frac{9mc}{f 2e^2 h} \frac{n}{(n^2 + 2)^2} k_m H \quad [\text{cm}^{-3}], \quad (7.2)$$

where m is the electron mass, c the light velocity, f the oscillator strength, e the electron charge, h Planck's constant, n the refractive index of the crystal at wavelength of maximum absorption coefficient k_m, and H the half-width of the F band in eV. The only unknown is the oscillator strength f. Determinations of f by chemical or photochemical means and by magnetic-resonance or magnetic-susceptibility methods vary from 0.66 to 1.17 for KCl and from 0.70 to 0.87 for NaCl.[10] The best practical correlation between n and f can be obtained by an independent (e.g., chemical) determination of n and graphical integration of the F band.

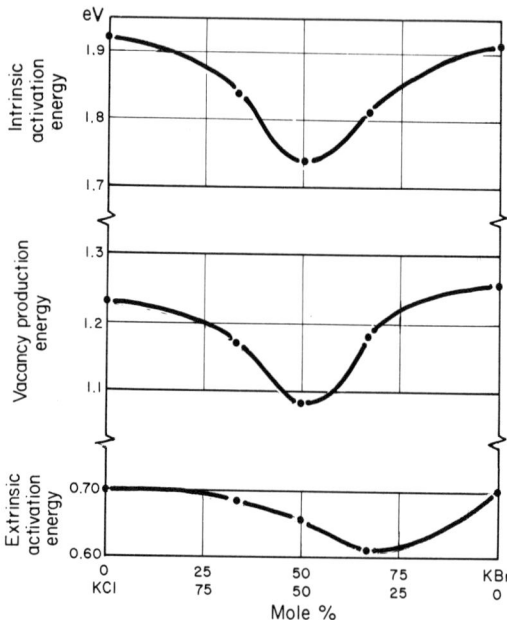

Fig. 7.11. Influence of composition on activation energies for KCl-KBr mixed crystals.

Objections have been raised against use of the Lorentzian line shape for the F band, because a Gaussian gives a better fit for the experimental curves.[11] This dilemma can be avoided by determining the integrated intensity $\int k d\varepsilon$ planimetrically. There is also some uncertainty about the proper electron mass: An effective mass m should be used in Eq. 7.2. According to Tibbs,[12] $m^* \sim m$, while Pekar[13] found $1.85m$ for KCl and $2.78m$ for NaCl. There is the further question

[10] W. T. Doyle, *Phys. Rev.* **111**, 1072 (1958).
[11] D. L. Dexter, *ibid.* **101**, 48 (1956).
[12] S. R. Tibbs, *Trans. Faraday Soc.* **35**, 1471 (1939).
[13] S. I. Pekar, *Untersuchungen über die Elektronentheorie der Kristalle*, Akademie-Verlag, Berlin, 1954.

whether a Lorentz or an Onsager effective-field expression is more appropriate.[14]

Figure 7.1 compares the F-center absorption produced by 10^5 rads of 3-MeV electrons in NaCl crystals grown from c.p. and from purified material. In the purified crystal (b) the concentration of vacancies appears three times lower than in the unpurified one (a). However, the coloration is temperature sensitive: The absorption is about eight times lower at $-190°$ than at $20°$ C, and the ratio unpurified to purified is only 1.6:1 (Fig. 7.2). The capture cross section for electrons depends strongly on lattice vibrations.

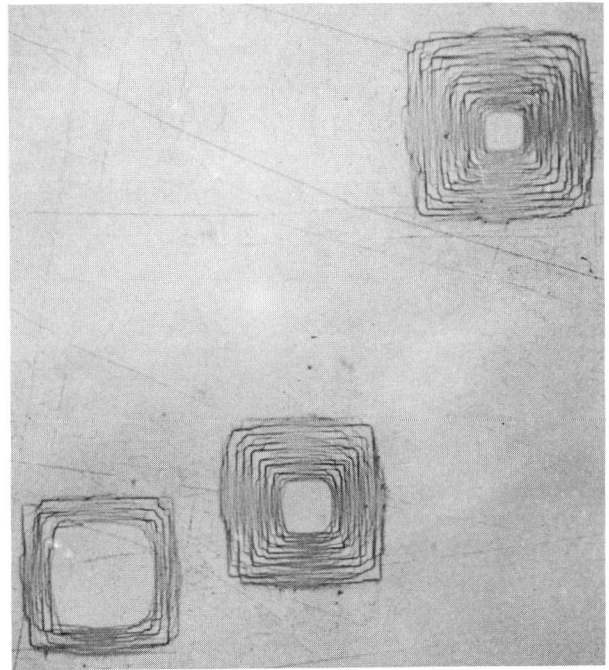

(a)

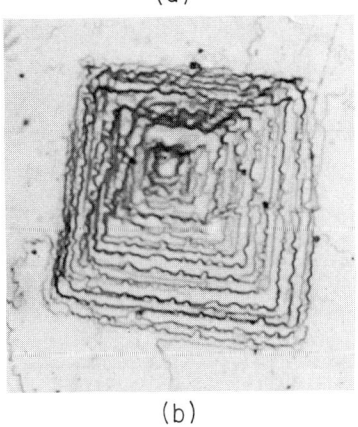

(b)

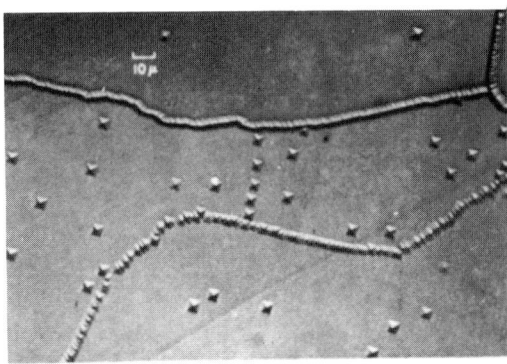

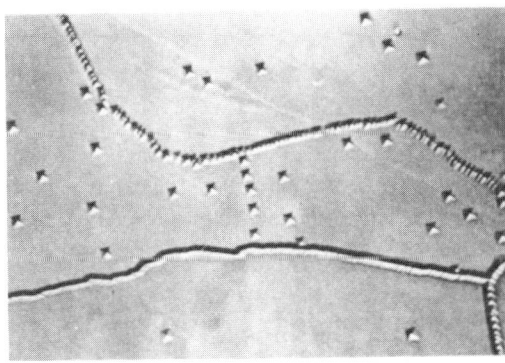

Fig. 7.12. Chemical etch pits on matched opposite sides of cleavage planes of LiF. (After Gilman and Johnston.[22])

Fig. 7.13. Thermal etch pits on a cleaved face of KBr: (a) regular shape, (b) irregular shape.

In solid solutions of ionic crystals one expects an increasing vacancy concentration and stronger trapping because of lattice distortions. Figures 7.3 and 7.4 show the influence of cation and of anion substitution, respectively, on electron trapping;[15] the outcome differs widely. The influence of ion substitution on binding energy, indicated by the spectral position of the F band, is illustrated in Fig. 7.5.

The decrease of the F-band intensity in the KCl-RbCl system may result from a higher recombination rate in solid solution. In the KCl-KBr system the situation is complicated by the much higher intensity of the F band in KBr.

Conductivity and crystal defects

Trapped electrons or holes can be released either thermally or photoelectrically; currents and in some cases light emission (glow curves) are observed. The number of traps and their energy can be ascertained from glow curves (cf. Chaps. 13 and 14).[16]

[14] R. H. Silsbee, *Phys. Rev.* **103**, 1675 (1956).
[15] A. Smakula, N. C. Maynard, and A. Repucci, *ibid.* **130**, 113 (1963).

[16] Cf., e.g., F. Matossi and S. Nudelman, "Luminescence," *Methods of Experimental Physics*, Vol. 6B, L. Marton, Editor-in-Chief, Academic Press, New York, 1959, p. 313.

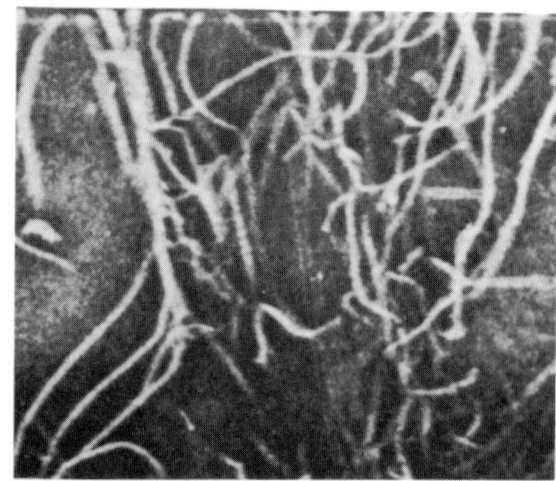

Fig. 7.14. Dislocations in Si single crystal, made visible by X-ray micrography.

Ionic-conductivity measurements are frequently used in studies of positive ion vacancies. A classical example is the change of conductivity in KCl by $CaCl_2$ addition (Fig. 7.9).[19]

There is an influence of the crystal-growth direction on the ionic conductivity and hence on defect distribution (Fig. 7.10).[20] In mixed crystals the composition influences not only the extrinsic but also the intrinsic region, as shown for the KCl-KBr system in Fig. 7.11.[21]

Dislocations

Dislocations are another type of defect extensively studied in recent years.[22] Chemical etching leads to the development of pits, revealing location and types of dislocations (Fig. 7.12). Thermal etching also produces pits of regular or irregular shape (Fig. 7.13), but no connection with chemically produced pits seems to exist. The most powerful tool for the study of dislocations is X-ray micrography, since it reveals dislocations within crystals (Fig. 7.14). The internal

When electrons or holes are released by thermal energy, the change of current with temperature provides information about the crystal defects. The conductivity of NiO (Fig. 7.6) and CoO (Fig. 7.7) is p-type

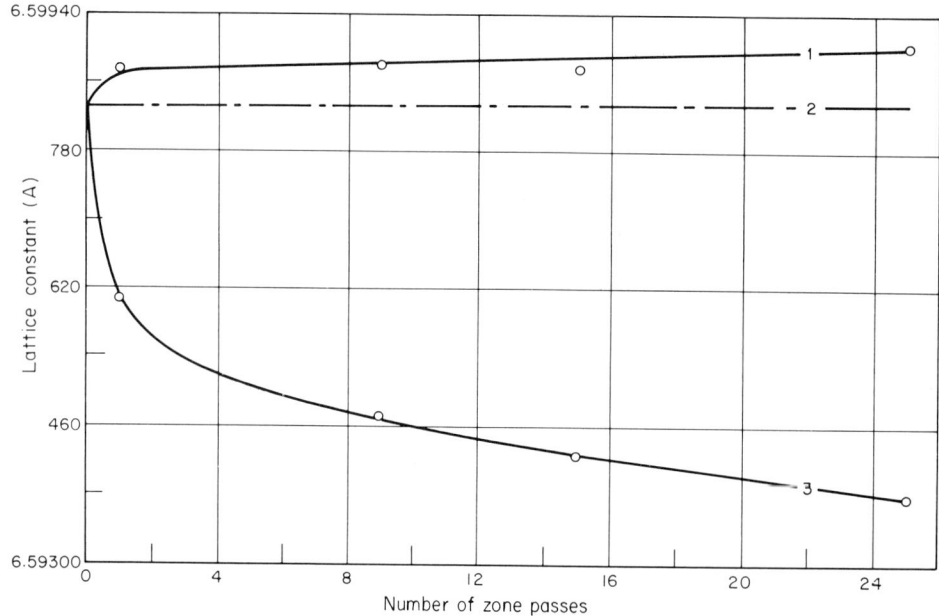

Fig. 7.15. Change of lattice constant caused by impurities in KBr: (1) purified, (2) original, (3) with concentrated impurities.

and comparatively low; in solid solution (molar composition 1:1) it is higher by several orders of magnitude (Fig. 7.8). Conductivity is caused by the presence of trivalent Ni or Co ions;[17,18] in solid solution the concentration of such trivalent ions is enhanced.

structure of crystals appears to be extremely distorted. The concentration of vacancies can be kept below

[17] R. W. Wright and J. P. Andrews, *Proc. Phys. Soc.* (London) *A62*, 446 (1949).
[18] F. J. Morin, *Phys. Rev. 93*, 1199 (1954).

[19] H. Kelting and H. Witt, *Z. Physik 126*, 697 (1949).
[20] V. Klemas, M.S. Thesis, Mass. Inst. Tech., May, 1959.
[21] D. L. Cannon, *Tech. Rep. 180*, Lab. Ins. Res., Mass. Inst. Tech., May, 1963.
[22] J. J. Gilman and W. G. Johnston, *J. Appl. Phys. 27*, 1018 (1956).

$1:10^6$ and that of dislocations below $1:10^{12}$. Although such concentrations seem very small, they are highly significant in relation to the number of ions. Still, they are negligible compared with the concentration of chemical impurities, which present the main problem in obtaining crystal perfection. In normal materials, the ion concentration of impurities can be of the order of 0.1 percent or more.

X-ray density versus macrodensity

Efficient methods for investigating impurities in crystals are high-precision lattice-constant and high-precision density determination; both can be accurate to within a few units in the fifth decimal place. Segregated impurities can be detected by X-ray diffraction only if their concentration amounts to several percent.

Figure 7.15 shows the change in lattice constant of KBr on purification for three types of samples.

The computation of density from the lattice constant depends on the molecular weight of the pure substance, disregarding impurities or vacancies. On the other hand, density determined by hydrostatic weighing or flotation includes both impurities and vacancies. The difference between X-ray and weighing density gives the average concentration of all impurities and vacancies, although unfortunately only up to $1:10^6$.

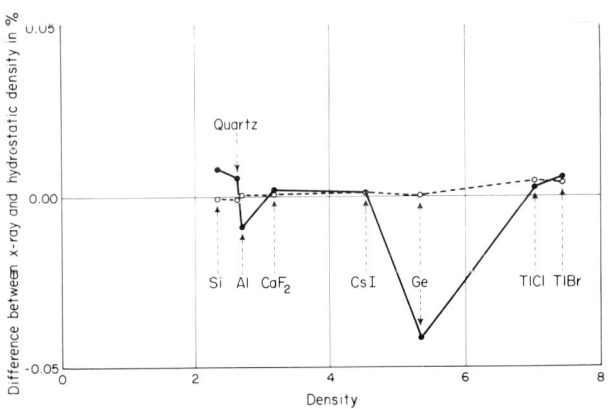

Fig. 7.16. "Apparent" density defect in Ge single crystals.

Three examples may illustrate the efficiency of density measurements. By comparing X-ray and weighing density we found[23] that the atomic weight of Ge as given by the International Commission is wrong: Instead of 72.59 it should be 72.627 (Fig. 7.16). The second example pertains to diamond. There are two

[23] A. Smakula, J. Kalnajs, and V. Sils, *Phys. Rev.* **99**, 1747 (1955).

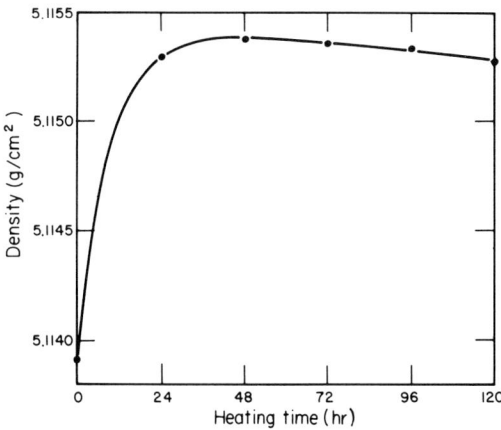

Fig. 7.17. Density variation of $SrTiO_3$ crystal by heat treatment in air at 1300° C.

distinct types of diamonds, as indicated in Table 7.1.[24]

Table 7.1. Differentiation between diamond types

	Type I	Type II
Infrared absorption	8 to 10 μ	none
Ultraviolet absorption edge	300 mμ	225 mμ
X-ray diffraction	extra spots	normal
Photoconduction	poor	good
Birefringence	present	absent

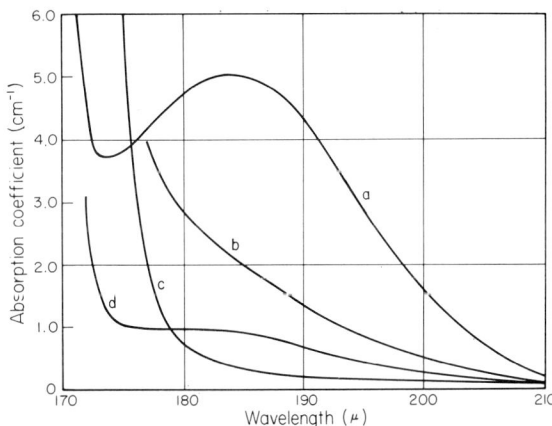

Fig. 7.18. Influence of impurities on absorption edge of NaCl crystal: (a) grown in Pt crucible in air, (b) grown in Vycor crucible in N_2, (c) natural crystal, (d) grown in Pt crucible in He.

The majority of diamonds are of Type I which, according to infrared and ultraviolet absorption, is more heavily contaminated than Type II. Kaiser and Bond[25] concluded that Type I is contaminated by

[24] G. B. B. M. Sutherland, D. E. Blackwell, and W. G. Simeral, *Nature* **174**, 901 (1954).
[25] W. Kaiser and W. L. Bond, *Phys. Rev.* **115**, 857 (1959).

nitrogen, in a concentration up to $1:10^3$, substitutionally bound in the lattice. Their evidence was based on an increase of the lattice constant from 3.56683 Å in Type II to 3.56725 Å in Type I, with no change in density up to the fifth decimal place. Our high-precision density determination by an improved flotation method on 33 diamonds gives the average values

Type I: 3.51537 ± 0.00004 g/cm^3,
Type II: 3.51507 ± 0.00004 g/cm^3.

A third example is the oxidation of SrTiO$_3$ grown by flame fusion. The crystal, originally dark, becomes colorless after treatment at high temperature in air. The hydrostatic density initially increases as expected for oxidation but later decreases (Fig. 7.17), indicating defect formation after prolonged heat treatment.

Optical absorption is one of the most sensitive methods for the investigation of crystal defects. Defects may cause additional absorption bands or modify existing ones, e.g., the absorption edge (Fig. 7.18).[26] Other methods, e.g., dielectric loss, electron-spin resonance, etc., are also frequently helpful. The main problem in the investigation of crystals at present is to understand their imperfections and to get them under control by close cooperation between chemists, physicists, and engineers.

[26] A. Smakula, *Optica Acta 9*, 205 (1962).

8 · DEFECTS IN METAL CRYSTALS

E. W. Müller

Introduction — Technique of field-ion microscopy — Observation of point defects — Complex defect structures — Conclusion

Introduction

Although it is difficult to calculate the strength of metals from first principles, it can be safely concluded that a metal crystal should be many hundred times stronger than our technical materials actually are. This discrepancy is due to lattice defects. Future designing of metals from the atomic level thus depends upon our knowledge of the nature of crystal lattice defects. The recent technique of transmission electron microscopy[1] has brought about great progress in understanding dislocation phenomena, precipitation, and other structural defects by making them accessible to direct observation. True atomic resolution microscopy, leading to the building stones of matter, has now been achieved with a new, still more powerful instrument, the field-ion microscope.[2] Now even point defects, vacancies, interstitials, and impurity atoms become directly visible, and the intimate atomic structure of dislocations, precipitations, slip bands, grain boundaries, and cold-work structure as well as displacement spikes due to radiation damage can be unfolded before the eye of the observer. It is the purpose of this chapter to outline the capabilities as well as the limitations of this new method.

[1] P. B. Hirsch, *J. Inst. Metals* **87**, 400 (1959).
[2] E. W. Müller, "Field Ionization and Field Ion Microscopy," *Advances in Electronics and Electron Physics*, Vol. 13, L. Marton, Ed., Academic Press, New York, 1960, p. 83.

Technique of field-ion microscopy

The field-ion microscope is a newer version of the field-electron microscope,[3] in which the specimen, the approximately hemispherical surface of a fine needle tip, is radially projected in more than millionfold magnification onto a phosphor screen (Fig. 8.1). By using positive ions,[4] preferably of helium or neon, the surface can be seen in a resolution of up to 2.5 Å. Field ionization occurs about 4 Å above the surface, dependent upon the local field strength of the order of 500 million volts/cm, which is modified by the protrusion of surface atoms. The imaging process is complicated by the process of accommodation of the gas atoms to the tip temperature, preceding ionization, and by the surface diffusion of gas atoms in the high field.[5] By using helium or neon, one of the most critical conditions for microscopy at the atomic level—prevention of contamination by adsorption—is automatically met when all contamination gases, having lower ionization potentials, ionize in space before reaching the surface and are then rejected by the field. This unique feature maintains absolutely clean surface conditions even in

[3] E. W. Müller, *Z. Physik* **106**, 541 (1937). For the history of field-emission microscopy see R. H. Good, Jr., and E. W. Müller, "Field Emission," *Handbuch der Physik*, Vol. 21, Springer-Verlag, Berlin-Göttingen-Heidelberg, 1956, p. 176.
[4] E. W. Müller, *Z. Physik* **131**, 136 (1951).
[5] E. W. Müller, *J. Appl. Phys.* **27**, 474 (1956).

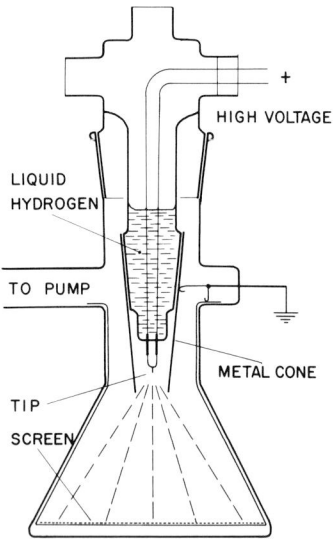

Fig. 8.1. Schematic diagram of field-ion microscope.

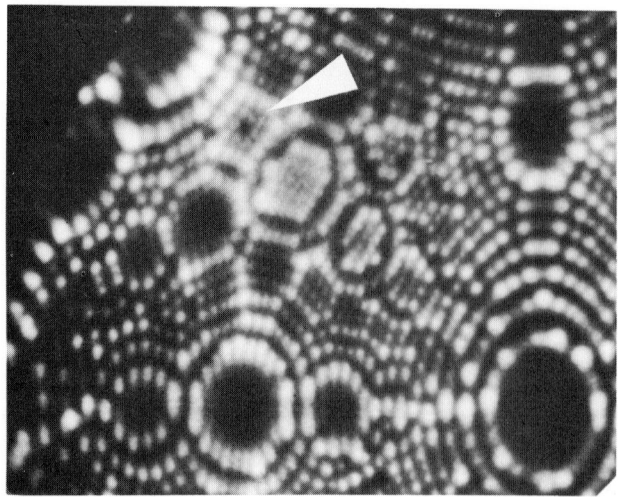

Fig. 8.2. Single vacancy in center of (203) plane of platinum crystal.

a microscope with the modest vacuum conditions of a demountable system.[6] The specimen surface is initially made absolutely clean by controlled low-temperature field evaporation. The theory of field evaporation[7,8] is still unsatisfactory because of the limitations of a one-dimensional surface model in a field of more than 5 volts/Å, which discredits classical image-force and polarization concepts. In practical field-ion microscopy, the onset of field evaporation below the field strength necessary for image formation limits applicability to metals with melting points above approximately 1000° C. Even when reduced surface stability is countered by the short exposure times made possible by electronic image intensifiers,[9] the mechanical field stress causes nucleation of disturbing dislocations in the softer metals.

Observation of point defects

Under favorable conditions vacancies, interstitials, and impurity atoms can be seen as individuals (Fig. 8.2). Most easily accessible are quenched-in vacancies in face-centered cubic metals such as Pt, where it was possible to determine a vacancy concentration of 5.9×10^{-4} by direct counting during controlled field evaporation to a depth of 70 atom layers.[10] In tungsten the vacancy concentration was found [11] to be well below 10^{-4} even when quenched from near the melting point at a rate of 450,000°/sec, thus indicating an energy of formation of more than 2.9 eV. Clusters of two to ten vacancies have been obtained by *in situ* bombardment of Pt, pile neutron bombardment of Mo,[11] or irradiation of W with 5-MeV protons. In no case was a predominance of divacancies observed when various heat treatments were applied to *in situ* bombarded materials. Copious surface vacancies are found in W and Pt as a result of irradiation or sputtering *in situ*, when Silsbee focusing collisions cause the ejection of surface atoms. For such observations the color-comparison technique[12] of photographs taken before and after bombardment is used advantageously. Interstitials might be expected to assume a normal lattice site when they arrive at the surface by either diffusion, focusing collisions, or by field evaporation, thus becoming undetectable. This, however, is not the case. It appears that the lattice strain relaxes when the interstitial arrives one atom layer below the surface, and the resulting surface deformation appears as a very bright dot, clearly distinguishable from individual surface atoms[2] (Fig. 8.3). Such spots have been produced in large numbers by α bombardment, medium-energy atom or negative ion bombardment, and cathode sputtering, all experiments being carried out *in situ* with the specimen at 21° K. At that temperature, shallow-lying interstitials arrive at the surface of W and Pt by a slight rise of field stress. Annealing experiments with

[6] E. W. Müller, "Examining the Atomic Structure of Metal Surfaces with the Field-Ion Microscope," *Proc. Fourth Interntl. Congress on Electron Microscopy*, Vol. 1, Springer-Verlag, Berlin, 1960, p. 820.
[7] E. W. Müller, *Phys. Rev. 102*, 618 (1956).
[8] R. Gomer, *Field Emission and Field Ionization*, Harvard University Press, Cambridge, 1961.
[9] S. B. McLane, Jr., E. W. Müller, and O. Nishikawa, *Rev. Sci. Instr.* 35, 1297 (1964).

[10] E. W. Müller, *Z. Physik 156*, 399 (1959).
[11] E. W. Müller, *J. Phys. Soc. Japan 18*, Suppl. 2, 1 (1963).
[12] E. W. Müller, *J. Appl. Phys. 28*, 1 (1957).

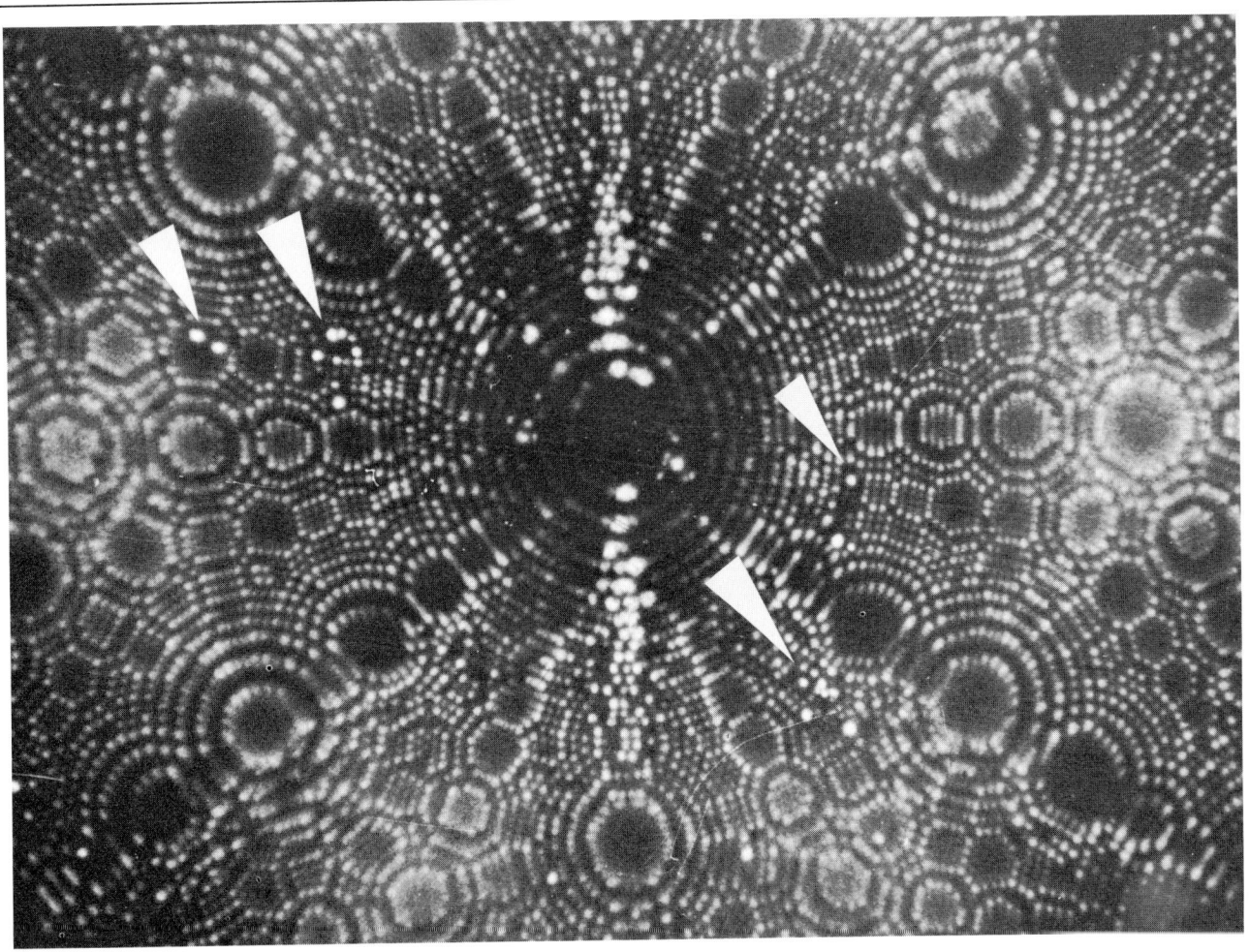

Fig. 8.3. Clusters of interstitials and vacancies on surface of α-bombarded tungsten tip.

W showed fast diffusion near 90° K,[13] and also an indication of interaction with vacancies at higher defect density.[11]

Impurity atoms as solutes are visible in some areas of the field-ion-microscope pattern when their size differs sufficiently from the atoms of the base metal. Such impurity atoms then show up in homogeneous atom rows by protruding either less or more than the surrounding atoms[2] (Fig. 8.4). Impurities in interstitial positions again deform the lattice and are then quite clearly visible as bright spots, randomly distributed or decorating dislocation lines.[11] Interstitial carbon appears in the tungsten pattern at a specific crystallographic area, on the one edge of the (132) plane toward (010), because only at this area can the small impurity atom displace a tungsten atom enough to make it protrude more.[14]

Complex defect structures

Dislocations with their long-range strain field can be easily detected by the more conventional techniques of microscopy, using either etch pits, thin-film transmission-electron microscopy, or X-ray methods. The field-ion microscope gives more intimate details of core structure, although the determination of the Burgers vector is often difficult, since only the intersection of the dislocation with the surface is seen. Such dislocations can be either intrinsically present in the specimen tip or they can be introduced, sometimes *in situ*, by stressing or cold work or by the collapse of vacancy clusters after stressing or particle bombardment. Unfortunately, some of the softer metals cannot always withstand the high stress exerted by the imaging field itself.

[13] E. W. Müller, "Observation of Radiation Damage with the Field Ion Microscope," *Reactivity of Solids*, Proc. 4th International Symposium on the Reactivity of Solids, J. H. de Boer, Editor-in-Chief, Elsevier Publishing Co., Amsterdam (Distributors: D. Van Nostrand Co., Princeton, N. J.), 1960, p. 682.

[14] E. W. Müller and Y. Yashiro, unpublished work.

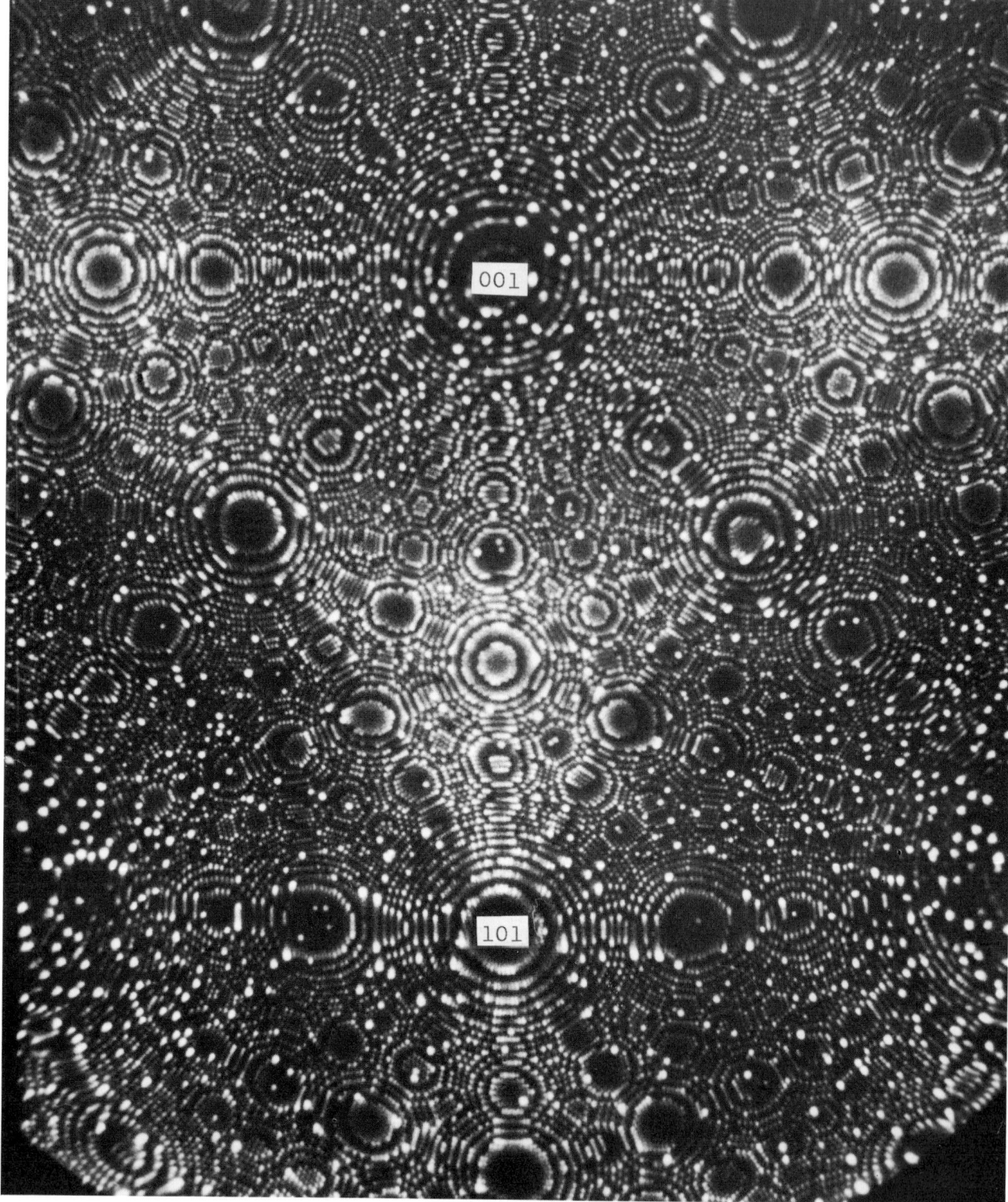

Fig. 8.4. Numerous impurity atoms (oxygen) appearing as scattered bright spots on surface of rhodium crystal.

Fig. 8.5. Various dislocations on platinum crystal (dislocation dipole in the upper left corner decorated by impurity atom; encircled (317)-plane dislocation consists of two inserted extra net planes.) (2,000,000X)

This outward stress, acting like a negative hydrostatic pressure of up to 100,000 atm, causes the development of dislocations near the (111) and (001) planes of molybdenum, or the (012) plane of nickel, as soon as these specimens are used beyond a minimum radius of about 300 Å. Grain boundaries were observed to have a very narrow contact area of imperfection,[11] which can be followed into the depth of the crystal by controlled field evaporation. The crystallographic orientation of such grain boundaries can be derived from the field-ion-microscope pattern (Fig. 8.5). The attraction of impurities by grain boundaries as well as their effect on focusing collisions during particle bombardment is easily observable.

No detailed studies have been made as yet of the process of precipitation, although the size and density of atom clusters acting as precipitation nuclei can be measured directly (Fig. 8.6). Guinier-Preston zones should be readily recognizable in suitable alloys of the Pt metals or of tungsten. Finally, the structure of alloys is a major area of future research with the field-ion microscope. Short-range order in dilute binary alloys should be detectable. Long-range order in 50% Pt-Co alloy has already been studied to some extent.[15] The high degree of strain resulting from the lattice deformation from face-centered cubic to a tetragonal structure during the ordering process as well as the structure of domain boundaries is directly evident in the ion micrograph (Fig. 8.7).

[15] E. W. Müller, *Bull. Am. Phys. Soc.* (II) **7**, 27 (1962).

Fig. 8.6. Precipitated cluster of impurity in platinum.

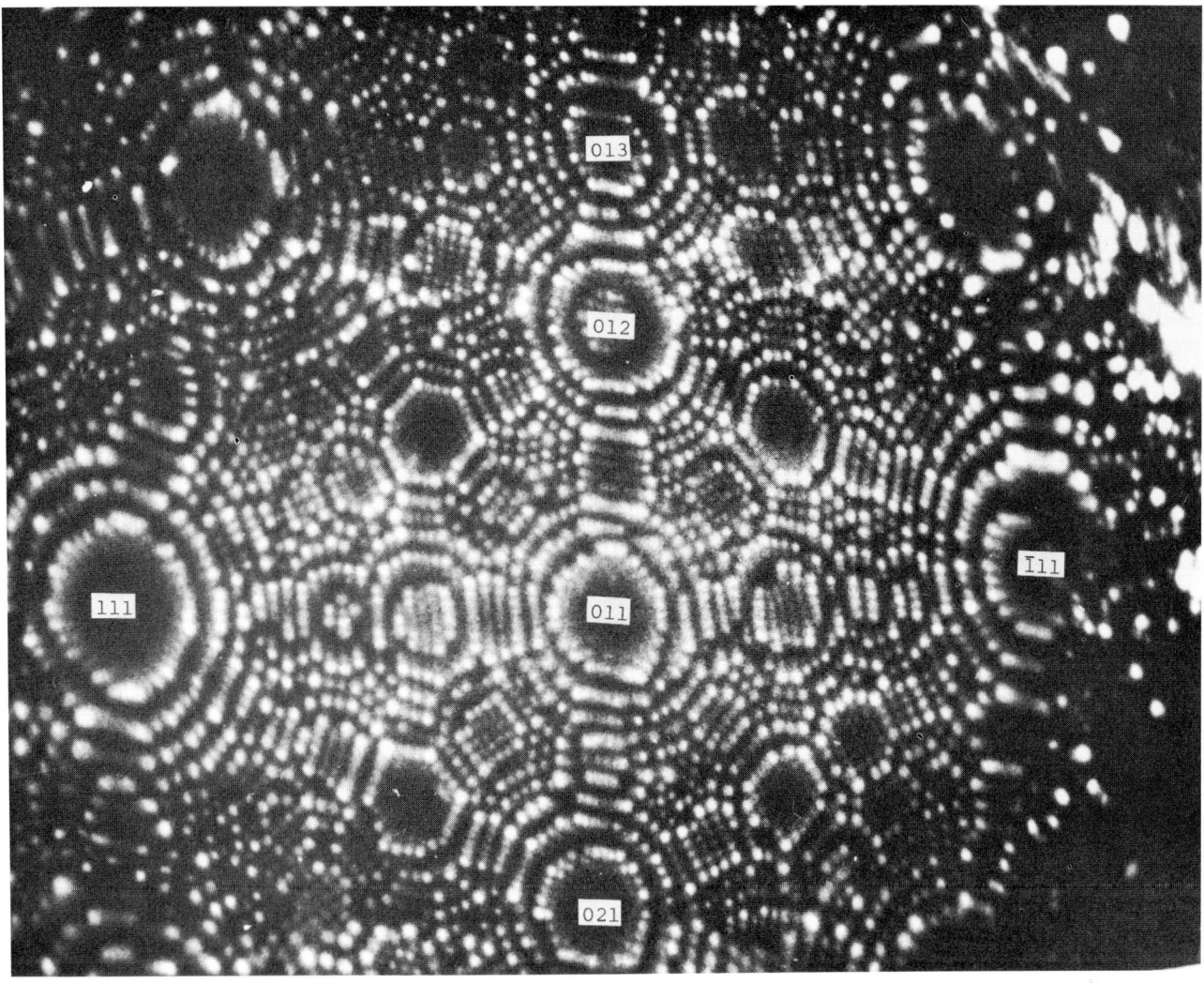

Fig. 0.7. 50% platinum-cobalt ordered alloy.

Conclusion

The conditions of visibility of various crystal imperfections with the field-ion microscope are now well established. Future efforts will aim at a quantitative determination of defect densities, activation energies of diffusion, or numerical data for the interaction of the imperfections. In addition, the effects of surface imperfections on such properties as the work functions, adsorption, nucleation, or epitaxial growth can now be studied with a more realistic image of the atomic structure.

9 · DEFORMATION AND FRACTURE OF SOLIDS

Ali S. Argon

Introduction — Elastic deformation — Viscous deformation — Plastic deformation by dislocations — Dislocation mills — Observation of dislocations — Deformation of polymers — Hardening of crystals — Solution hardening — Precipitation hardening — Work hardening — Yield strength of single crystals — Fracture — Conclusions

Introduction

All solids are, to a greater or lesser extent, reversibly and irreversibly deformable. Mathematical theories of elasticity, plasticity, and viscoelasticity have been developed over the past quarter century to a level where solutions to some problems of great complexity are now within reach. We shall not consider such phenomenological formalisms here but instead confine our attention to the physical processes that govern the mechanical behavior of materials.

Elastic deformation

The simplest mode of deformation is the reversible, small-strain elasticity exhibited by all solids at low temperatures and under low stresses. With the application of an external force, the atoms of a solid become displaced from their equilibrium positions until the internal stress resulting from the interatomic forces of extended bonds equals the externally applied force. For example, the energy of a sodium chloride crystal varies with interionic distance r as

$$U = -U_b \left[\frac{8}{7} \left(\frac{r_0}{r}\right) - \frac{1}{7} \left(\frac{r_0}{r}\right)^8 \right], \quad (9.1)$$

where $U_b = 180$ kcal/mole is the binding energy (lattice energy) and $r_0 = 2.81$ Å, the interionic distance of the unstressed lattice at atmospheric pressure. The average Young's modulus E can be found from the compressibility (the reciprocal of the bulk modulus B) through differentiation of the energy equation; i.e.,

$$B = \frac{E}{3(1-2\nu)} = -V\left(\frac{\partial p}{\partial V}\right)_T$$

$$= \left[V \left(\frac{\partial r}{\partial V}\right)^2 \left(\frac{\partial^2 U}{\partial r^2}\right) \right]_{r=r_0}, \quad (9.2)$$

where ν is Poisson's ratio. Disregarding the anisotropy of the crystal, we obtain

$$E = \frac{8}{3}(1-2\nu)\frac{U_b}{V_0} = 3.72 \times 10^{11} \text{ dynes/cm}^2, \quad (9.3)$$

with Poisson's ratio taken as 0.25. The stronger the binding of the crystal, the greater its stiffness.

The elasticity of rubber at room temperature stems from an entirely different mechanism. The long molecules of the structure are tied together at a few points only. The remainder of the molecule between these junctions acts like a kinking line, constantly changing its shape by thermal oscillations (Fig. 9.1). All along, however, the structure remains in a most probable state of disorder, with the configurational entropy maximized. Under uniaxial tension or compression

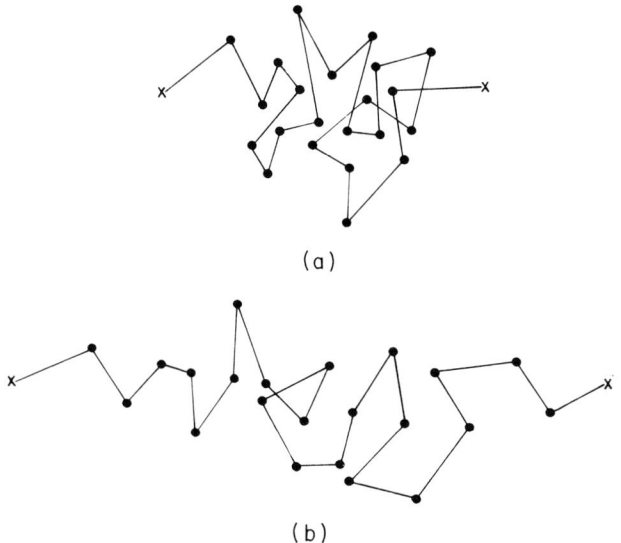

Fig. 9.1. Rubber molecule with random thermal kinks; (a) under no stress, (b) under uniaxial tensile stress.

the work of deformation is not stored as potential energy of extension or compression of individual bonds but is expended in straightening out the kinks against the randomizing tendency of thermal agitation. The result is a reduction in configurational entropy with practically no accompanying change in internal energy. When the load is removed, the entropy increases again, and the process is reversible. According to the first law of thermodynamics,

$$T\,dS = dU - f\,dl, \tag{9.4}$$

the force f becomes

$$f = -T\left(\frac{\partial S}{\partial l}\right)_T, \tag{9.5}$$

since $(\partial U/\partial l)_T = 0$, as already stated.

Under hydrostatic pressure, however, this mechanism cannot operate, and rubber behaves as an elastic solid with a high bulk modulus.

Viscous deformation

Under high nonhydrostatic stresses all solids undergo permanent, inelastic deformation if fracture does not intervene.

In an inorganic glass, which in many respects can be described as a supercooled liquid, the disordered structure makes it kinematically possible for ions to exchange positions between neighboring sites (Fig. 9.2). In jumping from one of these positions into the other, the ion not only must overcome an enthalpy barrier ΔH because of its state of binding to its neighbors but also alter the arrangement of these neighbors, i.e., change the configurational entropy by an amount ΔS. With no external stress there will be a steady back-and-forth shuffle of ions between such alternate positions of equilibrium, separated by a free-energy barrier $\Delta g = \Delta H - T\,\Delta S$, at a rate proportional to a Boltz-

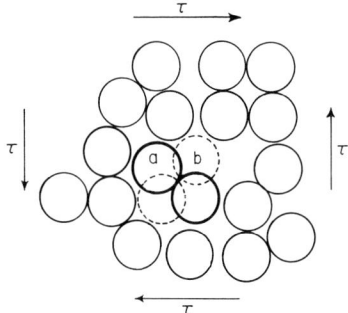

Fig. 9.2. Model for viscous deformation and delayed elasticity; (a) initial and (b) final configuration under stress.

mann factor. The application of a shear stress τ will give a net bias to this shuffle in the direction of the stress, resulting in an elementary shear strain $\gamma_a = b^3/V$ for each net switch, where b is an average interionic distance and V the total volume. If there is a constant concentration N of such sites per unit volume, the net shear strain rate $\dot\gamma$ would become (taking account also of reverse jumps)

$$\dot\gamma = NV\frac{b^3}{V}\nu$$
$$\times\left[\exp\left(-\frac{\Delta g - \tau b^3/2}{kT}\right) - \exp\left(-\frac{\Delta g + \tau b^3/2}{kT}\right)\right], \tag{9.6}$$

with ν the Debye frequency. In most cases,

$$\frac{\tau b^3}{2} \ll kT \ll \Delta g; \tag{9.7}$$

hence,

$$\dot\gamma = \frac{Nb^6\tau\nu}{2kT}\exp\left(-\frac{\Delta g}{kT}\right). \tag{9.8}$$

This is Orowan's[1] model of viscous deformation for a simple glass at elevated temperature; it leads to a viscosity coefficient

$$\mu = \frac{2kT}{Nb^6\nu}\exp\left(\frac{\Delta g}{kT}\right), \tag{9.9}$$

which (as expected) shows a very strong dependence on temperature.

[1] E. Orowan, *Proceedings of the First National Congress of Applied Mechanics*, American Society of Mechanical Engineers, New York, 1952, p. 453.

At lower temperatures the number N of sites switching in the direction of the applied stress cannot be maintained constant, independent of time. Instead, the total number of loose sites N will eventually be divided between lower and higher energy sites when the rates of transfer under the applied stress become equal; i.e.,

$$N_l \nu \exp\left(-\frac{\Delta g + \tau b^3/2}{kT}\right) = N_h \nu \exp\left(-\frac{\Delta g - \tau b^3/2}{kT}\right). \quad (9.10)$$

Since

$$N = N_l + N_h, \quad (9.11)$$

the steady-state value of N_l is

$$N_l = \frac{N}{[1 + \exp(-\tau b^3/kT)]} \cong \frac{N}{2}\left(1 + \frac{\tau b^3}{2kT}\right); \quad (9.12)$$

here the process comes to a standstill. This results in the familiar delayed elasticity of inorganic glasses.

The distinguishing feature of viscous or delayed elastic deformation is that the structure must have a certain amount of disorder, making it possible for individual ions to change positions. This fact, coupled with the condition that the work done during the switching operation be much smaller than the level of thermal-vibrational energy, makes viscous deformation highly temperature-sensitive. The stress serves only to give the process direction.

Plastic deformation by dislocations

Viscous deformation is not dominant in a crystalline material, primarily because the regularity of a lattice makes it very difficult for an atom to leave its regular site and be accommodated in its immediate vicinity. Although such movements do occur to some extent, producing vacant lattice sites and interstitial atoms, the low-temperature plasticity of crystals is due to the generation and motion of lattice dislocations.

A dislocation is a *line defect*, most commonly the boundary of a region in the crystal which has undergone relative shear on a crystallographic plane by one interatomic distance[2,3] (Figs. 9.3 and 9.4). Figure 9.4a illustrates the state of extension or distortion of atomic bonds in a plane of atoms immediately adjacent to the plane of relative translation (slip plane) in a hypothetical simple cubic crystal. The square dislocation ring bounding the slip patch is made up of two types of

[2] G. I. Taylor, *Proc. Roy. Soc.* (London) *A145*, 362 (1934).
[3] E. Orowan, "Dislocations and Mechanical Properties," *Dislocations in Metals*, M. Cohen, Ed., American Institute of Mining and Metallurgical Engineers, Institute of Metals Division, New York, 1954, p. 69.

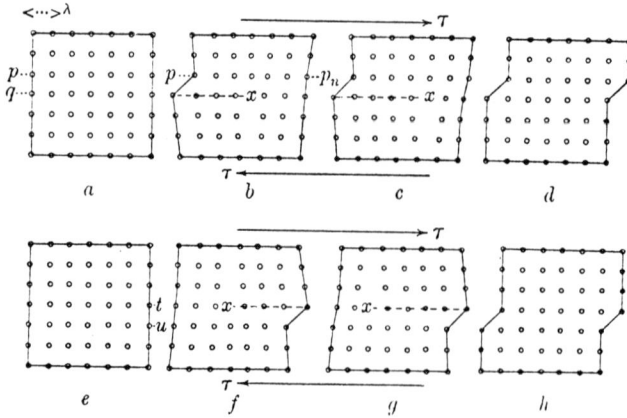

Fig. 9.3. Plastic deformation by motion of edge dislocations.[2]

pure dislocations (Fig. 9.4b). The characteristic feature of the dislocation normal to the direction of relative translation (slip direction) is an extra half plane (cf. Fig. 9.3); this is an *edge dislocation*. The dislocation parallel to the direction of translation produces a spiraling displacement; it is known as a *screw dislocation*.

Because dislocations are defects in a crystal lattice, the relative shear displacement necessary to generate both edge and screw dislocations must be related to a lattice identity vector (the *Burgers vector*) that measures the strength of the dislocation.

The Burgers vector of a dislocation line remains unaltered along its length as the dislocation snakes its way through the crystal.

Dislocations are internal stress singularities of considerable strain energy. Of the pure types, the screw

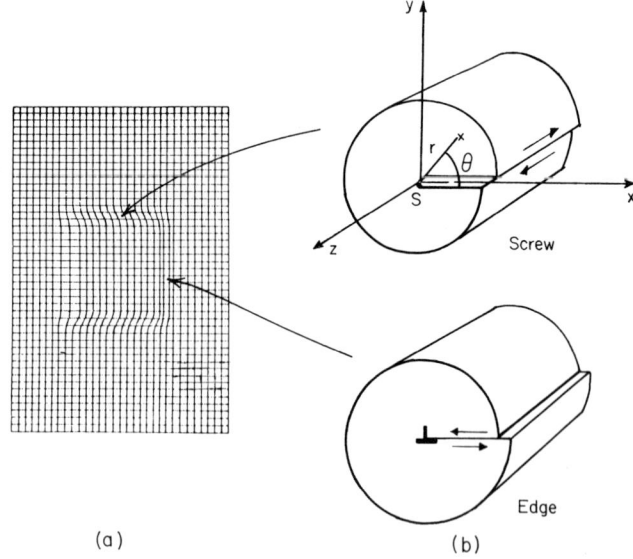

Fig. 9.4. Dislocation loop surrounding a slip patch; (*a*) displacement of atoms in layer immediately adjacent to the slip plane;[3] (*b*) the two types of pure dislocations: screw and edge.

dislocation is the simpler to analyze. In the coordinate system of Fig. 9.4b there is only one shear strain ($\gamma_{\theta z}$) caused by the screw nature of the displacements,

$$\gamma_{\theta z} = \frac{b}{2\pi r}, \qquad (9.13)$$

independent of the angle θ. It is related to a shear stress,

$$\tau_{\theta z} = \frac{Gb}{2\pi r}. \qquad (9.14)$$

Here b is the pitch (or strength) of the screw dislocation—the magnitude of the Burgers vector—and G the shear modulus. The stress field of the edge dislocation has both shear and normal stress components which have the same long-range dependence on the distance r as in the screw dislocation but in addition depend on θ. The general stress patterns around an edge dislocation are shown in Fig. 9.5[4] (compare orientation with Fig. 9.4b).

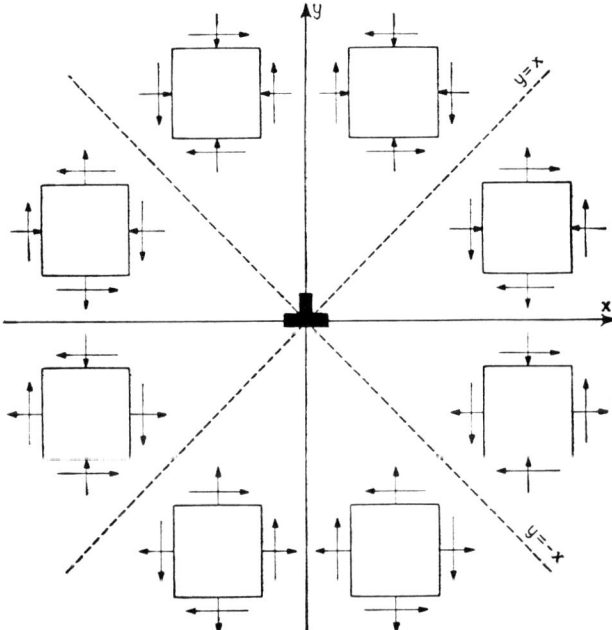

Fig. 9.5. Stress distribution around an edge dislocation in an elastically isotropic material. (The stress components change sign on the boundaries of the eight regions.[4])

Integration of the strain energy of a screw or edge dislocation over the cylindrical region of radius R allotted to one dislocation gives (excluding a small core region of radius $2b$, where linear elasticity does not hold)

$$U = \frac{Gb^2}{4\pi} \ln\left(\frac{R}{2b}\right) \qquad (9.15)$$

[4] T. W. Read, Jr., *Dislocations in Crystals*, McGraw-Hill Book Co., New York, 1953, p. 118.

per unit length of dislocation. The core energy is only about 10 percent of the strain energy of the dislocation outside the core and may be neglected in first-order calculations. In a real crystal with random dislocation distribution, the stress of a dislocation extends only as far as its nearest neighbor. Therefore, the radius of integration R is taken as half the mean distance between dislocations, and the energy per unit length becomes approximately

$$U = \frac{Gb^2}{2} \qquad (9.16)$$

in a crystal with a dislocation density of about 10^8 cm/cm^3. This line energy of a dislocation makes it behave as if it were under a tension of this magnitude.

The configurational entropy contribution to the free energy of a dislocation is negligible compared with the strain energy in any crystal of reasonable size. Hence, nearly all of the energy of a dislocation is free energy.

Since this free energy is proportional to the square of the magnitude of the Burgers vector, only dislocations of a strength equal to the smallest identity vector exist with any abundance in a crystal. Because of the inherent anisotropy of crystal lattices, there are in general some crystallographic planes containing the Burgers vector on which the free energy of dislocations is minimized.[5,6] The position of the very center of the dislocation in a lattice is governed by the energy of its core; it can be expected to be a minimum in some sym-

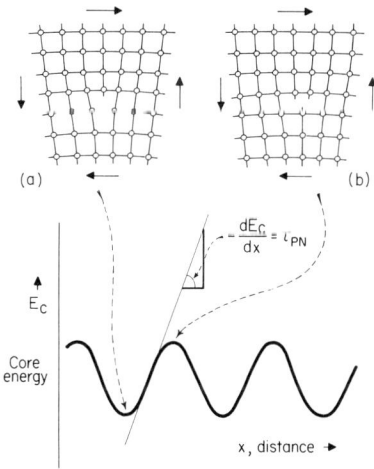

Fig. 9.6. Variation of the core energy of an edge dislocation as a function of position in a simple cubic lattice.

metry orientation (Fig. 9.6a) and a maximum in some other one (Fig. 9.6b). Estimates for simple lattices

[5] J. D. Eshelby, *Phil. Mag.* [7] **11**, 903 (1949).
[6] R. D. Heidenreich and W. Shockley, *Report of the Conference on Strength of Solids*, The Physical Society, London, 1948, p. 57.

indicate that the fluctuation of the core energy[5,7,8] of a dislocation is a minimum on the same planes on which the elastic energy is minimized. Therefore, when shear stress is applied, a dislocation can propagate more easily on these crystallographic slip planes than on others.

When dislocations move, the extra half plane confines the motion of the edge components to the slip plane; the screw components, having no such constraints, may move on any crystallographic plane that contains the slip direction parallel to the Burgers vector. This type of motion propagating the dislocation along slip planes is called *glide*. At high temperatures, above approximately half the absolute melting temperature, the extra half plane of a dislocation can be made to extend or contract by transport of vacancies or interstitials. This kind of motion of dislocations out of their slip plane is appropriately called *climb*. At such elevated temperatures, under the added degree of freedom provided by climb, dislocations will tend to annihilate each other, collect in low-energy configurations (such as low-angle grain boundaries), or leave the crystal through its surface, thereby increasing its perfection. This constitutes the mechanism of annealing.

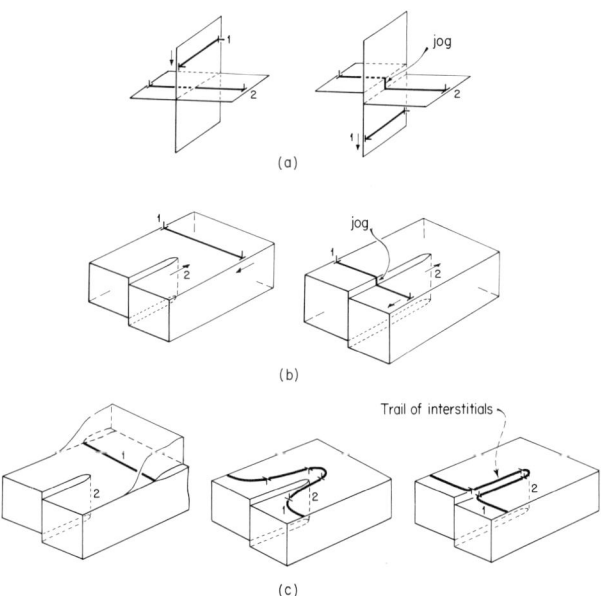

Fig. 9.7. Production of various types of jogs during the intersection of dislocations; (*a*) edge-edge intersection, (*b*) edge-screw intersection, (*c*) screw-screw intersection; note trail of interstitials.

During elastic deformation, dislocations intersect each other, thus producing defects on the dislocation lines themselves, called *jogs*. As shown in Fig. 9.7, jogs

[7] F. R. N. Nabarro, *Proc. Phys. Soc.* (London) **59**, 256 (1947).
[8] G. Leibfried and H.-D. Dietze, *Z. Physik* **131**, 113 (1951); H.-D. Dietze, *ibid.* **131**, 156 (1951).

arising from the intersection of edge dislocations or of edge and screw dislocations are simple and one identity distance long. The jogs generated by the intersection of two screw dislocations, on the other hand, are as long as the separation of the dislocations from each other and consist of a trail of vacancies or interstitials (cf. Fig. 9.7c).

If dislocations did not pre-exist in a crystal or were driven out by plastic deformation, new ones would have to be formed. This generation of dislocations inside the crystal under applied stress may proceed in slip patches outlined by closed loops of dislocations (Fig. 9.8). All parts of a dislocation loop attract each other;

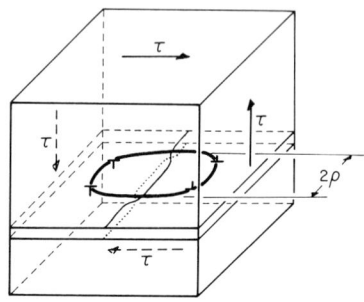

Fig. 9.8. Dislocation loop of radius ρ, outlining a slip patch in a prismatic crystal. (The planes immediately above and below the slip plane are shown.)

hence, a shear stress is required to prevent a slip patch from collapsing. A loop has a critical size for each shear stress τ, below which it would disappear spontaneously. The dislocation loop will be of critical size with radius ρ when the stress at any point of the dislocation loop resulting from the stress field of the rest of the loop is balanced by the applied stress τ, i.e., when

$$\frac{Gb}{4\rho\pi} \approx \tau. \qquad (9.17)$$

The circumference for the critical-size dislocation loop becomes $Gb/2\tau$, and its Gibbs free energy

$$\Delta g = \frac{Gb}{2\tau} \times \frac{Gb^2}{2}. \qquad (9.18)$$

For the shear stresses normally observed ($\tau < 10^{-3}G$) the free energy of a loop would be of the order of thousands of electron volts, four to five orders of magnitude beyond the realm of thermal activation at room temperature. Refinements of this calculation do not change the conclusion that dislocations cannot be generated by thermal fluctuations in a homogeneous manner. Thus, in contrast to viscous deformation, thermal activation does not play a dominant role in crystal plasticity.

Dislocation mills

The dislocations necessary for large amounts of plastic deformation are generated by dislocation mills operating under a critical shear stress. Dislocation mills are special configurations capable of producing large strains by continuously increasing the total length of a dislocation. One type of dislocation mill[9] (Fig. 9.9) consists of a short segment of dislocation in

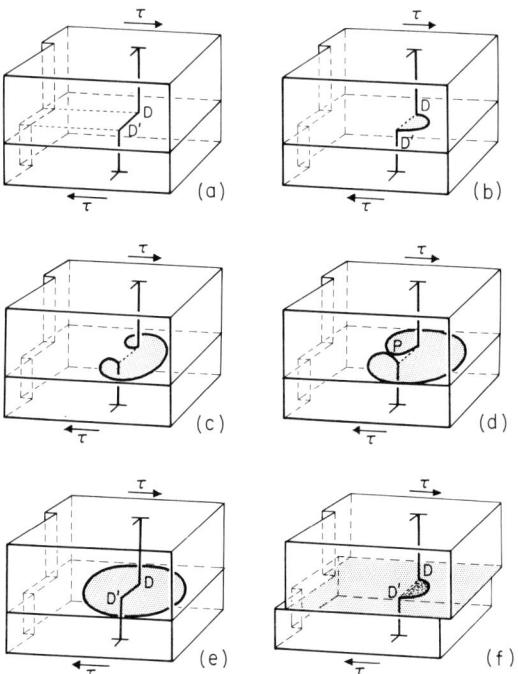

Fig. 9.9. Stages in the operation of a Frank-Read dislocation mill.

the slip plane, produced by a meandering dislocation. Under the proper shear stress this segment will move out as shown in the sequence of sketches of Fig. 9.9. When the dislocation has swept over the shaded region of the slip plane (Fig. 9.9d) and its two arms meet at point p, they will fuse together to give back the initial segment DD' and add a closed dislocation loop surrounding it. The latter in sweeping over the slip plane produces a shear displacement of one interatomic distance. Each successive revolution of the free segment DD' will contribute a new dislocation loop. Figure 9.10[10] shows a loop in such a dislocation mill in a silicon crystal.

Dislocation mills are produced in relative abundance during plastic deformation by double cross glide of a

[9] F. C. Frank and W. T. Read, *Phys. Rev.* **79**, 722 (1950).
[10] W. C. Dash, "The Observation of Dislocations in Silicon," *Dislocations and Mechanical Properties of Crystals*, J. C. Fisher et al., Eds., John Wiley and Sons, New York, 1957, p. 57.

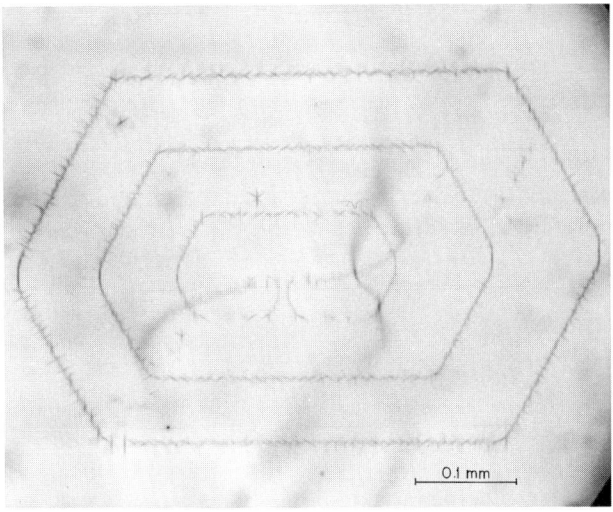

Fig. 9.10. A Frank-Read dislocation mill in silicon, made visible by decorating it with precipitates of copper.[10]

screw dislocation. In Fig. 9.11 a portion of a screw dislocation is shown shunted to a neighboring slip plane by gliding a short distance on a bridging plane. The resulting configuration on the neighboring plane is a dislocation mill.

A mill can be operated only by a shear stress which bows a dislocation out to an unstable semicircular configuration (Fig. 9.9b).

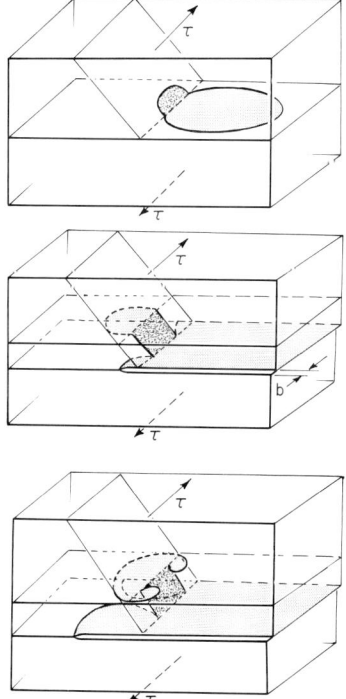

Fig. 9.11. Production of a Frank-Read dislocation mill during plastic deformation by the double cross glide of a segment of a screw dislocation.

Observation of dislocations

In addition to indirect methods, such as precision measurements of density, resistivity, etc., powerful methods are now available for the direct observation of dislocations: Impurity atoms can often be made to precipitate on dislocations by interaction of their mutual stress fields. Thus the dislocations become visible, e.g., in the dislocation mill (Fig. 9.10), where copper was precipitated on the dislocation loops and observed by infrared microscopy.

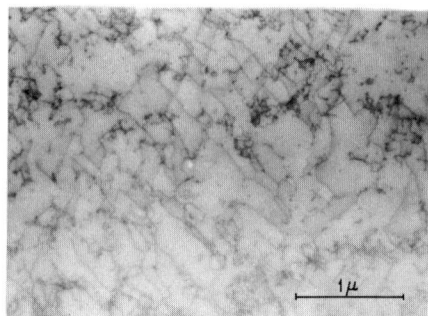

Fig. 9.12. Dislocations in a thin film of tungsten, prepared from a deformed single crystal oriented parallel to the cube edge. The observation is made in the electron microscope.

Dislocations can be seen directly in thin films by electron microscopy (Fig. 9.12). Thicker sheets can be studied by X-ray diffraction: The Bragg reflections are

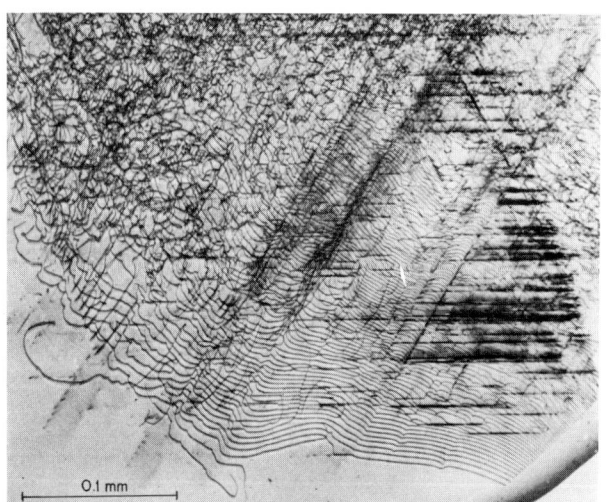

Fig. 9.13. Magnified X-ray image of a deformed silicon crystal showing dislocations.[11]

recorded on ultrahigh-resolution photographic plates and the images optically magnified (Fig. 9.13).[11] The

[11] A. R. Lang, *J. Appl. Phys* **30**, 1748 (1959).

dislocations become visible in both cases by the diffraction contrast produced by their stress field.[12]

In some crystals the sites at which dislocations reach the surface can be revealed by selective attack of an etchant producing etch pits (Fig. 9.14).[13]

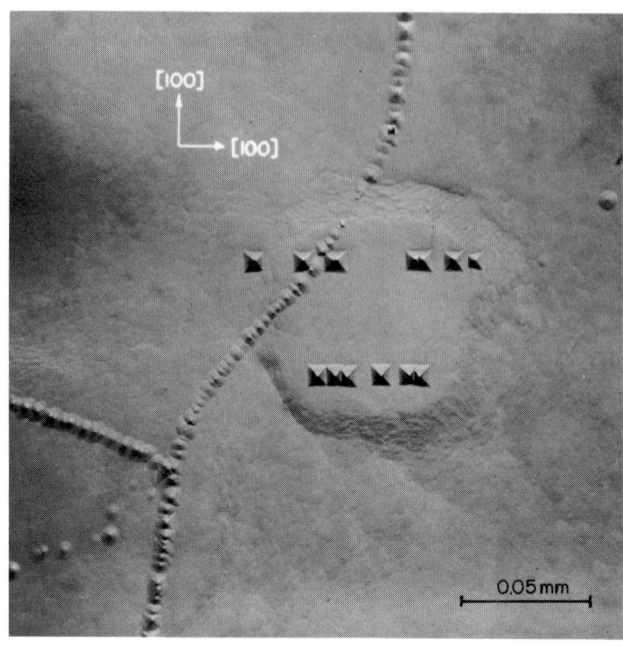

Fig. 9.14. Etch pits on the cube surface of a cleaved LiF crystal showing the sites where dislocations emerge. The different etching behavior of the dislocations making up the low-angle grain boundaries suggests that these dislocations are likely contaminated with impurity atoms.

Of these techniques, light-microscopic observation can measure dislocation densities up to 10^8 and electron microscopy, up to 10^{12} lines/cm^2.

Deformation of polymers

In polymers, inelastic deformation shows a great variety of forms ranging from viscous and rubbery to more stress-dependent types resembling plastic deformation in metals. The stress-dependent deformation is in part directly related to the crystalline content of a polymer. There are, however, certain interesting exceptions, such as the deformation of amorphous polystyrene[14] (Fig. 9.15): Here a significant portion of the

[12] P. B. Hirsch, A. Howie, and M. J. Whelan, *Phil. Trans. Roy. Soc. (London)* **A253**, 61 (1960).
[13] J. J. Gilman and W. G. Johnston, "The Origin and Growth of Glide Bands in Lithium Fluoride Crystals," *Dislocations and Mechanical Properties of Crystals*, J. C. Fisher et al., Eds., John Wiley and Sons, New York, 1957, p. 116.
[14] W. Whitney, *J. Appl. Phys.* **34**, 3633 (1963).

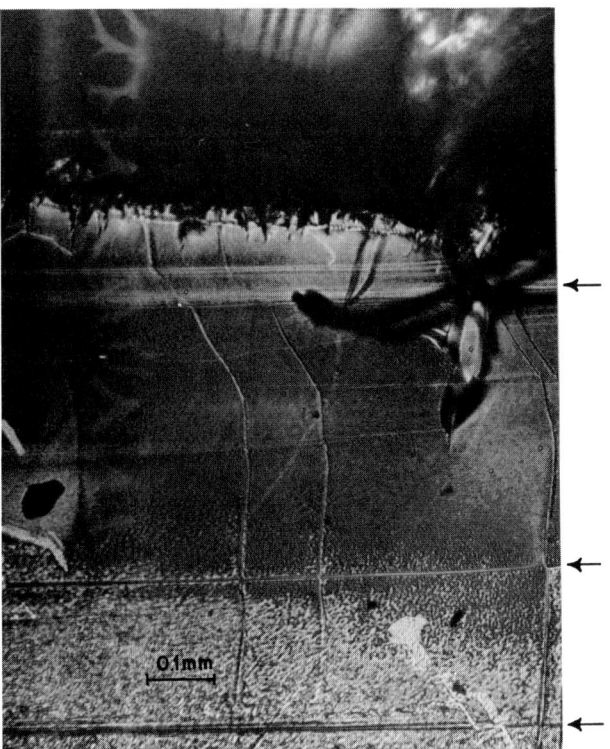

Fig. 9.15. Bands of shear deformation in polystyrene, revealed by cleavage.[14]

permanent deformation in compression is confined to well-defined bands closely resembling in appearance and behavior the slip bands in metals. The elucidation of the deformation mechanisms in polymers is one of the most interesting problems of microrheology still awaiting solution.

Hardening of crystals

A most active area in the study of the mechanical properties of solids over the past decade has been the hardening of crystals, with dislocation theory playing a central role. Although a full understanding of the mechanisms of hardening cannot be claimed, consistent theoretical interpretations can now be given for some phenomena concerning the strength of crystalline materials.

Solution hardening

One of the more successful theoretical interpretations concerns solution hardening, a process that had its earliest application when man mixed tin into copper and ushered in the bronze age. The introduction of atoms producing dilatations and/or distortions in a lattice will obstruct the motion of dislocations. In the simplest case a misfitting solute atom will interact with a dislocation if its volume differs from that of the matrix atom it displaces. The energy of interaction, the work done in introducing the misfitting sphere into the lattice against the stresses of a dislocation, can be calculated very simply.

From the theory of elasticity[15] it is known that the introduction of a rigid sphere into a somewhat smaller spherical hole in an infinite elastic body produces no dilatation anywhere outside the sphere; it merely displaces the boundaries of any closed surface enclosing the misfitting sphere by an amount that would increase its volume by the misfit volume of the rigid sphere. The stress field of an edge dislocation, on the other hand, produces a hydrostatic pressure p at a point re-

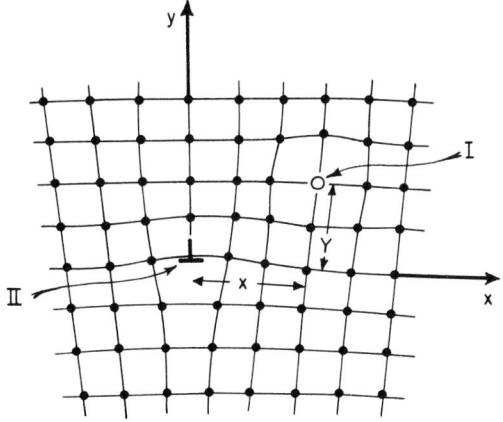

Fig. 9.16. Interaction of edge dislocation with substitutional solute atom (I, the solute atom; II, the dislocation).

moved by distances (x, Y) from the center of the dislocation (Fig. 9.16), which is

$$p = -\frac{\sigma_{xx} + \sigma_{yy} + \sigma_{zz}}{3} = \frac{\beta}{3\pi}\left(\frac{GbY}{x^2 + Y^2}\right). \quad (9.19)$$

Here β is a constant $(1 + \nu)/(1 - \nu)$, of order unity, involving Poisson's ratio. The volume change produced by the misfitting sphere would simply be

$$\Delta V = 4\pi a^3 \epsilon = \frac{\pi \epsilon b^3}{2}, \quad (9.20)$$

where $\epsilon = \Delta a/a$, the fractional difference between the radii of the rigid sphere and the spherical hole, is called the misfit parameter. The energy of interaction I will then be

$$I = p\, \Delta V = \frac{\beta}{6}\left(\frac{Gb^4 \epsilon Y}{x^2 + Y^2}\right); \quad (9.21)$$

i.e., the dislocation will resist the introduction of any misfitting sphere by its hydrostatic pressure by means

[15] J. D. Eshelby, *J. Appl. Phys.* **25**, 225 (1954).

of raising the energy of the crystal. The force between dislocation and solute atom will then become

$$F = -\frac{\partial I}{\partial x} = \beta \left[\frac{G\epsilon b^4 x Y}{3(x^2 + Y^2)^2}\right]. \quad (9.22)$$

As might have been expected, the dislocation will repel a solute atom situated relative to it, as shown in Fig. 9.16.

At low temperatures the solute cannot move because the required diffusion of vacant lattice sites cannot proceed with sufficient rapidity. As the dislocation is moved on its slip plane, therefore, it will encounter a varying interaction which is strongest at $x = Y/\sqrt{3}$, where the interaction force is maximized. This maximum force opposing the motion of the dislocation is

$$F_{\max} = \beta \frac{G\epsilon b^4}{Y^2} \cdot \frac{\sqrt{3}}{16}. \quad (9.23)$$

If Y is approximately equal to the mean distance of solute atoms randomly distributed in the lattice, the force on the dislocation per unit length becomes

$$f = \frac{F_{\max}}{Y} = \frac{\sqrt{3}\beta G\epsilon b}{16}\left(\frac{b}{Y}\right)^3 \equiv \frac{\sqrt{3}\beta}{16} G\epsilon b C; \quad (9.24)$$

C is the composition of the solute in atomic fractions. The motion of a dislocation through a unit cube will produce a shear strain $b/unity$, which makes an externally applied shear stress τ do work on the crystal to the extent τb. This work must appear on the slip plane as the displacement of the dislocation a distance unity against an opposing force of the magnitude given in Eq. 9.24. Thus the force given by Eq. 9.24 can be overcome by an external shear stress of

$$\tau = \frac{f}{b} = \frac{\sqrt{3}}{16}\beta G\epsilon C, \quad (9.25)$$

the shear strength of the solution-hardened crystal.

This analysis does not consider many essential refinements such as the effect of solutes situated below the slip plane, the flexibility of a dislocation line, the directional distortions produced by the solutes, and their inherent compressibility. It predicts, however, the often-observed linear dependence of the shear strength on composition (Fig. 9.17)[16] and a not too unreasonable magnitude. The refinements already mentioned have all been introduced into the theory [17-19] and should be

[16] J. O. Linde, B. O. Lindell, and C. H. Stade, *Arkif Fysik 2*, 89 (1950).
[17] N. F. Mott, "Mechanical Strength and Creep in Metals," *Imperfections in Nearly Perfect Crystals*, W. Shockley, J. H. Hollomon, and R. Maurer, Eds., John Wiley and Sons, New York, 1952, p. 173.
[18] R. L. Fleischer, *Acta. Met. 9*, 966 (1961).
[19] R. L. Fleischer, *ibid. 10*, 835 (1962).

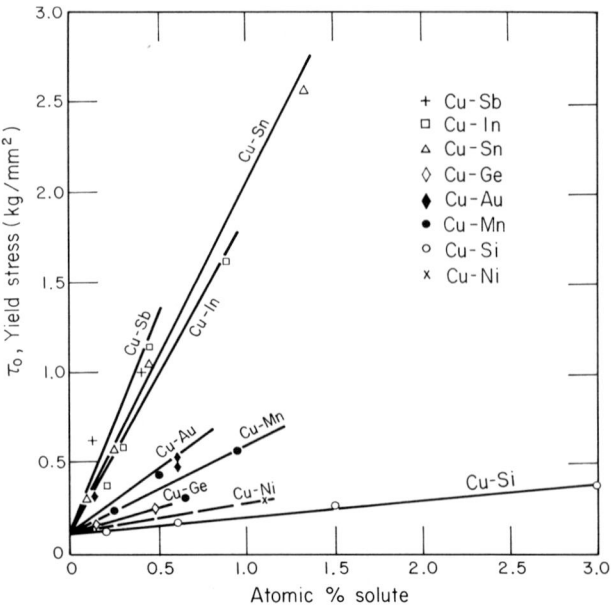

Fig. 9.17. Effect of solute concentration on the shear strength of copper.[16]

consulted for a more detailed understanding of solution hardening.

Precipitation hardening

Addition of constituents that remain in solid solution is not the only way of producing intrinsic hardening.

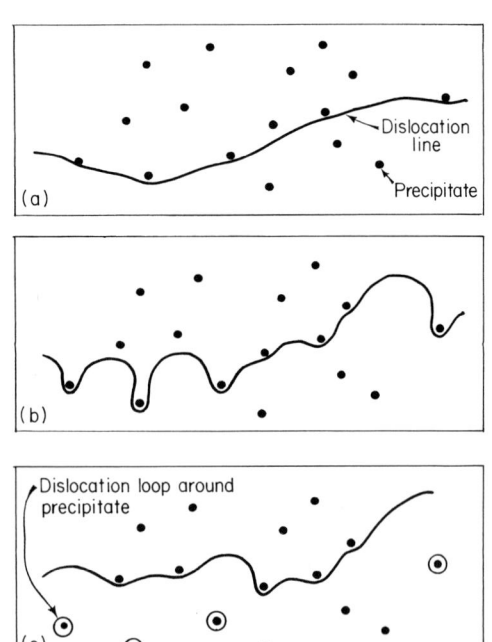

Fig. 9.18. Dislocation line in a slip plane held up by precipitates.

If the constituent goes out of solution, or if very small and insoluble particles are deliberately added to the melt to form a homogeneous distribution of dispersed precipitates in the solid state in the lattice, additional amounts of hardening can be produced. Such precipitates obstruct the motion of dislocations by their mere presence, much like pegs locking together the two parts of the crystal across a potential slip plane. If the precipitate fits incoherently into the host lattice, it will stop the motion of a dislocation locally (Fig. 9.18a). A dislocation, however, is not a rigid rod; it can snake its way through the gaps of precipitates (Fig. 9.18b), joining up with neighboring segments and leaving behind closed loops of dislocations around precipitates (Fig. 9.18c). The critical stage in this operation comes when the dislocation assumes a semicircular shape against the straightening tendency of its line energy (cf. Fig. 9.18b). The critical shear stress to hold the dislocation at this configuration is that given in Eq. 9.7. Thus, for a distance l between precipitate particles, one would expect a shear strength of

$$\tau = \frac{\alpha G b}{l}, \qquad (9.26)$$

with α about unity. This technique of hardening has been very successfully used in making high-strength and creep-resistant alloys.

Work hardening

Another, equally ancient method of hardening a metal as solution hardening is that which results from cold working. In pure metals where solution hardening is absent, this mode is one effective means of producing hardening.

Although the basic mechanisms of work hardening have long since been elucidated, a precise description of the sequence of events leading to detailed predictions of properties is still not possible.

It is known from experiments on individual dislocations[20] in a face-centered cubic metal that they can be set into short-range motion under shear stresses of approximately one ten millionth of the shear modulus. On the other hand, the shear strength of some face-centered cubic crystals of high melting temperatures can reach about one percent of the shear modulus after about one hundred percent of plastic straining.

It was Taylor,[2] one of the pioneers in the field of dislocations, who proposed that work hardening may be due to the interaction of the stress fields of dislocations. As a crystal is strained plastically, it becomes more dislocated and the increasingly closer stacking of dislocations will require a steady increase in the external stress. Elementary calculations of dislocation stress-field interactions of the type imagined by Taylor (Fig. 9.19) give a dependence between the density Λ of dis-

[20] F. W. Young, Jr., *J. Appl. Phys.* 32, 1815 (1961).

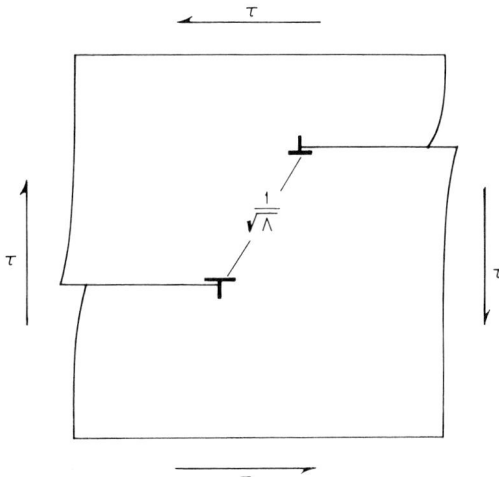

Fig. 9.19. Interaction of two edge dislocations of opposite type, resulting in hardening, according to Taylor's model.

locations threading through 1 cm² of area and the external stress τ required to hold them at this density:

$$\tau = \beta G b \sqrt{\Lambda}; \qquad (9.27)$$

β is a constant of order ≈ 0.1. Such a dependence has been experimentally confirmed in some cases.[21,22]

The fundamental element in this hardening process is the means by which an ever-increasing dislocation density inside a crystal can be held in equilibrium without collapsing by mutual annihilation or escaping through the free surfaces. Various ways have been proposed to account for this stabilization of a dislocation network. One frequent source in crystals of high symmetry is the production of segments of so-called "sessile" dislocations whose Burgers vectors are not in a crystallographically permitted slip plane. Such dislocations, formed by combination or dissociation of regular slip dislocations, cannot move easily through the lattice. These sessile dislocation segments thus serve to stabilize the remaining dislocation entanglement to which they are attached.

There can be another contribution to the increase in the shear strength. As mentioned earlier, when dislocations intersect and cut through each other—as they would have to when more than one slip system is active

[21] J. E. Bailey and P. B. Hirsch, *Phil. Mag.* [8] 5, 485 (1960).
[22] A. S. Keh, "Dislocation Arrangement in Alpha Iron during Deformation and Recovery," *Direct Observation of Imperfections in Crystals*, J. B. Newkirk and J. H. Wernick, Eds., Interscience Publishers (John Wiley and Sons), New York, 1962, p. 173.

—jogs are produced on the intersecting dislocations or other line defects formed between them if two screw dislocations intersect. Such defects raise the stored energy of the crystal and hence require additional expenditure of work on the part of the external stresses.[23] This contribution to the shear strength is probably not a major one: The total energy of such defects represents only a small part of the stored energy of the cold work, which in turn is only a small part of the work of plastic deformation.

There are still other contributions to the shear strength in work hardening. Our sketchy discussion cannot do full justice to this many-faceted subject, and the reader is referred to the literature[24] for further study.

Yield strength of single crystals

Although the elastic energy of a dislocation outside its very core is insensitive to its location in the lattice, the energy of the core itself must be a function of the nature of bonding of the crystal and of its position in the lattice (Fig. 9.6). Because of the periodicity of a lattice, the core energy is a periodic function of the lattice identity distance and gives rise to a force resisting the motion of a dislocation. The initial yield strength of pure crystals with strong directional bonds, such as silicon, or with ionic bonds and even of the body-centered cubic transition metals appears to be due to this cause.

The yield strength of pure close-packed metal crystals, such as copper, in single slip orientations appears to be governed by the rate of energy storage. In such single crystals yielding is accompanied by a very substantial rate of generation of new lengths of dislocations —having approximately a free energy per unit length as given by Eq. 9.16. Thus, at yielding, the rate of work done by the external forces must at least equal the rate of storage of internal free energy; i.e.,

$$\tau \, d\gamma = U \left(\frac{d\Lambda}{d\gamma} \right)_{\gamma = \gamma_0} d\gamma, \qquad (9.28)$$

where $(d\Lambda/d\gamma)_{\gamma = \gamma_0}$ is the initial rate of generation of dislocation lengths per cubic centimeter per shear-strain increment at yielding. Thus

$$\tau = U \left(\frac{d\Lambda}{d\gamma} \right)_{\gamma = \gamma_0}. \qquad (9.29)$$

For copper, $U \cong 1.5 \times 10^{-4}$ erg/cm, and the measured values of $(d\Lambda/d\gamma)_{\gamma = \gamma_0} \cong 3 \times 10^{10}$ cm/cm³,[25] giving for the initial shear strength a value of 45 g/mm², the required magnitude of actually observed values.

Fracture

The deformation process in tension is normally terminated by fracture. Most often this interrupts prematurely a hardening or forming process or, in some structural applications, intervenes before local stress concentrations can be leveled down by plastic deformation. In the first instance fracture can be frustrating, in the second, catastrophic.

The fracture of brittle materials is conceptually the simplest of all; it has been understood for the past forty years. Griffith[26] proposed, and experimentally showed, that a glassy material cannot exhibit its ideal strength of about a tenth of the Young's modulus, because it normally contains many surface defects that concentrate the nominal stress hundred- to thousandfold (Fig. 9.20).[27]

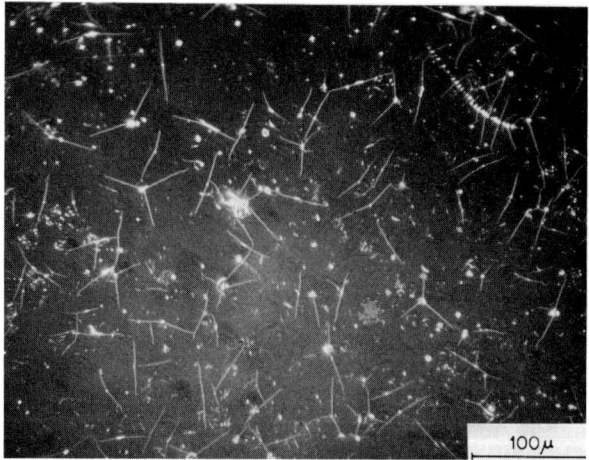

Fig. 9.20. Strength-impairing cracks on the surface of Pyrex glass revealed by a sodium vapor etch that initiates the star-shaped fissures from them.[27]

Fracture in a plastically deforming metal, on the other hand, is much more complicated. In most practical cases, commercially useful metals have weak interfaces between precipitates and the matrix material, slag pockets, and the like. As demonstrated by Puttick,[28] cracks may arise at such interfaces due to stress concentrations resulting from inhomogeneities.

---

[23] P. B. Hirsch and D. H. Warrington, *Phil. Mag.* [8] 6, 735, (1961).

[24] L. M. Clarebrough and M. E. Hargreaves, *Progr. Metal Physics.* 8, 1 (1959).

[25] M. J. Hordon, *Acta Met.* 10, 999 (1962).

[26] A. A. Griffith, *Phil. Trans. Roy. Soc. (London)* A221, 163, (1920).

[27] A. S. Argon, *Proc. Roy. Soc. (London)* A250, 472 (1959).

[28] K. E. Puttick, *Phil. Mag.* [8] 4, 964 (1959).

They lead to the formation of embryonic holes which may enlarge under further deformation, primarily on account of the hydrostatic tensile stresses present in any type of loading except pure shear.[29] Eventually such holes are bridged as the ligaments between them rupture; the effect of the discontinuity as a strain concentration then becomes too severe and leads to greatly accelerated plastic deformation and shearing off. This produces the familiar cup-and-cone fracture in the tension experiment (Fig. 9.21).

Fig. 9.21. Double-cup ductile fracture in a strained specimen of high-conductivity copper.

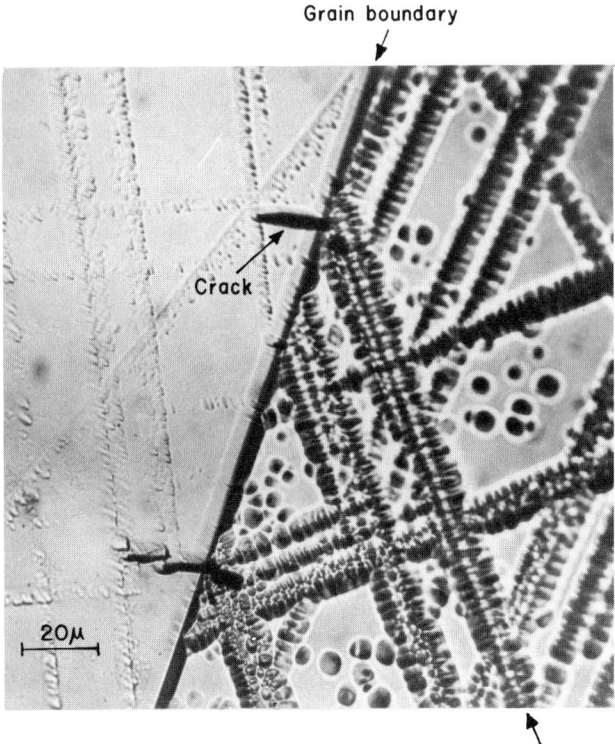

Fig. 9.22. Formation of cracks at the boundary in a bicrystal by the dislocation pileup mechanism of Zener.[30]

In pure materials, where such relatively trivial surfaces of weakness are eliminated and all hardening agents are kept in solution, other more inherent mechanisms may become limiting. Thus, for instance, as conceived by Zener,[30] if the progress of a set of edge dislocations on a slip plane is obstructed by a hard obstacle such as a grain boundary, their tensile stresses can reinforce each other and concentrate the applied shear stress in direct proportion to the number of dislocations in the pileup. This may nucleate a crack if the tensile stress at the tip of the pileup can be amplified to the level of the ideal tensile strength, provided the stress concentration cannot be relieved by deformation in the immediate vicinity. Johnston, Stokes, and Li[31] observed cracks nucleated by this mechanism at the boundary of a bicrystal of magnesium oxide (Fig. 9.22).

In hexagonal close-packed crystals where the base plane is the only good slip plane, slip dislocations of one type on parallel slip planes tend to attract each other to form a dislocation boundary that produces a lattice tilt between the two sides of the crystal it separates. Such so-called tilt boundaries, if abruptly terminated as a

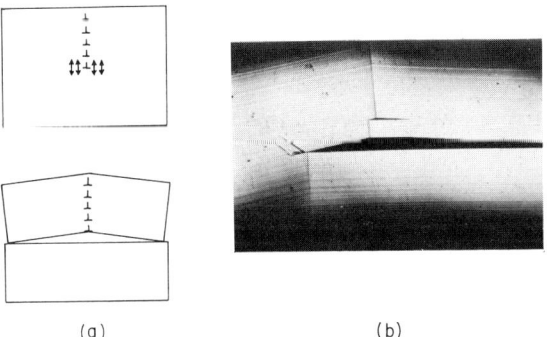

Fig. 9.23. (a) Production of a crack by the formation of a finite tilt boundary in inhomogeneous deformation;[3] (b) a crack of this type observed in a zinc crystal.[32]

result of inhomogeneous deformation, can lead to crack nucleation (Fig. 9.23).[32] Cracks of this type have been observed in crystals of many hexagonal metals.

In body-centered cubic metals that can undergo

[29] F. A. McClintock, in press.
[30] C. Zener, *Fracturing in Metals*, American Society for Metals, Cleveland, 1949, p. 3.
[31] T. L. Johnston, R. J. Stokes, and C. H. Li, *Phil. Mag.* [8] 7, 23 (1962).
[32] J. J. Gilman *Trans. AIME*, Metals Branch, 200, 621 (1954).

twinning at low temperatures in preference to slip, the intersection of two twin lamellae with noncollinear twinning directions can set up very severe stress concentrations, primarily because of the large twinning shear strain in this particular lattice. Once again, if such stress concentrations cannot be substantially relieved by slip deformation inside or around the twins, cracks can be nucleated. This is shown in Fig. 9.24 for

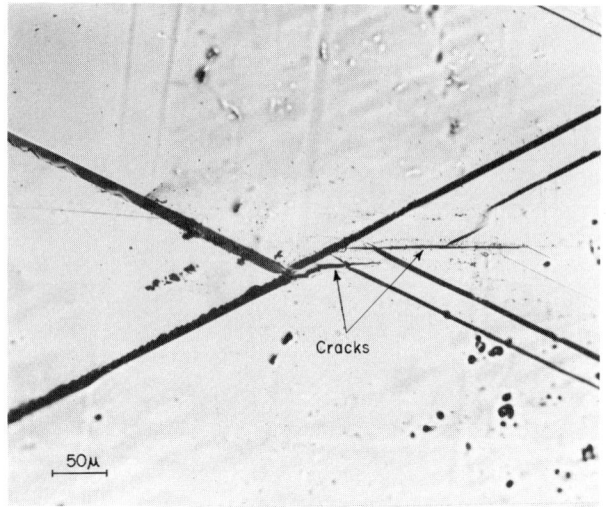

Fig. 9.24. Formation of cracks by twin band intersections in a single crystal of magnet iron.[33]

the case of a crystal of magnet iron containing 3 percent silicon.[33]

In pure materials at temperatures where the stress concentrations due to pileups, twin intersections, etc., can be relieved by slip processes, fracture may not occur, and the part may instead rupture by thinning down to a point or knife-edge (Fig. 9.25).

Mechanisms of crack nucleation by plastic deformation are of great interest, since they turn an inherently continuous material into a discontinuous one by such cracks. Under certain conditions of low temperature and high strain rate the presence of cracks can lead to brittle crack propagation and brittle behavior in a potentially ductile material. The ductile-to-brittle transition of the mode of fracture in mild steel is the best known example.[34]

The subject of fracture is a broad one to which the

[33] D. Hull, *Acta Met. 8*, 11 (1960).
[34] A. N. Stroh, *Advances in Physics 6*, 418 (1957).

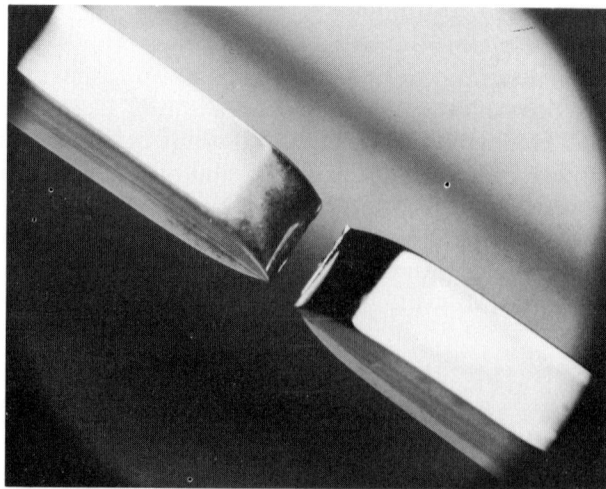

Fig. 9.25. A tungsten single crystal ruptured at room temperature in tension by thinning down to a knife-edge.

sketchy coverage given here does no justice. For further involvement the interested reader is referred to proceedings of fracture conferences which are now held almost yearly.[35]

Conclusions

As the detailed understanding of the mechanisms of deformation and fracture becomes more complete, it will be increasingly possible to tailor materials for specific uses.

Some of the present successes can serve as a guide of what can be expected in the future: the "Chemcor" process of Corning of setting up extremely high compressive stresses in thin surface layers by ion-exchange techniques, imparting to ordinary glass an extreme resistance to mechanical damage; the production of composite materials of hard phases held together by deformable matrices having both high strength and remarkable ductility; the production of metals of very high purity, alleviating the low-temperature brittleness of some of them; the production of polymers with outstanding strength and wear properties; etc.

The possibilities in the future are many and are only limited by the imagination and ingenuity of the engineer.

[35] *Fracture of Solids*, D. C. Drucker and J. J. Gilman, Eds., Interscience Publishers, New York, 1963.

10 · STRUCTURE IN LIVING SYSTEMS*

Francis O. Schmitt

Introduction — Self-replicating polymers: a monumental step in evolutionary advance — The proteins — Molecular assemblies: an essential structural hierarchy — Cell membranes — The molecule ↔ cell, component ↔ system reciprocal control — The molecule ↔ neuron ↔ brain ↔ mind problem

Introduction

Life, as manifested on this planet, requires—in addition to solar energy, water, minerals, and low-weight organic metabolites—a complex array of high-molecular-weight constituents, assemblies, and hierarchies of structures at the colloidal level of size. Endowed with high specificity and interacting with still only dimly perceived perfection of spatial and temporal ordering, these molecular and supramolecular systems structure the cells and tissues of living organisms. In this chapter we shall outline some of the facts, discovered mostly since World War II, about the molecular basis of fundamental life processes.

Through the application of physical and chemical principles and techniques, a body of knowledge, designated as "molecular biology," has been accumulated. The development of subdisciplines, such as molecular genetics, molecular immunology, and molecular neurology, exemplifies the necessity and fruitfulness of interdisciplinary research.[1,2]

Self-replicating polymers: a monumental step in evolutionary advance

To encode information that can be transferred biochemically so as to direct the synthesis of the body's constituents, each at its appropriate place and time in development, and to pass such information on to successive generations require a molecular mechanism for storing, transferring, and retrieving an enormous amount of data. A great step forward in evolution was achieved with the development of large copolymers of purine and pyrimidine bases, two of each kind, giving rise to the nucleic acid polymer deoxyribonucleic acid (DNA). The material was called nucleic acid because it was thought to occur only in the cell nucleus.

---

* Acknowledgment is gratefully given for grants from the National Institute of Neurological Diseases and Blindness (NB-00024-15) and the National Institute of General Medical Sciences (GM-10211-03) of the National Institutes of Health, U. S. Public Health Service; from the National Aeronautics and Space Administration (NsG-462); for contracts between the Office of Naval Research, Department of the Navy, and the Massachusetts Institute of Technology (Nonr-1841(27) and Nonr(G)-00089-64); and for support from The Rogosin Foundation; Neurosciences Research Foundation, Inc.; The Louis and Eugenie Marron Foundation; and from the Trustees under the Wills of Charles A. and Marjorie King.

[1] *Biophysical Science—A Study Program*, J. L. Oncley, F. O. Schmitt, R. C. Williams, M. D. Rosenberg, and R. H. Bolt, Eds., John Wiley and Sons, New York, 1959.

[2] F. O. Schmitt, *Biophysics, Wet and Dry*. First Klopsteg Lecture, Technological Institute, Northwestern University, Evanston, Ill., 1962.

Later another kind of nucleic acid, ribonucleic acid (RNA), differing from DNA by only one atom in the sugar residue, was found in both the nucleus and the cytoplasm of the cell.

With a molecular alphabet of four letters, the information-encoding capacity of DNA depends on the length of the polymer and on the monomer sequence. Such polymers must be capable of precise replication within the primeval organic aggregates which, at that stage of evolution, hardly merit the term "cell." With copolymers having n monomers of four types, the variants possible equal 4^n. The DNA of some organisms contains more than 10^5 monomers and hence has a theoretical capacity to form $4^{100,000}$ variants. On such encoding capacity rest the heredity, the development, and probably even the mental life of man.

Proof of DNA's role in molecular genetics and in biological processes of self-replication is exemplified in the viral infection of bacteria. Infection occurs when the DNA of the virus (bacteriophage) is injected into the victimized bacterial host through the tail canal of the 'phage (Fig. 10.1).[3] After injection, the protein

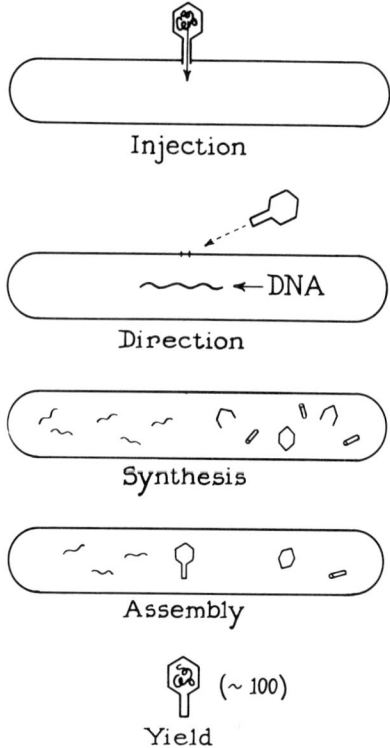

Fig. 10.1. Bacterial virus reproduction. (After Horsfall.[3])

"clothing" of the virus is discarded. The injected DNA takes command of the biochemical economy of

[3] F. L. Horsfall, Jr., *Proc. Am. Phil. Soc. 102*, 442 (1958).

the infected cell and directs its synthesis so as to replicate the invading DNA code and also the viral protein "body." Thousands of new viruses may thus be synthesized from a single infecting virus; the cell disintegrates and releases the newly formed viruses. The fact that the DNA exists as a single thread is illustrated dramatically in Fig. 10.2, an electron mi-

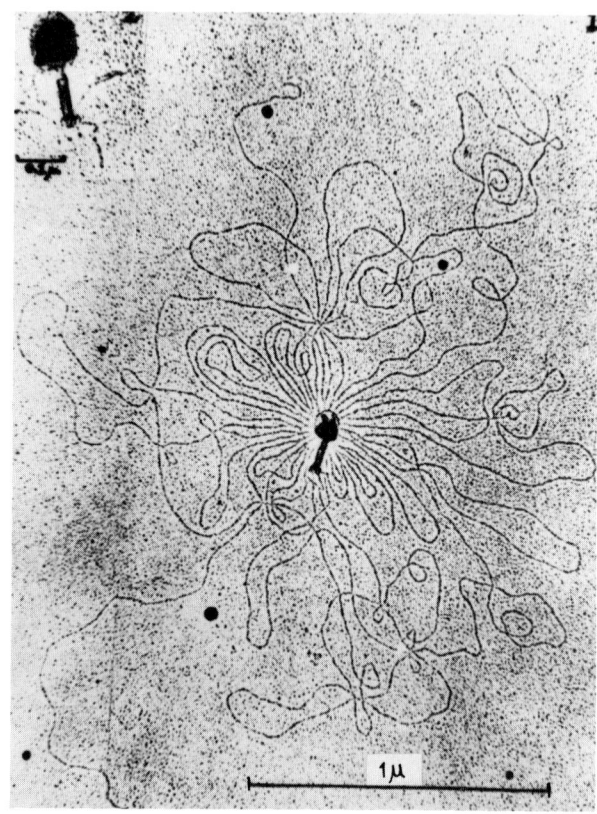

Fig. 10.2. Electron micrograph of continuous DNA thread liberated by osmotic shock from a single T_2-bacteriophage. (From Kleinschmidt et al.[4])

crograph showing the DNA liberated by osmotic shock from a single T_2-bacteriophage.[4]

DNA was proved to be a double-strand helix by the crystallographic studies of Watson and Crick,[5] for which they were awarded the Nobel Prize. The complementary pairs of purine bases adenine (A) and guanine (G) and of pyrimidine bases cytosine (C) and thymine (T) hydrogen-bond to each other in double strands, A with T, and C with G (Figs. 10.3[6] and 10.4[7]). In replication, after separation of the strand pairs,

[4] A. D. Kleinschmidt, D. Lang, D. Jacherts, and R. K. Zahn, *Biochim. Biophys. Acta 61*, 857 (1962).
[5] J. D. Watson and F. H. C. Crick, *Nature 171*, 964 (1953).
[6] I. Menzies, *Boston Globe*, May 12 and 19, 1963.
[7] M. Calvin and G. J. Calvin, *Proc. Am. Phil. Soc. 108*, 73 (1964).

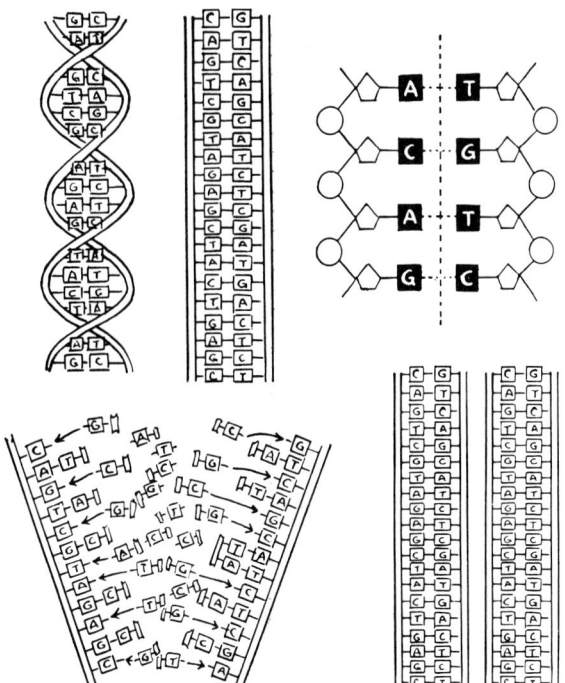

Fig. 10.3. Structural model of DNA and its replication from monomer pool. (Courtesy of the *Boston Globe*.[6])

purine and pyrimidine bases from the pool in the cell's nucleus attach to the single-strand template and are covalently bonded under the influence of specific synthesizing enzymes.

Particular contiguous regions along the DNA correspond to particular genes that direct cellular biosynthesis through intermediary polymers closely related to DNA, the *messenger* RNA. For such readout of the genetic code, the DNA double-helix must open

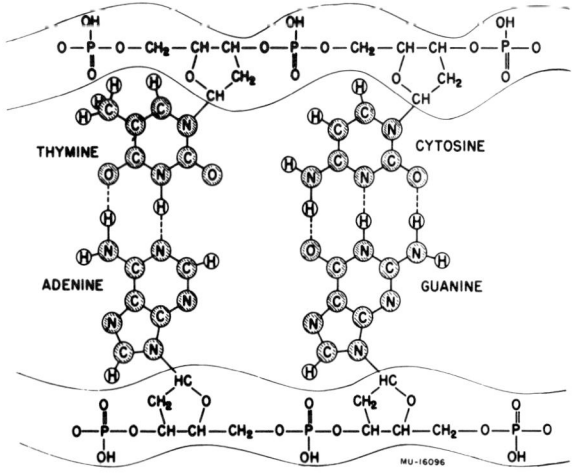

Fig. 10.4. Pairing of purine and pyrimidine bases by hydrogen bonding across polynucleotide chains of DNA. (From Calvin.[7])

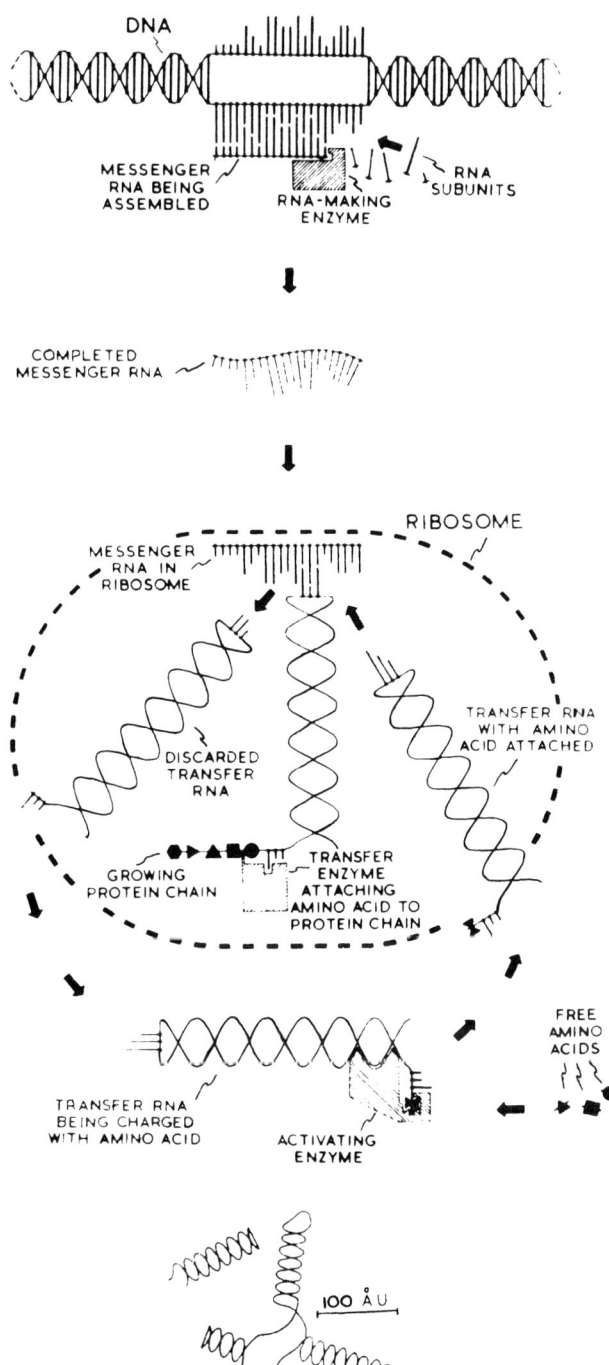

Fig. 10.5. Role of DNA, messenger RNA, transfer RNA, and ribosomes in protein biosynthesis. (From Spencer.[8])

(Fig. 10.5),[8] and the mononucleotides in the nuclear pool form upon complementary base pairs to produce single strands of RNA. In RNA, uracil (U) replaces T as the pyrimidine, hydrogen-bonding partner of A. The single-strand messenger RNA, synthesized on a

[8] M. Spencer, *New Scientist* **16**, 554 (Fig. 3) (1962).

particular portion of the DNA as template (molecular weight 10^5 to 10^6), directs the synthesis of protein in both nucleus and cytoplasm in combination with the RNA-ribosome complex ("polysomes"); it assembles the amino acids in a sequence conforming with the sequence of bases in RNA. The translation process, by which a sequence of 4 bases dictates a permuted sequence of 20 amino acids, occurs through mediation of *transfer* RNA of low (ca. 25,000) molecular weight (Fig. 10.6). A sequence of 3 bases, a so-called "codon,"

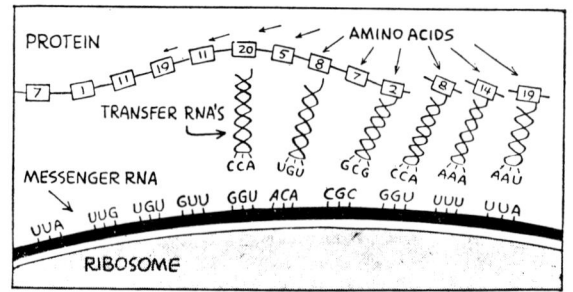

Fig. 10.6. Role of messenger RNA codons, transfer RNA, and ribosomes in the synthesis of protein chains. (Courtesy of the *Boston Globe*.[6])

dictates the selection of 1 specific amino acid. The *ribosomes*, dark-staining granules seen in the electron microscope and frequently attached to the cytoplasmic membranes, are especially abundant in actively synthesizing cells.

This topology facilitates passage of the synthesized product out of the cell through the membrane of the endoplasmic reticulum. Through low-molecular-weight transfer-RNA molecules, each bearing at one end a specific amino acid and at the other a triplet of purine or pyrimidine bases, the message carried by the messenger RNA is read out. The triplet code has been partly deciphered; in many cases it is now known which triplet codes which amino acid. Through attachment by complementary base pairs of transfer RNA upon messenger RNA, the amino acids bound at the ends of the transfer RNA are brought together. Action of synthesizing enzymes joins the amino acids to form a polypeptide (protein) chain (cf. Fig. 10.6)[6]; the transfer-RNA molecules are liberated for recycling. The coded transfer of information seems highly precise, and the sequence of amino acids in the synthesized protein remains constant from one molecule to the next.

The proteins

The *primary* polypeptide chain, synthesized by the mechanism already described, because of chemical properties of its main chain and side chains, achieves minimal energy by assuming higher-order configurations. The *secondary* conformation is in some cases uniformly helical; such proteins are fibrous. The alpha helix, characteristic of muscle, keratin, and fibrin, and the triple helix, characteristic of the collagen class, are shown in Fig. 10.7.

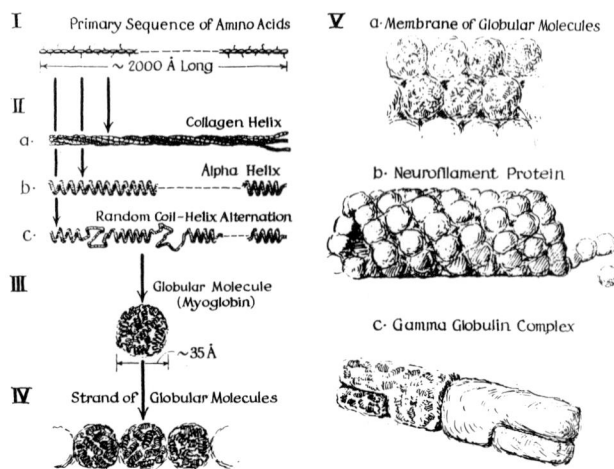

Fig. 10.7. Formation of conformational levels of protein molecules from the genetically determined primary sequence of amino acids.

For certain sequences of amino acids in the primary chain, helically ordered and random coil (disordered) regions form alternately along the chain. The secondary configuration achieves stability only with the formation of a *tertiary* conformation in which the long chain is folded into a compact globular molecule. The internal structure of the globular tertiary configuration, nevertheless, is also precisely determined by the primary sequence of amino acids. It is possible to determine the structure by X-ray crystallographic analysis. The monumental work of Kendrew and Perutz (for which they have been awarded the Nobel Prize) reveals the structure of the myoglobin[9] and the hemoglobin molecules, respectively.

Enzymes usually occur as globular molecules with highly specific secondary and tertiary conformation. The enzymatically active site may be determined by a few amino acid residues in the same or in contiguous folds of the chain. Sequence analysis, together with crystallographic and biochemical data, permits accurate determination of molecular configuration and enzymatically active sites. Figure 10.8[10] shows the sequence and tertiary structure characteristic of the enzyme ribonuclease, which depolymerizes RNA. Figure 10.7c depicts the structure of gamma globulin, which rep-

[9] J. C. Kendrew, *Sci. Am.* **205**, 98 (1961).
[10] W. H. Stein, *Federation Proc.* **23**, No. 3, Pt. I, 599 (1964).

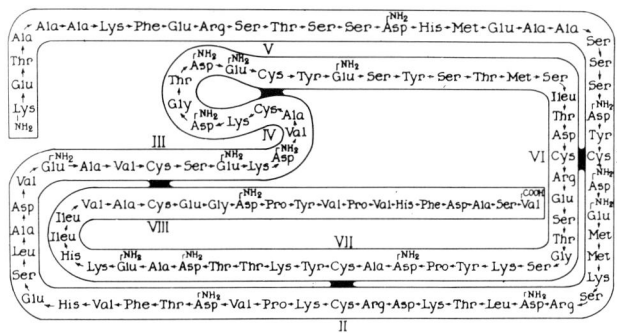

Fig. 10.8. Sequence and tertiary conformation of ribonuclease. (From Stein.[10])

resents the antibody molecules of mammalian immunocytes. Gamma-globulin molecules are composed of two heavy and two light subunits bonded by disulfide and secondary linkages; the four subunits, under appropriate conditions, can be dissociated reversibly. The immunological specificity is apparently not lost after reversible dissociation and reassociation.

Molecules of a single species may combine on a *quaternary* conformation level. This process is illustrated in Fig. 10.7 as a strand of globular molecules. Such specific aggregation can give rise to structures such as to the neurofilament protein or the tobacco mosaic virus protein, filaments 100 to 200 Å in width. Still another type of aggregation may occur in the formation of two-dimensional membranes through interbonding of component globular molecules at *several* sites on the surface of the globular molecules. A single amino acid change in a protein may result in a change in conformation, and this in turn may produce drastic physiological or pathological consequences. One such single amino acid change in hemoglobin molecules[11] produces strands of altered molecules; this quasi-crystalline orientation causes the red blood corpuscle to assume the elongate form characteristic of sickle-cell anemia, which Pauling[12] calls a "molecular disease."

It is becoming increasingly evident that conformational changes in molecules and in molecular interaction due to infinitesimal alterations, frequently in single amino acids or upon the side chains of single amino acids, may effect significant physiological changes, not merely at the site of the change but also, through an allosteric effect, at other regions of the molecule. Configurational changes may thus underlie vital mechanical or electrical phenomena in the cell. They may also serve information transfer in the brain, hence be significant for understanding the physical basis of memory, learning, and other mental processes.

Contractile tissue, such as skeletal, visceral, and cardiac muscle, is composed of a parallel alignment of highly asymmetrical myosin threads, ca. 100 Å in diameter, interspersed with thinner (ca. 40 Å) strands of actin (Fig. 10.9).[13] The former have an α-helix

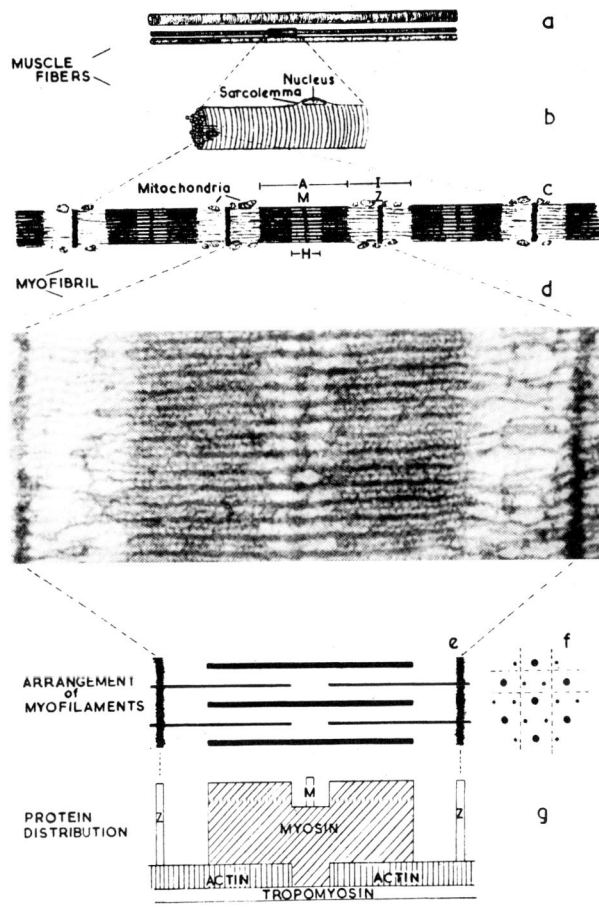

Fig. 10.9. Muscle structure at several levels of size. (From Finean, J. B., *Chemical Ultrastructure in Living Tissues*, 1961. Courtesy of Charles C Thomas, Publisher, Springfield, Ill.)

tertiary conformation; the latter are globular molecules joined in quaternary fibrous strands. Changes in interaction between myosin and actin are thought to underlie shortening and tension production in muscle (Fig. 10.10).[14] Adenosine triphosphate (ATP), produced by muscle mitochondria, decreases the interaction, while splitting of ATP by an enzyme situated on the myosin leads to increased interaction.

One typical representation of fibrous tissue proteins

[11] V. M. Ingram, *Hemoglobin and its Abnormalities*, Charles C Thomas, Springfield, Ill., 1961.

[12] L. Pauling, *Harvey Lectures*, Ser. 49, 1953–1954, Academic Press, New York, 1954, pp. 216–241.

[13] J. B. Finean, *Chemical Ultrastructure in Living Tissues*, Charles C Thomas, Springfield, Ill., 1961.

[14] P. Stubbs, *New Scientist* 19, 602 (Fig. 2) (1963).

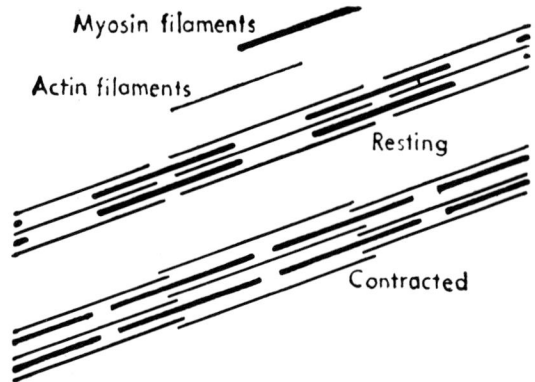

Fig. 10.10. Myosin-actin interaction in muscle contraction, showing how the two kinds of muscle filament slide between one another during contraction. (After Stubbs.[14])

that bestow mechanical properties is the *collagen* class. Comprising about 30 percent of the body's protein, collagen is the protein characteristic of skin, tendons, bone, and connective tissue generally. The macromolecules of the collagen precursor, *tropocollagen*, are very asymmetric (28×2800 Å in size) as seen directly with the electron microscope.[15] Such elongate macromolecules, synthesized in the connective tissue cells, appear extracellularly as monomer tropocollagen dissolved in the intercellular ground substance (a complex colloidal system containing polysaccharides, proteins, salts, and other substances) of the connective tissue. The monomers aggregate in characteristic, linearly quarter-staggered arrays to form fibers of collagen. This native fiber shows an axial-repeat pattern of about 700 Å (i.e., one quarter of the molecular length) and the wide-angle X-ray-diffraction pattern

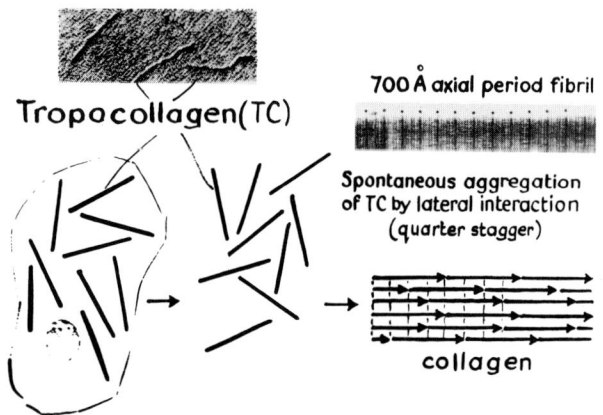

Fig. 10.11. Role of tropocollagen macromolecules in the formation of collagen fibers.

of the three-strand helix that comprises the elongate tropocollagen monomeric unit (Fig. 10.11). The quar-

[15] C. E. Hall and P. Doty, *J. Am. Chem. Soc.* **80**, 1269 (1958).

ter-staggered configuration (which might be called a quaternary level of conformation) is spontaneously assumed when dissolved tropocollagen macromolecules are brought into an appropriate chemical environment experimentally or naturally in the course of development or wound repair.[16–18] This particular intermolecular relationship is stabilized by covalent cross bonds as the collagenous tissue matures to form insoluble fibers. A full understanding of the molecular biology of collagen is important not only because in its own right it comprises a major class of proteins, but also because it bears an intimate relationship with rheumatoid and certain kinds of cardiovascular diseases and with aging processes.

Molecular assemblies: an essential structural hierarchy

Another important evolutionary advance has been the emergence of specific assemblies of macromolecules, particularly enzymes.[19,20] Certain enzymes, composed of a number of constituent protein and nonprotein subunits, are now being examined in the electron microscope. An excellent study of this kind describes the pyruvate dehydrogenase complex of *Escherichia coli*.[21]

Biochemical investigations show that in certain fundamental cellular processes (e.g., the synthesis of "energy-rich" molecules of ATP) a number of enzymes act seriatim on the carbohydrate substrate molecules to complete a cycle (e.g., the Krebs cycle) in oxidative phosphorylation and in electron transport via the cytochrome carriers; the final outcome is reduction of oxygen and concomitant production of ATP (Fig. 10.12).[22] If the enzyme catalyzing the various steps in such cyclic and linear sequential processes were freely dissolved and if it were necessary between each step for the substrate to diffuse in the cytoplasm until

[16] F. O. Schmitt, *Bull. New York Acad. Med.*, 2nd Ser., **36**, 725 (1960).

[17] A. J. Hodge and F. O. Schmitt, "The Tropocollagen Macromolecule and its Properties of Ordered Interaction," *Macromolecular Complexes*, M. V. Edds, Jr., Ed., Ronald Press, New York, 1961, pp. 19–51.

[18] F. O. Schmitt, *Federation Proc.* **23**, No. 3, Pt. I, 618 (1964).

[19] F. O. Schmitt, "Molecule-cell, component-system reciprocal control as exemplified in psychophysical research." Robert A. Welch Foundation Conference on Chemical Research. V. *Molecular Structure and Biochemical Reactions*, W. O. Milligan, Ed., R. A. Welch Foundation, Houston Texas, 1961, pp. 33–39.

[20] F. O. Schmitt, *Dev. Biol.* **7**, 546 (1963).

[21] H. Fernández-Morán, L. J. Reed, M. Koike, and C. R. Willms, *Science* **145**, 930 (1964).

[22] A. L. Lehninger, "Evaluative Summary: NRP work session on cell membranes," *Bull. Neurosci. Res. Program* **2**, No. 2 (March–April), 5 (1964).

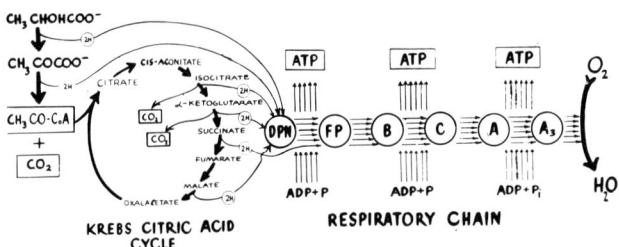

Fig. 10.12. Assemblies of enzymes and carriers in energy mobilization. (After Lehninger.[22])

collision occurred with the next enzyme in the chain, the time for completion of cycles and sequences would be far too great to be compatible with organismic function. The kinetics of the reaction would be markedly accelerated if all the enzymes participating in particular cyclic and sequential processes were organized as solid-state devices with enzymatically active groups exposed in topochemically appropriate fashion.

Assemblies of a dozen or more enzymes are large enough to be seen in the electron microscope. Under conditions of negative staining, the membranes of mitochondria appear covered with mushroomlike structures (Fig. 10.13)[23,24] that may be the structural manifestation of the oxidative phosphorylation and of electron transport-controlling enzyme assemblies. Subsequent investigations[22,25] suggest that these structures may pre-exist within, rather than upon, the mitochondrial membranes, but their existence can hardly be questioned. It has been suggested that the sequence of cytochrome carriers in the electron-transport mechanism may be only partially located on these particles;[26] this controversy is still in flux.

The study of macromolecular assemblies might be termed *solid-state macromolecular biology*; its vast possibilities are only now becoming obvious.

Ribosomes, previously mentioned in connection with RNA-controlled biosynthesis of protein, are such assemblies, which again may group into polyribosomes (briefly polysomes), apparently functioning seriatim in the synthesis of large protein chains. The maximum molecular weight of protein synthesized on a single ribosome is believed to be ~5000. A large protein would therefore require a group of 20 to 30 ribosomes, possibly functioning as a unit. Ribosome grouping has been observed in the electron microscope.[27]

Another example of a macromolecular assembly is the *quantasome*, proposed by Calvin[7] as the unit structure of the chloroplast, containing all the molecular equipment needed for utilizing the energy of absorbed light quanta for photosynthesis. Figure 10.14 shows diagrammatically the components of such a unit.[28] Figure 10.15, an electron micrograph from the same investigator's laboratory, shows ordered arrays of quantasomes within a single granum of a spinach chloroplast lamella.[7]

Cell membranes

Membranes are thin layers, 50 to 100 Å thick, surrounding cells and cellular constituents and partitioning intracellular phases functionally as well as structurally.

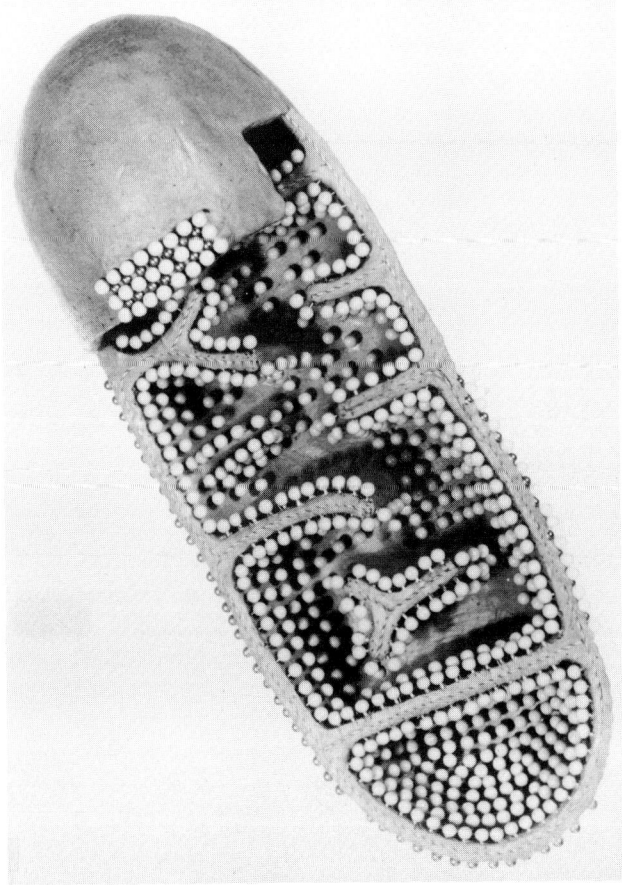

Fig. 10.13. Diagrammatic representation of enzyme assemblies on membranes of mitochondria subjected to "negative" staining. Precise relationship of particles to membranes before fixation and staining is unknown. (Courtesy of H. Fernández-Morán.)

[23] H. Fernández-Morán, T. Oda, P. V. Blair, and D. E. Green, *J. Cell Biol.* **22**, 63 (1964).
[24] H. Fernández-Morán, *J. Roy. Miscroscop. Soc.* **83**, 183 (1964).
[25] A. L. Lehninger, *The Mitochondrion*, W. A. Benjamin, Inc., New York, 1964.
[26] B. Chance, D. F. Parsons, and G. R. Williams, *Science* **143**, 136 (1964).
[27] J. R. Warner, A. Rich, and C. E. Hall, *ibid.* **138**, 1399 (1962).
[28] M. Calvin, "Energy Reception and Transfer in Photosynthesis," *Biophysical Science—A Study Program*, J. L. Oncley et al., Eds., John Wiley and Sons, New York, 1959, pp. 147–156.

The general ultrastructure of the typical membrane consists of bimolecular layers of mixed lipids, considered either as continuous layers, as in Robertson's[29] "unit membrane" theory (Fig. 10.16), or as a patchwork of discontinuities or micelles serving as a substructure for organized enzymes and other macromolecular effectors. Classically, the membranes surrounding the cell and the nucleus were thought to provide structural separation between cytoplasm and extracellular space and between cytoplasm and nucleoplasm, respectively. Electron-microscopic studies reveal pores in the nuclear membrane through which molecular traffic from nuclear via the pores of the nuclear membrane, contains the ribosomes and mitochondria. A second colloidal phase, relatively poor in solids and possibly connecting with extracellular space through evanescent pores in the outer cell membrane, is contained within the cytoplasmic membranes and called the *endoplasmic reticulum*. As illustrated in Fig. 10.17, "extracellular space" might be thought of as extending through the cytoplasm, via endoplasmic reticulum, all the way to the nuclear membrane, and "nuclear space" might be thought of as extending from the interior of the nucleus, via endoplasmic reticulum, all the way to the outer cell membrane. This diagram by Robertson,[30] admittedly rather extreme in its portrayal of a concept, also illustrates the author's view that many, if not all, types of cellular organelles, including mitochondria and Golgi apparatus as well as the more obvious endoplasmic reticulum and the cell and nuclear membranes, are clear DNA and RNA can flow to the cytoplasmic factories, the ribosomes or polysomes, and back to the DNA to establish feedback control. The cytoplasm, continuous with the nucleoplasm as a colloidal phase

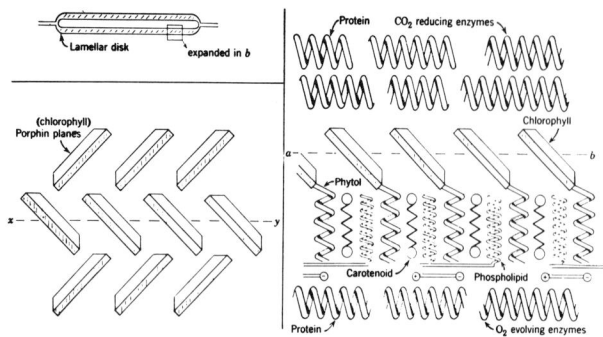

Fig. 10.14. Molecular componentry of photosynthetic mechanism. (From Calvin.[28])

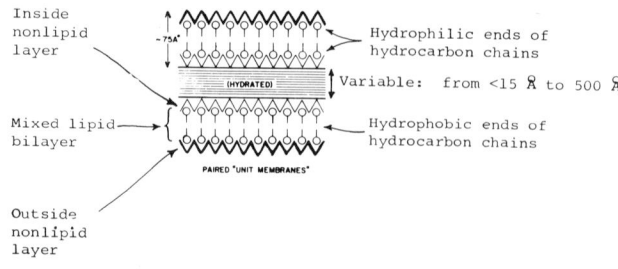

Fig. 10.16. J. D. Robertson's "unit membrane" concept. (From Lehninger.[22])

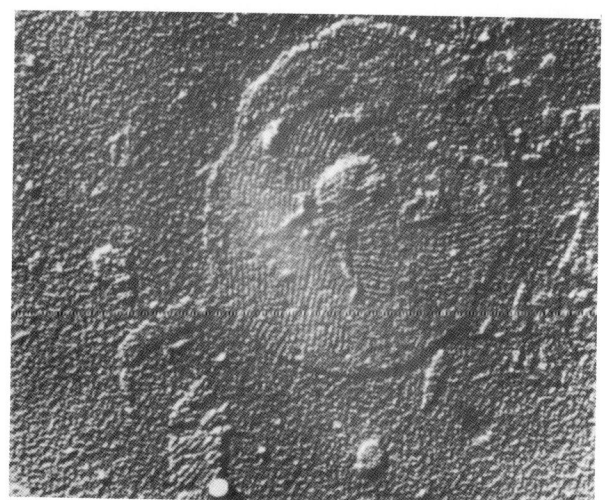

Fig. 10.15. Quantasome of spinach chloroplast. (From Calvin.[7])

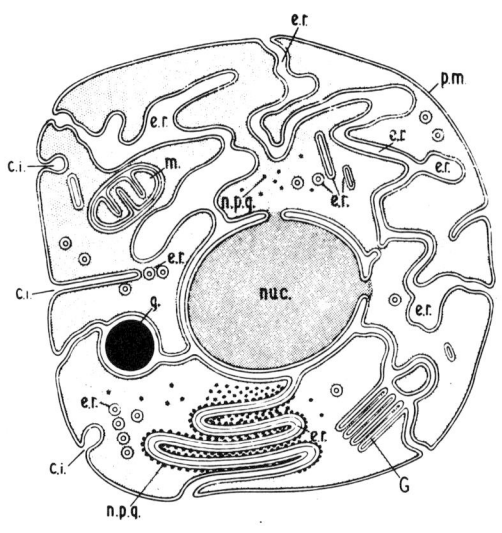

Fig. 10.17. Diagram of hypothetical cell structure illustrating origin of cell organelles from membrane precursor. (After J. D. Robertson.[30])

[29] J. D. Robertson, *Progr. Biophys.* 10, 343 (1960).

[30] J. D. Robertson, *Biochem. Soc. Symposium* 16, 3 (1959).

membrane derivatives arising from a common-type process of biogenesis and unfolding.

The mitochondrion is the energy producer of the cell, for it generates the ATP which, in the presence of appropriate enzymes, can liberate high-energy phosphate groups; scission of each of the two terminal phosphates provides ~10,000 calories of energy available for driving cellular processes, such as muscle contraction, biosynthetic, bioelectric, and other endergonic processes. The mitochondrion[22,25] requires much "floor space" for its enzymatic equipment that produces most, if not all, of the cell's requirement of ATP. As we have seen, this process involves certain cycles and sequences of enzymes and carriers acting in assemblies (cf. Fig. 10.12) that catalyze oxidative phosphorylation and cytochrome-mediated electron transport by which oxygen is reduced and ATP liberated. Lehninger estimates that membranes comprise some 60 percent of the mass of the "average" cell, while mitochondrial membranes represent 40 to 45 percent of the total cell-membrane material. Upon or within this membrane, chiefly as intramitochondrial "cristae," the enzyme assemblies are situated.

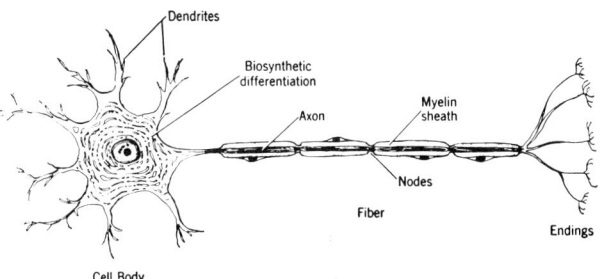

Fig. 10.18. Diagrammatic representation of a myelinated nerve fiber. Note cell body with dendrites and cytoplasmic organelles, axon, myelin sheath with Schwann cells and nodes of Ranvier, and terminal twigs making synaptic connection with peripheral tissues. (From Schmitt.[31])

to ~100 m/sec, is made possible in evolution by insulating the conducting axon with fatlike (lipid) material over its entire extent except at small nodal regions, ~1 to 5μ long, at the junction of the myelin-producing Schwann cells that, in tandem array, cover the axon. At the nodal discontinuities, where the axon surface membrane is not completely covered by lipid myelin layers, current can flow from inner, axo-

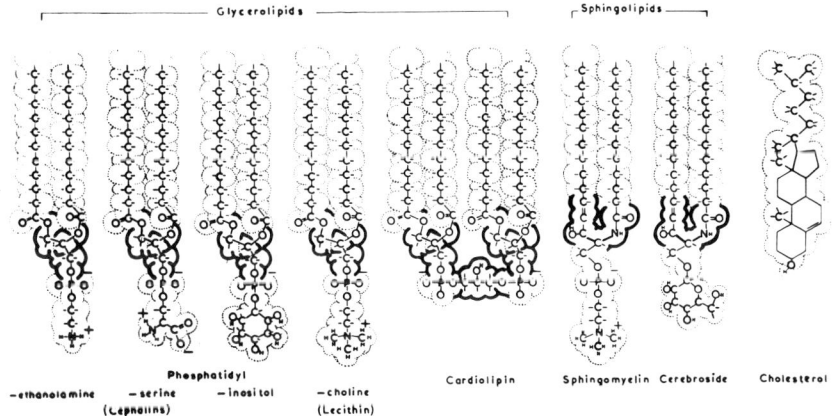

Fig. 10.19. Chemical formulae of structurally important lipids adapted to the approximate spatial outlines of the molecular configurations. (From Finean, J. B., *Chemical Ultrastructure in Living Tissues*, 1961. Courtesy of Charles C Thomas, Publisher, Springfield, Ill.)

Another example in which differentiation of membranous structure accomplishes an important physiological mission is the development of the myelin sheath around nerve fibers or axons (Fig. 10.18).[31] The nerve impulse, arriving from the synaptic junctions on the dendrites of the cell body, is transmitted over the axon to the axon terminals either at synapses with other neurons or at endings upon muscles, glands, or blood vessels. Rapidity of impulse propagation, up

[31] F. O. Schmitt, "Molecular Organization of the Nerve Fiber," Ref. 28, p. 456.

plasmic fluid to outer, extra-axonal fluid, thereby producing a regenerative action current involving successive nodal membranes as the impulse passes down the axon.

Polarization-optical and X-ray-diffraction evidence has demonstrated that the highly birefringent myelin sheath is composed of concentric layers 170 to 190 Å thick; that the lipid molecules, as mixtures of several classes of lipid material, extend with their long paraffin chains perpendicular to the surfaces of the concentric membranes; and that between bimolecular layers of mixed lipid there are thin, possibly monomolecular, layers of nonlipid material, chiefly protein.[13,32] The nature of the protein is still not well known. The lipid types are shown in Fig. 10.19. When thin-sectioning techniques made it possible to examine structures like nerve myelin with the electron microscope, the predictions from polarization optics and X-ray diffraction were confirmed: The myelin was seen to consist of alternating dark- and light-staining layers, consistent with the presence of lipid bimolecular layers of mixed lipids and of protein, or other nonlipid material (Fig. 10.20).[33,34]

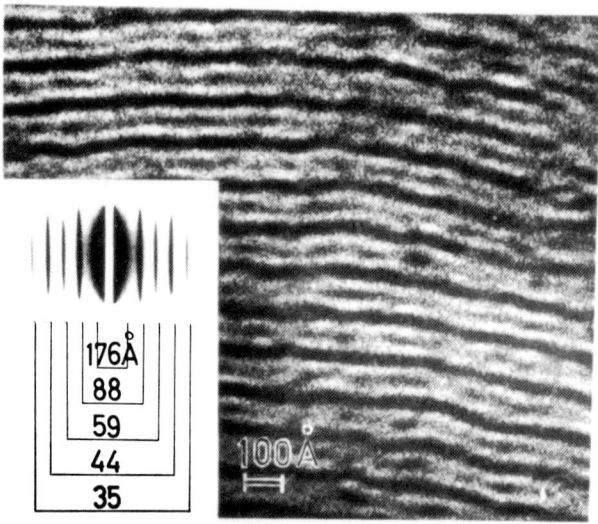

Fig. 10.20. Electron micrograph and small-angle X-ray pattern providing evidence of layered structure of nerve myelin. (From Fernández-Morán.[33])

It had long been known that myelin is produced through some kind of cooperative interaction between the naked axons, growing out from their cell bodies during development, and other, so-called Schwann cells migrating toward the outgrowing fibers from the crest region of the neural ectoderm. Geren[35] solved the structural problem of myelogenesis by showing that the Schwann cells, after attaching themselves seriatim along the outgrowing axon, wrap their rapidly growing surface membranes about the axon (Fig. 10.21). After a few to more than a hundred double cell membranes, depending on fiber type, have been wrapped about the axon, extrusion of cytoplasmic contents from between the layers occurs, permitting the latter to condense into compact myelin capable of showing five orders of the fundamental 170- to 190-Å repeating period in the small-angle X-ray diffraction pattern (cf. Fig. 10.20). A similar process has been demonstrated in the brain and the central nervous system, where the counterparts of the Schwann cells, represented by certain kinds of "glial" cells, perform the same function. The degree of their membrane production and wrapping around axons (sometimes several adjacent axons per glial cell) portrays in a fantastic manner the way in which the

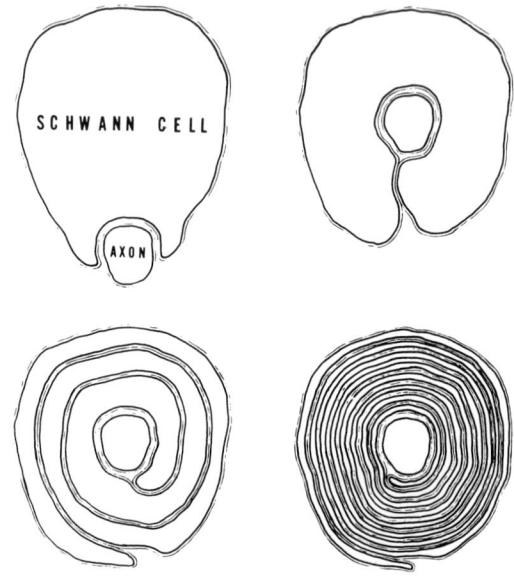

Fig. 10.21. Origin of myelin by wrapping of Schwann cell membrane about axon. (From Geren.[35])

biosynthetic processes of a cell type (in this case the spectacular production of outer cell membrane by the Schwann or glial cells) may be channeled to accomplish a developmental or physiological mission.

Muscle cells equally spectacularly produce elongate protein macromolecules of myosin and actin; mesodermal cells produce elongate tropocollagen macromolecules in enormous quantities in such processes as wound repair. Such specification of types of products synthesized by particular cell types is believed to result from readout of different portions of the genetic code of the chromosomal DNA.

Little is known about the enormously complex feedback processes that determine which portions of the DNA code shall be activated to accomplish specific missions by particular cells in particular parts of the body at any moment, from the fertilization of the egg cell at conception throughout the life of the organism. This important area of biology, remaining to be

---

[32] F. O. Schmitt, R. S. Bear, and K. J. Palmer, *J. Cell Comp. Physiol.* **18**, 31 (1941).

[33] H. Fernández-Morán. "Fine Structure of Biological Lamellar Systems," Ref. 28, p. 319.

[34] H. Fernández-Morán, "Lamellar Systems in Myelin and Photoreceptors as Revealed by High-Resolution Electron Microscopy," Ref. 17, pp. 113–159.

[35] B. B. Geren, *Exptl. Cell Res.* **7**, 558 (1954).

explored, must be further cultivated by biologists working at the cell, tissue, or organismic levels before it can be effectively harvested by molecular biologists.

The molecule ↔ cell, component ↔ system reciprocal control

In the foregoing descriptions, individual cellular components and organelles have been considered at or near the molecular level of size. The physiological role played by these components has been deduced frequently with the help of information obtained from partial systems or purified substances derived from cells. Large extrapolations and generalizations are made or implied by biochemists about function in the tissues of higher organisms, from experiments with viruses and micro-organisms. To what extent these generalizations will prove tenable remains for future work to prove.

Recent successes of molecular biology have accentuated the tendency of investigators to think that molecular types exert a sort of hierarchical chain of control in cell function. The DNA-RNA-protein trilogy illustrates such a chain of command. A similar "control" in biological systems is frequently imputed in the biomedical literature to very small amounts of certain highly active substances, such as hormones or pharmacodynamic agents. Preoccupation with the task of isolating and characterizing the biologically active molecules and molecular complexes tends to make investigators lose sight of the fact that the properties of life and living reside in the *system*—the cell or organism—rather than in the components. It is in the *ordered interaction*, in space and time, of the specific intracellular molecules and supramolecular complexes that life resides.[19] As pointed out by Weiss,[36] "'Controlling' molecules have themselves acquired their specific configurations, which are the key to their power of control by virtue of their membership in the population of an organized cell, hence are under 'cellular control.' . . . the cell is not just a playground for a few almighty masterminding molecules, but is a *system*, a hierarchically *ordered* system of mutually interdependent species of molecules, molecular groupings, and supramolecular entities, . . . life depends on the *order* of their interactions."

The molecule ↔ neuron ↔ brain ↔ mind problem

Neurons (cf. Fig. 10.18) conduct the transient electrochemical membrane process called the "impulse" from dendrite and cell body, via the axon, to axon terminal and across the synaptic junction to the dendritic receptors of the next neuron. This phenomenon brings information from sensory receptors, which transduce the sensed modalities into bioelectric impulses, via sensory neuronal nets, to appropriate regions of the brain where further routing occurs: either to motor effectors or to higher cortical centers where the information may eventually be stored as permanent memory "engrams." The physiological and eventually biochemical and biophysical processes that bring about such storage are still almost entirely unknown. Impulses arriving from sense organs may be switched to other circuits, which trigger recognition of the sensory information, involving retrieval of stored information and giving rise to cognitive processes. Long-term memory, hence also learning and other mental processes more complex than memory, must have correlates in enduring molecular structures having stabilities guaranteed by covalent bonds, as well as in action waves coursing in reverberating circuits over particular neuronal networks, however complex. Identification of these memory-encoding molecules or molecular systems is a primary task of the newly emerging discipline of molecular neurology. Possibly the memory encoding is to be sought in some manifestation of the DNA-RNA-protein system in the neuronal nets of the brain.[37-40] Efforts are being made to test this hypothesis by applying ingenious microanalytical tests of great sensitivity to individual cells obtained from brain regions involved in learning a task.[41,42] If the chemical composition of a substance such as RNA, which occupies a strategic place in the chain of biochemical command of biosynthetic systems, is altered by the learning experience, this may signify the establishment of a chemical correlate of memory in brain cells.

Proteins are, in general, the effectors of cell function, as is obvious in tissues such as muscle, glands, or connective tissue. It seems reasonable that proteins play a key role also in brain function. Yet little is known about the composition and properties of the proteins of neurons and their satellite cells, the glia in the central nervous system.

Investigations in this laboratory, extending over several decades, have isolated, identified, and partially

---

[36] P. Weiss, "From Cell to Molecule," *The Molecular Control of Cellular Activity*, J. M. Allen, Ed., McGraw-Hill Book Co., New York, 1962, p. 1.

[37] J. Gaito, *Psychol. Rev. 68*, 288 (1961).
[38] M. H. Briggs and G. B. Kitto, *ibid. 69*, 537 (1962).
[39] W. Dingman and M. B. Sporn, *Science 144*, 26 (1964).
[40] F. O. Schmitt, *Bull. Neurosci. Res. Program 2*, No. 3, 43 (1964).
[41] H. Hydén, "Biochemical and Functional Interplay between Neuron and Glia," *Recent Advances in Biological Psychiatry*, Vol. 6, J. Wortis, Ed., Plenum Press, New York, 1964, p. 31.
[42] H. Hydén, *Bull. Neurosci. Res. Program 2*, No. 3, 23 (1964).

characterized the principal protein of axoplasm, the contents of the nerve axon.[40,43,44] This is true at least of the axons of the nerve type having a diameter large enough to permit obtaining the protein uncontaminated by nonneural materials, e.g., the giant nerve fiber of the squid. To obtain quantities of axoplasm sufficient for isolation and characterization of the protein, we have utilized the 4- to 6-foot squid that abound in the Humboldt Current off the western coast of South America. Our laboratories are at the Chilean Marine Laboratory at Montemar, a suburb of Valparaiso.

This fibrous protein, present in all neurons, is called the neurofilament protein. Bundles of neurofilaments (neurofibrils) resolvable by the light microscope (Fig. 10.22)[45] course undivided down the axon to the ter-

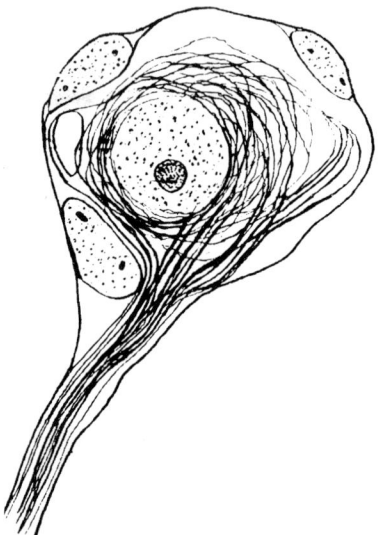

Fig. 10.22. Neurofibrils in living neuron of chick embryo in tissue culture. (From Weiss and Wang.[45])

minal twigs. They may be very long indeed in neurons that originate in the spinal cord in the regions of the small of the back and terminate in the toes—distances of a meter or more! A curious differentiation of neurofilaments of unknown function which take the form of a toroid[46] is shown in Fig. 10.23. As indicated in Figs. 10.7 and 10.24,[47] the neurofilament (width about 100 Å) is a conformational variant of the protein whose tertiary subunit molecules have a weight of 20,000 to

[43] P. F. Davison and E. W. Taylor, *J. Gen. Physiol.* 43, 810 (1960).

[44] F. O. Schmitt and P. F. Davison, "Biologie moléculaire des neurofilaments," *Actualités neurophysiologiques*, 3ième série, A.-M. Monnier, Ed., Masson et Cie, Paris, 1961, p. 355.

[45] P. Weiss and H. Wang, *Anat. Record* 67, 105 (1936).

[46] B. B. Boycott, E. G. Gray, and R. W. Guillery, *Proc. Roy. Soc. (London)* B154, 151 (1961).

[47] M. Maxfield, *J. Gen. Physiol.* 37, 201 (1953).

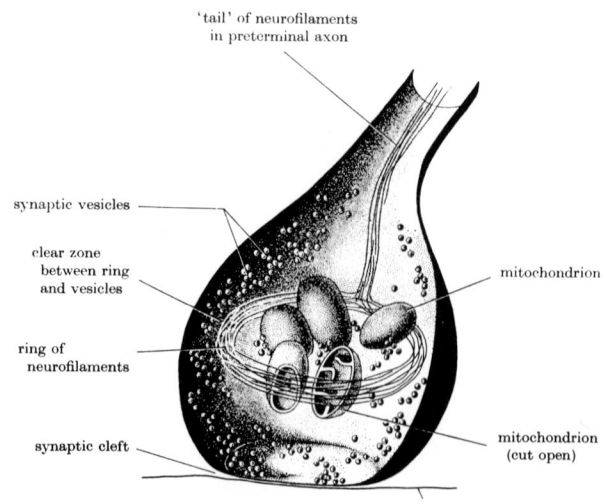

Fig. 10.23. Toroid of neurofilaments; schematic three-dimensional representation of axon presynaptic terminal. (From Boycott, Gray, and Guillery.[46])

30,000.[43,44] No physiological role has yet been demonstrated for this unique structure, nor indeed for any of the proteins of axoplasm, but recent immunoneurological experiments[48-50] suggest interesting possibil-

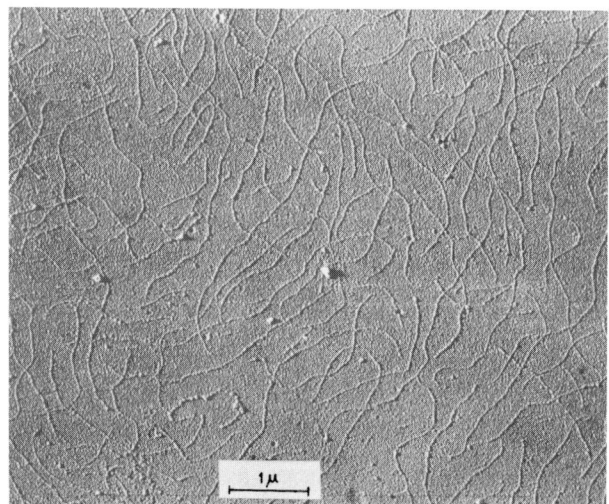

Fig. 10.24. Neurofilaments as seen in electron micrograph of purified axon fibrous protein. (From Maxfield.[47])

ities both in relation to conduction of the impulse in nerve fibers and to the establishment of patterns of neuronal circuits in the brain, on which may depend higher mental function.

[48] F. Huneeus-Cox, *Science* 143, 1036 (1964).

[49] F. O. Schmitt and P. F. Davison, *Ber. Bunsenges. physik. Chem.* 68, 887 (1964).

[50] L. Mihailović and B. D. Janković, *Nature* 192, 655 (1961).

However much success molecular neurologists may eventually achieve in laying a physical, molecular basis for the establishment of long-term memory or other mental processes, the entity known as "mind" has the logic, not of a group of individual components, but of a *system*.[19] Its functioning must be considered as a total interacting system, rather than as a set of particular molecules, molecular complexes, networks of neurons, or other partial systems. In the discovery of the true nature of this system lies a challenge far greater than that of deciphering the molecular componentry of brain cells and their circuits. It is a field the development of which requires and deserves the keenest interest and participation of many types of scientists: mathematicians, physicists, chemists, and engineers, as well as life scientists of all kinds.

11 · MAGNETIZATION PROCESS

David J. Epstein

Introduction — The fundamental equation — Magnetization of the infinite crystal — Domains and domain walls — Fine particles — Dynamic excitation — Domain-wall dynamics — Magnetic losses — Diffusion damping

Introduction

A systematic examination of the properties of ordered magnetic systems starts from fundamental atomic considerations and arrives at a description of the system in terms of field variables that characterize the bulk behavior. At the atomic beginnings we are in the realm of the *microscopic*, concerned with the origin of the basic magnetic moments and with the interaction among moments that leads to ordered magnetic states. The problems encountered are the same as those arising in atomic multiplet structure and the band theory of solids. At the terminus of the journey we are in the *macroscopic* world. Here we treat the magnetic material as completely continuous and describe its properties in terms of field vectors **M** (magnetization per unit volume), **B** (magnetic induction), and **H** (magnetic field), obeying the familiar relation (in the rationalized mks system)

$$\mathbf{B} = \mu_0(\mathbf{H} + \mathbf{M}), \quad (11.1)$$

with μ_0 the permeability of free space ($4\pi \times 10^{-7}$ henry/m).

Between micro- and macroscopic is an intermediate realm we might call *mesoscopic*. Here we are concerned with phenomena that at the smallest end of the size scale involve a magnetic moment and its immediate neighbors (a cluster extending over several atomic spacings) and at the upper end include systems of many moments covering a region several millimeters in extent. To understand the magnetization process, that is, the response of an ordered magnetic system to an applied magnetic field, we must focus attention primarily on the mesoscopic realm.

We shall formulate in semiclassical terms the fundamental mesoscopic equation that must be solved for a quantitative analysis of the magnetization process. A rigorous solution has been obtained only for a relatively few cases. Fortunately, many situations arise in which the exact solution can be safely bypassed and the analysis carried out in terms of "domain theory." We shall look into a number of magnetization problems, some treated exactly, others handled by the domain approximation.

The fundamental equation

The two basic sources of magnetic moments are circulation of electrons in orbits and magnetic moments associated with electron spins. In a solid the orbital moments are "quenched" by the electrostatic field of the crystal lattice, and it is mainly the resultant spin moment of the atoms that is left for cooperative interaction. If we designate the angular momentum of the resultant spin of the ith atom by $\mathbf{S}_i$, the atom carries a resultant magnetic moment

$$\mathbf{m}_i = \gamma \mathbf{S}_i. \quad (11.2)$$

The factor γ, the magnetomechanical ratio, has the value

$$\gamma = \frac{ge}{2m}, \quad (11.3)$$

with e the electronic charge, m the electron mass, and g the so-called Landé factor. Normally g has a value in the vicinity of 2. Since γ is a negative number, $\mathbf{m}_i$ is antiparallel to $\mathbf{S}_i$.

The isolated spin $\mathbf{S}_i$, subjected to an external magnetic field $\mathbf{H}_a$, is acted on by a torque

$$\mathbf{T}_a{}^i = \mathbf{m}_i \times \mu_0 \mathbf{H}_a = \gamma \mathbf{S}_i \times \mu_0 \mathbf{H}_a. \quad (11.4)$$

In a solid, as a member of a system of ordered magnetic moments, this same spin is acted on by a variety of additional torques.

The very fact that we can have ordered magnetic states suggests that there must be internal torques. They arise from quantum-mechanical spin interactions because certain wave-function symmetries must be imposed in order to satisfy the Pauli exclusion principle. These so-called exchange torques are of such short range that only nearest-neighbor interactions need be considered. The exchange torque on a spin $\mathbf{S}_i$ is given by

$$\mathbf{T}_E{}^i = 2J \mathbf{S}_i \times \sum_j \mathbf{S}_j, \quad (11.5)$$

where the sum on j is to be taken over the spins on atoms that are nearest neighbors of $\mathbf{i}$.[1] The strength of the interaction, given by the exchange integral J, is determined by carrying out an appropriate integration over wave functions of atom i and its neighbors. If J is positive, the torque favors parallel spin alignment. This is the situation in ferromagnetism.

To understand the significance of Eq. 11.5, we consider torques in a one-dimensional chain of spins. If all spins are aligned (Fig. 11.1a), the cross product $\mathbf{S}_i \times \mathbf{S}_{i-1} = 0$, and there is no torque between adjacent spins. If the angular variation of spin $d\theta/dx = $ constant (Fig. 11.1b), there is no *net* torque on $\mathbf{S}_i$, because the spin $i-1$ exerts a torque exactly counterbalanced by the opposing torque of spin $i+1$. There will be a resultant torque on $\mathbf{S}_i$ only when $d\theta/dx = f(\theta)$, i.e., for $d^2\theta/dx^2 \neq 0$.

When the variation of spin orientation is small compared to lattice spacing, we can treat the spin distribution as a continuous function of the space coordinates. If the sum in Eq. 11.5 is now carried out, we find

$$\mathbf{T}_E{}^i = \mathbf{S}_i \times C \nabla^2 \mathbf{S}_i, \quad (11.6)$$

where C involves J, the lattice constant a, and a numerical factor which depends on the specific geometry of the spin lattice. For a simple cubic lattice, $C = 2Ja^2$. The operation ∇^2 is the Laplacian operating on the rectangular components of $\mathbf{S}_i$. The general condition for a nonvanishing torque, $\nabla^2 \mathbf{S}_i \neq 0$, reduces to $d\theta^2/dx^2 \neq 0$ for a one-dimensional chain.

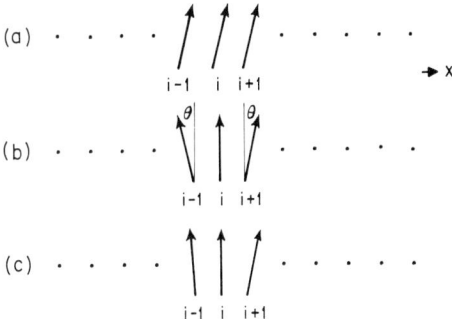

Fig. 11.1. Exchange torques in a one-dimensional chain of spins. A torque on i exists only in (c). In (a) $d\theta/dx = 0$; in (b) $d^2\theta/dx^2 = 0$; in (c) $d^2\theta/dx^2 \neq 0$.

It is convenient to replace the variable $\mathbf{S}_i$ by the spin density $\mathbf{S}^v$ (spin angular momentum per unit volume) or by $\mathbf{M}_s$, the local magnetization per unit volume. The exchange torque density becomes

$$\mathbf{T}_E{}^v = \mathbf{S}^v \times a^3 C \nabla^2 \mathbf{S}^v \quad (11.7a)$$

(assuming one spin $\mathbf{S}_i$ per unit cell, volume a^3) or alternatively

$$\mathbf{T}_E{}^v = \frac{a^3 C}{\gamma^2} \mathbf{M}_s \times \nabla^2 \mathbf{M}_s = C' \mathbf{M}_s \times \nabla^2 \mathbf{M}_s. \quad (11.7b)$$

The spins are also acted on by lattice torques that tend to align the spins along certain preferred crystallographic directions. This coupling between spin system and lattice can be represented phenomenologically by an anisotropy energy

$$E_K = f(\alpha_1, \alpha_2, \alpha_3), \quad (11.8)$$

where the α's are direction cosines of $\mathbf{M}_s$ relative to reference directions in the crystal. The precise form of E_K must be consistent with crystal symmetry. Normally, one writes E_K as a power series in the α's, eliminating terms inconsistent with symmetry and carrying the expansion only as far as is needed to fit the data. For a cubic crystal the anistropy energy density has the form

---

[1] The spin vectors $\mathbf{S}_i$ and $\mathbf{S}_j$ must actually be regarded as quantum-mechanical operators. We, however, shall treat them as ordinary vectors. It is this approximation that makes the treatment semiclassical.

$$E_K = K_1(\alpha_1^2\alpha_2^2 + \alpha_2^2\alpha_3^2 + \alpha_3^2\alpha_1^2) + K_2\alpha_1^2\alpha_2^2\alpha_3^2 + \cdots, \quad (11.9)$$

where α_1, α_2, α_3 are direction cosines of $\mathbf{M}_s$ referred to the cubic axes.

For a uniaxial crystal

$$E_K = K_1\alpha^2 + K_2\alpha^4 + K_3\alpha^6 + \cdots, \quad (11.10)$$

where α is the cosine of the angle between $\mathbf{M}_s$ and the plane perpendicular to the unique axis. Normally, the data can be adequately described in terms of only the first and second-order anisotropy constants K_1 and K_2.

The torque on $\mathbf{M}_s$ is related to the derivatives of the anisotropy energy as

$$\mathbf{T}_K^v = -\mathbf{a}_s \times \sum_j \mathbf{a}_j \frac{\partial E_K}{\partial \alpha_j}, \quad (11.11)$$

with $\mathbf{a}_s$ a unit vector along $\mathbf{M}_s$, $\mathbf{a}_j$ a unit vector along one of the crystal reference directions.

From Eq. 11.1 and the relation $\nabla \cdot \mathbf{B} = 0$, it follows that $\nabla \cdot \mathbf{H} = -\nabla \cdot \mathbf{M}_s$, where $-\nabla \cdot \mathbf{M}_s$ defines a magnetic pole density that acts as a source of magnetic field. This field, in turn, exerts a torque on $\mathbf{M}_s$ which tends to reorient the local magnetization so that the magnetic pole density is reduced. For example, when the magnetization in a long cylinder is tilted from the axis, a surface divergence of $\mathbf{M}_s$ is produced and, accordingly, magnetic poles are created on the surface

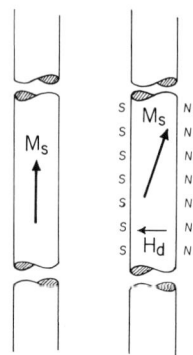

Fig. 11.2. Deviation of $\mathbf{M}_s$ from axis of cylinder leads to surface poles which create a demagnetizing field $\mathbf{H}_d$.

(Fig. 11.2). The result is a demagnetizing field $\mathbf{H}_d$ that exerts a torque

$$\mathbf{T}_d^v = \mathbf{M}_s \times \mu_0 \mathbf{H}_d, \quad (11.12)$$

seeking to return the magnetization to the axial direction. A divergence of $\mathbf{M}_s$ *within* a magnetic material will give rise to similar demagnetizing fields and attendant torques.

The total torque per unit volume acting on the local magnetization is the sum of these contributions:[2]

$$\mathbf{T}^v = \mathbf{M}_s \times \left[\mu_0(\mathbf{H}_a + \mathbf{H}_d) + C'\nabla^2 \mathbf{M}_s - \frac{1}{M_s} \sum_j \mathbf{a}_j \frac{\partial E_K}{\partial \alpha_j} \right]. \quad (11.13)$$

In equilibrium this torque must be equal to the rate of change of angular moment:

$$\mathbf{T}^v = \frac{d\mathbf{S}^v}{dt} = \frac{1}{\gamma} \frac{d\mathbf{M}_s}{dt}. \quad (11.14)$$

This is our fundamental equation.

The importance of this equation for understanding the magnetization process has been emphasized by W. F. Brown, Jr.[3] Solutions of the static equation

$$\mathbf{T}^v = 0 \quad (11.15)$$

have been investigated in some detail.[4,5]

Equation 11.14, as Brown has pointed out, represents in fact a set of two differential equations; the independent variables are the two angles that describe the orientation of $\mathbf{M}_s$. These nonlinear equations must be solved subject to two subsidiary conditions: one relating $\mathbf{H}_d$ to $\nabla \cdot \mathbf{M}_s$, the other a boundary condition on $\mathbf{M}_s$ which Brown has taken to be

$$\mathbf{M}_s \times \frac{\partial \mathbf{M}_s}{\partial n} = 0. \quad (11.16)$$

The operation $\partial/\partial n$ denotes differentiation along an outward normal to the boundaries of the sample. This equation signifies that a spin on the surface, lacking an exterior neighbor, can only be in equilibrium parallel to its interior neighbor.

Magnetization of the infinite crystal

The problem easiest to handle is the quasi-static magnetization of a crystal infinite in extent. Since we can ignore all boundary conditions, the only torques present are those due to the applied field and the anisotropy energy. The requirement that the free energy be a minimum demands that the exchange energy be zero, i.e., the spins parallel.

---

[2] Equation 11.13 applies only to the undeformed crystal. If the crystal is free to deform, there are additional torques caused by magnetoelastic coupling.

[3] Brown has coined the term "micromagnetics" to describe magnetic studies based on a more or less rigorous solution of Eq. 11.14. In terms of our nomenclature, micromagnetics and domain theory would constitute two branches of "mesoscopic magnetism."

[4] W. F. Brown, Jr., *J. Appl. Phys.*, Suppl., *30*, No. 4, 62S (1959).

[5] W. F. Brown, Jr., *Magnetostatic Principles in Ferromagnetism*, Interscience Publishers (John Wiley and Sons), New York, 1962.

As an example, consider the magnetization of a uniaxial crystal (Fig. 11.3) with the anisotropy energy

$$E_K = K_1 \cos^2 \theta. \quad (11.17)$$

Without an applied field the condition for minimum free energy is satisfied if $\mathbf{M}_s$ lies either along $+x$ or $-x$. Suppose the magnetization is initially oriented

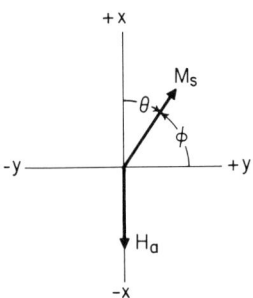

Fig. 11.3. Geometry for investigating magnetization of uniaxial crystal.

along $+x$ and we apply a field $\mathbf{H}_a$ in the $-x$ direction. The equilibrium position assumed by the magnetization vector is determined by the torque balance

$$\mu_0 \mathbf{M}_s \times \mathbf{H}_a - (\mathbf{a}_s \times \mathbf{a}_y) \frac{\partial E_K}{\partial \cos \phi} = 0, \quad (11.18a)$$

$$-\mu_0 M_s H_a = 2K_1 \cos \theta. \quad (11.18b)$$

This equation can be rewritten

$$-\frac{H_a}{\frac{2K_1}{\mu_0 M_s}} = \frac{M_s \cos \theta}{M_s}. \quad (11.19)$$

Usually we measure the magnetization parallel to the applied field,

$$M_H = M_s \cos \theta. \quad (11.20)$$

If we define $2K_1/\mu_0 M_s \equiv H_K$, Eq. 11.19 can be written as

$$\frac{H_a}{H_K} = -\frac{M_H}{M_s}. \quad (11.21)$$

In the M_H-vs.-H_a plane this equation gives a straight line of negative slope passing through the origin (Fig. 11.4). Equation 11.21 must, however, be bounded in extent, because M_H can never exceed M_s in magnitude.

Let us examine the magnetization curve resulting when we decrease the applied field from very large positive to very large negative values. For $\mathbf{H}_a$ very large and parallel to M_s, $M_s = M_H$; the torque balance (Eq. 11.18a) is satisfied because both $\mathbf{M}_s \times \mathbf{H}_a$ and T_K^v are zero for $\theta = 0$. This condition of balance will be satisfied until the field is reduced to $-H_K$.

Here the torque balance is described by Eq. 11.21. However, it is now clear that Eq. 11.21 must represent a metastable equilibrium, for, if the field is monotonically decreased, we never enter the branch 3-6.

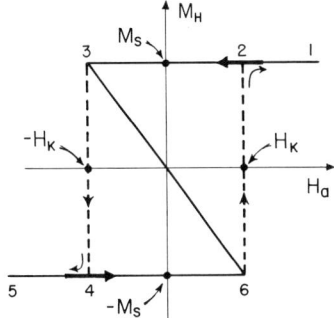

Fig. 11.4. Hysteresis loop for uniaxial crystal.

Instead, the magnetization flips over and we trace out the segment 3-4. On reversing the field we return to 1 via the vertical branch 6-2 and we trace out the segment 3-4.

If the field $\mathbf{H}_a$ is applied at an angle to the x axis, the shape of the hysteresis loop is modified (Fig. 11.5).

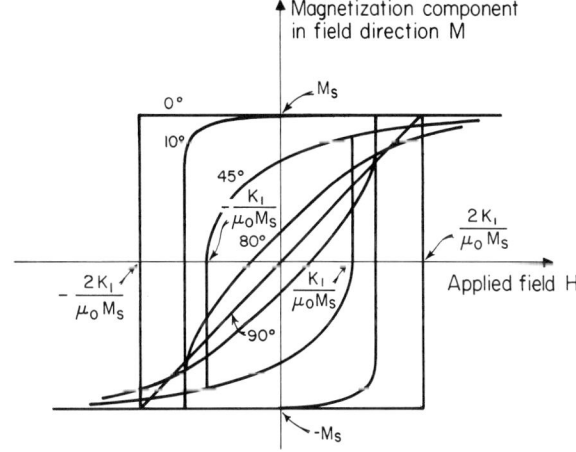

Fig. 11.5. Hysteresis loops for uniaxial crystal for various orientations of applied field relative to magnetic axis.

With increasing angle the loop width progressively decreases until finally, when the field is applied at $\theta = 90°$, the loop collapses to a straight line through the origin.

Domains and domain walls

If we examine a uniaxial crystal of finite size in the absence of an external field, we find that the crystal, in its state of lowest free energy, is subdivided into regions (domains) in which the local magnetization is directed along $+x$ or $-x$ (Fig. 11.6); the overall arrangement produces an average magnetization $\mathbf{M} = 0$. This situation is quite different from that for the infinite crystal, and it is instructive to inquire into the physical origin of the difference.

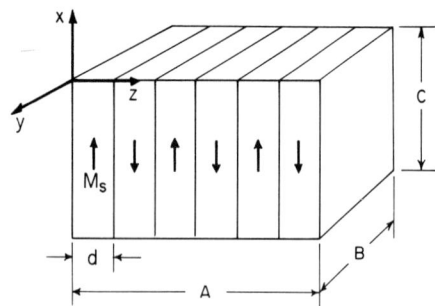

Fig. 11.6. Domains in uniaxial crystal.

In the infinite crystal there were no boundaries; consequently, we were not troubled by demagnetizing fields produced by $\nabla \cdot \mathbf{M}_s$. In the finite crystal such fields and their torques cannot be ignored. Consequently, if we wish to determine the magnetic state of the finite crystal rigorously, we have to solve Eq. 11.15 in its full form. Unfortunately the nonlinear character of the equation has thus far proved too formidable a hurdle. Instead, the following approximate procedure has been adopted. Suppose we consider an infinite uniaxial crystal composed of two domains. At $z = -\infty$, the local magnetization $\mathbf{M}_s$ is directed along $+x$; at $z = +\infty$, it is directed along $-x$. Our problem is to determine how the magnetization varies as a function of z. The torques in this problem arise from exchange and anisotropy only, and we can solve Eq. 11.15 subject to the appropriate boundary conditions.

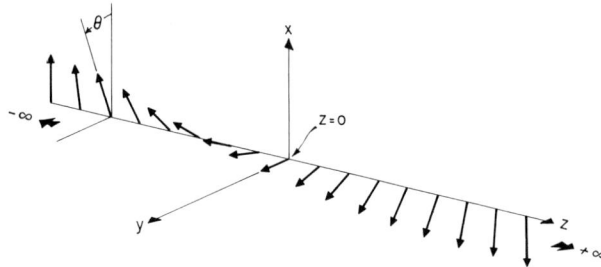

Fig. 11.7. Domain wall.

When we do so, we find that the orientation of the local magnetization varies with z position as (Fig. 11.7)

$$z = \left(\frac{A}{K_1}\right)^{1/2} \ln \tan \frac{\theta}{2}, \quad (11.22)$$

where θ is the angle between $\mathbf{M}_s$ and $+x$, and $A = 2JS_i^2/a$.[6] The quantity $\delta = \sqrt{A/K_1}$ defines the domain-wall thickness parameter. Of the total 180° angular variation, 60 percent occurs within the interval $-\delta \leq z \leq \delta$.

One can show that compared with the situation of parallel spins the two-domain configuration has an extra energy $4(AK_1)^{1/2}$ per unit area,[7] customarily referred to as the domain-wall energy per unit area. One half of the energy is due to anisotropy, i.e., to the deviation of the local magnetization from the preferred crystallographic direction. The other half results from the work performed by the exchange torques. That such torques must exist in this problem follows from Eq. 11.22, because $d\theta/dz \neq$ constant.

Let us return to the multidomain situation of Fig. 11.6 and determine the domain structure of the crystal in its equilibrium state. A reasonable starting point is to assume that the domain wall will have nearly the same structure as in the two-domain problem. Under these assumptions we can construct a reasonably accurate free-energy function for the multidomain crystal by using only two terms: (a) the magnetostatic energy resulting from the alternating magnetic poles at the top and bottom surfaces of the crystal; (b) the domain-wall energy. Solution of the magnetostatic problem gives an energy per unit surface[8]

$$u_M = \frac{0.85 \mu_0 M_s^2 d}{(4\pi)^2} = a' M_s^2 d. \quad (11.23)$$

If u_w is the energy per unit area of domain wall and A/d approximately the total number of walls, the total free energy of the crystal is

$$U(d) = 2a' M_s^2 \, d \, AB + \frac{u_w ABC}{d}. \quad (11.24)$$

Minimizing $U(d)$ with respect to d, we obtain the equilibrium domain dimension

$$d_0 = \frac{1}{M_s} \left(\frac{u_w C}{2a'}\right)^{1/2}. \quad (11.25)$$

The important feature of this result is that the domain dimension depends on the size (and shape) of the crystal.

The application of a field along $+x$ will cause the domains parallel to the field to grow at the expense of the antiparallel domains. For a perfect crystal this growth would proceed in infinitesimally small fields

---

[6] This is correct in magnitude only for spins distributed over a simple cubic lattice; for more complex lattices the numerical factor 2 changes slightly.

[7] The value of energy as given is correct only for the 180° wall in a uniaxial crystal. In a cubic crystal the energy, although similar in form, is different in magnitude.

[8] C. Kittel, *Revs. Modern Phys.* **21**, 541 (1949).

(magnetization curve in Fig. 11.8a). Actually, imperfections in the crystal impede the rearrangement of the domain structure, and the magnetization curve broadens into a loop (Fig. 11.8b). For fields at an angle to the $+x$ direction the magnetization changes will occur approximately in a two-step process: a change from multidomain to single-domain configuration with magnetization along that preferred crystallographic direction which makes the smallest angle with the applied field, followed by a rotation of the single-domain magnetization into alignment with the field (Fig. 11.8c).

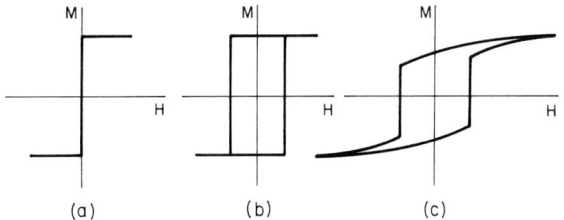

Fig. 11.8. Magnetization curves for multidomain crystal (idealized): (a) no crystal imperfections, field parallel to domain walls; (b) imperfections present, field parallel to domain walls; (c) imperfections present, field at angle to domain walls.

This elementary example illustrates the approximate analytical procedure of "domain theory." The starting point is always a judicious guess about the general character of the domain structure. The goal of the analysis is the evaluation of the free energy for the assumed configuration. If the configuration involves magnetic poles, the appropriate magnetostatic problem must be solved in order to find the magnetic contribution to the free energy. The calculation of wall energy is based upon the assumption that the structure of the wall is unchanged from that in an appropriate two-domain problem. It is, of course, not sufficient to have evaluated the free energy for some particular configuration. We must determine which one of all the possible configurations possesses the *lowest* free energy. If the problem is sufficiently simple, we can construct a family of configurations in terms of one or more parameters of variation (in our example the parameter d) and minimize the energy. In general, the problem is too complicated to permit such a straightforward procedure. The more usual method is to make a number of guesses as to possible configurations, calculate the free energy for each choice—if this can be done—and compare results. By this procedure we determine the most stable among several reasonable domain structures but are never certain that we have found the correct equilibrium state. The "exact" solution to the problem is, of course, embedded in Eq. 11.15, which proves tractable only in a limited number of cases. We shall consider one solvable "class" in the following section.

Fine particles

The approximations of domain theory break down when the dimensions of the sample become very small, but for this situation certain limited solutions of the fundamental equation have been found.

A case examined is that of a long cylindrical crystal with the preferred magnetic axis along the cylinder axis. The diameter is assumed comparable to and smaller than a domain wall. With the magnetization initially pointing in the $+x$ direction, we wish to determine the response of the magnetization to a static field applied in the $-x$ direction.

For small departures from equilibrium, Eq. 11.15 can be linearized and solved. The results give the initial mode of magnetization reversal and the threshold field for its stability (i.e., the coercive field required to produce reversal). One finds that two modes are possible: a curling of the magnetization and a buckling (Fig. 11.9). The cylinder diameter determines which

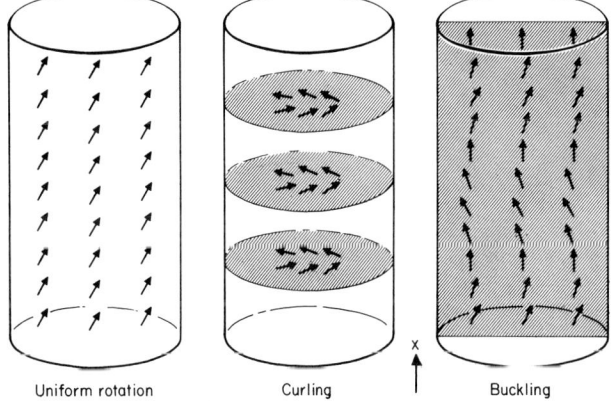

Fig. 11.9. Magnetization reversal modes in cylindrically shaped fine particles.

of these two modes is the less stable (Fig. 11.10). At the lower end of the diameter scale the reversal starts via the buckling mode; at the high end the curling mode is the first to go unstable. In Fig. 11.10 we also show a curve representing the coercive field for reversal by coherent rotation. Actually this mode, although it has been postulated in some analyses, does not satisfy Eq. 11.15.[9]

[9] The uniform mode would satisfy Eq. 11.14 only if no magnetic poles were allowed to develop at the surface of the sample. This situation will of course be achieved only for an infinite sample (when the boundaries are suitably closed at infinity).

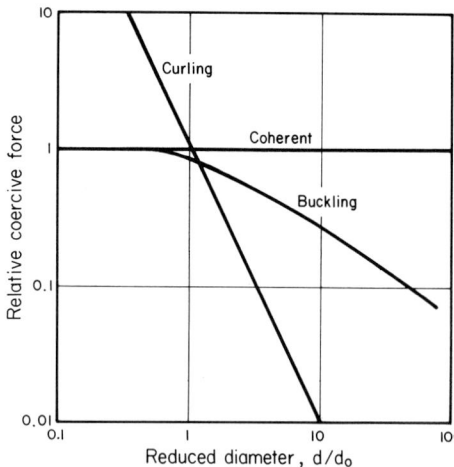

Fig. 11.10. Coercive fields for cylindrically shaped fine particles.

Calculations of the coercive force of small particles are of considerable technical importance in connection with the development of some newer materials for permanent magnets. In these materials one attempts the molecular designing of permanent magnets by making fine-particle compacts with particle shape and concentration as design parameters in addition to the intrinsic properties of the individual particle.

In the design of these materials[10] there remains as yet considerable disparity between experiment and theory, because the actual particles deviate from the shapes amenable to analysis and because particle interactions can be treated only approximately.

Dynamic excitation

Thus far we have considered spin configurations under static or quasi-static conditions. The question now is the dynamic response of the spins to a time-dependent excitation. To obtain some physical insight into this problem, let us consider the following simple example. Let the magnetization vector $\mathbf{M}_s$ be subjected to the torque of a d-c field $\mathbf{H}_a = \mathbf{a}_z H_z$ (Fig. 11.11). The equilibrium state clearly will be one in which $\mathbf{M}_s$ is aligned parallel to $\mathbf{H}_a$. Let us now apply a small d-c field $\mathbf{h}(\mathbf{h} \ll \mathbf{H}_a)$ at right angles to $\mathbf{H}_a$. The equilibrium position of $\mathbf{M}_s$ shifts to line up with the direction of the resultant field $\mathbf{H}_a + \mathbf{h}$; that is, the vector deviates by some small angle θ from its initial direction. If now $\mathbf{h}$ is suddenly reduced to zero, how will $\mathbf{M}_s$ return to $\theta = 0$?

The motion of the magnetization vector is, of course, governed by Eq. 11.14,

$$\gamma \mathbf{T}^v = \frac{d\mathbf{M}_s}{dt}.$$

[10] F. E. Luborsky, *J. Appl. Phys.* **32**, No. 3, 171S (1961).

In our present example $\mathbf{T}^v = \mu_0 \mathbf{M}_s \times \mathbf{H}_a$ with this torque directed at right angles to the plane established by $\mathbf{M}_s$ and $\mathbf{H}_a$. It is evident that such a torque can drive only the magnetization around $\mathbf{H}_a$ and into line with $\mathbf{H}_a$. This result is of course to be expected; the

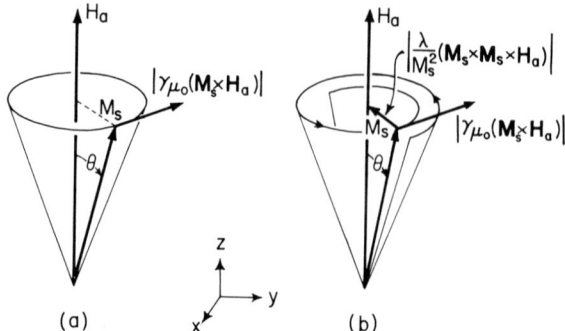

Fig. 11.11. In (a) the precessional motion H undamped; in (b) the damping torque closes the angle θ.

magnetization vector, because of its associated angular momentum, exhibits a gyroscopic precession. In the absence of losses the angle of inclination will remain fixed at θ, and $\mathbf{M}_s$ precesses about $\mathbf{H}_a$ at a natural angular frequency

$$\omega_0 = -\gamma \mu_0 H_z. \qquad (11.26)$$

In reality the magnetization must end up in the state of lowest energy, i.e., alignment of $\mathbf{M}_s$ with the z axis. It is evident that the equation of motion is not complete without the inclusion of a suitable damping torque. There are several different phenomenological forms for the damping that have been used. In the Landau-Lifshitz[11] form the equation of motion becomes

$$\frac{d\mathbf{M}_s}{dt} = (\mathbf{M}_s \times \mu_0 \mathbf{H}_a) - \frac{\lambda}{M_s^2} \mathbf{M}_s \times (\mathbf{M}_s \times \mathbf{H}_a), \qquad (11.27)$$

where λ is a damping factor having the dimensions of frequency.

The damping term exerts a torque that tends to close the angle θ (cf. Fig. 11.11). The resultant motion of the spin system is a damped precessional spiral.

Extremely interesting is the response of the magnetization vector to combined d-c and a-c fields. Let $\mathbf{H}_a$ remain a d-c field but $\mathbf{h}$ be a small a-c field $\mathbf{h} = \mathbf{h}_0 e^{j\omega t}$ directed at some arbitrary angle to $\mathbf{H}_a$. If we solve now Eq. 11.14 (the undamped case) in a linearized approximation, we obtain for the resultant change in magnetization $\mathbf{m}$ (assuming θ small)

[11] For a discussion of this and other forms of damping, see B. Lax and K. Button, *Microwave Ferrites and Ferrimagnetics*, McGraw-Hill Book Co., New York, 1962, Chap. 4.

$$m_x = -\frac{\gamma\mu_0 M_s}{\omega_0^2 - \omega^2}(\omega_0 h_x - j\omega h_y), \quad (11.28a)$$

$$m_y = -\frac{\gamma\mu_0 M_s}{\omega_0^2 - \omega^2}(j\omega h_x + \omega_0 h_y), \quad (11.28b)$$

$$m_z = 0. \quad (11.28c)$$

The relation between **m** and **h**

$$\mathbf{m} = \underset{\leftrightarrow}{\chi}\mathbf{h} \quad (11.29)$$

defines a tensor magnetic susceptibility which written in matrix form is

$$|\underset{\leftrightarrow}{\chi}| = \begin{vmatrix} \chi_{xx} & \chi_{xy} & 0 \\ \chi_{yx} & \chi_{yy} & 0 \\ 0 & 0 & 0 \end{vmatrix}, \quad (11.30)$$

where

$$\chi_{xx} = \chi_{yy} = \frac{\omega_0 \omega_m}{\omega_0^2 - \omega^2}, \quad (11.31a)$$

$$\chi_{xy} = \chi_{yx} = \frac{j\omega \omega_m}{\omega_0^2 - \omega^2}, \quad (11.31b)$$

$$\omega_m = -\gamma\mu_0 M_s. \quad (11.31c)$$

The susceptibility tensor is antisymmetric and characterized by a resonance denominator that has a singularity at $\omega = \omega_0 = -\gamma\mu_0 H_a$. With damping included, the singularity is suppressed, but the susceptibility will ordinarily still show resonance character in the vicinity of ω_0.

Other d-c torques can be included by replacing H_a by an appropriate effective field. Suppose, for example, we are dealing with resonance in a uniaxial single crystal in the form of an ellipsoid of revolution, the axis of revolution and the preferred magnetic axis coinciding with the applied field H_a. The resonance frequency is no longer ω_0 but modified by additional torques due to anisotropy and demagnetizing fields. For this specific example Kittel[12] has shown that resonance occurs at

$$\omega_r = -\gamma\mu_0 \sqrt{\begin{array}{l}[H_z + H_K + (N_y - N_z)M_s] \\ \times [H_z + H_K + (N_x - N_z)M_s]\end{array}}, \quad (11.32)$$

where N_x, N_y, and N_z are the demagnetizing coefficients along the principal axes of the ellipsoid and $H_K = 2K_1/\mu_0 M_s$ is the anisotropy contribution to the effective field.

In a spherical sample the interaction between magnetization and surface poles is independent of the orientation of M_s, and consequently there are no torques due to demagnetizing fields. In consequence, for a sphere $N_x = N_y = N_z = \frac{1}{3}$, and the terms $N_x - N_z$, $N_y - N_z$ vanish. If, further, $K_1 = 0$, Eq. 11.32 reduces to our previous result $\omega_r = \omega_0$.

[12] C. Kittel, *Phys. Rev.* **73**, 155 (1948).

Domain-wall dynamics

In a real crystal the motion of a domain wall is impeded by crystal imperfections which locally trap the wall. A schematic representation of the trapping energy as a function of wall position is shown in Fig. 11.12. In the absence of a field the wall is trapped at a locally stable position such as a, b, or c.

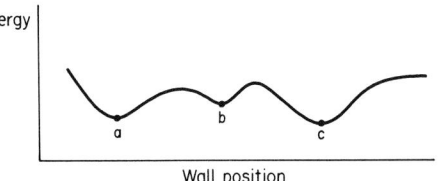

Fig. 11.12. Wall-trapping energy as a function of wall position (schematic).

Application of a field parallel to the wall exerts a pressure $2\mu_0 H M_s$ which tries to drive the wall away from the local minimum. For small displacements from equilibrium the local trap potential provides a springlike restoring force Kz per unit area opposing displacement,

$$2\mu_0 H M_s = Kz. \quad (11.33)$$

For small wall velocities one can formally account for energy dissipation by adding a viscous damping term

$$2\mu_0 H M_s = Kz + \beta\dot{z}, \quad (11.34)$$

where β is a damping parameter.[13] For a sinusoidal field the displacement is complex,

$$z = \frac{2\mu_0 M_s/K}{1 + j\omega\tau}, \quad (11.35)$$

with $\tau = \beta/K$. The magnetic susceptibility χ, defined as $\Delta M/H$, becomes

$$\chi = \frac{2M_s z}{H} \quad (11.36)$$

and has a frequency dependence

$$\chi = \frac{\chi_0}{1 + j\omega\tau}, \quad (11.37)$$

where

$$\chi_0 = \frac{4\mu_0 M_s^2}{K} \quad (11.38)$$

is the susceptibility at $\omega = 0$.

The occurrence of a relaxation spectrum for small-signal domain-wall susceptibility has been established

[13] The wall also has an effective (dynamic) mass, and therefore, a term $m_{\text{eff}}\ddot{z}$ should be added, but this term is ordinarily small and may be omitted.

in a number of experiments. Most experimental work, however, has been carried out on multidomain samples, where the observed relaxation cannot be described in terms of a single time constant but by a distribution of τ's.

The magnetic permeability[14] spectrum of a ceramic sample of yttrium iron garnet is shown in Fig. 11.13.

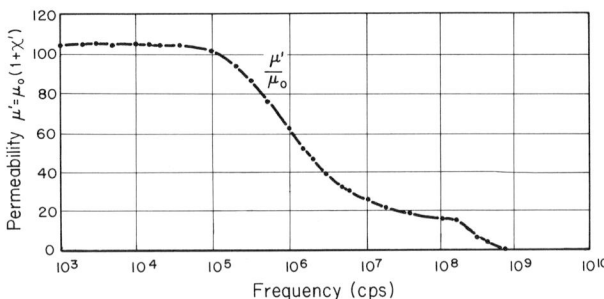

Fig. 11.13. Permeability spectrum for yttrium iron garnet.

The spectrum is characterized by two distinct regions of dispersion. The lower one results from relaxation of the domain-wall contribution. The dispersion at approximately 300 Mc/sec is due to gyromagnetic resonance of the spin system in an effective field produced by anisotropy and internal demagnetizing fields. Fairly wide variations in the effective field from point to point within the material make the over-all gyromagnetic "resonance" a distribution of individual resonances. A broad dispersion instead of a sharp resonance results.

Magnetic losses

Any dynamic excitation of an ordered spin system is accompanied by the dissipation of energy, but what is the basic mechanism for loss? In metals, of course, losses arise principally from the flow of eddy currents induced by the time rate of change of magnetization. In magnetic insulators, e.g., ferrites, garnets, where the resistivities can be extremely high,[15] this eddy current mechanism is not operative.

Fundamental studies on loss in magnetic insulators have, for the most part, dealt with resonance experiments, where the sample is magnetized to saturation by a large d-c field and simultaneously subjected to a small transverse a-c field. Thus excited, the magnetic system exhibits resonant response at a frequency $\omega_r = \gamma \mu_0 H_{\text{eff}}$ (Eq. 11.32). With the external field nor-

[14] The permeability μ and susceptibility χ are related as follows: $\mu = \mu_0(1 + \chi)$.

[15] In yttrium iron garnet, for example, the resistivity at room temperature can exceed 10^{12} ohm-cm.

mally required to produce saturation, the resonant frequency occurs at microwaves. In this region the limited bandwidth of the experimental arrangement usually dictates that the a-c frequency be held fixed and the resonant line be traced out by varying the external d-c magnetic field (which ordinarily is the major component of H_{eff}). A satisfactory theory of magnetic loss must explain the line width observed in this kind of resonance experiment.

Theoretical investigations have focused on two mechanisms. In one of them the precessing spin system loses energy directly to the lattice vibrations (spin-lattice relaxation). This mechanism has been ruled out because quantitative calculations predict a line width about four orders of magnitude narrower than observed. The second mechanism involves transfer of energy from the uniform precessional excitation to nonuniform thermal excitations of the spin system (spin-spin relaxation). In effect, these higher-order spin excitations, known as spin waves, act as an energy reservoir in contact with both the uniform precession and the lattice vibrations. The dissipative energy extracted from the a-c exciting field first enters the uniform precessional mode, is then transferred to the spin-wave system, and ultimately shared by the spin waves and the lattice vibrations. The limiting step in this over-all transfer process is the coupling between uniform precession and spin waves.

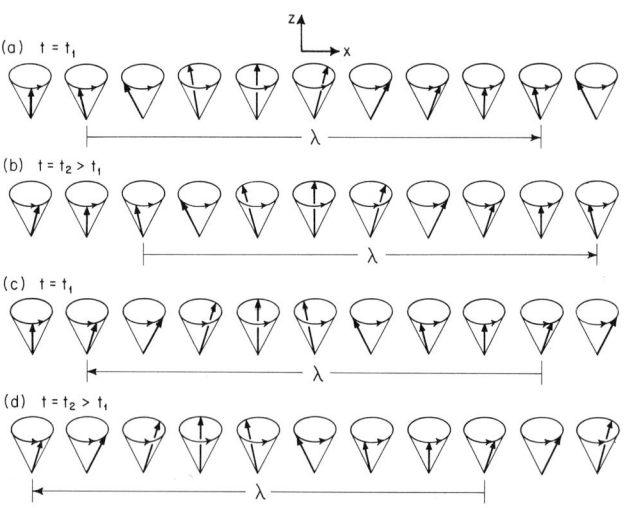

Fig. 11.14. Traveling spin wave in a one-dimensional lattice of magnetic moments. In (a) a spin wave traveling in the $+x$ direction is shown at $t = t_1$; the same wave is shown at a later time t_2 in (b). A spin wave traveling in the $-x$ direction is shown at two successive instants in time in (c) and (d).

The structure of a typical spin wave is illustrated in Fig. 11.14; for the sake of simplicity we show only a one-dimensional lattice of spins. The essence of the spin wave is the existence of a progressive change in precessional phase between adjacent spins. In Fig. 11.14 all spins are precessing about the z axis at the same angular frequency, but if at some instant $t = t_1$ we examine the phase of precession, we find a progressive phase lag as we proceed in the $+x$ direction. The shortest distance between two spins of equal phase defines the wavelength of the excitation. If we follow a point of fixed phase, we find that as time evolves this point moves with a velocity which defines the phase velocity of the spin waves.[16]

The spin-wave-spectrum characteristic of a magnetic lattice is obtained by looking for those solutions of Eq. 11.14 which represent small-amplitude, time-harmonic magnetic deviations from the perfectly aligned state. For an ellipsoid of revolution magnetized to saturation along the z axis, the allowed spin waves were shown[17] to cover a spectrum of frequencies[18] (Fig. 11.15).

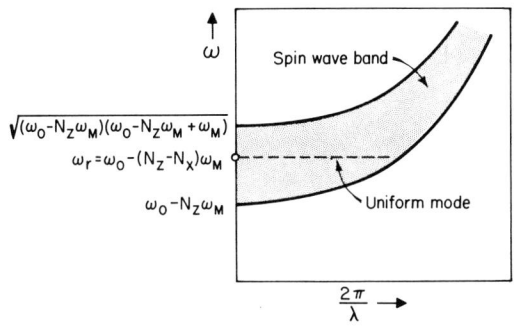

Fig. 11.15. Spin-wave spectrum.

The important feature of this spectrum is the existence of spin-wave frequencies degenerate with the uniform precessional mode. There is now ample evidence to demonstrate that the losses in magnetic resonance result principally from the transfer of energy from the uniform mode to these degenerate spin waves.[19]

Recently Janak[20] has shown that this same mechanism can quantitatively account for the domain-wall damping in certain ferrites. The analysis basically follows the one used in treating the resonance problem, but there is considerable difference in detail. The spin-wave spectrum for the wall turns out to be quite different from the one occurring in the resonance problem. However, the fundamental mode of the wall in translation is found to be degenerate with the wall spin-wave spectrum so that again we have an effective energy-transfer mechanism.

Diffusion damping

The presence of mobile impurities in magnetic solids can sometimes give rise to very unusual magnetic loss characteristics. The situation is typified by the behavior of carbon as an impurity in α-iron.[21]

The carbon atoms are incorporated into the iron lattice at interstitial sites located on the cube edges midway between iron atoms. For a crystal that is rigorously cubic, these sites (labeled 1, 2, 3 in Fig. 11.16)

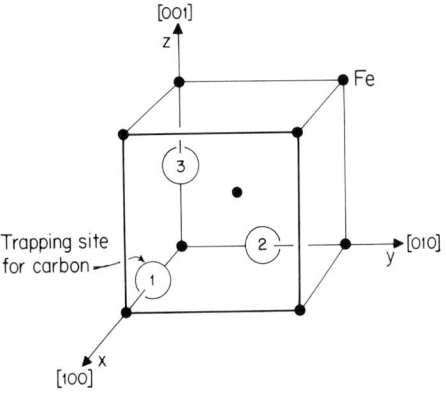

Fig. 11.16. Carbon impurity sites in iron.

are equivalent, and the probability of occupancy is consequently the same for each site. However, we are dealing with a crystal that has a magnetic axis and is therefore only pseudocubic, so that the sites are in fact nonequivalent. As a result, the impurity atoms show a statistical preference for certain sites, a preference determined by the orientation of the magnetic axis relative to the crystallographic axes. It has been found[22] that when the magnetic axis lies in the $+x$ direction, Site 1 becomes energetically unfavorable compared to 2 and 3; similarly, when the magnetic axis coincides with $\pm y$ (or $\pm z$), Site 2 (or 3) is less favored relative to the remaining two.

From examination of the energetics involved (Fig. 11.17) it is easy to see that this coupling between magnetization and impurity ordering must lead to magnetic loss. Suppose that in an initial equilibrium state the magnetization lies along $+x$. The interstitial carbons, distributed according to Boltzmann statistics, settle

---

[16] In the three-dimensional lattice the phase velocity is obtained by observing the motion of a surface of constant phase.

[17] A. M. Clogston, H. Suhl, L. R. Walker, and P. W. Anderson, *J. Phys. Chem. Solids* **1**, 129 (1956).

[18] The calculation that gives this spectrum does not include anisotropy torques. The shape of the spectrum is determined by torques originating in exchange coupling and in internal demagnetizing fields, usually referred to as dipolar torques.

[19] For further discussion, see Ref. 11, Chap. 5.

[20] J. F. Janak, *Phys. Rev.* **134**, A411 (1964).

[21] Typical carbon concentrations are of the order of 1 percent.

[22] G. DeVries, D. W. van Geest, R. Gersdorf, and G. W. Rathenau, *Physica* **25**, 1131 (1959).

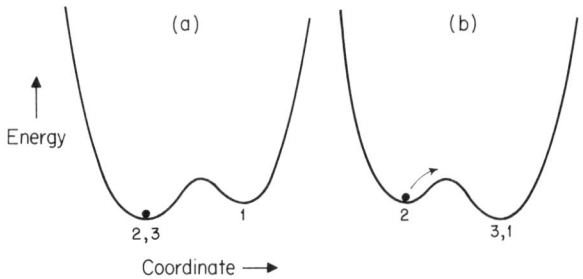

Fig. 11.17. Energy diagram (schematic) for carbon impurities in iron.

preferentially in Sites 2 and 3. If by applying an external field we shift the magnetic axis to $+y$, the preferred sites now become 1 and 3. The initial impurity distribution no longer satisfies the requirement for equilibrium, and consequently the carbon atoms undergo rearrangement by preferential diffusion from Site 2 to 3. Thus, we have a mechanism whereby energy is extracted from the applied magnetic field and dissipated to the lattice.

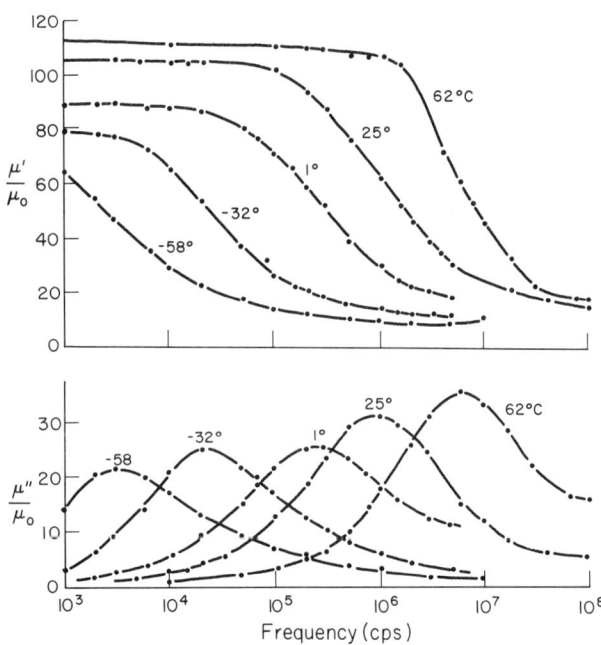

Fig. 11.18. Domain-wall permeability spectra at several temperatures for a sample of YIG containing divalent iron; μ' is real part of permeability, μ'' is imaginary part.

Diffusion damping shows up dramatically in certain spinels and garnets containing small concentrations of Fe^{2+} cations. Here the diffusion particle is not an atom but an electron that, as the magnetic axis changes, transfers by diffusion from one Fe^{3+} cation to another in an equivalent lattice position.[23] Formally, this

[23] B. W. Lovell and D. J. Epstein, J. Appl. Phys. 34, 1115 (1963).

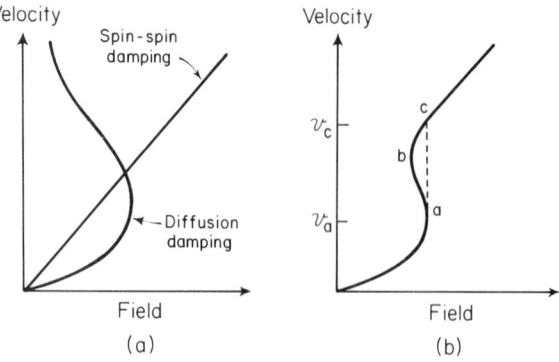

Fig. 11.19. Steady-state wall velocity as a function of amplitude of the step-function of driving field. In (a), diffusion and spin-spin damping are plotted separately; in (b), the two are added.

electron rearrangement is equivalent to a field-induced redistribution of Fe^{2+} cation "impurities" in an Fe^{3+} sublattice.

Because diffusion is a temperature-activated process, the magnetic losses are similarly activated, as seen in the domain-wall permeability spectrum (Fig. 11.18).

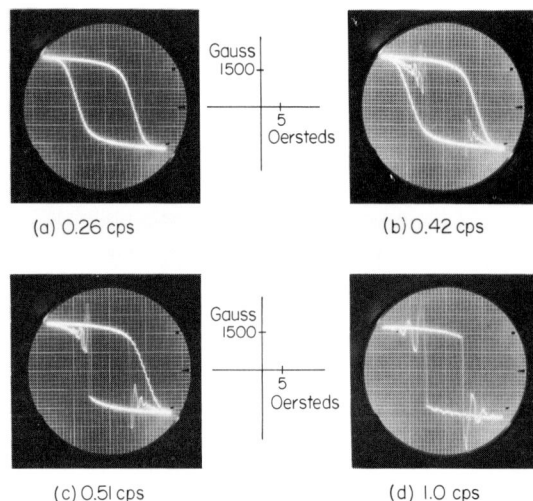

Fig. 11.20. Hysteresis loops for yttrium iron garnet at 77° K, showing transition between normal and "square" behavior.

Diffusion damping may be thought of as exerting a viscositylike drag on the moving wall. As the temperature reduces and diffusion becomes more difficult, the "viscosity" becomes "stickier" and the walls fail to follow the a-c driving field to as high a frequency.

For the large-scale wall displacements that occur in magnetic switching, the drag imposed by diffusion is highly nonlinear. For the 180° wall pictured in Fig. 11.7, calculations[24] show that the steady-state wall velocity for a step-function excitation H varies as indicated in Fig. 11.19a. When we add to the diffusion

[24] J. F. Janak, J. Appl. Phys. 34, 1119 (1963).

drag the always present spin-spin relaxation, we obtain the v-H characteristic shown in Fig. 11.19b. The segment a-b clearly represents an unstable solution, for in this region a decrease in driving field leads to an increase in wall velocity. Thus, the entire segment a-b-c normally is not accessible; when the field is increased above point a, we should expect a discontinuous change in velocity from v_a to v_c. The occurrence of such instabilities is illustrated by the sequence of hysteresis loops (Fig. 11.20) obtained in a sample of yttrium iron garnet at low temperatures, where diffusion damping is particularly pronounced. With triangular wave excitation at 0.26 cps, the loop appears quite normal (Fig. 11.20a); however, as the frequency of excitation, and therefore the wall velocity, is increased, irregularities appear. In Fig. 11.20b, for example, one cycle is normal, another abnormal; a similar effect appears in Fig. 11.20c. At 1.0 cps (Fig. 11.20d), the instabilities occur repeatedly, and the loop achieves a state of "dynamic squareness."

12 · CONDUCTION AND DIFFUSION IN IONIC CRYSTALS

Carl Wagner

Introduction — Conduction in ionic crystals — Defects in stoichiometric binary crystals — Defects due to nonstoichiometry — Change of conductivity by doping — Conduction in ternary ionic crystals — Self-diffusion — Diffusion in substitutional solid solutions — Diffusion in binary crystals involving a gradient of metal-to-nonmetal ratio — Diffusion in ternary compounds — Transport phenomena with diffusion in different phases — Nonisothermal transport — The use of ionic conductors for thermodynamic measurements — The use of ionic conductors in fuel cells

Introduction

Electrical conduction connotes motion of charged particles, i.e., ions and electrons, in a phase of uniform composition under a gradient of the electrical potential. On the other hand, diffusion connotes the transport of matter in a phase of nonuniform composition without a flow of net current. In many cases, these phenomena are closely interrelated, since diffusion may be effected by the motion of charged particles that are operative in conduction. This justifies considering conduction and diffusion phenomena in the same chapter.

Conduction in ionic crystals

Electronic conduction in semiconductors is due to motion of electrons and/or electron holes. Likewise, ionic conduction is due to motion of ions and/or ion vacancies. An interstitial ion may either jump to an adjacent empty interstice (interstitial mechanism) or displace a lattice ion by forcing it into an adjacent empty interstice (interstitialcy mechanism) (Fig. 12.1).

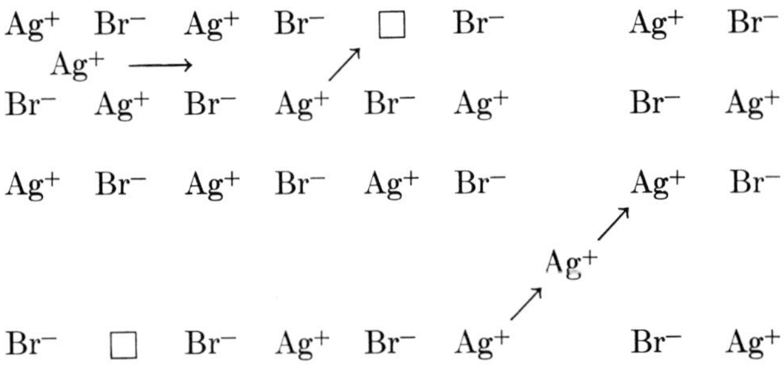

$$
\begin{array}{ccccccccc}
Ag^+ & Br^- & Ag^+ & Br^- & \square & Br^- & & Ag^+ & Br^- \\
& Ag^+ \longrightarrow & & & \nearrow & & & & \\
Br^- & Ag^+ & Br^- & Ag^+ & Br^- & Ag^+ & & Br^- & Ag^+ \\
Ag^+ & Br^- & Ag^+ & Br^- & Ag^+ & Br^- & & Ag^+ & Br^- \\
& & & & & & \nearrow & & \\
& & & & & Ag^+ & & & \\
Br^- & \square & Br^- & Ag^+ & Br^- & Ag^+ & & Br^- & Ag^+ \\
\end{array}
$$

Fig. 12.1. Lattice defects in AgBr.

An ion may also migrate by jumping from a regular lattice site into a vacancy, thereby creating another vacancy.

Defects in stoichiometric binary crystals

In a stoichiometric binary ionic crystal (e.g., AgCl, KCl, or CuO) various types of disorder involving pairs of majority defects may occur.

If only electronic defects are present, excess electrons

and holes must be in equal concentrations (condition for an intrinsic semiconductor, e.g., CuO).

If only ionic defects are present, the defects with positive and negative excess charges must balance. The following limiting cases are noted:[1]

  a. Equal concentrations of interstitial cations and cation vacancies (Frenkel disorder type, cationic conduction[2]).

  b. Equal concentrations of interstitial anions and anion vacancies (anti-Frenkel disorder type, anionic conduction).

  c. Equivalent concentrations of cation and anion vacancies (Schottky disorder type[1]).

  d. Equivalent concentrations of interstitial cations and anions (anti-Schottky disorder type).

In general, the concentrations of defects are determined by thermodynamic equilibrium conditions, analogous to those governing electrolytic dissociation in aqueous solution, and may be calculated by means of equations of statistical mechanics and from the enthalpy changes corresponding to the formation of the various pairs of defects listed. These enthalpy changes may be estimated by taking into account Coulomb energies and polarization effects. Thus it can be determined theoretically which disorder type is most likely to be found in a particular compound.

Alternatively, the prevailing defects in a particular compound can be deduced from experimental data, e.g., a comparison between dilatometric observations and the change of the lattice parameter with temperature according to X-ray observations, or the change of electrical conductivity upon doping (e.g., AgBr doped with Ag_2S or $CdBr_2$).[3-10]

Substances close to case a are the silver and the copper halides (free of excess halogen), while the alkali halides are close to case c. As an example for case b, CaF_2 may be cited.[9] No example is known for case d, and it seems unlikely that one will be found because of the high energy required for the formation of interstitials in close-packed lattices of ions of nearly equal size.

A modified Schottky disorder type has recently been proposed for $PbCl_2$: The predominant defects are supposedly anion vacancies and associates consisting of a cation vacancy and an anion vacancy.[10] Conduction takes place mainly by motion of anions via anion vacancies not linked to cation vacancies.

In addition to the majority defects that characterize the disorder types already listed, there may be minority defects in smaller concentrations (e.g., excess electrons and electron holes in AgBr).[11,12]

Defects due to nonstoichiometry

Excess metal in a binary crystal causes excess electrons and either interstitial cations or anion vacancies. In ZnO with excess Zn at 600° C, for example, the predominant defects are supposed to be interstitial Zn^+ ions and excess electrons in virtually equal concentrations.[13] Since the mobility of an electron is in general much greater than that of an interstitial ion, electronic n-type conduction prevails. In KCl with excess metal, most of the excess electrons are fixed at anion vacancies. These are the F centers (color centers) according to the nomenclature introduced by Pohl.[14] In part, the F centers are dissociated into ordinary anion vacancies and excess electrons, which produce n-type conduction.

A binary crystal with excess nonmetal contains electron holes and either interstitial anions or cation vacancies in addition to the lattice defects present at ideal stoichiometry. In Cu_2O with excess oxygen at 1000° C, the predominant defects are supposed to be Cu^+ vacancies and holes (represented by Cu^{2+} ions in an environment of Cu^+ ions as the normal cationic constituent) in virtually equal concentrations.[15] Since the mobility of a hole is in general much greater than that of a cation vacancy, electronic p-type conduction prevails.

In many binary crystals, deviations from ideal stoichiometry are very small. Cuprous oxide,[16,17] for example, is known to have a homogeneity range between $Cu_{1.9999}O$ and $Cu_{1.9973}O$ at 1000° C. At lower temperatures, the homogeneity range is still smaller. Nevertheless, the electrical conductivity has been found to vary widely, between about 0.9 and 6 ohm^{-1} cm^{-1}

---

[1] W. Schottky, *Z. physik. Chem.* **B29**, 335 (1935).

[2] J. Frenkel. *Z. Physik* **35**, 652 (1926).

[3] E. Koch and C. Wagner, *Z. physik. Chem.* **B38**, 295 (1937).

[4] J. Teltow, *Ann. Phys.* [6] **5**, 63, 71 (1950); *Z. physik. Chem.* **195**, 213 (1950).

[5] H. Kelting and H. Witt, *Z. Physik* **126**, 697 (1949).

[6] C. Wagner and P. Hantelmann, *J. Chem. Phys.* **18**, 72 (1950).

[7] H. W. Etzel and R. J. Maurer, *ibid.* **18**, 1003 (1950).

[8] Y. Haven, *Rec. trav. chim. Pays-Bas* **69**, 1259, 1471 (1950).

[9] R. W. Ure, *J. Chem. Phys.* **26**, 1363 (1957).

[10] G. Simkovich, *J. Phys. Chem. Solids* **24**, 213 (1963).

[11] B. Ilschner, *J. Chem. Phys.* **28**, 1109 (1958).

[12] C. Wagner, *Z. Elektrochem.* **63**, 1027 (1959).

[13] G. Heiland, E. Mollwo, and F. Stöckmann, "Electronic Processes in Zinc Oxide," *Solid State Physics*, Vol. 8, F. Seitz and D. Turnbull, Eds., Academic Press, New York, 1959, p. 193.

[14] R. W. Pohl, *Proc. Phys. Soc. (London)* **49**, 4 (1937) (extra part).

[15] H. Dünwald and C. Wagner, *Z. physik. Chem.* **B22**, 212 (1933).

[16] C. Wagner and H. Hammen, *ibid.* **B40**, 197 (1938).

[17] M. O'Keefe and W. J. Moore, *J. Chem. Phys.* **36**, 3009 (1962).

at 1000° C, and as much as 9 orders of magnitude at room temperature. Since small deviations from ideal stoichiometry are difficult to determine with sufficient precision, one measures the electrical conductivity of oxides such as Cu_2O at elevated temperatures as a function of the oxygen partial pressure in a coexisting gas atmosphere rather than as a function of composition. At 1000° C the electrical conductivity of Cu_2O varies proportionally to the seventh root of the oxygen partial pressure.[15]

By and large, the properties of a binary ionic crystal, in particular its ionic and the electronic conductivity, have well-defined values only if in addition to total pressure and temperature either the deviation of the composition from the ideal stoichiometric ratio or the activity of one component is given. It is often convenient to consider the concentrations of the various defects and the corresponding partial conductivities as a function of the equilibrium partial pressure of the nonmetallic component, e.g., the Br_2 partial pressure in the case of AgBr (Fig. 12.2). Diagrams for binary ionic crystals have been compiled and discussed by Kröger and Vink.[18]

Change of conductivity by doping

The conductivity of a binary ionic compound may be markedly changed by dissolution of another compound with a different cation-anion ratio, thereby creating additional defects. By introducing $CdBr_2$ into AgBr, the cation/anion ratio becomes less than unity: The solid solution of AgBr-$CdBr_2$ contains more cation vacancies and fewer interstitial cations than pure AgBr. Upon dissolving $CdBr_2$ in AgBr, the ionic conductivity first decreases somewhat and then rises about proportionally to the $CdBr_2$ content.[3,4]

Changes of the electronic conductivity are also observed. For instance, the conductivity of ZnO at 600° C and an oxygen partial pressure of 1 atm increases by a factor of 50 on addition of 1 wt % Ga_2O_3.[19] Since Ga_2O_3 contains only $\frac{2}{3}$ cation for each anion, there will be cation vacancies in a solid solution of Ga_2O_3 in ZnO. Cation vacancies favor the uptake of excess zinc in form of zinc ions and excess electrons. Thus the electronic conductivity of the solid solution ZnO + Ga_2O_3 is higher than that of ZnO at a given oxygen partial pressure. Similar effects have been obtained with

[18] F. A. Kröger and H. J. Vink, "Relations between the Concentrations of Imperfections in Crystalline Solids," *Solid State Physics*, Vol. 3, F. Seitz and D. Turnbull, Eds., Academic Press, New York, 1956, p. 310.
[19] C. Wagner, *J. Chem. Phys.* 18, 62 (1950).

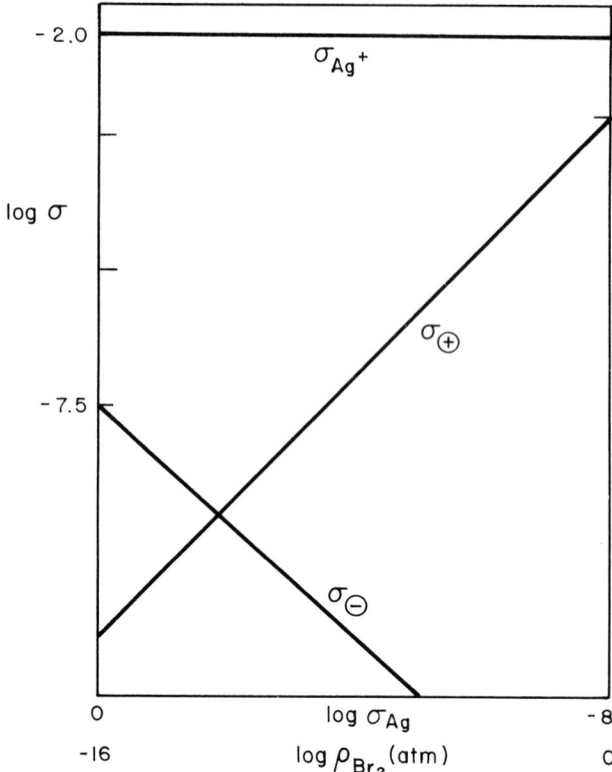

Fig. 12.2. Partial conductivities $\sigma_{Ag^+}, \sigma_-, \sigma_+$ for Ag^+ ions, excess electrons, and electron holes in AgBr as function of bromine pressure. (After Wagner.[12])

Al_2O_3 and Cr_2O_3 as solutes; the addition of Li_2O has the opposite effect.[20]

Upon dissolving Cr_2O_3 in NiO, the cation/anion ratio becomes less than unity, and the concentration of cation vacancies increases accordingly. Thus the uptake of excess oxygen,

$$\frac{1}{2}O_2(g) + Ni^{2+} \text{ (regular lattice site)}$$
$$+ 2 \text{ electrons (valence band)} \quad (12.1)$$
$$= NiO + \text{cation vacancy} + 2 \text{ electron holes,}$$

is diminished according to the law of mass action. Consequently, the electrical conductivity of NiO with Cr_2O_3 at 1000° C in an atmosphere producing a given oxygen partial pressure is less than that of NiO without Cr_2O_3.[21] The opposite is true for NiO with Li_2O, which is a so-called controlled-valency semiconductor with one Ni^{3+} ion (corresponding to an electron hole) for each Li^+ ion.[22] The conductivity of NiO containing Li_2O is thus essentially proportional to the Li_2O

[20] K. Hauffe and A. L. Vierk, *Z. physik. Chem.* 196, 160 (1950).
[21] K. Hauffe, *Ann. Phys.* [6] 8, 201 (1950).
[22] E. J. W. Verwey, P. W. Haayman, and F. C. Romeyn, *Chem. Weekblad* 44, 705 (1948); E. J. W. Verwey, *Bull. soc. chim. France* 1949, D 93.

content and independent of the oxygen partial pressure of the surrounding gas phase.[23]

Conduction in ternary ionic crystals

In ternary ionic crystals such as spinels, conduction phenomena are determined by the same defects as in binary ionic crystals. In addition, a crystal such as $MgAl_2O_4$ may contain defects due to "substitutional disorder" corresponding to an exchange of cations between different cation sublattices. Such defects do not contribute to electrical conduction but must be considered in the formulation of the law of mass action and stoichiometric relations.[24]

According to Gibbs, the characteristic properties of a ternary phase such as the concentrations of lattice defects and the electrical conductivity depend on four independent variables: pressure, temperature, and two mole fractions or two activities.[24,25] Thus the properties of a spinel, e.g., $CoCr_2O_4$, may be considered as functions of either the Co/Cr ratio and the metal/oxygen ratio in terms of equivalents, or of the Co/Cr ratio and the oxygen activity given by the oxygen partial pressure in a coexisting gas atmosphere at elevated temperature, or else of the activities of two components, such as CoO and oxygen.

The spinel $CoCr_2O_4$ exhibits mostly electronic conduction. The conductivity at 1005° C as function of oxygen partial pressure (obtained with the help of CO_2-CO mixtures) shows a distinct minimum at $p_{O_2} \sim 10^{-8}$ atm with predominant p-type conduction at higher, and n-type conduction at lower oxygen partial pressures.[26] In contrast, the dependence of this conductivity on the activity of CoO is much smaller. It varies only by a few per cent in the homogeneity range of the phase $CoCr_2O_4$ determined by coexistence with CoO or Cr_2O_3. The variability of the partial ionic conductivities is more significant, as follows from theoretical considerations and self-diffusion measurements.

Self-diffusion

Consider a AgBr crystal doped with radioactive ^{110}Ag and brought in contact with a normal AgBr crystal. The Ag/Br ratio is supposed to be equal to unity throughout the diffusion couple. Exchange diffusion takes place; i.e., equal amounts of inactive and radioactive silver move in opposite directions until a uniform distribution of radioactive ^{110}Ag is reached. This type of diffusion is called self-diffusion according to von Hevesy.

Self-diffusion takes place by the same elementary processes that are operative in ionic conduction. Thus a close relation between the rate of self-diffusion of Ag and the cationic conductivity may be expected. If there were only jumps of interstitial Ag^+ ions to adjacent empty interstices, essentially an independent motion of individual Ag^+ ions would result if the concentration of defects is small. Under these conditions, the self-diffusion coefficient D^*_{Ag} can be related by the Nernst-Einstein equation to the average mobility of Ag^+ ions in AgBr and thus to the cationic conductivity σ_{Ag^+} as

$$D^*_{Ag} = \frac{kT}{Ne^2}\sigma_{Ag^+}, \quad (12.2)$$

where k is the Boltzmann constant, T the temperature, N the total number of silver ions per unit volume, and e the electronic charge.

Actually there are also displacements of cations on normal lattice sites by interstitial ions and jumps of cations from regular lattice sites into vacancies. Under these conditions, the Nernst-Einstein relation holds only with a correlation coefficient f_{Ag} on the right-hand side of Eq. 12.2. The value of the correlation coefficient has been calculated theoretically for special models and found to be in general of the order of unity[27,28] (e.g., between 0.33 and 2.3). Much smaller values may occur in a body-centered cubic lattice, e.g., in TlCl.[28,29]

In principle, self-diffusion is possible by a direct interchange of cations on adjacent regular lattice sites in a single step (or by ring diffusion). If such a direct interchange were predominant, a much greater value of the self-diffusion coefficient would result than calculated from the Nernst-Einstein relation. This, however, has not been observed in ionic crystals. Direct interchange of adjacent ions on adjacent regular lattice sites seems to be negligible.

Since self-diffusion takes place via defects, the self-diffusion coefficient varies upon increasing or decreasing the number of such defects (by adding excess metal or nonmetal, small additions of other compounds involving a greater or smaller cation/anion ratio, etc.) The following examples may serve as illustration.

---

[23] K. Hauffe and J. Block, *Z. physik. Chem.* **198**, 232 (1951).
[24] H. Schmalzried and C. Wagner, *ibid.* (*N.F.*) **31**, 198 (1962).
[25] J. S. Prener, *J. Appl. Phys.*, Suppl., **33**, 434 (1962).
[26] H. Schmalzried, *Z. Elektrochem.* **67**, 93 (1963).
[27] A. N. Lidiard, "Ionic Conductivity," *Encyclopedia of Physics*, Vol. 20, Springer-Verlag, Berlin-Göttingen-Heidelberg, 1957, pp. 246 ff.
[28] R. J. Friauf, *J. Appl. Phys.*, Suppl., **33**, 494 (1962).
[29] R. J. Friauf, *J. Phys. Chem. Solids* **18**, 203 (1961).

The self-diffusion coefficient of iron in wüstite, $Fe_{1-\delta}O$, increases with the iron deficit δ (ranging from 0.05 to 0.13 at 900° C) in accord with the motion of iron ions via vacancies.[30]

The self-diffusion coefficient of iron in magnetite Fe_3O_4 at 1115° C increases with the oxygen partial pressure of the ambient gas phase by a factor of about 150, although there is only a very minor change in the composition of the solid. This also indicates a vacancy mechanism.[31]

The self-diffusion coefficients of the cationic components in spinels AB_2O_4 such as $CoCr_2O_4$, $CoAl_2O_4$, and $NiCr_2O_4$ have been determined in the spinel phase coexisting with the oxide of the divalent metal AO as well as in the spinel coexisting with the oxide of the trivalent metal B_2O_3. From the ratios of the self-diffusion coefficients for different activities of AO and B_2O_3, the majority defects in spinel phases have been deduced.[32]

Diffusion in substitutional solid solutions

Diffusion in substitutional solid solutions of ionic compounds of the same valence type, e.g., in the system AgCl-CuCl, corresponds closely to self-diffusion of cations. In solutions of ionic compounds of different valence type, the following points are noteworthy:[27]

1. In the system $AgCl-CdCl_2$ or $NaCl-CdCl_2$, interdiffusion exchange of a divalent cation (and a cation vacancy) for two monovalent cations is facilitated by the presence of extra cation vacancies.

2. Because of higher charge, however, the jump of a divalent cation to an adjacent cation vacancy requires a higher activation energy than that of a monovalent cation and therefore occurs at a lower rate.

3. The variability of the interdiffusion coefficient with composition changes the mathematics of diffusion. As a limiting case, the interdiffusion coefficient in the system $AgCl-PbCl_2$ may be set proportional to the concentration of $PbCl_2$ as the solute[33] except for very low concentrations of $PbCl_2$, where the intrinsic Frenkel disorder of AgCl prevails.

Diffusion in binary crystals involving a gradient of metal-to-nonmetal ratio

When Cu_2O samples involving different Cu deficiencies are brought in contact, diffusion takes place until a uniform Cu deficit is attained. This type of diffusion has not been studied widely, because deviations from the ideal stoichiometric ratio are normally small and therefore difficult to measure. Diffusion in binary compounds involving a gradient of composition, however, is an important phenomenon and instrumental during oxidation of metals at elevated temperatures.

The rate of oxidation of a metal is frequently inversely proportional to the instantaneous thickness of the oxide layer. Accordingly, the thickness increases in proportion to the square root of time,[34,35] indicating that the rate is essentially diffusion controlled. If Cu is oxidized in air at 1050° C, only Cu_2O is formed. At the Cu/Cu_2O interface, Cu_2O has nearly ideal stoichiometric composition. At the Cu_2O/gas interface, Cu_2O contains a small excess of oxygen corresponding to a deficit of Cu. Thus, there is a gradient in the metal/oxygen ratio corresponding to a gradient of the activity of metal or oxygen as a necessary condition for diffusion. In view of the results of the electrochemical investigations discussed earlier, one may interpret[36-38] oxidation of Cu to Cu_2O in terms of migration of Cu^+ ions via Cu^+ vacancies and of electrons via holes from the metal through the oxide layer to the outer oxide surface where the reaction with gaseous O_2 takes place:

$$4Cu^+ (\rightarrow \text{in } Cu_2O) + 4e^- (\rightarrow \text{in } Cu_2O) + O_2(g) = 2Cu_2O(s). \quad (12.3)$$

If this mechanism is assumed, one may calculate the rate of oxidation from conductivity and transference data as functions of the oxygen partial pressure. The experimental rate has been found essentially in accord with the calculated rate,[38] thus confirming that mechanism.

The transport rates of Cu^+ ions and electrons across a Cu_2O layer must be virtually equal, because each volume element must remain neutral. Since electronic conduction prevails, the rate of oxidation of Cu is essentially determined by the motion of the slow Cu^+ ions via vacancies, caused in part by a concentration gradient and in part by a diffusion potential resulting from the tendency of the electrons to move ahead.

Similarly, one may interpret other phenomena of high-temperature oxidation of metals (dry corrosion)

---

[30] L. Himmel, R. F. Mehl, and G. E. Birchenall, *Trans. Am. Inst. Mining Engrs.* 197, 827 (1953).

[31] H. Schmalzried, *Z. physik. Chem.* (N.F.) 31, 184 (1962).

[32] A. Morkel and H. Schmalzried, *ibid.* (N.F.) 32, 76 (1962).

[33] C. Wagner, *J. Chem. Phys.* 18, 1227 (1950).

[34] G. Tammann, *Z. anorg. u. allgem. Chem.* 111, 78 (1920).

[35] N. B. Pilling and R. E. Bedworth, *J. Inst. Metals* 29, 529 (1923).

[36] C. Wagner, *Z. physik. Chem.* B25, 21 (1933).

[37] C. Wagner, "Diffusion and High Temperature Oxidation," *Atom Movements*, American Society for Metals, Cleveland, Ohio, 1951, p. 153.

[38] C. Wagner and K. Grünewald, *Z. physik. Chem.* B40, 455 (1938).

in terms of migration of cations (or anions) and electrons. The same is true for reactions of metals with sulfur, selenium, chlorine, etc.[36,37].

In AgBr, cationic conduction prevails. Thus, the formation of AgBr on solid Ag in bromine gas is essentially determined by the motion of electrons via holes.[39]

Additions with a different cation/anion ratio may change the defect concentrations in an oxide significantly. Accordingly, significant changes of the oxidation rate of metals in the presence of alloying constituents have been observed.[40]

Diffusion in ternary compounds

The state of a ternary compound such as $CoAl_2O_4$ or $AgSbS_2$ at a given pressure and temperature is determined by two independent concentrations or activities. Accordingly, the diffusion rates in a ternary compound are determined by two independent concentration or activity gradients, as the following cases illustrate.

The rate of formation of a spinel AB_2O_4 from the oxides AO and B_2O_3 at a constant oxygen activity has been investigated by many authors.[41] After a short induction period, the reactants are separated from each other by a layer of the reaction product AB_2O_4. Accordingly the reaction can proceed only as diffusion in AB_2O_4 can take place. In general, the oxygen activity is constant, and there is only an activity gradient for AO (and likewise for B_2O_3). If only migration of the smaller cations is considered, the formation of a spinel may be formulated in terms of counterdiffusion of A^{2+} and B^{3+} ions in equivalent amounts (Fig. 12.3a). Other simple limiting cases involving the diffusion of cations A^{2+} or B^{3+} and O^{2-} ions are shown in Figs. 12.3b and 12.3c. Furthermore, especially in powder mixtures of AO and B_2O_3, either A^{2+} ions and electrons or B^{3+} ions and electrons may possibly migrate in solid AB_2O_4, while oxygen is transported through the gas phase in form of O_2 molecules (Fig. 12.3d and 12.3e). Assuming independent migrations of the various ions, one may calculate diffusion rates from conductivity data and the free energy of formation of the spinel from AO and B_2O_3 as the driving force.[42,43] Instead of conductivity data, self-diffusion data related to conductivity data may be used.

Recent experimental investigations[43] are available

[39] C. Wagner, Z. physik. Chem. B32, 447 (1936).
[40] K. Hauffe, Progress in Metal Physics 4, 71 (1952).
[41] J. A. Hedvall and R. Lindner, Einführung in die Festkörperchemie, Friedrich Vieweg und Sohn, Braunschweig, 1952, pp. 162 ff.
[42] C. Wagner, Z. physik. Chem. B34, 309 (1936).
[43] H. Schmalzried, ibid. (N.F.) 33, 111 (1962); H. Schmalzried and W. Rogalla, Naturwissenschaften 50, 593 (1963).

$$AO(s) \mid AB_2O_4(s) \mid B_2O_3(s)$$

(a) $\xrightarrow{3A^{2+}}$ $\xleftarrow{2B^{3+}}$

$2B^{3+} + 4AO$ $3A^{2+} + 4B_2O_3$
$= 3A^{2+} + AB_2O_4$ $= 2B^{3+} + 3AB_2O_4$

(b) $\xrightarrow{A^{2+},O^{2-}}$

AO $A^{2+} + O^{2-} + B_2O_3$
$= A^{2+} + O^{2-}$ $= AB_2O_4$

(c) $\xleftarrow{2B^{3+},3O^{2-}}$

$2B^{3+} + 3O^{2-} + AO$ B_2O_3
$= AB_2O_4$ $= 2B^{3+} + 3O^{2-}$

(d) $\xrightarrow{A^{2+},2e^-}$

AO $A^{2+} + 2e^-$
$= A^{2+} + 2e^- + \frac{1}{2}O_2(g)$ $+ \frac{1}{2}O_2(g) + B_2O_3$
    $= AB_2O_4$

(e) $\xleftarrow{2B^{3+},6e^-}$

$2B^{3+} + 6e^-$ B_2O_3
$+ \frac{3}{2}O_2(g) + AO$
$= AB_2O_4$ $= 3B^+ + 6e^- + \frac{3}{2}O_2(g)$

Fig. 12.3. Mechanism of the reaction $AO(s) + B_2O_3(s) = AB_2O_4(s)$.

for the formation of $NiAl_2O_4$, $CoAl_2O_4$, $CoCr_2O_4$, and Co_2TiO_4. The dependence of the rate and the mechanism on the composition of the gas atmosphere has been investigated especially for the reaction

$$2CoO(s) + TiO_2(s) = Co_2TiO_4(s), \quad (12.4)$$

with single crystals of TiO_2 embedded in CoO powder at 1200° C. With the help of Pt markers it was found that the mechanism in an inert gas atmosphere is essentially a counterdiffusion of cations. In oxygen of atmospheric pressure, Co^{2+} ions and electrons migrate in solid Co_2TiO_4, and oxygen is transported through the gas phase.

The mechanism of formation of double halides and sulfides from the constituent simple compounds is analogous to that of the double oxides. The formation of K_2SrCl_4 from KCl and $SrCl_2$ at 530° C proceeds essentially by virtue of migration of K^+ and Cl^- ions in K_2SrCl_4,[44] while that of Ag_2HgI_4 from AgI and HgI_2 at 65° C is due to counterdiffusion of Ag^+ and Hg^{2+} ions in Ag_2HgI_4.[45] Likewise, the formation of $AgSbS_2$ from Ag_3SbS_3 and Sb_2S_3 at 400° C,

$$Ag_3Sb_3(s) + Sb_2S_3(s) = 3AgSbS_2(s), \quad (12.5)$$

[44] H. Schmalzried, Z. physik. Chem. (N.F.) 33, 129 (1962).
[45] E. Koch and C. Wagner, Z. physik. Chem. B34, 317 (1936).

is due to counterdiffusion of Ag^+ and Sb^{3+} ions.[46] The rate is essentially determined by the motion of Sb^{3+} ions of a mobility much lower than that of Ag^+ ions because of the triple charge.

When ternary compounds such as $CoFe_2O_4$ or $AgSbS_2$ are formed during the scaling of an alloy at elevated temperatures, there arises a gradient of the activity of the nonmetallic component. To define conditions completely, a second activity gradient must be given. Instead, one may use the gradients of the activities of the two metals as the decisive variables. Experimental investigations are available for $AgSbS_2$.[46] When a Ag-Sb alloy is exposed to sulfur, a rather complex scale is formed, one of the constituent phases is $AgSbS_2$. In view of the rather high mobility of Ag^+ ions and electrons in $AgSbS_2$, there is a practically constant activity of Ag and only a gradient of the activity of Sb. The rate of diffusion of Sb in form of Sb^{3+} ions and electrons is essentially determined by the rate of migration of Sb^{3+} ions (supported by a potential gradient) and therefore closely related to the rate of reaction (Eq. 12.5), in accord with experimental results. On the other hand, one may conduct investigations with a gradient of the activity of Ag with rather rapid diffusion of Ag in form of Ag^+ ions and electrons, e.g., experiments involving the displacement reaction

$$3Ag \text{ (in } Ag_3Sb) + 2Sb_2S_3(s)$$
$$= 3AgSbS_2(s) + Sb(s). \quad (12.6)$$

The gradient of the activity of Sb and likewise the transport of Sb in $AgSbS_2$ are immaterial because the mobility of Sb^{3+} ions is very low.

Transport phenomena with diffusion in different phases

To demonstrate the possibility of local cell action during high-temperature corrosion of metals and alloys, the following experiment was made:[47] A Ta wire covered with Ag at one end was heated at 174° C in I_2 vapor at 575 torr. A AgI layer was observed to grow not only on the Ag surface but also to spread rapidly along the surface of the Ta wire (1.8 cm within 79 minutes, while the AgI layer on Ag reached a thickness of only 0.046 cm). This observation is explained by the mechanism shown schematically in Fig. 12.4 with Ag^+ ions migrating sidewise in the AgI layer and electrons migrating in the Ta wire (and across the AgI layer). The observed rate of the sidewise growth of the AgI

[46] H. Rickert and C. Wagner, *Z. Elektrochem.* 64, 793 (1960); 66, 502 (1962).
[47] C. Ilschner-Gensch and C. Wagner, *J. Electrochem. Soc.* 105, 198 (1958).

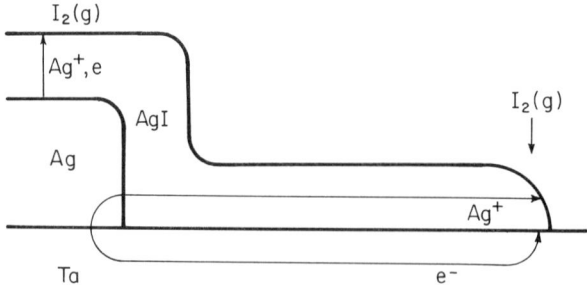

Fig. 12.4. Reaction of $I_2(g)$ with silver on tantalum.

layer on Ta is in accord with the rate calculated theoretically from the free energy of formation of AgI as the driving force and the Ag^+ conductivity of AgI.

A similar phenomenon took place when the formation of $AgSbS_2$ from Ag_2S or Ag_3SbS_3 and Sb_2S_3 tablets pressed against each other under sulfur vapor was investigated.[48] Under these conditions, $AgSbS_2$ is formed not only between the Ag_2S (or Ag_3SbS_3) and the Sb_2S_3 tablet but also spreads along the Sb_2S_3 surface by virtue of migration of Ag^+ ions in the $AgSbS_2$ film, migration of electrons in $AgSbS_2$ and Sb_2S_3, and transport of sulfur in the gas phase.

Nonisothermal transport

When a Ag_2S rod is heated above 180° C in a furnace involving a temperature gradient in the presence of sulfur vapor, some Ag_2S disappears at the hot end of the sample while the same amount of Ag_2S is formed at the cold end.[49] This transport phenomenon may be described in terms of migration of Ag^+ ions and electrons in Ag_2S and migration of sulfur molecules in the gas phase. The rate of this process may be calculated with the help of equations of irreversible thermodynamics and measurements of the gradient of the activity of Ag in a nonisothermal Ag_2S sample under steady-state conditions, where transport in the gas phase is excluded and therefore no transport in Ag_2S takes place. In view of the high mobility of Ag^+ ions and electrons in Ag_2S, this kind of transport is fairly rapid. For a temperature difference between 250° and 350° C over a distance of 3.5 cm and a cross section of 0.11 cm², a transport of 0.43 g Ag_2S within 18 hours has been observed.

The use of ionic conductors for thermodynamic measurements

Measurements of the emf of galvanic cells are very helpful in determining thermodynamic data, such as

[48] H. Rickert and C. Wagner, *Z. Elektrochem.* 66, 502 (1962).
[49] H. Rickert and C. Wagner, *ibid.* 67, 621 (1963).

free energies of formation or activities. In particular, many data have been obtained with the help of cells involving aqueous solutions at room temperature. Likewise, cells with liquid or solid electrolytes may be employed at elevated temperatures. Solid electrolytes offer special advantages, because cells can be set up in a very compact form.

As an example, consider the cell[50]

$$Ag(s) \mid AgI(s) \mid Ag_2S(s) \mid C(s), S(l), \quad (12.7)$$

in which AgI is essentially an ionic conductor. Thus on passing 1 faraday, 1 mole Ag is transferred from the pure state into Ag_2S coexisting with liquid sulfur. This corresponds to the reaction

$$Ag(s) + \tfrac{1}{2}S(l) = \tfrac{1}{2}Ag_2S(s). \quad (12.8)$$

Upon equating the electrical work done on Cell 12.7 and the change in the Gibbs free energy ΔF^0 of Reaction 12.8, one has

$$\Delta F^0 = -E\mathfrak{F}, \quad (12.9)$$

where E is the voltage of Cell 12.7 and $\mathfrak{F}$ the Faraday constant.

With the help of cells analogous to Cell 12.7, the standard free energies of formation of other metal compounds, such as oxides or halides, may be determined.[50] Likewise one may determine the free-energy changes of displacement reactions,[50] such as $Co(s) + Cu_2O(s) = 2Cu(s) + CoO(s)$, or the formation of spinels[51] and silicates[52,53] with the help of galvanic cells. For this purpose it is necessary to use ionic compounds with mostly ionic and practically negligible electronic conduction (< 0.1 percent). One electrolyte used in many investigations is ZrO_2 with the addition of about 10 mole percent CaO providing anion vacancies. The emf of a cell with different oxygen partial pressures p_{O_2}' and p_{O_2}'' at the electrodes,

$$\text{electrode I}(p_{O_2}')\mid ZrO_2(s)\mid\text{electrode II}(p_{O_2}'') \quad (12.10)$$

is given by the Nernst equation

$$E = \frac{RT}{4}\ln\frac{p_{O_2}''}{p_{O_2}'}. \quad (12.11)$$

Assume that p_{O_2}'' is known, e.g., by using a Pt lead in oxygen of atmospheric pressure at the right-hand side of Cell 12.10, or by using a metal-oxide mixture, e.g., Co-CoO. Here one can measure E and calculate the oxygen partial pressure p_{O_2}' at the left-hand side

[50] K. Kiukkola and C. Wagner, *J. Electrochem. Soc.* **104**, 308, 379 (1957).
[51] H. Schmalzried, *Z. physik. Chem.* (*N.F.*) **25**, 178 (1960).
[52] R. Benz and H. Schmalzried, *ibid.* **29**, 77 (1961).
[53] R. Benz and C. Wagner, *J. Phys. Chem.* **65**, 1308 (1961).

of Cell 12.10. In this manner, e.g., the oxygen equilibrium pressure for phases of the system U-O as a function of the mole fraction of oxygen has been determined.[54] Weissbart and Ruka[55] described a simple gauge consisting of a ZrO_2 tube with an inner and an outer Pt lead. Satisfactory tests have been reported for oxygen pressures ranging from 10^{-5} to 1 atm between 600° and 750° C. According to Peters and Möbius[56] and Schmalzried,[57] a gauge of this type can also be used to determine the oxygen-equilibrium partial pressure in CO_2-CO or H_2O-H_2 mixtures down to about 10^{-24} atm at 750° C. Under still more reducing conditions, however, n-type electronic conduction in ZrO_2 becomes noticeable and the Nernst equation (Eq. 12.11) is no longer obeyed.[57,58] This limitation may be overcome to some extent by using $ThO_2(+Y_2O_3)$ as a solid electrolyte.[59]

The use of ionic conductors in fuel cells

In recent years, ZrO_2 (doped with CaO) and other oxides have been tested as solid electrolytes in fuel cells.[60] The conductivity reaches a sufficiently high value only above 800° C.

Upon passing current across the cell

$$Pt(s), CO(g), CO_2(g)\mid ZrO_2(s)\mid Pt(s), O_2(g), \quad (12.12)$$

one has the anodic reaction

$$CO(g) + O^{2-} (\leftarrow \text{in } ZrO_2)$$
$$= CO_2(g) + 2e^- \text{ (in Pt)} \quad (12.13)$$

and the cathodic reaction

$$\tfrac{1}{2}O_2(g) + 2e^- (\leftarrow \text{in Pt}) = O^{2-} \text{ (in } ZrO_2). \quad (12.14)$$

Even at 1000° C, no undesirable changes in the structure and the geometry of ZrO_2 components are likely to occur. In spite of some disadvantages, a high operating temperature is attractive, because polarization of the electrode processes is low, and especially the anodic oxidation of cheap fuels, such as CO or hydrocarbons, is easier to accomplish than at lower temperatures.

[54] S. Aronson and J. Belle, *J. Chem. Phys.* **29**, 151 (1958).
[55] J. Weissbart and R. Ruka, *Rev. Sci. Instr.* **32**, 593 (1961).
[56] H. Peters and H. H. Möbius, *Z. physik. Chem.* (*N.F.*) **209**, 298 (1958).
[57] H. Schmalzried, *Z. Elektrochem.* **66**, 572 (1962).
[58] H. Schmalzried, *Z. physik. Chem.* (*N.F.*) **38**, 81 (1963).
[59] C. B. Alcock, private communication.
[60] *Fuel Cells*, G. J. Young, Ed., Vols. 1 and 2, Reinhold Publishing Corp., New York, 1960 and 1963, with contributions by G. H. J. Broers and J. A. A. Ketelaar, G. H. J. Broers and M. Schenke, E. B. Shultz, Jr., K. S. Vorres, L. G. Marianowski and H. R. Linden, J. Weissbart, and R. Ruka.

13 · LUMINESCENCE IN SOLIDS

James H. Schulman

General aspects of luminescence — Luminescence efficiency — Types of activators — Configuration-coordinate diagrams — Temperature dependence of absorption and emission — Derivation of configuration-coordinate diagrams — Polarized luminescence of centers — Systems involving energy transfer — Resonance transfer — Forbidden transitions — Other aspects of energy transfer

General aspects of luminescence

A discussion of luminescence might well be subtitled: "From the Lightning Bug to the Laser." Such a title would express the wide variety of phenomena and systems that have a claim to be considered under the heading of luminescence. From the complex biochemical system that nature has evolved over the ages for the lightning bug to the comparatively simpler physical systems that have been devised as a result of the recent rapid development of the laser, the field of luminescence has been intimately concerned with the constitution and structure of inorganic and organic materials in the gas, liquid, and solid phases, and with their spectroscopy from X rays to at least the infrared.[1-9] Luminescence is therefore of interest not only because it has led to technically important devices such as the cathode-ray tube, the fluorescent lamp, and the scintillation counter, but because it is an interdisciplinary study that rests upon and contributes to our understanding of the structure of matter and its interaction with energy.

In a short survey such as this it is clearly impossible to explore all the aspects of luminescence that are deserving of discussion. Since we are primarily interested in the molecular designing of materials and devices, we shall have little to say about the lightning bug; in the present state of the art the know-how (and the motivation) for designing this particular device is the province only of other lightning bugs. The laser, on the other hand, merits special attention, which is given to it elsewhere in this volume.

Our discussion will be focused on the class of inorganic solid phosphors. These are insulating crystals (like the alkaline-earth oxides, silicates, phosphates),

---

[1] D. Curie, *Luminescence in Crystals* (Translated by G. F. J. Garlick), John Wiley and Sons, New York, 1963.

[2] F. A. Kröger, *Ergeb. exakt. Naturw.* 29, 62 (1956).

[3] C. C. Klick and J. H. Schulman, "Luminescence in Solids," *Solid State Physics*, Vol. 5, F. Seitz and D. Turnbull, Eds., Academic Press, New York, 1957, p. 97.

[4] F. E. Williams, "Solid State Luminescence," *Advances in Electronics and Electron Physics*, Vol. 5, L. Marton, Ed., Academic Press, New York, 1953, p. 137.

[5] H. W. Leverenz, *Luminescence in Solids*, John Wiley and Sons, New York, 1950.

[6] G. F. J. Garlick, *Luminescent Materials*, Clarendon Press, Oxford, 1949.

[7] P. Pringsheim, *Fluorescence and Phosphorescence*, Interscience Publishers, New York, 1949.

[8] *Luminescence of Organic and Inorganic Materials* (Proc. Internatl. Conf. on Luminescence), H. P. Kallman and G. M. Spruch, Eds., John Wiley and Sons, New York and London, 1962.

[9] T. Förster, *Fluoreszenz organischer Verbindungen*, Vandenhoek und Ruprecht, Göttingen, 1951.

or semiconducting crystals (like zinc and cadmium sulfides and selenides). The vast majority of these materials owe their luminescence to small concentrations of selected chemical impurities, structural defects, or combinations of these two; these entities are called "activators." The luminescence that we shall discuss is a spontaneous incoherent light emission which can be excited by various means, including light absorption or particle bombardment. These emissions occur in the visible and in the near ultraviolet and infrared regions of the spectrum. Most of the emission spectra consist of rather broad bands, with spectral widths of a few tenths of an electron volt.

The immediate task of luminescence research is to understand how the efficiency, spectral distribution, polarization, and time dependence of the light emission from a system are determined by its composition, structure, and state of aggregation; by the type and intensity of the excitation energy; and by certain external variables, particularly temperature.

Luminescence efficiency

Of the various properties enumerated, the one that concerns us first is the efficiency, for it is the extraordinary efficiency of certain luminescent systems that attracts our attention and leads to useful applications. On the other hand, the low efficiency of most materials raises a number of provocative questions.

At least two steps are involved in luminescence: the absorption of energy and its re-emission as light. In many systems these two processes do not take place in the same "center" (i.e., in the same atom, ion, molecule, or atomic cluster). In these cases, further processes are involved—the transport of the energy from the absorbing center to the site of the emitting center, and the transfer of the energy to the emitter. Obviously, the net luminescence efficiency will depend on the effectiveness with which energy is protected from loss in other ways at every stage of events—from its incidence upon the system up to and including its residence on the emitting center. Since our primary interest is in the efficiency of the luminescent center itself, we shall consider the simplest case, in which the center is directly excited by a photon; in this event we do not have to contend with energy losses in the absorption or excitation process, nor in processes of energy transport and transfer.

Even in this elementary case we almost invariably find that radiationless processes compete with the emissive process as a means of disposing of the absorbed energy. The luminescence efficiency is determined by this competition. As a simple illustration, we may leave the range of optical frequencies for a moment and consider photoluminescence in the X-ray region. When an X-ray photon ejects an electron from the K shell of an atom, this atom, now ionized in the inner shell, is the luminescent center, and the emission that it is potentially able to emit is, of course, the characteristic K X-ray radiation of the atom.

Even under this ideal circumstance—where the electron involved in the luminescent transition lies in a deep inner shell and is therefore well shielded from interaction with external atoms—an important internal dissipative process exists which cuts down the fluorescence yield. This process is the Auger effect.[10] Instead of a radiative transition of an L electron to fill the hole in the K shell, the $L \rightarrow K$ transition energy can be used to expel a second L electron from the atom, leaving the atom doubly ionized in the L shell. The transition probability for this dissipative (or "internal conversion") process is almost independent of the atomic number of the element, while the transition probability for a radiative transition increases approximately with the fourth power of the atomic number. The photofluorescence yield, therefore, is very low for the lighter elements, and comes close to unity only for the heavier ones. As we go to somewhat lesser shielding from the surroundings, matters get considerably worse. The total yield for M X-ray fluorescence in uranium is only 6 percent, although here we are still exciting the atom deep inside a quite well-shielded shell of electrons.

If we now return to consider the optical region of the spectrum, where the radiative lifetime is of the order of 10^{-8} to 10^{-9} sec instead of 10^{-15} sec for X-ray emission, we see that many other types of competitive radiationless processes will have time to intervene and cut down the luminescence yield. Furthermore, if the luminescent centers are not isolated atoms but atoms crowded together under pressure in the gas phase by combination into molecules or by condensation into the liquid and solid states of aggregation, we find many new paths for radiationless de-excitation: excitation of vibrational and rotational motion of the molecules or atoms comprising the luminescent center, transfer of the excess energy to other atoms or molecules by collision or similar processes, or chemical dissociation. It should not surprise us, therefore, that efficient luminescence is the exception rather than the rule. Nor should it be surprising, although it is admittedly disappointing, that there is no single pattern of molecular architecture that can be relied upon to build a structure into which we can deposit the energy of an optical-

---

[10] L. H. S. Burhop, *The Auger Effect*, Cambridge University Press, Cambridge, 1952.

frequency photon with the assurance that it can be retrieved more or less in one piece.

Types of activators

One of the architectural design principles often applied in the attempt to achieve efficiency is the use of activators such as the rare-earth ions, whose electronic transitions occur in well-shielded inner electronic shells. Although, as we have noted in the case of X-ray fluorescence, this principle does not guarantee success, it often does work; in the successful cases the luminescence has properties that might be expected to result from electronic transitions of this type. The emission spectra of a given rare-earth ion in a wide variety of host matrices resemble each other closely, as do the absorption spectra; and they often consist of rather sharp lines that are closely related to the line spectra of the free ion. But the principle is not universally effective, and there are many matrices in which the luminescence of the rare-earth ions is quenched.

Furthermore, there are a large number of efficient activators in which the transitions involve electrons in outer shells, unshielded from interaction with the electrons of the neighboring atoms to which they are bonded. Perhaps the most striking illustration of such a case is the F center in the alkali halides.[11] This center consists of an electron trapped at a negative ion vacancy in the crystal, with a wave function that spreads out over more than its nearest neighboring ions. One could hardly conceive of a more poorly shielded situation; the absorption spectrum of the F center is broad and depends on the lattice constant of the crystal and on the temperature, thus betraying the strong interaction of the trapped electron with its environment. Yet, under appropriate circumstances (low concentrations and low temperatures) the F center luminesces with high efficiency. With this system, as with many others, the balance between emissive and dissipative processes is so delicate that the luminescence is quenched if two F centers are situated closer than 14 lattice spacings from each other.[12]

We see, therefore, that the existence of inner-shell transitions is neither a necessary nor a sufficient condition for efficient luminescence. In general, both experiment and theory show that even with a given class of luminescent solids, the specific details of the coupling between the emitting atom and its surroundings have a strong influence on the processes that follow the excitation of the atom, and hence on the luminescence yield.

Configuration-coordinate diagrams

A great deal has been learned about the factors that influence the efficiency and other properties of a luminescent system from a study of so-called configuration coordinate diagrams, which are frequently used[1-9] to describe the luminescent center. These diagrams are attempts to represent the potential energy of the various electronic states of the center in terms of a single coordinate that characterizes the center. If the center consists of a vibrating diatomic molecule, for example, the configuration-coordinate diagram is identical with the ordinary potential energy diagram, in which the potential energy in the ground state and in the various electronically excited states is plotted as a function of the internuclear distance. In this case the internuclear distance is the configuration coordinate. In a more complex center, consisting of a central positively charged impurity ion and its nearest shell of anion neighbors in a solid, there are several modes of vibration of the center (Fig. 13.1).[13] If it is assumed

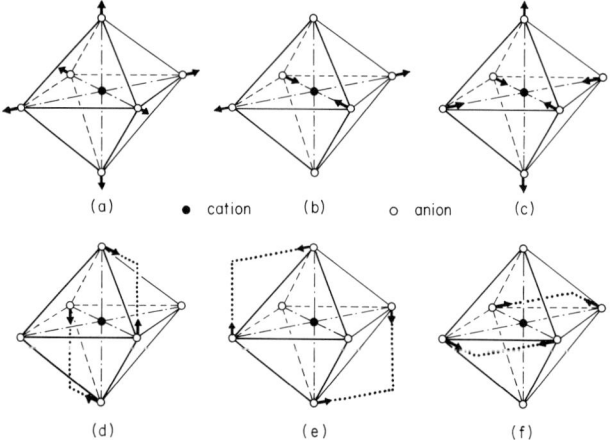

Fig. 13.1. Vibrational modes of a luminescent center, a central cation octrahedrally surrounded by anions.[23]

that one of these modes, i.e., the radial or "breathing" mode (Fig. 13.1a), is the one which has the greatest influence on the energy of the center, the configuration coordinate can be taken as the distance between the central ion and any one of its equidistant anion neighbors.

[11] J. H. Schulman and W. D. Compton, *Color Centers in Solids*, A Pergamon Press Book, The Macmillan Company, New York, 1962.

[12] A. Miehlich, a paper presented at the International Symposium on Color Centers in the Alkali Halides, Stuttgart, Germany, 1962.

[13] H. Kamimura and S. Sugano, *J. Chem. Phys. Soc. Japan* 14 (11), 1612 (1959).

Figure 13.2 shows the configuration-coordinate diagram of such a center. The minimum in the lower curve, which describes the potential energy of the system in its electronic ground state, occurs at the equilibrium separation between the central ion and its neighbors. At all temperatures above absolute zero, thermal vibrations will cause displacements of the order of kT from the equilibrium positions. The upper curve shows the potential energy of the first electronically excited state of the center. Because the interatomic forces in the excited state are generally weaker than those in the ground state, the top curve has correspondingly less curvature; its minimum is also generally displaced with respect to the minimum of the ground-state curve.

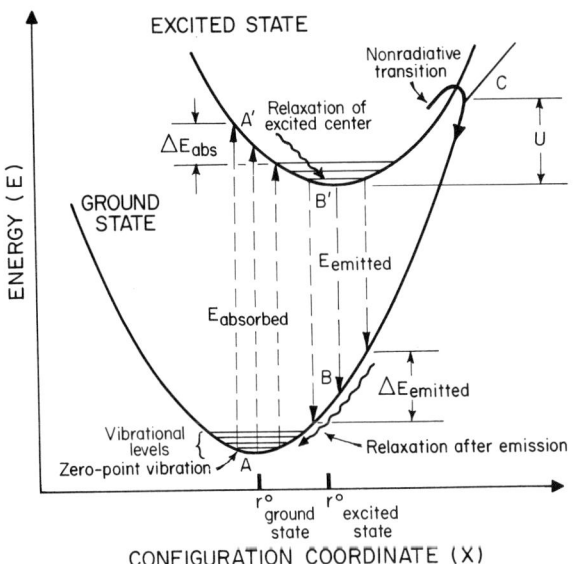

Fig. 13.2. Schematic configuration-coordinate diagram of a luminescent center.

$E_{B'} = E_A = E_0$; $r_{\text{excited state}} - r_{\text{ground state}} = X_0$.

To describe the luminescence process with the aid of this diagram, we must recall the Franck-Condon principle, which states that an electronic transition involving absorption or emission of a photon takes place in a time interval much shorter than that of nuclear motions. Such absorptive or emissive transitions are therefore represented by vertical lines on the diagram. The process of photoluminescence in the center is then as follows: Absorption of a photon, $h\nu_a$, excites the system from a point such as A on the ground-state curve to a point A' on the excited-state curve. Since the system at A' is not in its equilibrium configuration, it will dissipate its excess vibrational energy nonradiatively to neighboring molecules, thereby arriving at the point B' on the curve. Here it will remain for a time that depends on the probability of the optical transition $B'B$. When this transition occurs, a photon $h\nu_f$ is emitted. The system is now back in the ground electronic state, but with an excess of vibrational energy. This excess energy is also dissipated to the surroundings, and the system returns to its original condition.[14]

From this representation it is clear that the energy of the emitted photon is less than that of the absorbed photon, the difference being due to the energy dissipated as heat in the relaxations of the center from $A' \to B'$ and $B \to A$. The corresponding relationship between the wavelengths of the emitted light and the absorbed light was first observed experimentally by Stokes more than one hundred years ago and is now known as Stokes's law.[15]

The diagram also shows how a radiationless transition can take place from the excited state to the ground state by an "internal conversion" process. The curves for these two states approach very closely (or "cross") at point C. If the activation energy $U (= B'C)$ is provided, the center can make this transition without emitting radiation, after which it relaxes to the minimum of the ground-state curve by giving its excess vibrational energy to the surroundings. All of the absorbed energy is thus dissipated nonradiatively. The probability of such a radiationless transition, P_{diss}, is proportional to $\epsilon^{-U/kT}$, while the radiative transitional probability, P_{lum}, is independent of temperature. The quantum efficiency of luminescence is thus

$$\eta = (1 + \text{const } \epsilon^{-U/kT})^{-1}. \qquad (14.1)$$

The dissipative transition is therefore favored by an increase and the radiative transition by a decrease of temperature. This general result is confirmed by experiment; a great many systems that are nonluminescent at room temperature luminesce quite efficiently at liquid-nitrogen or liquid-helium temperature.[16]

From measurements of the quantum efficiency as a function of temperature one can obtain the activation energy U. A good activator has a high value of U. When the doping agent or defect has a low value of the activation energy, the absorbed energy is easily degraded as heat, and the center is called a "killer" center or "poison."

[14] A. von Hippel. *Z. Physik* **101**, 680 (1936).
[15] G. Stokes, *Cambridge Phil. Soc.* **9**, 8 (1851).
[16] F. A. Kröger, *Some Aspects of the Luminescence of Solids*, Elsevier Publishing Co., Amsterdam, 1948.

Temperature dependence of absorption and emission

From configuration-coordinate curves some general ideas can be obtained about the shape and temperature dependence of the absorption and emission bands of a center. From Fig. 13.2 it can be seen that the spread of energies in an optical transition increases with the amplitude of oscillation of the center. On a purely classical picture, for an assumption that the center can be represented by a harmonic oscillator, it follows rather simply[17] that the bandwidth in absorption and emission is proportional to $T^{1/2}$. This predicts that all absorption and emission spectra should narrow to sharp lines at very low temperature. Experimentally the $T^{1/2}$ law is approached at higher temperatures but is seriously in error at low temperatures. Experiment shows that the bands do not narrow as fast as this law would predict at low temperatures, and that the width remains as high as a few tenths of an electron volt even when the temperature is lowered to near absolute zero. This behavior has been explained on the basis of a quantum-mechanical description of the center[18]. Because of the zero-point energy of a quantized oscillator, the lowest occupied vibrational level of the center lies $\frac{1}{2} h\nu_{vib}$ above the minimum of the classical curve (cf. Fig. 13.2).

If the ground and excited states are parabolic (harmonic oscillator), the classical formula for absorption or emission spectra can be written in terms of the constants of the curves

$$P(E) = \left(\frac{K}{2\pi kT}\right)^{1/2} \exp\left(-\frac{KX^2}{2kT}\right) \frac{dX}{dE}, \quad (13.2)$$

where X is the displacement from the minimum in the initial state, E the energy difference between the two curves at the configuration coordinate X, and K the force constant for the initial state. This equation can be used in the quantum-mechanical case[19] if T is replaced by an effective temperature

$$T_{eff} = \frac{h\nu}{2k} \coth\left(\frac{h\nu}{2kT}\right), \quad (13.3)$$

with ν the vibrational frequency of the center in the initial state. At high temperature, T_{eff} reduces to T; at low temperature, it approaches the constant value $h\nu/kT$. The expression for the width of the absorption band is

$$W_{abs} = (\text{const}) \left(\frac{h\nu_{ground}}{2k}\right) \coth\left(\frac{h\nu_{ground}}{2kT}\right)^{1/2}; \quad (13.4)$$

the emission bandwidth is given by a similar expression with $\nu_{excited}$ replacing ν_{ground}.

If an analysis is carried through with initial and final states of the center quantized, low-temperature absorption or emission spectra are predicted which consist of a large number of closely spaced lines. Such fine structure is not generally observed. Because the transition usually takes place to a portion of the final-state curve where the quantum-mechanical curve approaches the classical curve (i.e., to an energy many vibrational levels above the minimum), reasonable accuracy is achieved in most cases by treating only the initial state as a quantized harmonic oscillator and using the classical curve for the final state.

Derivation of configuration-coordinate diagrams

Considerable effort has been spent on the difficult task of deriving the curves theoretically. Starting with the energy levels of the free activator ion, one can compute how these are modified by interaction with the host lattice in which the ion is incorporated. This approach has been pursued very extensively with the thallium-activated alkali halide phosphors. The framework of the theoretical investigations of these systems was established by Seitz,[20] who proposed that the various absorption bands in the KCl:Tl phosphor were due to transitions between the ^1S and the $^3P_1$ and $^1P_1$ states of the thallium ion. Williams[21] and others have carried out quantitative calculations, but so far without completely satisfying results. This problem was not made any easier by the unavailability of good wave functions for the thallium ion, which necessitated the introduction of a certain amount of empirical data.[22] Another source of difficulty is that continued experimental work has unearthed new facts about this system, which the original theory cannot account for.[23,24]

An empirical approach to configuration-coordinate curves, pioneered by Klick,[25] has been much less ambitious in its objective but has nevertheless had important and useful consequences. One can readily derive the relationship between the configurational coordinate parameters (E_0, X_0, K_g, K_e, etc.) and such experimentally measurable quantities as absorption and emission bandwidths and peak positions. The configu-

---

[17] N. F. Mott and R. W. Gurney, *Electronic Processes in Ionic Crystals*, Oxford Univ. Press, London, 1940.
[18] M. Schön, *Ann. Phys.* [6] *3*, 343 (1948).
[19] F. E. Williams and M. H. Hebb, *Phys. Rev.* *84*, 1181 (1951).
[20] F. Seitz, *J. Chem. Phys.* *6*, 150 (1938).
[21] Cf. Ref. 8, pp. 306–309; also P. D. Johnson and F. E. Williams, *Phys. Rev.* *117*, 964 (1960).
[22] R. S. Knox and D. L. Dexter, *ibid.* *104*, 1245 (1956).
[23] D. A. Patterson and C. C. Klick, *ibid.* *105*, 401 (1957).
[24] C. C. Klick and W. D. Compton, *J. Phys. Chem. Solids* *7*, 170 (1958).
[25] C. C. Klick, *Phys. Rev.* *85*, 154 (1952).

ration-coordinate diagram can thus be constructed from experimentally measured quantities. With empirical diagrams of this sort one can study systematically how macroscopic variables such as the host-lattice composition or the concentration of the activator, for example, affect the microscopic properties of a given type of center, e.g., the vibrational frequencies and activation energies for radiationless transitions.

Furthermore, different experimental measurements can often be related to the same configurational coordinate parameter, thus providing a check on the internal consistency of the scheme and of the validity of the assumptions and approximations employed. For example, the ratio of the vibrational frequencies in the excited and ground states can be determined from this empirical approach in two ways: from the variation in the bandwidths of absorption and emission at high temperatures, and also from the ratio of the low-temperature bandwidths. For the KCl:Tl phosphor, one gets 1.28 for this ratio by the first method, and 0.65 by the second method.[23] The results are thus not internally consistent, implying that the simple configuration-coordinate treatment may be insufficient and that a multidimensional configuration-coordinate diagram may be necessary to represent the center.

Polarized luminescence of centers

This implication is further emphasized by the discovery that when KCl:Tl is excited by polarized light at liquid-helium temperature, the luminescence emission is polarized.[24] It is difficult for two reasons to reconcile this observation with the simple configuration-coordinate scheme: on the assumption of a symmetrical, radially vibrating $TlCl_6$ center excited to one of three equivalent p states, one would not expect to observe polarized emission; and second, configuration-coordinate diagrams indicate that the center is excited to some 40 vibrational levels above the minimum of the excited-state curve. Its local temperature at this point is thus many thousands of degrees Kelvin, and it executes many vibrations in relaxing to the minimum of the excited-state curve. Under these conditions, it is hard to see how the center can "remember" the original orientation of excitation.

The first difficulty has led to the suggestion[24] that the degeneracy of the p states is lifted by a distortion of the center according to the Jahn-Teller effect; i.e., the mode of vibration of the center in the excited state is no longer the radial or "breathing" mode but one of lower symmetry. Under these circumstances, however, the configuration coordinate that suffices to characterize the ground state is no longer a pertinent coordinate for the description of the excited state, and a multidimensional configuration-coordinate description[13] is needed.[26]

The second difficulty mentioned earlier, that the thallium ion has a "memory" of its excitation polarization, can be minimized if one postulates that the center interacts with the lattice vibrations as a whole rather than with the local vibrational modes of the center. In this "continuous dielectric" model,[27] many phonons are simultaneously created throughout the lattice when the thallium ion is excited. The lattice thus acts as a "thermostat," the center itself is not heated to a local high temperature, and memory of the polarization can more easily be preserved by the thallium ion. However, the simplest form of this theory ignores any relaxation of the lattice around the center, and it consequently predicts that the emission and absorption bands should have an identical shape; this is in disagreement with the experimental facts. Also on the basis of this model, the center should interact with the longitudinal optical vibrations of the lattice, whereas the observed values of ν_{ground} and $\nu_{excited}$ are much less than the lattice frequency. Although the continuous dielectric model may be applicable in cases where the center is not strongly localized, as in materials of high dielectric constant, it is difficult to accept the long-range interaction mechanism for phosphors of the KCl:Tl type.

In summary, considerable understanding has been achieved of the constitution of luminescent centers and of the processes involved in direct excitation of, and emission from, such centers. The theory, however, is not yet fundamental or detailed enough to permit synthesis of luminescent materials from first principles. Attempts to "tailor-make" new luminescent inorganic solids for specific purposes require recourse to the extensive body of empirical data on luminescent systems, with present-day theory serving primarily as a qualitative guide. This state of affairs arises because the processes that follow the absorption of energy by a luminescent center are very sensitive to the details of

---

[26] A similar conclusion is suggested by recent studies of the F center [F. Lüty and W. Gebhardt, *Z. Physik* **169**, 475 (1962); and C. C. Klick, D. A. Patterson, and R. S. Knox, *Phys. Rev.* **133**, A1717 (1964)]. Again the configuration-coordinate curves obtained from analysis of absorption data are not consistent with those obtained from emission data. Theoretical results of Wood and Korringa [*Phys. Rev.* **123**, 1138 (1961)] on the lattice relaxation accompanying the excitation of the F center in LiCl indicate that two of the Li neighbors are pushed out by about 15 percent from their ground-state equilibrium positions. The symmetrical octahedral arrangement of atoms around the trapped electron in the ground state is then replaced by a tetragonal configuration in the excited state.

[27] S. I. Pekar, *Untersuchungen über die Elektronentheorie der Kristalle*, Akademie-Verlag, Berlin, 1954.

the coupling between the activator atom and its surroundings. A complete description of these details presents the hardest problem in the field of luminescence.

Systems involving energy transfer

We now turn our attention to more complicated systems, in which energy is initially absorbed by a center or by the host material and then transferred to the luminescent center. Greater theoretical progress toward tailor-making materials has been achieved in such complex systems than in the simpler systems just discussed. The reason for this apparent anomaly lies in the different nature of the questions asked in the two cases. In dealing with problems of energy transfer, we are no longer concerned with the detailed structure of the centers; we may not even know the proper models of the centers, let alone details of their configuration-coordinate diagrams. Success in dealing with questions of energy transfer arises because of the more limited problems that we set ourselves and because we employ an experimental description of the absorption and emission properties of the centers.

What mechanisms of energy transfer can be visualized between two centers? The simplest mechanism, of course, is a photon-cascade process in which the luminescence from the first center is reabsorbed by the second center, which then luminesces with its own characteristic emission. This rather trivial process sometimes plays a significant role and has to be excluded before physically more interesting cases can be identified. In certain instances this is not at all easy to do. One feature that helps to characterize this process is the dependence of luminescence yield on the system's geometry.

A more important mechanism is resonance transfer of energy without the actual emission and reabsorption of photons.[28] In many Mn^{2+}-activated phosphors, the luminescence of Mn^{2+} cannot be excited—or can be excited only feebly—by direct irradiation into the Mn^{2+} absorption bands.[16] These bands are very weak; they correspond to a forbidden ($^4G \rightarrow {}^6S$) transition in the $3d^5$ shell of Mn^{2+}. But if an impurity like Pb^{2+}, which has strong characteristic absorption bands, is incorporated into the phosphor along with Mn^{2+}, excitation with light absorbed by the Pb^{2+} center will result in an intense emission characteristic of Mn^{2+}.[29,30] The Pb^{2+}, in this case, is a "sensitizer" for the Mn^{2+}. Conversely, since the Pb^{2+} would emit its own luminescence in the absence of Mn^{2+}, the latter acts as a "poison" or "quencher" for the Pb^{2+}. It can readily be shown that the transfer of energy from Pb^{2+} to Mn^{2+} is due to a resonance and is not a simple cascade process.[29]

This behavior suggests the possibility of tailor-making a class of phosphors in which certain weakly absorbing activators are brought into a photon-excitable condition, and in which excitation and emission frequencies are controlled—the former by the absorption spectrum of the sensitizer, the latter by choice of the activator. Figure 13.3 gives an example

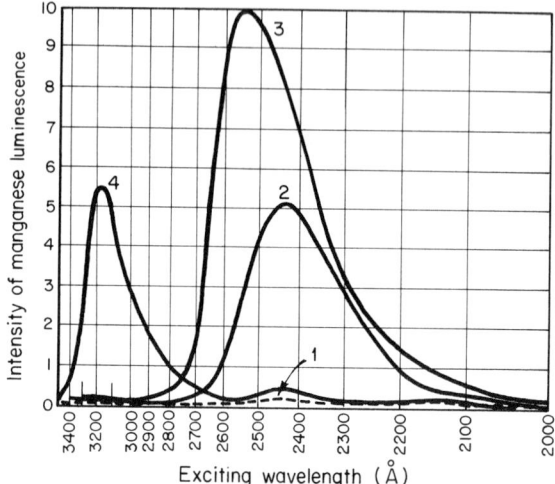

Fig. 13.3. Excitation spectra for Mn^{2+} emission of $CaCO_3$ phosphor triggered with different sensitizers:[31] (1) Without sensitizer; (2) Tl^+ sensitizer; (3) Pb^{2+} sensitizer; (4) Ce^{3+} sensitizer.

of excitation-spectrum control for Mn^{2+}-activated calcium carbonate.[29] The excitation peak may be placed at 2540 Å by Pb^{2+} as a sensitizer, at 2470 Å by Tl^+, or at 3200 Å by Ce^{3+}. In order to perform this manipulation effectively, we have to know the efficiency of energy transfer between sensitizer and activator in relation to concentration and other properties of the individual centers.

Resonance transfer

Pertinent relationships can be derived from essentially classical considerations, as in the following treatment originated by Förster.[28] Classically, an atom is coupled to the radiation field by its oscillating electric charge and to the neighboring atoms by interactions of their oscillating electric dipole fields (Fig. 13.4). At large separation distances the direct interaction is weak and the excited atom emits its energy as radiation; if an unexcited distant center reabsorbs this

---

[28] T. Förster, *Naturwissenschaften* 6, 166 (1946).
[29] J. H. Schulman, L. W. Evans, R. J. Ginther, and K. J. Murata, *J. Appl. Phys.* 18, 732 (1947).
[30] T. P. J. Botden, *Philips Research Repts.* 7, 197 (1952).

radiation, we have energy transfer by a photon. At close approach the interaction allows some direct energy transfer by coupling between the atoms, as in a system of coupled mechanical oscillators.

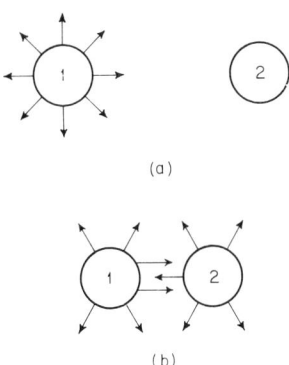

Fig. 13.4. Emission from two oscillators (schematic diagram).[30] (a) Separation $\gg d_c$: photon emission from initially excited oscillator 1; (b) separation $\ll d_c$: energy transfer leading to equal probability of emission from oscillator 1 or 2.

The radiative lifetime of an individual oscillator (angular frequency ω, dipole moment μ) is

$$\tau_{\text{rad}} \sim \frac{\hbar c^3}{\mu^2 \omega^3}. \tag{13.5}$$

(Planck's constant is introduced to simplify the formulation.) The interaction energy U of two oscillators separated by a distance d is

$$U \sim \frac{\mu^2}{d^3}, \tag{13.6}$$

and in a time of the order $1/\omega$ this amount of energy is transferred from one oscillator to the other. In the case of perfect resonance, these energy increments add, so that the energy $U\omega$ is transferred per unit time. To transfer the total energy $\hbar\omega$ from one to the other, the time required is

$$\tau_{\text{transfer}} = \frac{U_{\text{total}}}{U\omega} = \frac{\hbar\omega}{U\omega} \sim \frac{\hbar d^3}{\mu^2}. \tag{13.7}$$

At some critical distance d_0, $\tau_{\text{transfer}} = \tau_{\text{rad}}$, or

$$d_0 \sim \frac{c}{\omega} = \frac{\lambda}{2\pi}.$$

If $\lambda = \lambda_{\text{absorbed}} = \lambda_{\text{emitted}} = 6000$ Å, we find a critical distance of ~ 1000 Å. This applies, of course, to the case of perfect resonance between two identical dipole oscillators.

Even if we are dealing with the same species of luminescent center, we normally do not have perfect resonance. The oscillators are not characterized by a single frequency but by a band of frequencies, and the absorption band does not coincide with the emission band because of the Stokes shift. The coupling between the oscillators is therefore usually poorer than perfect resonance prescribes. For the imperfect reso-

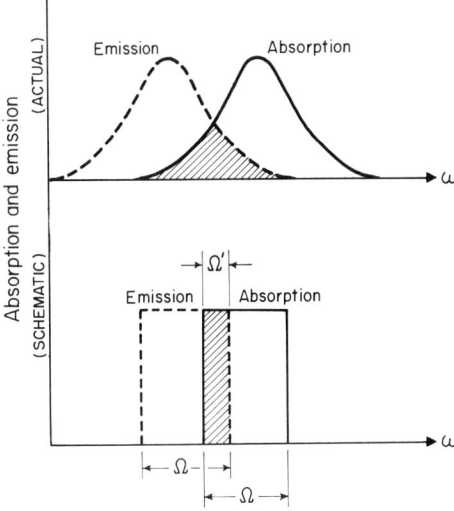

Fig. 13.5. Idealized absorption and emission of two dipole oscillators with spectral overlap (Ω').[30]

nance case shown schematically in Fig. 13.5, the critical distance is

$$d_0 = \frac{\lambda}{2\pi} \sqrt[6]{\frac{\Omega'}{\tau_{\text{rad}}\Omega^2}}, \tag{13.8}$$

where Ω is the half-width of the absorption and emission bands (taken for simplicity to be equally broad) and Ω' the overlap of these two spectra.

With reasonable values for these parameters, the critical transfer distance now proves smaller; taking the half-widths as about 5 percent of the peak frequency and assigning 10 percent overlap between absorption and emission bands and a radiative lifetime of 5×10^{-9} second, we get $d_0 \sim 75$ Å for $\lambda = 6000$ Å. For oscillators separated by this distance, resonance transfer of the energy is as probable as photon emission. Resonance transfer becomes more probable for shorter distance and larger overlap. The temperature dependence of transfer efficiency is caused by the variation of bandwidths with temperature, which affects the overlap.

By this resonance process of energy transfer from center to center, energy absorbed by the ions of the host lattice can migrate through the crystal. This is one picture of the "exciton," an entity that transports energy through a crystal without transporting charge. In a similar way energy can be transferred from one of the host-lattice ions to an impurity center, or from one species of impurity center to another species. When the absorbing atoms are those of the host material, we speak of host-sensitized luminescence; when the absorber is an impurity center, we speak of impurity-sensitized luminescence. An exact quantum-mechanical treatment of resonance transfer processes has also been given by Förster.[9]

Forbidden transitions

So far we assumed that the initially excited atom and the receptor atom are both electric dipole oscillators, i.e., that, in quantum terms, both the sensitizer and the activator operate with allowed transitions. In most sensitized inorganic phosphors, however (cf. the Mn^{2+}-activated systems), the activator has a forbidden transition. To complete the theory, one must find out how the transfer probability is affected by that type of transition. This analysis was carried out by Dexter,[31] who extended the treatment from electric dipole–dipole coupling to electric dipole–electric quadrupole and electric dipole–magnetic dipole cases, and to energy transfer by quantum-mechanical exchange.

The critical distance for dipole–quadrupole transfer is approximately one third as large as the distance for dipole–dipole coupling; so there is effective sensitization even when the activator has a forbidden transition of the electric quadrupole type. For the electric dipole–magnetic dipole case the transfer probability is very low, and at the small critical distances required for effective coupling the transfer takes place more rapidly by quantum-mechanical exchange. The latter mechanism requires the activator to be no farther away from the sensitizer than about the third nearest-neighbor shell in order to have effective transfer.

Other aspects of energy transfer

A new and very interesting phenomenon in resonance transfer has been discovered by Varsanyi and Dieke:[32] The cooperative absorption of one photon by two atoms, which can occur when the energy of the photon equals the sum of the energies needed to excite the individual atoms. It may be expected that this effect will be exploited to produce a number of interesting new luminescent systems.

The resonance processes for energy transfer do not depend directly on the periodicity of the lattice; hence, the effects leading to sensitization, quenching, and depolarization of luminescence occur in gases, liquids, and glasses as well as in crystalline solids. Resonant transfer processes underlie many luminescent systems[1–3,16] and also have an important bearing on other fields, such as photosynthesis and other areas of photochemistry.[9,28]

We have yet to consider one widely prevalent mechanism of energy transfer: the migration of energy by the movement of free electrons and holes. This important subject is treated in the following chapter by H. Klasens; it deals with phosphors in which free charge carriers play a major role, not only in energy transfer but in the luminescence process itself.

[31] D. L. Dexter, *J. Chem. Phys.* **21**, 836 (1953).

[32] F. Varsanyi and G. H. Dieke, *Phys. Rev. Letters* **7**, 442 (1961).

14 · PHOTOCONDUCTING PHOSPHORS

Hendrik A. Klasens

Introduction — Intrinsic and extrinsic luminescence — Activators and capture cross sections — Equilibrium states — Effects of excitation intensity — Quenching of fluorescence — Luminescent centers in ZnS — n-type conductivity and automatic compensation — Vacancies and center structure — Paramagnetic studies

Introduction

Any material, hence also the phosphors discussed by Dr. Schulman in Chap. 13, will show free carriers when exposed to high-energy radiation. However, while such carriers may be an intermediary between the absorption of the exciting radiation and the excitation of the activator, the actual luminescence discussed thus far occurs between the excited and ground states of this activator.

In the phosphors of our present concern, the light emission is a direct result of recombination between two free carriers or between one free and one trapped carrier.

Phosphors of this type have already provided entertainment since Roman days. Livius writes that the bacchanalian priestesses at Rome used to carry torches at night, containing sulfur and lime, which did not extinguish when plunged into the Tiber. No doubt this is the first report on the phosphorescence of CaS. A similar phosphor was made twelve centuries later by Japanese priests from oyster shells and sulfur. In 1600, during the days of alchemy, a Bolognese shoemaker tried to make silver by heating the mineral $BaSO_4$ (which has a glossy appearance) with coal. The experiment was bound to fail; the disappointed shoemaker threw the product (BaS) out of the window, but, much to his surprise, the lump of material kept glowing in the dark.

The first contribution to the molecular designing of this material was made by Homberg at Paris, who in 1694 at first did not succeed in repeating the preparation of phosphorescent BaS until by accident he replaced his iron mortar by a brass one and concluded rightly that traces of copper are required. Only at the turn of the last century, Klatt and Lenard[1] undertook a systematic study starting from purified materials and adding traces of impurities. Their starting materials, however, were not sufficiently pure, and many conclusions they drew are erroneous. But even nowadays, when starting materials and preparative methods are more refined, the nature of the luminescent centers is still an object for speculation.

The photoconducting phosphors most extensively studied are those of the ZnS type, because properly activated ZnS and (ZnCd)S phosphors have by far the highest cathode-ray efficiencies; furthermore, CdS and CdSe are the most sensitive photoconductors known today. The molecular designing of such materials aims

[1] V. Klatt and P. Lenard, *Wied. Ann. Phys. Chem.* 38, 90 (1889); P. Lenard and V. Klatt, *Ann. Phys.* [4] 15, 225, 425, 633 (1904); P. Lenard, *ibid.* [4] 31, 641 (1910).

at controlling preparation and composition to such an extent that optimum performance is obtained for any requirement specified.

Intrinsic and extrinsic luminescence

As already stated, the luminescence accompanies the recombination between free carriers or between a free and a trapped carrier (Fig. 14.1). In the second case,

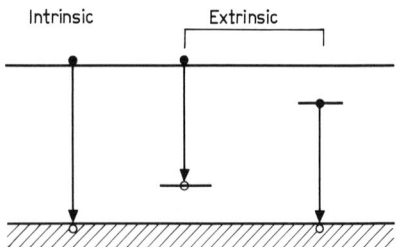

Fig. 14.1. Three types of recombination processes between carriers in photoconducting phosphors.

the free carrier can be either an electron or a hole. In the field of semiconductors, the radiation stemming from the recombination between free carriers is called "intrinsic" and that between free and trapped carriers "extrinsic." Typical examples of both types of fluorescence are found in silicon.[2]

When a p-n junction of silicon doped with boron and arsenic is biased in the forward direction (Fig. 14.2),

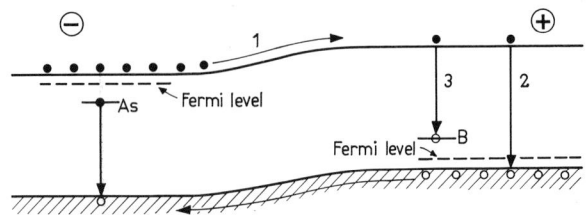

Fig. 14.2. Model of p-n junction of silicon doped with boron and arsenic and biased in the forward direction.

electrons are injected from the n-type into the p-type region (1) where they recombine with the free holes (2) and at low temperatures also with holes trapped in the acceptor boron (3).

The fluorescent spectrum is shown in curve I of Fig. 14.3. The peak at 1.1 eV which corresponds to the band gap in silicon is due to intrinsic recombination. The small peak at 1.04 eV is due to extrinsic recombination via boron; the other small peak probably stems from extrinsic radiation between an injected hole and an electron trapped at the donor state of arsenic. The

[2] I. R. Haynes and W. C. Westphal, *Phys. Rev.* **101**, 1676 (1956).

boron emission appears only at low temperatures. At room temperature the holes cannot be kept long enough at the acceptor levels, which are rather close to the top of the full band. In the absence of other impurity levels only intrinsic radiation can then be expected. At lower temperatures a competition between the intrinsic and the extrinsic process takes place. To promote the chances for the latter, a fair number of acceptors must be present; also the temperature must be low enough or the acceptor level high enough above the full band to keep the hole trapped sufficiently long. These prerequisites are illustrated by the other curves of Fig. 14.3. In the In-doped silicon, where the ac-

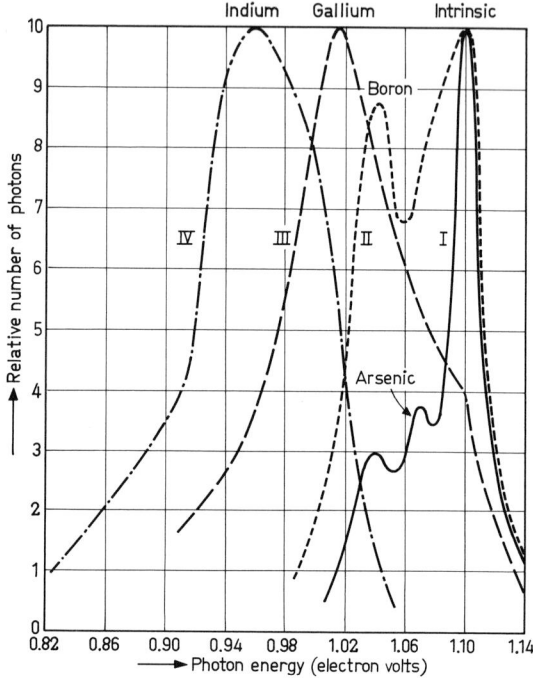

Fig. 14.3. Recombination radiation from a p-n junction of silicon doped with arsenic and various acceptors and biased in the forward direction (77° K): I. some boron added; II. 50 times higher boron content; III. with gallium; IV. with indium. (After Haynes and Westphal.[2])

ceptor level is farthest away from the full band, no further intrinsic recombination is seen at liquid-air temperature.

The positions of most impurity levels in silicon are well known from measurements of conductivity and can be unambiguously defined because there is practically no Franck-Condon shift (at least for the shallow acceptor levels in silicon). In other words, there is hardly any difference between thermal and optical activation energies for the electronic transitions. Consequently the emission peak can be calculated by subtracting the acceptor depth from the band gap and making a small correction for the average kinetic energy of the moving electron.

This is probably no longer true for deeper-lying levels in silicon and certainly not for the levels in ZnS-type phosphors that are responsible for the fluorescent emission. When discussing the position of an impurity level in such phosphors in terms of the band picture, one must be aware that different positions with respect to the band edges have to be allocated to an impurity level, depending on whether optical or thermal processes are considered.[3,4] Because of the Franck-Condon shift one should also be careful in deducing energies by carrying out simple additions or subtractions with the band gap as sum.

Activators and capture cross sections

In silicon we have considered the competition between an intrinsic and extrinsic process, both of which produce light; in zinc sulfide phosphors the competition is nearly always between radiative and nonradiative extrinsic recombination processes. Activators are incorporated to promote the radiative processes, because without them the recombination usually occurs via nonradiative recombination centers, such as lattice defects, dislocations, surface states, undesired impurities, etc., which are present in most materials. The purpose of introducing activators is to steer the recombination away from these unwanted "killers" to luminescent centers.

This can be done despite the fact that luminescent centers in ZnS-type phosphors have extremely small capture cross sections for electrons. The green copper center, the blue silver center, and the blue zinc vacancy center in ZnS have capture cross sections[4,5] between 10^{-19} and 10^{-21} cm^2, many orders of magnitude smaller than the geometrical size of the impurity atom. This is perhaps not surprising, since the electrons must not only meet the impurity but also lose a large amount of energy. Large capture cross sections can be expected only when there is a Coulomb attraction.

In silicon and germanium, capture cross sections range from 10^{-15} to 10^{-12} cm^2 if there is a Coulomb attraction between the carrier and the recombination center. Many neutral centers have capture cross sections of 10^{-17} to 10^{-15} cm^2,[6] but much smaller values have also been reported. For example, for the radiative recombination between an electron and a neutral In center in silicon, a capture cross section as low as

4.10^{-22} cm^2 at 80° K and even lower at higher temperatures has been measured.[7]

How can an activator of very low capture cross section compete with other recombination centers in ZnS of much larger capture cross sections? Let us consider a model with two recombination centers, one radiative and one nonradiative (Fig. 14.4). The wide-band

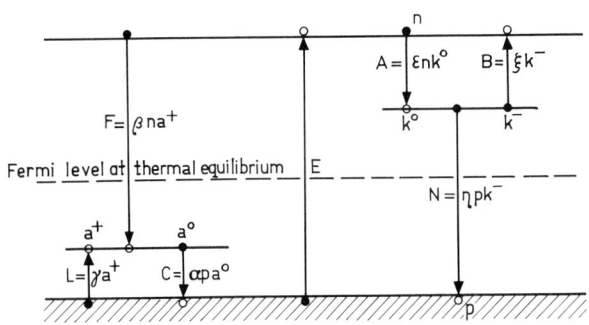

Fig. 14.4. Transitions, parameters, and symbols used in discussion of the competition between fluorescent recombination F and nonradiative process N.

material ($E = 3.6$ eV) has the Fermi level at thermal equilibrium somewhere near the middle of the gap, in keeping with the fact that most ZnS phosphors are highly insulating in the dark. The activator levels are chosen 0.5 eV above the full band, well below the Fermi level. They are completely occupied by electrons when the material is not excited. The nonradiative recombination level is assumed well above the Fermi level, about 0.7 eV below the empty band. We assume about 10^{16} of the latter centers per cm^3 to have capture cross sections of 10^{-16} and 10^{-15} cm^2 for electrons and holes, respectively. The corresponding cross sections for the activator levels shall be 10^{-19} and 10^{-15}.

Equilibrium states

When radiation injects electrons from the full band into the conduction band, a stationary state will be reached, characterized by

$$a = a + a^+, \quad (14.1)$$

$$k = k^0 + k^-, \quad (14.2)$$

$$a^+ + p = k^- + n \quad \text{(neutrality equation)}, \quad (14.3)$$

$$E = F + N, \quad (14.4)$$

$$C = L + F, \quad (14.5)$$

$$A = B + N, \quad (14.6)$$

---

[3] A. von Hippel, *Z. Physik* **101**, 680 (1936).
[4] M. Schön, *Z. Naturforsch.* **6a**, 287 (1951).
[5] W. Hoogenstraaten, *Philips Research Repts.* **13**, 624 (1958).
[6] M. Lax, *Phys. Rev.* **119**, 1502 (1960).
[7] Y. E. Pokrovsky and K. J. Svistunova, "Recombination Processes in Silicon Doped with Ga, In and Sb," *Proc. International Conference on Photoconductivity* (New York, August 1961), H. Levinstein, Ed., *J. Phys. Chem. Solids* **22**, 39 (1961).

where a, k, n, and p are densities of activators, killers, free electrons, and free holes, respectively; a^0 and a^+ are the densities of occupied and empty activator states, and k^0 and k^- those of empty and occupied killer states, respectively;[8] F, N, E, C, L, A, and B are transition probabilities as indicated in Fig. 14.4.

These six equations enable us, for example, to solve for the six unknowns (n, p, a^0, a^+, k^0, and k^-) and obtain the various transition rates of electrons as function of activator content for a given set of rate constants (α, β, γ, ϵ, ζ, and η), and prescribed values of excitation density E and killer content k. Unfortunately, the solution of this relatively simple set of equations is rather complex and leads to high-order equations of no particular help for a better understanding of the underlying processes. We will therefore make use of a method first introduced by Brouwer[9] for solving Kröger-Vink equations and by Duboc[10] for handling kinetic problems as here discussed and extended by this writer and others[11,12] to analyze intensity and temperature dependence of photoconducting phosphors.

The procedure consists of neglecting systematically and in proper sequence all terms but one on each side of all equations. This is done in a graphical way as illustrated by the following example: First we have to select relevant values for the rate constants (see Fig. 14.4) and chose $\alpha = 10^{-8}$, $\epsilon = 10^{-9}$, $\eta = 10^{-8}$, $\beta = 10^{-12}$ cm^3/sec^{-1}; $\gamma = 10^2$, $\zeta = 10^{-2}$ sec^{-1}. Furthermore we have put $k = 10^{16}$ cm^{-3} and $E = 10^{19}$ cm^{-3} (which is rather high).

Next we find by trial and error which of the terms in the preceding equations can be neglected when a is very low. The proper set of simplified equations[13] is found to be

$$a = a_0^+, \qquad (14.1a)$$

$$k = k, \qquad (14.2a)$$

$$p = k^-, \qquad (14.3a)$$

$$E = N, \qquad (14.4a)$$

$$C = L, \qquad (14.5a)$$

$$A = N. \qquad (14.6a)$$

One can easily check if the neglected terms are indeed small by solving these equations and find that for $a \leq 10^{12}$ the given set is the correct one. All entities and transition rates are simple power functions of a; in log-log scale, straight lines are obtained

---

[8] The + sign does not necessarily mean that the activator has an effective positive charge. It merely indicates that the center has one electron less than in the unexcited state.

[9] G. Brouwer, *Philips Research Repts.* **9**, 366 (1954).

[10] C. A. Duboc, *Brit. J. Appl. Phys.*, Suppl., **4**, 107 (1955).

[11] H. A. Klasens, *J. Phys. Chem. Solids* **7**, 175 (1958).

[12] H. G. Grimmeiss and H. Koelmans, *Phys. Rev.* **123**, 1939 (1961).

[13] The author[11] has previously introduced a code system of great help in identifying a given set of simplified equations. The relevant code numbers are given in Fig. 14.5.

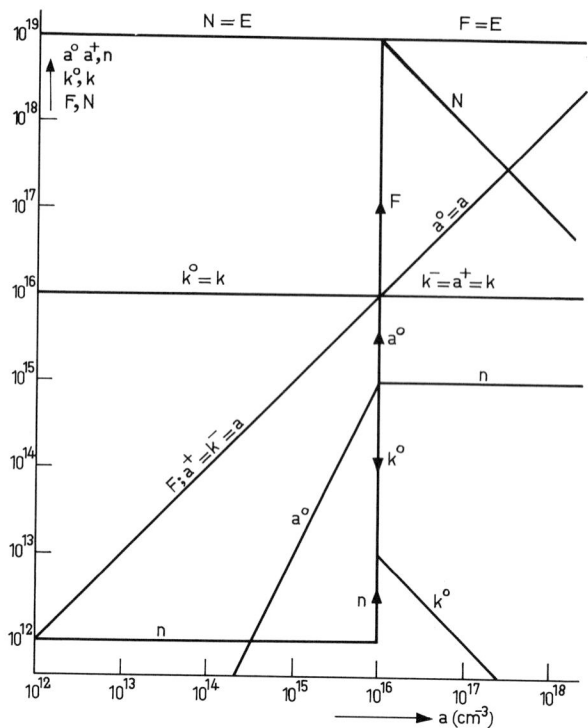

Fig. 14.5. Behavior diagram showing to a first approximation the effect of activator density A on a ZnS material with 10^{11} killers, electron density n, transition rates F and N per cm^3 at an excitation density E of 10^{13} cm^{-3} and with $\epsilon = 10^{-9}$, $\eta = 10^{-8}$, $\beta = 10^{-12}$, and $\alpha = 10^{-8}$ cm^3 sec^{-1}; $\gamma = 10^2$ and $\zeta = 10^{-2}$ sec^{-1}.

(cf. left part of Fig. 14.5). When following the line for a^+, one notices that it cuts the line $p = k^-$ at $a = 10^{13.5}$. This means that for $a > 10^{13.5}$ another set of equations must be used, obtained by replacing the simplified neutrality equation 14.3a by a new one:

$$a^+ = k^-. \qquad (14.3b)$$

With this new set (211212), all entities and rates can again be calculated easily and new lines drawn. For activator densities below 10^{16}, the killers still remain practically empty ($k^- \ll k^0$), while all activators have captured a hole ($a^+ - a$). At an activator content of 10^{16}, however, there is a drastic change in the situation. Here k^- has become equal to k, and in our set of equations Eq. 14.2a must be replaced by

$$k = k^-. \qquad (14.2b)$$

With this set of equations n and k^0 cannot be solved separately. They are indefinite, and their product is determined and equal to E/ϵ. Since k^0 can become only smaller, n must now increase in a vertical direction (in our first approximation a remains constant). At the same time $F (= \beta na)$ moves upward along the line $a = 10^{16}$. The probability F first catches up with C, necessitating the replacement of Eq. 14.5a by

$$C = F. \qquad (14.5b)$$

There is again a singular condition: F continues to rise vertically until equal to E. Simultaneously a^0 jumps from 10^{15} to 10^{16} and therefore remains equal to a. We have entered a new region in which practically all killers remain occupied ($k^0 \ll 10^{13}$

for $a > 10^{16}$). They cannot capture any more electrons from the conduction band. In fact the recombination via the killers has become blocked by electrons originating from the activator states, and the recombination is forced to proceed via the activators $(E - F)$.

At the same time there is a sudden rise by several orders of magnitude in the density of free electrons n. If these electrons have a high mobility, as in CdS (100 cm² volt⁻¹ sec⁻¹), we have now turned the material into a sensitive photoconductor.

Effects of excitation intensity

One has to keep in mind, especially when the activator states are near the full band, that the holes should stay trapped at these states and not leak away by thermal agitation. If the supply of holes is too small (e.g., for too low excitation rates), this may easily happen with the result shown in Fig. 14.6. Here E

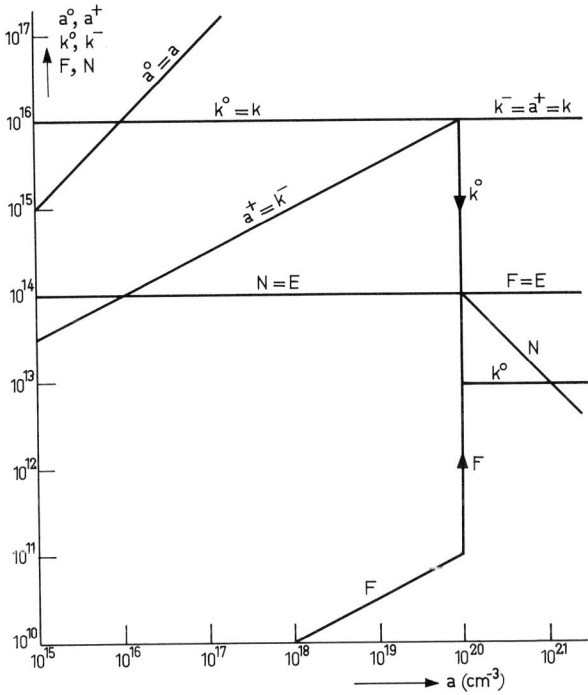

Fig. 14.6. Material as in Fig. 14.5 but with excitation density E lowered from 10^{19} to 10^{14} cm⁻³.

has been lowered from 10^{19} to 10^{14}. Although the neutrality equation is still $a^+ = k^-$, only 1 percent of the activator states is empty at $a = 10^{16}$, and the killers are not getting blocked until the activator density is increased to 10^{20}.

Let us now keep the activator density at 10^{18} cm⁻³ and vary the excitation E. At low excitation density the efficiency is very low, because the rate of generation of holes is too small to keep the majority of them trapped in the activator state. The killers are not blocked, and recombination proceeds through them. On increasing the excitation, however, this situation changes rather abruptly: The efficiency increases in a narrow range from a low to a high value. This is the explanation for superlinearity in the intensity dependence.

A classical example, observed by Nail, Urbach, and Pearlman,[14] is shown in Fig. 14.7. The nonradiative

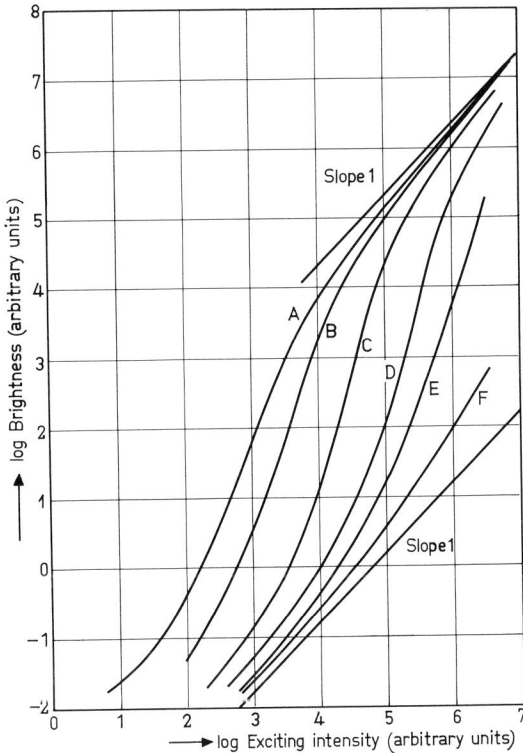

Fig. 14.7. Fluorescence in intensity as f (exciting intensity) for (Zn, Cd)S activated with silver and containing various amounts of nickel.[12] A: 0.0; B: 0.3; C: 1.0; D: 3.0; E: 5.0; F: 20.0 ppm.

recombination center here is nickel. The higher the nickel content, the more difficult to block these centers, hence the more intense the excitation has to be to reach the high-efficiency region. Simultaneously with the shift to high efficiency, the photoconductivity rises sharply. Superlinear photoeffects have also been found in CdSe-Cu (Fig. 14.8) and explained similarly by Bube.[15] More recently a strong superlinear dependence was found by Grimmeiss and Koelmans[12] in the pn emission of a GaP diode (Fig. 14.9).

[14] N. R. Nail, F. Urbach, and D. Pearlman, *J. Opt. Soc. Am.* **39**, 690 (1949).
[15] R. H. Bube, *J. Phys. Chem. Solids* **1**, 234 (1957).

Quenching of fluorescence

Returning to the model of Fig. 14.5, it is clear that the temperature can have a large effect on the luminescence because of the strongly temperature-dependent rates L and B. If the activator state is near the full band, the holes do not remain trapped at the activator when the temperature is increased too much. The recombination will again shift to the killer; this is "temperature quenching" of fluorescence.

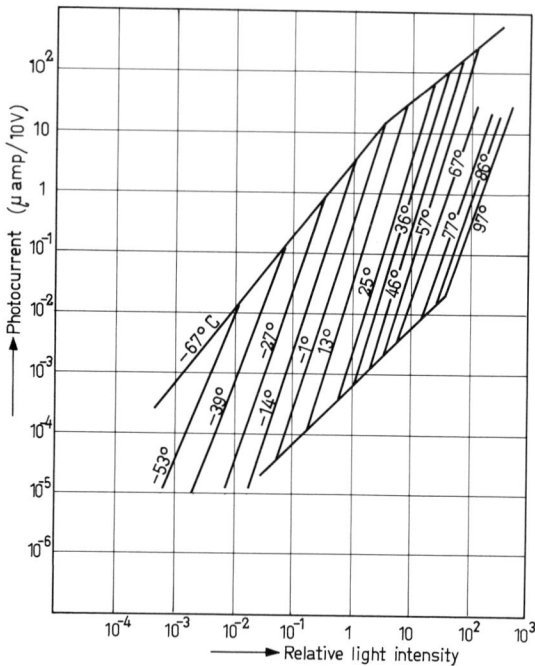

Fig. 14.8. Change of photocurrent with excitation intensity at various temperatures (°C) for CdSe crystal activated with Cu.[15]

Another way to liberate holes from the activator and quench the fluorescence is irradiation with infrared, a process known as "infrared quenching." The reverse "infrared stimulation" may occur when the recombination proceeds largely through a killer and exposure by infrared stimulates process B (cf. Fig. 14.4).

Luminescent centers in ZnS

Let us consider in more detail how activators are incorporated into the lattice of ZnS and what the real nature of a luminescent center is.

The best-known activators for zinc sulfide are copper and silver. However, firing pure zinc sulfide with Cu_2S in a neutral atmosphere or in H_2S does not result in the well-known green copper emission but often in a grayish-looking powder with weak reddish luminescence. As known for many years, to get a good phosphor a third substance, a "flux," must be added. Materials with a low melting point were chosen, but, as Rothschild[16] pointed out, only materials with specific halogen anions proved suitable.

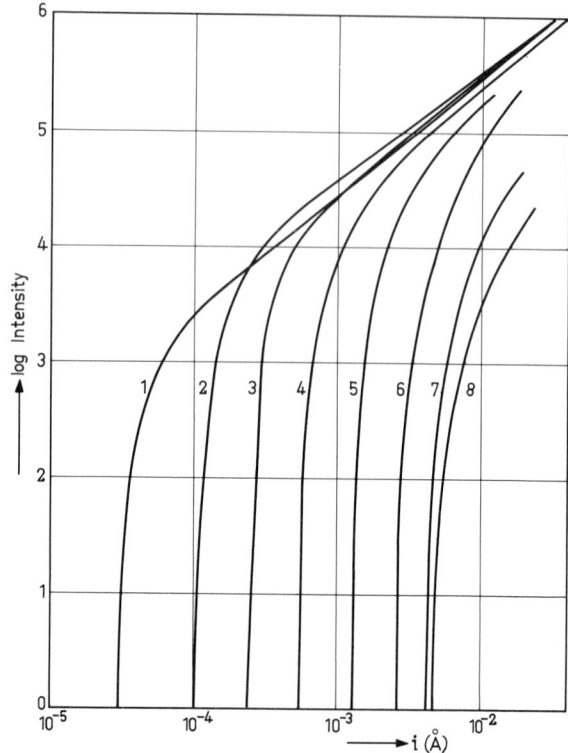

Fig. 14.9. Red luminescence (7000 Å) of GaP p-n diode (forward biased) as function of current and temperature:[14] (1) −175° C; (2) −160° C; (3) −146° C; (4) −130° C; (5) −107° C; (6) −70° C; (7) −20° C; (8) +20° C.

Kröger[17] concluded that Ag_2S is not very soluble or, if dissolved, gives rise to unwanted complications (e.g., vacancies); the halogen is needed to get a solid solution of CuCl in ZnS. To maintain neutrality, a S^{2-} ion has to be replaced by a Cl^- ion for every Zn^{2+} ion replaced by a monovalent Cu^+ ion. He next tried to achieve the same result by replacing a second Zn^{2+} ion by a trivalent ion, such as Al^{3+}, Ga^{3+}, etc. This proved successful.[18] Very efficient zinc-sulfide phosphors can be obtained without any flux by firing, for example, pure zinc sulfide with a small addition of copper and aluminum.

Originally it was thought that elements such as the halogens or aluminum, often referred to as coactivators, are necessary for the incorporation of the activator. Although the solubility is indeed increased in some

[16] S. Rothschild, *Trans. Faraday Soc.* **42**, 635 (1946).
[17] F. A. Kröger and J. E. Hellingman, *J. Electrochem. Soc.* **93**, 156 (1948).
[18] F. A. Kröger and J. A. M. Dikhoff, *Physica* **16**, 297 (1950).

cases,[19] activators alone are soluble to some extent in zinc sulfide.

It is not so much the quantity of activator but the manner by which it is incorporated that is determined by the coactivator (or rather by the ratio activator/coactivator). If ZnS is fired with different amounts of Cu and Ga, phosphors with different emissions result (Fig. 14.10).[20] The characteristic green

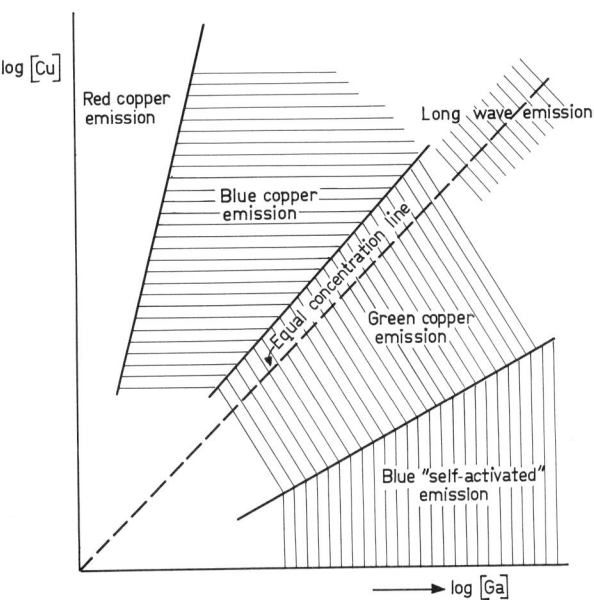

Fig. 14.10. Schematic representation of fluorescent emissions obtainable for ZnS activated with Cu and Ga.[20]

emission is obtained along the line of equal but low concentrations. At higher concentrations longer wave bands begin to appear. At Cu/Ga ratios larger than unity the blue Cu emission appears, and for very low coactivator content a red emission is observed. Small quantities of gallium alone produce the blue "self-activated" emission. This name goes back to the time when the influence of the flux was not yet fully recognized. Firing ZnS with a halide without activator, one obtained the blue emission and called it "self-activated."

Let us return to the green emission. It is due, as we will see presently, to an isolated copper atom on a zinc site. In the periodic table, copper and silver are in the column to the left of Zn and Cd. They are one electron short for completing the binding and therefore act as acceptors, just like Zn and Cd in GaAs and Ga in Ge. Similarly, the elements of the column to the right of Zn and Cd must act as donors when replacing

Zn in ZnS. The same is true for a halide replacing S in ZnS. Using equal concentrations of donors and acceptors results in a compensated semiconductor.

Copper accepts an electron from the donor elements and becomes negatively charged. The energy level of this electron corresponds to the lower level (activator level) in Fig. 14.4. If the copper loses its electron again by excitation, it becomes neutral; this explains its small capture cross section for electrons.

The empty donor levels can act as traps. Most of them are so shallow, however, that they are not noticeable at room temperature and certainly do not act as recombination centers. These levels have therefore not been drawn in Fig. 14.4. Only the heavier trivalent elements, such as indium, produce deeper traps.[21]

If we follow the equal-concentration line (Fig. 14.10) to higher concentrations, one obtains new emission bands. Their positions depend on the nature of both the activator and the coactivator and must therefore be caused by associations between these two.[22,23]

n-type conductivity and automatic compensation

If one fires ZnS with a donor like gallium alone, the result need not be an n-type semiconductor but may be a compensated material. This "automatic" compensation arises because it is often energetically economical for the system to produce a vacancy as a compensating center.[24] We will follow the method of Kröger and Vink[25] to see which factors determine whether the addition of a donor such as Ga to ZnS produces n-type conductivity or automatic compensation.

The donor level of Ga substituting for Zn in ZnS is so shallow that all gallium is ionized at the temperature of preparation. It thus has an effective positive charge and (following the notation of Kröger and Vink) the center is designated as Ga_{Zn}^+, indicating that a Ga atom replacing a Zn atom has lost an electron. One assumes furthermore that a zinc vacancy (V_{Zn}) acts as a double acceptor. Figure 14.11 indicates the situation schematically. The level position determines the thermal equilibria between the free electrons in the

[19] H. C. Froelich, *J. Electrochem. Soc.* **100**, 497 (1953).
[20] W. van Gool, *Philips Research Repts.*, Suppl. No. 3, 1961.
[21] W. Hoogenstraaten, Ref. 5, p. 597.
[22] W. van Gool, A. P. Cleiren and H. J. M. Heyligers, *Philips Research Repts.* **15**, 254 (1960).
[23] E. F. Apple and F. E. Williams, *J. Electrochem. Soc.* **106**, 224 (1959).
[24] J. S. Prener and F. E. Williams, *J. Chem. Phys.* **25**, 361 (1956).
[25] F. A. Kröger and H. J. Vink, "Relations between the Concentrations of Imperfections in Crystalline Solids," *Solid State Physics*, Vol. 3, F. Seitz and D. Turnbull, Eds., Academic Press, 1956, p. 3.

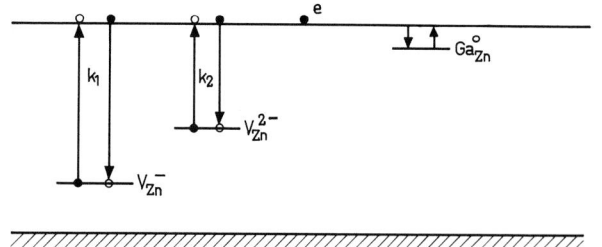

Fig. 14.11. Possible energy levels in ZnS activated with Ga.

conduction band and the trapped electrons in the acceptor levels V_{Zn}^- and V_{Zn}^{2-}:

$$V_{Zn}^- \rightleftarrows V_{Zn}^0 + e \quad \text{or} \quad \frac{n \cdot [V_{Zn}^0]}{V_{Zn}^-} = k_1;$$

$$V_{Zn}^{2-} \rightleftarrows V_{Zn}^- + e \quad \text{or} \quad \frac{n \cdot [V_{Zn}^-]}{[V_{Zn}^{2-}]} = k_2. \quad (14.7)$$

During preparation the material is in equilibrium with vapor in which a certain Zn pressure is maintained. Zn vacancies are produced by the reaction

$$Zn_{Zn} \rightleftarrows Zn_{vapor} + V_{Zn}. \quad (14.8)$$

In thermal equilibrium,

$$P_{Zn} \cdot [V_{Zn}^0] = k_r, \quad (14.9)$$

where k_r is a measure for the tendency of the material to produce Zn vacancies.

Finally, we have to maintain electric neutrality:

$$n + [V_{Zn}^-] + 2[V_{Zn}^{2-}] = [Ga_{Zn}^+]. \quad (14.10)$$

From these equations the values of n, V_{Zn}^-, and V_{Zn}^{2-} can be calculated as function of total Ga content and Zn vapor pressure, once the position of the relevant energy levels and the chemical reaction constant k_r are known. The result of such calculation is that the $P_{Zn}:[Ga]$ plane can be divided into three regions,[20] in each of which one of the negative species in the neutrality equation is dominant (Fig. 14.12). At high Zn pressure and low Ga content gallium acts as a straight donor: For every Ga atom one free electron is produced,

$$[Ga_{Zn}^+] = n. \quad (14.11)$$

The chemical composition of the material can then be described as a solution of GaS in ZnS.

For CdS this situation can easily be realized. In ZnS the bandgap is wider and the value of k_1 and k_2 much lower. The triple point $[Ga] = k_2/2$ and $P_{Zn} = k_r/k_1$ lies at a much higher metal pressure and much lower dope for the same value of k_r. This is probably the reason that the n-type region in ZnS cannot be realized as easily. Instead one has either the mechanism

$$[Ga_{Zn}^+] = 2[V_{Zn}^{2-}], \quad (14.12)$$

i.e., in chemical terms a solution of Ga_2S_3 in ZnS, or, at low Ga content,

$$[Ga_{Zn}^+] = [V_{Zn}^-], \quad (14.13)$$

a solution of GaS_2.

Vacancies and center structure

Since nearly the same emission was obtained with Cl, Br, Al, or Ga, it was first assumed that the luminescent center must be an isolated Zn vacancy; but luminescence data alone cannot decide which of the two vacancies (V_{Zn}^- or V_{Zn}^{2-}) is responsible for the observed

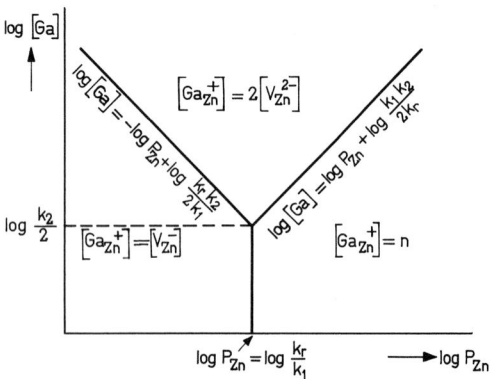

Fig. 14.12. Possible incorporation mechanisms of Ga in ZnS.

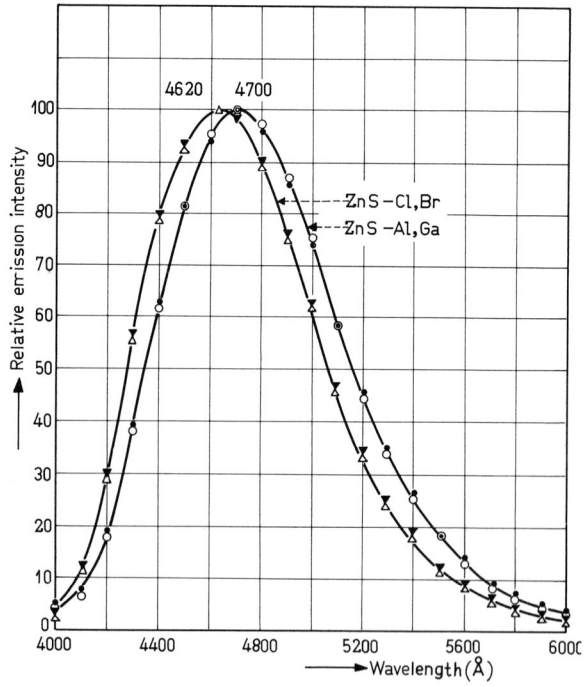

Fig. 14.13. Emission spectra for 3650-Å excitation of ZnS, activated with Cl, Br, Al, or Ga (room temperature).[26]

blue luminescence. Later an influence of the donor on the spectra distribution was found[26] (Fig. 14.13). This points to a possible association between donor and vacancy, as Prener and Williams[24] suggested. However, these are not the only emission bands found in donor-activated phosphors. Longer wave emissions had already been found by Kröger and Dikhoff in ZnS activated by Sc, Ga, and In.[18] Later, other bands have been added at wavelengths both longer and shorter than that of the original blue emission.[27,28]

Whereas Zn vacancies are present in donor-activated phosphors, sulfur vacancies must be expected to be involved in emissions (e.g., the blue and red copper bands) found in phosphors doped with an acceptor.

In summary, ZnS phosphors show an embarrassing number of emissions. One knows that vacancies occur, that associations have to be reckoned with, and that interstitial atoms cannot be excluded. Obviously many possible centers can be constructed from these elements and have indeed been proposed. It is clear, however, that without corroborating evidence all descriptions of centers based on optical observations alone remain mere conjecture. Possible exceptions are two well-known and widely used emission bands: the blue Ag band and the green Cu band. They occur in compensated ZnS at low dope and are not affected by the nature of the donor.[29] Therefore one can be reasonably certain that for these emissions no vacancies or associations are responsible and that the luminescent center is a single filled acceptor.

Paramagnetic studies

As far as the other centers are concerned, conclusive evidence is still needed. It may come from a study of paramagnetism, especially of paramagnetic resonance.

Larach and Turkevich[30] as well as Bowers and Melamed[31] already pointed out that the center in self-activated zinc sulfide is not paramagnetic; this fact alone rules out several proposed structures. For example, the occupied activator state cannot be V_{Zn}^-, as suggested by Kröger and Vink,[32] because such a center would have an unpaired spin.

[26] J. S. Prener and D. J. Weil, *J. Electrochem. Soc. 106*, 409 (1959).
[27] R. W. A. Gill and S. Rothschild, *Electrochem. Soc. Electronics Div. Abstr. 9*, 72 (1962).
[28] H. Koelmans, *J. Phys. Chem. Solids 17*, 69 (1960).
[29] W. v. Gool and G. Diemer, "Association of Centers in Zinc Sulfide," *Luminescence of Organic and Inorganic Materials* (Proc. International Conference on Luminescence), John Wiley and Sons, New York, 1962, p. 39L.
[30] S. Larach and J. Turkevich, *Phys. Rev. 98*, 1015 (1955).
[31] R. Bowers and N. T. Melamed, *ibid. 99*, 1781 (1955).
[32] F. A. Kröger and H. J. Vink, *J. Chem. Phys. 22*, 250 (1954).

Paramagnetic-resonance measurements, particularly if carried out on single crystals, can reveal many more details. The deviation of g from theoretical value for a free spin can give information on the nature of a trapped carrier in a shallow state. The anisotropy of g reveals the symmetry of the center. The atoms of the center and its immediate surrounding may manifest themselves in the hyperfine structure.

Single crystals of donor-activated zinc sulfide were investigated by a number of authors[33-35] with the following results.

No signals were found in unexcited phosphors, but on excitation at 77° K two signals appear. One of these has an isotropic g of 1.883 in Cl-, Br-, and I-doped samples and 1.885 in Al-doped crystals. This low value is typical for trapped electrons. Its symmetry indicates a single impurity; this evidence strongly supports the presence of diamagnetic Cl_S^+ or Al_{Zn}^+ centers which, after trapping an electron, become neutral and paramagnetic. No hyperfine structure was found, although expected; this has been attributed to motion of the trapped carrier along a shallow donor band.[36] The other signal has an anisotropic g factor with axial symmetry. In halogen-doped crystals the axis of symmetry lies along the [111] directions of the cubic crystals. The g values are greater than 2, indicating a trapped hole. In Al-activated crystals the axial symmetry is lacking. Again hyperfine structure that might give more conclusive evidence is absent, but so far the data are in good agreement with a center originally proposed by Prener and Williams:[24] a Zn vacancy with a halogen atom in a nearest sulfur position or a trivalent element in a nearest zinc position.

The incorporation mechanism then seems to be

$$[(V_{Zn}Cl_S)^-] = [Cl_S^+] \quad \text{or} \quad [Al_{Zn}^+]. \quad (14.14)$$

This is equivalent to saying that Cl is incorporated as $ZnCl_2$, and Al as Al_2S_3. For every two donor atoms one vacancy is created, which then associates with one donor atom. This associate complex is an acceptor and accepts an electron from the remaining donor atom.

Hyperfine structure was found for the transition elements Cr, Mn, and Fe. ZnS-Mn, the first semiconductor investigated with paramagnetic resonance[37] shows the characteristic fine structure of the d^5 configuration of the Mn^{2+} ion ($^6S_{5/2}$) and a hyperfine structure arising from a nuclear spin of $\frac{5}{2}$. There is no

[33] P. H. Kasai and Y. Otomo, *Phys. Rev. Letters 7*, 17 (1961).
[34] A. Räuber, J. Schneider, and F. Matossi, *Z. Naturforsch. 17a*, 654 (1962).
[35] A. Räuber and J. Schneider, *Phys. Letters 3*, 230 (1963).
[36] K. A. Müller and J. Schneider, *ibid. 4*, 288 (1963).
[37] E. E. Schneider and P. S. England, *Physica 17*, 221 (1951).

evidence that this state is represented by an impurity level within the forbidden zone. The other two elements also give signals in compensated phosphors but only upon excitation. The observed resonance in both cases is characteristic for the d^5 configuration. Since both elements are probably incorporated as a bivalent element, Cr must first trap an electron[38] and Fe a hole[39] before the d^5 signals show up. We thus know that there are levels in the forbidden zone corresponding to Cr_{Zn}^- and to Fe_{Zn}^0. In uncompensated donor-activated phosphors the Fermi level can be raised so high that the Cr_{Zn}^- center (or Cr^+ ion) shows up in the dark, indicating that Cr in ZnS can be either divalent or monovalent.

It is to be expected that many more interesting facts will become known in the near future about the nature of other impurity centers in ZnS by use of the paramagnetic-resonance technique in combination with measurements of other physical properties.

[38] J. Dieleman, R. S. Title, and W. V. Smith, *Phys. Letters 1*, 334 (1962).

[39] A. Räuber and J. Schneider, *Z. Naturforsch. 17a*, 226 (1962).

General Reference: H. K. Henisch, *Electroluminescence*, Pergamon Press, London, 1962.

15 · ELECTRONS AND HOLES IN SEMICONDUCTORS

Herbert J. Zeiger

Introduction — Electrons and holes in bands — Landau levels and cyclotron resonance — Excitons — Donor and acceptor impurities — Impurities and excitons in a magnetic field — Interband magneto-optical phenomena — Impurity states and stimulated emission

Introduction

In the past two decades a revolutionary development in our understanding of holes and electrons in semiconductors has taken place. This has been due in large part to the development of pure single-crystal semiconducting materials for devices (impurity concentrations from 10^{13} to 10^{18}/cm^3), the development of experimental techniques for the study of such single crystals, and the sharpening of theoretical concepts based on these new experimental results. In this chapter some of these developments will be summarized.

Electrons and holes in bands

In the one-electron band model,[1] the energy of one outer (or valence) atomic electron in a crystal is calculated in an average periodic potential due to the atomic cores and the remaining valence electrons. The resulting energy spectrum is characterized by a quantum number specifying the bands, and a **k** vector for electrons in the bands. If there are N atoms in the crystal and m valence states per atom, then the total number of states to which electrons may be assigned is $m \cdot N$. The number of possible **k** vectors times the number of bands must then be $m \cdot N$, a large but finite number. One of possible **k** vectors that uniquely and nonredundantly defines the states of the crystal falls within the first Brillouin zone in k space.[1]

In a pure semiconductor crystal like Ge or Si there are just enough electrons present to fill one set of bands (the valence bands) completely and none left over to occupy higher energy (conduction) bands. The valence and conduction bands are separated from each other by an energy gap, so that at low temperatures the valence bands are filled and no electrons are present in conduction bands for current transport. At higher temperatures electrons can be excited into conduction bands, and current can flow through the semiconductor.

When electrons are either thermally or optically excited into conduction bands in a semiconductor, they tend to collect about energy minima, while the empty states or holes left behind in the valence bands tend to collect about maxima in the valence bands, giving the lowest energy consistent with the temperature of the crystal. The holes in the valence band behave like particles of positive charge in their interaction with electric and magnetic fields.

In the cases of Ge and Si, the conduction electron minima are away from **k** = 0, and the valence band maxima are at **k** = 0 in the first Brillouin zone (Fig. 15.1). Optical transitions from the top of the valence

[1] F. Seitz, *Modern Theory of Solids*, McGraw-Hill Book Co., New York, 1940.

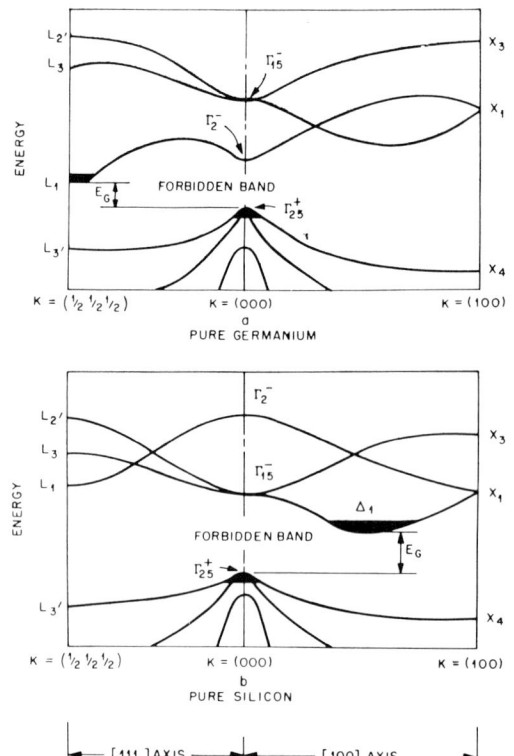

Fig. 15.1. Conduction and valence bands in [100] and [111] directions of Si and Ge first Brillouin zones. The symbols L, Γ, Δ, and X refer to symmetry points in the Brillouin zone (cf. Chap. 16). The dark regions show the normal positions of electrons in conduction and of holes in valence bands.

band to the lowest conduction-band minima are indirect transitions and must be accompanied by the emission or absorption of lattice-vibration quanta, or phonons. Transitions from the valence-band maxima to the conduction-band minima at $\mathbf{k} = 0$ are direct transitions and are much stronger, because they do not require phonons to conserve $\mathbf{k}$.

The motion of holes and electrons in semiconductors, subjected to applied electric and magnetic fields, are described by their "effective mass" properties, which are related to the curvatures and shapes of the energy maxima or minima as a function of $\mathbf{k}$.[2] For the electron minima in Ge, for example, the relation of "kinetic energy" (relative to the band edge) and $\mathbf{k}$ is of the form

$$\mathcal{E}_{\text{kin}} = \hbar^2 \left(\frac{k_z^2}{2m_\parallel} + \frac{k_x^2 + k_y^2}{2m_\perp} \right). \quad (15.1)$$

The constant-energy surfaces are elongated ellipsoids of revolution with axis k_z along [111] directions in the crystal. The kinetic energy of holes in the valence band is given by a more complicated set of warped energy surfaces.[2]

The transport properties of semiconductors are described in terms of the motion of holes and electrons in magnetic and electric fields, governed by the effective mass properties of the carriers. Motion of the charge carriers is interrupted by scattering with lattice vibrations, charged and neutral centers, or other imperfections in the lattice. Such a model has given a detailed picture of Hall effect, magnetoresistance, and other transport properties of semiconductors.[3]

Landau levels and cyclotron resonance

When a uniform d-c magnetic field is applied to a semiconductor, the holes and electrons present occupy quantized orbital levels called Landau levels. The quantized motion of the carriers corresponds to the fact that the classical motion of a charged particle in a d-c magnetic field is periodic. For a charge carrier of mass m^*, the energy of the carrier in a magnetic field is

$$\mathcal{E} = (n + \tfrac{1}{2})\hbar\omega_c + \frac{\hbar^2 k_z^2}{2m^*}, \quad (15.2)$$

where

$$\omega_c = \frac{eH}{m^*c}, \quad (15.3)$$

k_z is the component of $\mathbf{k}$ along the field direction, and n is a quantum number. The cyclotron angular frequency ω_c is the same as the classical value.

The Landau quantization of carriers in a magnetic field has been observed in a number of different experiments, in one of which interband absorption is observed in a magnetic field.[4] Oscillations are observed in absorption as a function of frequency, due to transitions to the Landau levels (Fig. 15.2).[5]

Transitions between adjacent Landau levels in one band have been observed directly in microwave cyclotron-resonance experiments.[6] These transitions

[2] W. Kohn, "Shallow Impurity States in Silicon and Germanium," *Solid State Physics*, Vol. 5, F. Seitz and D. Turnbull, Eds., Academic Press, New York, 1957, p. 257.

[3] H. Y. Fan, "Valence Semiconductors, Germanium and Silicon," *Solid State Physics*, Vol. 1, F. Seitz and D. Turnbull, Eds., Academic Press, New York, 1955, p. 284; H. Brooks, "Theory of the Electrical Properties of Germanium and Silicon," *Advances in Electronics and Electron Physics*, Vol. 7, L. Marton, Ed., Academic Press, New York, 1955, p. 87; F. J. Blatt, "Theory of Mobility of Electrons in Solids," *Solid State Physics*, Vol. 4, F. Seitz and D. Turnbull, Eds., Academic Press, New York, 1957, p. 200.

[4] S. Zwerdling, "Magneto-Absorption Experiments in Semiconductors," *Solid State Physics in Electronics and Telecommunications*, Vol. 3, Pt. 1, M. Désirant and J. L. Michiels, Eds., Academic Press, London and New York, 1960, p. 526.

[5] S. Zwerdling, B. Lax, and K. J. Button, *Phys. Rev.* **114**, 80 (1959).

[6] C. Kittel, *Introduction to Solid State Physics*, John Wiley and Sons, New York, 1956.

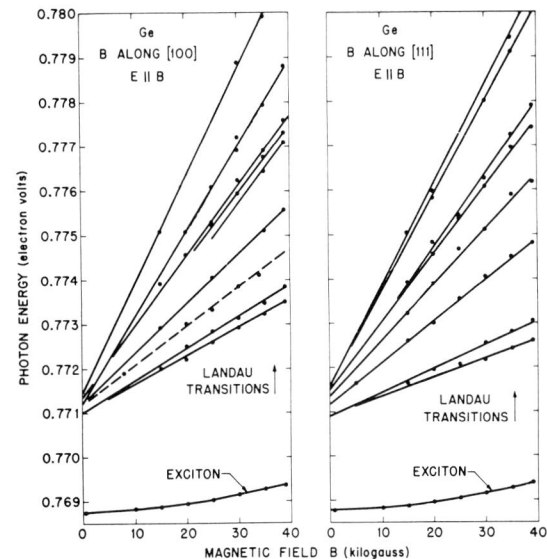

Fig. 15.2. Observed direct optical transitions in Ge in a d-c magnetic field.[5] The straight lines correspond to transitions from the valence band to quantized Landau levels in the conduction band; the curve below represents creation of an exciton.

are produced by microwave electric fields at frequency ω_c, transverse to the d-c field H. A classical description of this same effect pictures the absorption as due to the acceleration of carriers by the circularly polarized component of the microwave electric field rotating in step with the carriers.

Excitons

Free holes and electrons in a semiconductor attract each other and can form bound hole-electron pair structures, called excitons, which spread over many atoms and are the crystalline analogues of positronium. They are described basically by kinetic energy terms characteristic of the effective mass properties of the hole and electron, and a Coulomb interaction between hole and electron modified by the low-frequency polarizability of the crystal.[7] The Coulomb interaction is of the form $V = -e^2/\kappa' r$, where κ' is the low-frequency dielectric constant. The properties of excitons have been studied in many crystals, and this model is now well established experimentally.[4,8]

Donor and acceptor impurities

A free electron can also be attracted by a positively charged impurity in a semiconductor to form a structure analogous to a hydrogen atom. Again, the kinetic energy behavior is described by the effective mass properties of the electron, and the force of attraction is screened by the dielectric constant of the semiconductor.[2] The energies of the s-like donor states, in which the electron spends much of its time close to the impurity, may be modified somewhat from the predictions of this simple model. An interesting generalization of the behavior of a hydrogen atom arises in donor states in some semiconductors because of anisotropic effective mass properties of the electron. This gives rise to a lifting of degeneracy of p states, as would occur for a real hydrogen atom in anisotropic space.

The acceptor state is the analogue of the antihydrogen atom. Its structure is described by the effective mass behavior of holes and is further complicated, in the case of semiconductors like Ge and Si, by the fact that the valence bands are warped and degenerate.

Donor and acceptor states play a central role in the operation of semiconducting devices. In particular, they are the source of excess positive and negative charge in transistors and related devices.

Impurities and excitons in a magnetic field

The description of an impurity or an exciton state in a magnetic field is given by the effective mass behavior of holes and electrons, with the magnetic interaction terms also modified by the effective mass properties. Excited p-like states of donors tune with a linear

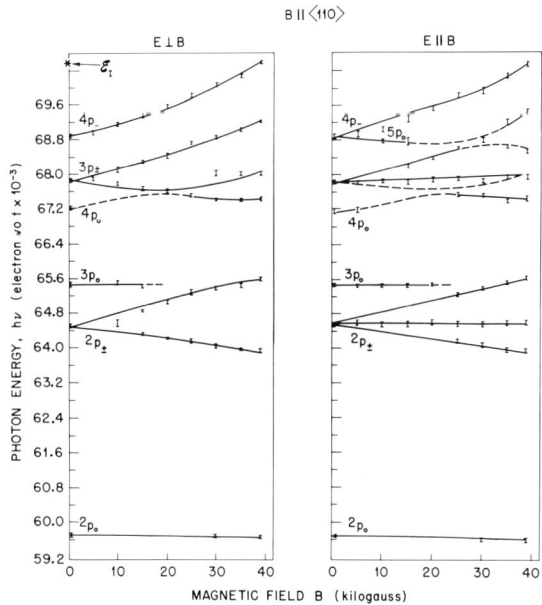

Fig. 15.3. Absorption peak versus magnetic field for noninteracting Bi donor states in Si.[9] ε_i is the ionization energy of the Bi donors. The notations $2p$, $3p$, $4p$, $\cdots$ designate hydrogenic energy levels. Note lifting of degeneracy of the p states due to anisotropy of the conduction electron masses in Ge.

[7] G. Dresselhaus, *J. Phys. Chem. Solids* **1**, 14 (1956).
[8] J. J. Hopfield and D. G. Thomas, *Phys. Rev.* **122**, 35 (1961).

Zeeman effect in a magnetic field, dominated by the large orbital moment of the electron for low effective masses (Fig. 15.3).[4,9] The ground states are s-like and show initially a quadratic Zeeman effect in a magnetic field. Donor impurities are of special interest, because it is possible in many cases to make the magnetic interaction energy much larger than the donor-binding energy in easily attainable fields,[10] in contrast with the case of the free hydrogen atom, where unattainably high fields ($> 10^9$ gauss) would be necessary to produce this situation.

Interband magneto-optical phenomena

Impurity and exciton states have been studied extensively in recent years (both by absorption and fluorescent emission) with the aid of applied d-c magnetic fields. Exciton states have been observed as absorption peaks just below the band edge.[8] This transition is analogous to the creation of a hole-electron pair in a bound positronium state and is possible in a solid but not in the free-vacuum case. Absorptions have also been observed, corresponding to the creation of an exciton bound to an impurity.[11] Fluorescence spectra have been observed from the annihilation of excitons and bound excitons,[12] as well as the radiative transitions of electrons from donor to a neighboring empty acceptor state.[13] The elucidation of many of these phenomena has been helped by the study of their Zeeman effects.

Impurity states and stimulated emission

In semiconductors with a concentration of impurities high enough that wave-function overlap takes place, interaction of impurities begins to play an important role and changes the properties of the crystals significantly. The impurity states begin to conduct electricity and form a smear of states below the conduction band edge in the case of donor states.[14] The emission of radiation in direct-gap semiconducting diodes such as GaAs diodes, has been studied in a magnetic field.[15] These and other experiments indicate that interacting impurity states play a major role in the efficient spontaneous emission of these devices.

By sending a strong enough current through a semiconducting diode, it has been possible to produce population inversion of interacting donor states relative to acceptor states in the vicinity of the diode junction. When the stimulated emission produced is great enough to overcome absorption processes in the lossy regions bordering the region of inversion, coherent emission of radiation occurs (Fig. 15.4).[16] Such coherent emission

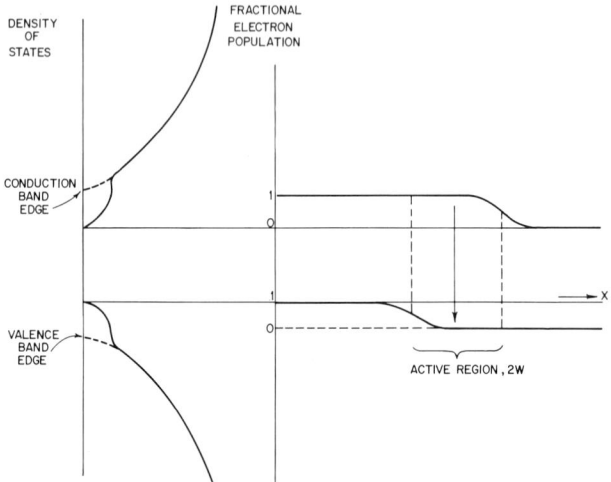

Fig. 15.4. Electron population in impurity states associated with conduction and valence bands in semiconducting diode laser (schematic). Electron flow is to the right from n to p region. Width $2W$ is region of population inversion.

in a very narrow frequency bandwidth has now been observed in a number of semiconducting diode systems, including GaAs,[16] InP,[17] InAs,[18] InSb;[19] and alloys of GaAs-GaP,[20] InAs-InP,[21] and InAs-GaAs.[22]

[9] S. Zwerdling, B. Lax and K. J. Button, *Phys. Rev.* **118**, 975 (1960).

[10] Y. Yafet, R. W. Keyes, and E. N. Adams, *J. Phys. Chem. Solids* **1**, 137 (1956).

[11] D. G. Thomas and J. J. Hopfield, *Phys. Rev. Letters* **7**, 316 (1961); E. J. Johnson, I. Filinski, and H. Y. Fan, *Proc. Exeter Conf. on the Phys. of Semiconductors*, A. C. Stickland, Ed., The Institute of Physics and The Physical Society, London, 1962, p. 375 (distributors: Chapman and Hall, Ltd., London).

[12] J. R. Haynes, M. Lax, and W. F. Flood, *J. Phys. Chem. Solids* **8**, 392 (1959).

[13] J. J. Hopfield, D. G. Thomas, and M. Gerchenzon, *Phys. Rev. Letters* **10**, 162 (1963).

[14] C. S. Hung and J. R. Gliessman, *Phys. Rev.* **79**, 726 (1950).

[15] F. Galeener, G. B. Wright, W. E. Krag, T. M. Quist, and H. J. Zeiger, *Phys. Rev. Letters* **10**, 472 (1963).

[16] R. N. Hall, G. E. Fenner, J. D. Kingsley, T. J. Soltys, and R. O. Carlson, *Phys. Rev. Letters* **9**, 366 (1962); M. I. Nathan, W. P. Dumke, G. Burns, F. H. Dill, Jr., and G. Lasher, *Appl. Phys. Letters* **1**, 62 (1962); T. M. Quist, R. H. Rediker, R. J. Keyes, W. E. Krag, B. Lax, A. L. McWhorter, and H. J. Zeiger, *ibid.* **1**, 91 (1962).

[17] K. Weiser and R. S. Levitt, *ibid.* **2**, 178 (1963).

[18] I. Melngailis, *ibid.* **2**, 176 (1963).

[19] R. J. Phelan, A. R. Calawa, R. H. Rediker, R. J. Keyes, and B. Lax, *ibid.* **3**, 143 (1963).

[20] N. Holonyak and S. F. Bevaqua, *ibid.* **1**, 82 (1962).

[21] F. B. Alexander, V. R. Bird, D. R. Carpenter, G. W. Manley, P. S. McDermott, J. R. Peloke, H. F. Quinn, R. J. Riley, and L. R. Yetter, *ibid.* **4**, 13 (1964).

[22] I. Melngailis, A. J. Strauss, and R. H. Rediker, *Proc. IEEE* **51**, 1154 (1963).

16 · ELECTRONS IN METALLIC SYSTEMS

Gert W. Rathenau

Metals and energy bands — Properties caused by high Fermi energy of typical metals — The Hume-Rothery rule — Band calculations with one-particle wave functions — Specific heat and magnetoresistance — The de Haas–van Alphen effect and cyclotron resonance — Cohesion — Transition metals — Magnetic interaction through conduction electrons — Transition metal solute atoms in alloys — Exchange fields and Knight shift — Electron correlation in plasma oscillations — Superconductivity — Model for soft superconductors — Occupation function and gap parameter

Metals and energy bands

The interaction between the ionic cores of a crystalline solid and one single electron is presented by a periodic potential V. If we restrict ourselves to one dimension, the wave function of the Bloch type reads

$$\psi_k = u_k(x)e^{ikx}. \qquad (16.1)$$

The wave vector **k** is related to the particle momentum p; in the limiting case of very small V,

$$\mathbf{k} = \frac{p}{\hbar}. \qquad (16.2)$$

A positive **k** signifies a wave traveling to the right, a negative one traveling to the left. The function $u(x)$ has the periodicity of the lattice potential V.

If N (the number of atoms per unit length in the lattice) is large, the possible states k represent a sequence with very small change Δk between neighboring states

$$\Delta k = \frac{2\pi}{Na}, \qquad (16.3)$$

where a is the lattice constant. When we scan the sequence of states with k growing from small values, something special happens at $k = \pm \pi/a$. The amplitude of the wave at neighboring lattice points differs only in sign. At these values of k the *traveling waves* moving to the right $(+k)$ and left $(-k)$ combine to *two independent standing-wave solutions*

$$\psi_1 = u_{\pi/a}(x) \sin \frac{\pi x}{a}; \qquad \psi_2 = u_{\pi/a}(x) \cos \frac{\pi x}{a}. \qquad (16.4)$$

This means that an electron wave with $k = \pm \pi/a$ is constructively reflected by all atoms (Bragg reflection). An analogous situation arises for all $k = \pm n\pi/a$, with n an integer.

The two solutions belonging to standing waves of the same k have different energies, because the charge density $(-e/|\psi_k|^2)$ is either concentrated at the nuclei or between them (Fig. 16.1).[1] The energy difference for

[1] C. Kittel, *Introduction to Solid State Physics*, John Wiley and Sons, New York, 1953.

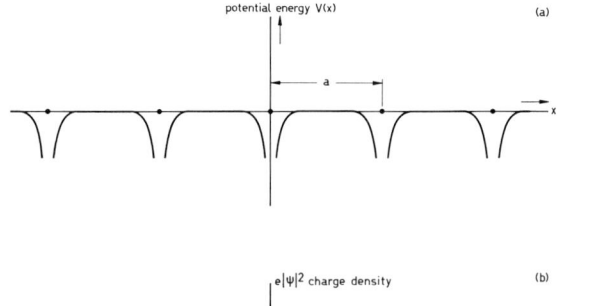

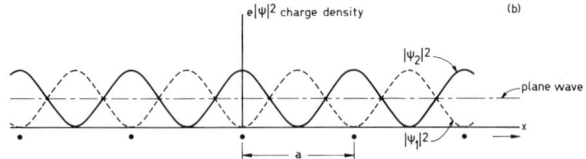

Fig. 16.1. (a) Potential energy due to interaction of positive ion cores of monatomic linear lattice and the conduction electrons. (b) Charge density for $|\psi_1|^2 \sim \sin^2 \pi x/a$ and $|\psi_2|^2 \sim \cos^2 \pi x/a$ as compared to charge density of traveling plane waves. (After Kittel.[1])

the two standing-wave solutions for $|k| = |n\pi|/a$ is the nth Fourier component of the lattice potential:

$$\Delta\varepsilon\left(\frac{\pi n}{a}\right) \cong 2|V_n|. \tag{16.5}$$

The energy dependence on k of our one free electron in the one-dimensional solid can be depicted as follows: At first ε as function of k increases monotonically, e.g., as $\varepsilon \cong \hbar^2 k^2/2m^*$, similarly to that of free electrons. At $k = +\pi/a$ and $-\pi/a$ (the wave numbers bounding the first Brillouin zone), a discontinuity in the energy occurs, and similarly for $k = \pm n\pi/a$ with integers $n > 1$. Such discontinuities make themselves felt already within the zones or energy bands (Fig. 16.2).

In Fig. 16.2a the relationship between energy ε of the crystal electrons and their wave vector is schematically drawn for the one-dimensional case. Figure 16.2b gives an equivalent representation, with all relevant information compacted in the interval $-\pi/a \leq k \leq \pi/a$ and each band defined by a quantum number n. Since the states for extreme k at the right and left are actually identical, one can also extend the reduced zone scheme of Fig. 16.2b to present the relationship between energy and wave number as being periodical. This leads to the repeated zone scheme of Fig. 16.2c. From the graphs we immediately read the group velocity

$$v = \left(\frac{1}{\hbar}\right)\left(\frac{\partial \varepsilon}{\partial k}\right) \tag{16.6}$$

of the electron wave packets.

If we are allowed—and within certain limits we are—to extend this scheme from one to three dimensions and from nuclei with one common electron to nuclei, say with one "free" electron each, we can fill the Brillouin zones or energy bands in k space with elec-

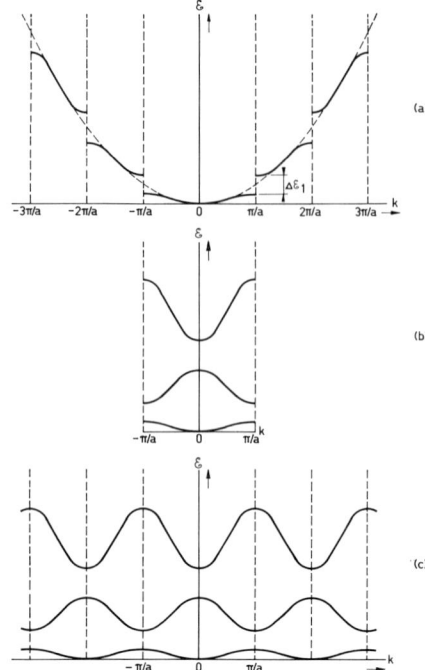

Fig. 16.2. Electron energy within crystal plotted against wave number k. (a) k Defined in range $-\infty < k < +\infty$ (extended zone scheme); (b) all information given in interval $-\pi/a \leq k \leq \pi/a$ (reduced zone scheme); (c) repeated zone scheme.

trons. These zones are related in shape to the crystal structure.

Electrons obey Fermi-Dirac statistics; accordingly each state of k can be filled with at most two electrons, one with spin up and one with spin down. Figure 16.3

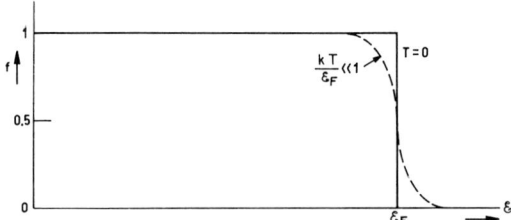

Fig. 16.3. Fermi-Dirac distribution function.

shows the occupation probability in k space at low temperatures as prescribed by those statistics.

Metals then are solids in which at least one Brillouin zone (energy band) is partly empty, and for which the k states are filled up to about the Fermi energy ε_F, removed from bottom and top of the band at a distance $\gg kT$. This description of metals assumes that the interaction between electrons can be taken into account essentially by a reasonable choice of the potential function.

Properties caused by high Fermi energy of typical metals

The properties discussed under this heading were explained about thirty years ago[1,2] and brought the early quantum theory to triumph.

The low electron heat capacity and small spin paramagnetism (Pauli magnetism) of metals stem from the fact (cf. Fig. 16.3) that thermal excitation of the electrons can occur only for electrons over energy distances of the order kT from the Fermi-level located at 50-percent occupancy ($f = \frac{1}{2}$). The electron heat capacity therefore will be by a factor

$$\frac{T}{T_F} = \frac{kT}{\mathcal{E}_F} \qquad (16.7)$$

smaller than for particles that do not obey the Pauli exclusion principle ($T_F \sim 10^{5\circ}$ K). A magnetic field **H** displaces the parts of the band for spin up and down relative to each other by $2\mathbf{m}_B\mathbf{H}$, where $\mathbf{m}_B$ represents the Bohr magneton. For large $\mathcal{E}_F$, the magnetic susceptibility χ_p will be temperature independent and small (Fig. 16.4), according to the Pauli formula:

$$\chi_p \simeq \frac{\mathbf{m}_B}{3\mathcal{E}_F} \cdot \mathbf{m}_B N = \frac{\mathbf{m}_B{}^2 N}{3\mathcal{E}_F}, \qquad (16.8)$$

with N the number of electrons per unit volume.

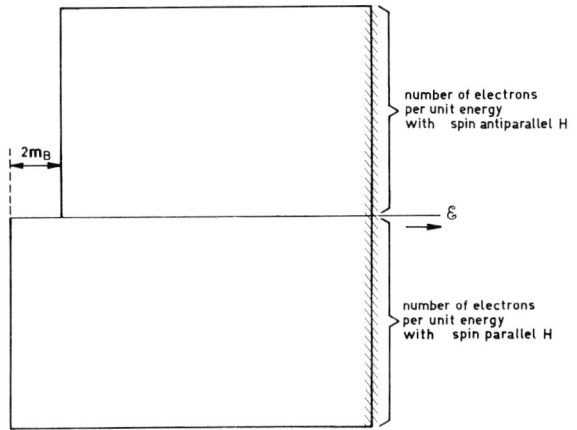

Fig. 16.4. Relative shift of subbands for spin up and down, respectively, in an applied electric field.

The Hume-Rothery rule

One of the properties that can be explained qualitatively is the crystal structure of metals and alloys at low temperature. Metals generally crystallize in simple structures. An empirical rule of Hume-Rothery shows that for example in Cu and Ag alloys, for which the ratio n (number of electrons/number of atoms) can be varied in a continuous fashion by substitution, the low-temperature crystal structure is essentially determined by n (Table 16.1).

Table 16.1. Electron/atom ratios for copper alloys

Alloy	F.c.c. Boundary α	B.c.c. Boundary β	γ Boundaries	H.c.p. Boundaries ϵ
Cu-Zn	1.38	1.48	1.58–1.66	1.78–1.87
Cu-Al	1.41	1.48	1.63–1.77	—
Cu-Ga	1.41	—	—	—
Cu-Si	1.42	1.49	—	—
Cu-Ge	1.36	—	—	—
Cu-Sn	1.27	1.49	1.60–1.63	1.73–1.75
Inscribed Fermi Sphere Contacts Brillouin Zone Surface	1.36	1.48	1.54	1.69

To explain this rule, two assumptions were made: (1) The external electrons of Cu and of the added elements are accommodated in one common valence band; (2) the Fermi boundary (which in k space confines all electrons up to the Fermi energy $\mathcal{E}_F$) has a spherical shape. As electrons are filled into the first Brillouin zone by alloying with elements containing more valence electrons than Cu, the Fermi sphere expands into k space and at a certain n touches the Brillouin zone boundary. Thereafter, because of the energy gap, the density of available states $N(\mathcal{E})$ as function of the energy $\mathcal{E}$ sharply decreases. At this critical composition a naive interpretation of the rule says the structure will give way to a different one in which the Fermi sphere can contain more electrons per atom prior to touching its Brillouin zone.

The rule became suspect when it was ascertained that already in the pure metals Cu and Ag the boundaries are touched. Closer inspection,[3] however, shows that the internal energy of one phase relative to another has its minimum not at the concentration where the Fermi sphere touches the boundary but for higher concentrations n of electrons. This is indicated in Fig. 16.5, where the Fermi gas in the crystal is compared to

---

[2] Cf. N. F. Mott and H. Jones, *The Theory of the Properties of Metals and Alloys*, Oxford University Press, London, 1936, also Dover ed.; F. Seitz, *Modern Theory of Solids*, McGraw-Hill Book Co., New York, 1940; C. Kittel, *Introduction to Solid State Physics*, John Wiley and Sons, New York, 1953.

[3] H. Jones, *J. phys. radium* **23**, 637 (1962).

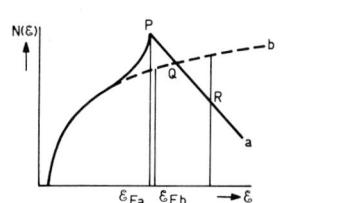 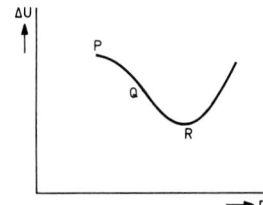

Fig. 16.5. *Left:* density-of-states curve in crystal (*a*) as compared with that of free-electron gas, and (*b*) with the same density as electrons in real space. The energies of the Fermi level are generally different, ε_{Fa} and ε_{Fb}, as shown for the case where the crystal zone boundary is just touched (*P*). At *R*, where the bands are more filled, $\varepsilon_{Fa} = \varepsilon_{Fb}$. *Right:* difference of internal energy of crystal and electron density in real space. The letters *P, Q, R* indicate the energy of the Fermi crystal level in the left-hand figure.

a free-electron gas with the same density in real space. It thus seems that the general content of the rule remains valid even though a close fit between experiment and theory may be fortuitous.

Band calculations with one-particle wave functions

All properties of metals might be understood if the correct electronic wave functions of the many-particle system were known. Such wave functions, however, are unobtainable. Instead, one-particle functions in a periodic potential that accounts both for the ionic attraction and (through the Wigner-Seitz and the Hartree-Fock approximations) also for part of the electronic repulsion are combined to form the model wave function of standard band theory.[4] Clearly, because of electron interaction, this procedure is not rigorous, and properties of metals will not be adequately represented if they involve collective effects. The one-particle approximation actually works very well.[5] There seem to exist a sharp Fermi energy ε_F and Fermi surface, although the theoretical reasons for this are not yet quite clear.[6]

During the last years enormous progress has been made in the numerical evaluation of the dispersion law

$$\varepsilon = \varepsilon(\mathbf{k}) \qquad (16.9)$$

—energy as function of wave vector—by prescribing a suitable crystal potential or by using refined calculation methods based on atomic spectroscopic constants. Re-

cent band calculations have been made, e.g., of the alkali metals,[7] of noble metals[8] as Cu and of transition metals.[9] Let us consider some of their results and the experimental facts they explain. The right side of Fig. 16.6 shows the first Brillouin zone for the b.c.c.

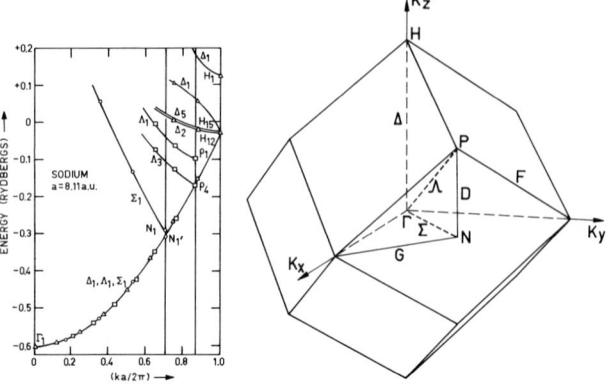

Fig. 16.6. *Left:* sodium energy bands in *k* space. (After Ham.[7]) *Right:* first Brillouin zone of b.c.c. lattice.

lattice of Na (at the boundaries of this zone, energy discontinuities are expected). On the left the dispersion law $\varepsilon(\mathbf{k})$ for Na is represented. Points of symmetry on the Brillouin zone boundary (*N, H, P*) and the lines joining them with the center (Γ) are explained on the right and used in the left-hand figure ($\Sigma \equiv [110]$; $\Delta \equiv [100]; \Lambda \equiv [111]$). Evidently there is a very small energy gap $N_1 - N_1'$ at point *N*. The electrons fill the zone only to 0.88 of the *k* value belonging to N_1, i.e., only to 0.88 (the distance from Γ to *N*). The Fermi surface is very nearly a sphere, which made it easy to draw ε versus $\mathbf{k}$ along the [100], [110], and [111] directions in one figure by branching off from a common curve. Apparently the free-electron approximation holds very well for Na.

For Cu it does not hold nearly as well (Fig. 16.7). Again the first Brillouin zone of this f.c.c. metal is shown on the right and the dispersion law drawn at the left along the [111] direction Λ (from the center Γ of the zone to the center *L* of the hexagonal face). The free-electron dispersion law (shown by dashes) obviously is not obeyed. This fact is connected with the presence of the 3*d* shell of electrons, which is filled just below the conduction band. Here Γ_{12} and $\Gamma_{25'}$ refer to *d*-like wave functions, while Γ_1 is related to *s*-like

---

[4] J. Callaway, "Electron Energy Bands in Solids," *Solid State Physics*, Vol. 7, F. Seitz and D. Turnbull, Eds., Academic Press, New York, 1958, p. 99.

[5] N. F. Mott, *Nature 178*, 1205 (1956).

[6] J. M. Luttinger, "Theory of the Fermi Surface," *The Fermi Surface*, W. A. Harrison and M. B. Webb, Eds., John Wiley and Sons, New York, 1960, p. 2.

[7] F. S. Ham, *Phys. Rev. 128*, 82 and 2524 (1962).

[8] F. S. Segall, *ibid. 125*, 109 (1962); G. A. Burdick, *ibid. 129*, 138 (1963).

[9] J. G. Hanus, Quart. Prog. Rep. No. 43, p. 73, 1962, Solid State and Molecular Theory Group, Mass. Inst. Tech.; L. F. Mattheiss, *ibid.* No. 48, 1963, p. 5; H. Ehrenreich, H. R. Philip, and D. J. Olechna, *Phys. Rev. 131*, 2469 (1963).

Specific Heat and Magnetoresistance

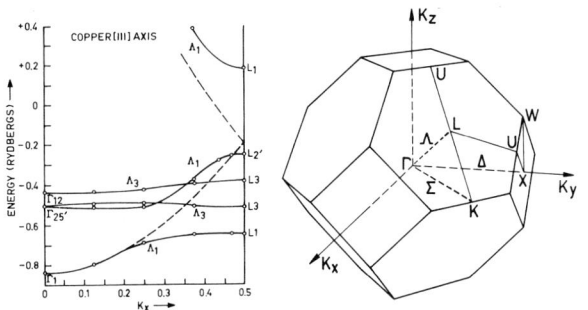

Fig. 16.7. *Left:* energy bands for Cu along the [111] direction in k space. (After Segall.[8]) *Right:* first Brillouin zone of f.c.c. lattice.

wave functions. The zone boundary is contacted at L_2', since the Fermi level lies just above $\varepsilon = -0.2$ rydberg. Figure 16.8 shows the dispersion law for Cu along other directions in k space (for reference points, cf. the Brillouin zone polyhedron of Fig. 16.7).

For the $3d$ transition elements (Fig. 16.9), one sees the interplay between the $4s$ and the $3d$ conduction bands; the latter narrows as the nuclear charge increases. Obviously such band calculations should be able to account for experimental results. Let us see whether or not they do.

Specific heat and magnetoresistance

It was already mentioned that the specific heat C_e due to metal electrons is small. If $N(E_F)$ is the density of states at the Fermi surface, one finds

$$C_e = (\tfrac{2}{3})\pi^2 k^2 N(E_F) T \equiv \gamma T. \qquad (16.10)$$

Table 16.2 gives some values relative to a free-electron

Table 16.2. Ratio of electronic specific heat to electronic specific heat for free electrons (γ/γ^0)

Na	Cu	Pd	
1.27	1.38	27	exper.
1.00	1.12		theor.
1 electron atom	1 electron atom	0.55 conduct. electron/atom	

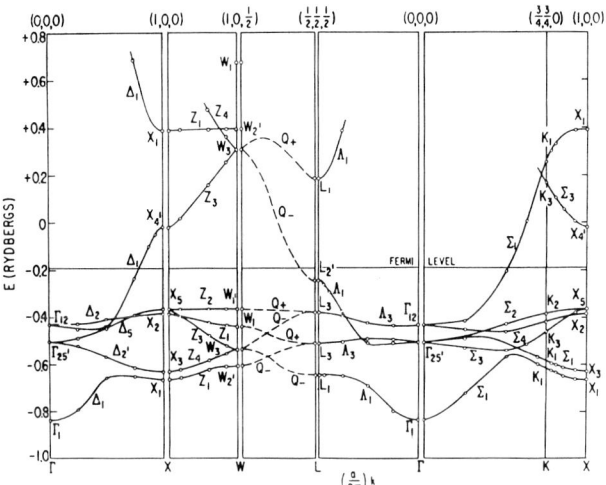

Fig. 16.8. Energy bands for Cu along various directions in k space. (After Segall.[8])

Fermi gas. (The experimental result for Na might have been influenced by a martensitic phase transformation at low temperature.)

A more severe test object of band calculations is the shape of the Fermi surface. It can be determined experimentally, e.g., from measurements on single crystals

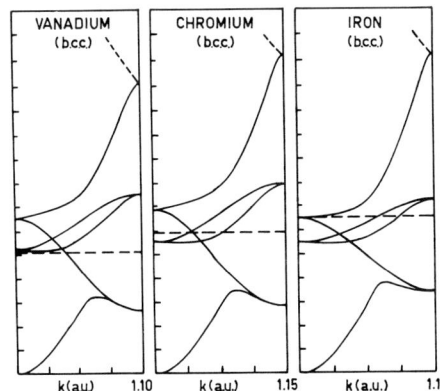
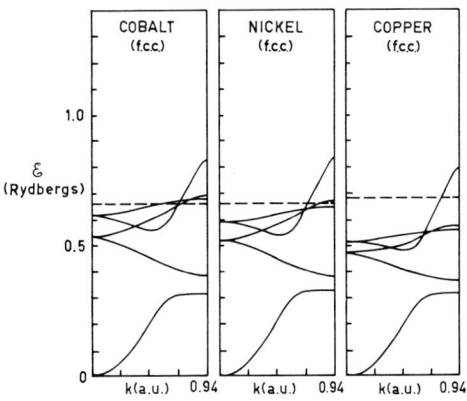

Fig. 16.9. Energy bands for V, Cr, Fe, and f.c.c. Co, Ni, Cu along the [100] direction in k space. (After Mattheiss.[9])

in magnetic fields.[10] Such fields exert forces upon the electrons which easily surpass those of externally applied electric fields. The energy of the electrons is not changed by the presence of the magnetic field but their motion is. The Lorentz force makes electrons move on orbits in k space perpendicular to the magnetic field, according to the force equation

$$\hbar \frac{d\mathbf{k}}{dt} = e(\mathbf{E} + \mathbf{v} \times \mathbf{B}). \quad (16.11)$$

For a closed spherical Fermi surface, as in Na, the electrons in a magnetic field (in the absence of scattering) move on closed orbits in k space and on similar closed orbits or helices in real space.

If a high magnetic field and a small electric one are applied simultaneously, conduction is heavily restricted except in the direction parallel to B. However, the resistivity in a plane perpendicular to B does not approach infinity with increasing B. It saturates, because at constant current the Hall voltage, increasing linearly with B, supplies the driving force on the electrons. This is not generally the case if one measures transverse magnetoresistance for a metal like Cu, where the Fermi surface is not simply connected in the repeated zone scheme.

As depicted schematically in Fig. 16.10, sectioning reveals two kinds of closed orbits on the Fermi surface:[11] electron orbits such as A, which enclose filled states, and hole orbits such as the "dog's bone" C, which enclose empty states. The velocity vectors given by the gradient of ε (Eq. 16.6) point outward for an electron orbit and inward for a hole orbit. An electron therefore traverses electron and hole orbits in opposite directions. In addition, open orbits occur, as indicated by the interconnected area B in Fig. 16.10.

If an open orbit exists in a plane perpendicular to the induction B, the magnetoresistance increases to infinity with B^2, except when the current is parallel to the direction in direct space of the open orbit. This behavior is connected with the fact that the Hall voltage no longer supplies a driving force on the electrons in high fields. It is short-circuited by the open orbit. By measuring magnetoresistance of single crystals in high fields it has been possible[12] to show that Na has a closed spherical Fermi surface, and Cu one that touches the

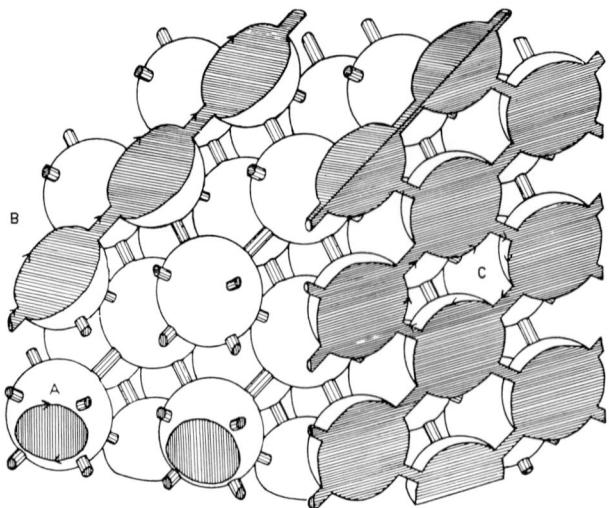

Fig. 16.10. The Fermi surface of Cu in the repeated zone scheme showing closed electron (A) and hole (C, dog's bone) orbits and open orbits (B). (After Ziman.[11])

Brillouin boundary and allows open orbits, as has been previously calculated.[13]

The de Haas–van Alphen effect and cyclotron resonance

Other modern techniques permit a more detailed comparison between theory and experiment. One is the de Haas–van Alphen effect, the occurrence (at low temperature) of periodic variations in the magnetic susceptibility as the magnetic field strength increases. Electrons in a plane normal to a magnetic field move in quantized orbits (Landau levels). When the magnetic field (thought parallel to the z direction) is increased in strength, quantum states with a high quantum number of orbital motion move in k space normal to z outward through the Fermi surface. This causes a periodic redistribution and excitation of electrons along the Fermi surface, observed as oscillations of the diamagnetic susceptibility. Regions of extremal area normal to z contribute the decisive contribution. The period of the oscillations, e.g., the field strengths for which, say, maxima of χ occur, is given by

$$\Delta \left(\frac{1}{H}\right) = 2\pi \left(\frac{e}{\hbar A_0}\right), \quad (16.12)$$

where A_0 is an extremal area of the Fermi surface cross section perpendicular to z. Table 16.3 compares the radii of A_0 calculated from band theory and measured

---

[10] R. G. Chambers, "Magnetoresistance," *The Fermi Surface*, W. A. Harrison and M. B. Webb, Eds., John Wiley and Sons, New York, 1960, p. 100; A. B. Pippard, "The Dynamics of Conduction Electrons," *Low-Temperature Physics*, C. DeWitt, B. Dreyfus, and P. G. de Gennes, Eds., Gordon and Breach, New York, 1962, p. 149.

[11] J. M. Ziman, *Electrons and Phonons*, Clarendon Press, Oxford, 1960.

[12] F. Garcia Moliner, *Proc. Phys. Soc.* (London) 72, 996 (1956).

[13] R. G. Chambers, Ref. 10; I. M. Lifshitz and V. G. Peschanskii, *Sov. Phys.—JETP* 8, 875 (1959); N. E. Alekseevski and Yu. P. Gaidukov, *ibid.* 10, 481 (1960); J. R. Kaluder and J. E. Kunzler, *J. Phys. Chem. Solids* 18, 256 (1961).

Table 16.3. Dimensions of Fermi surface in units of 10^8 cm^{-1} (from de Haas–van Alphen experiments)

	K	Cu		
Exper.	0.744	1.38 $H \parallel [111]$ 1.40 $H \parallel [100]$	exper.	average belly radius of area A_0
		1.33 ± 0.01	theor.	
		0.26 – 0.28	exper.	neck radius
		0.28 ± 0.03	theor.	
Free-Electron Value	0.745	1.365		free-electron value

with the help of the de Haas–van Alphen effect. Figure 16.11 of the Cu Fermi surface helps to visualize dog's-bone orbit, neck, and belly.[14]

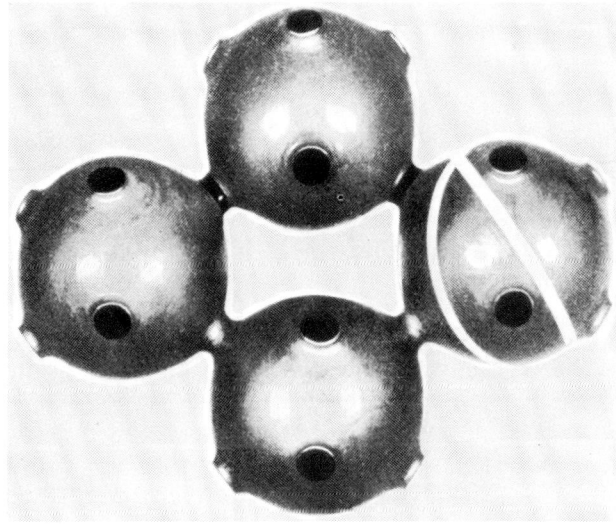

Fig. 16.11. Fermi surface of Cu showing dog's bone, belly, and neck. (After Shoenberg.[14])

Another effect capable of checking Fermi surface calculations is cyclotron resonance. It has been adapted to real metals by Azbel' and Kaner.[15] In this experiment, both the constant magnetic (H_z) and the high-frequency electric field are parallel to the metal surface and normal to each other. The electrons

[14] D. Shoenberg, "The de Haas–Van Alphen Effect," *The Fermi Surface*, W. A. Harrison and M. B. Webb, Eds., John Wiley and Sons, New York, 1960, p. 74.

[15] M. I. Azbel' and E. A. Kaner, *Sov. Phys.—JETP* **3**, 772 (1956); **5**, 730 (1957). See also B. Lax and J. G. Mavroides, "Cyclotron Resonance," *Solid State Physics*, Vol. 11, F. Seitz and D. Turnbull, Eds., Academic Press, New York, 1960, p. 261.

periodically surface from below the penetration depth of the electric field and then dive down. The cyclotron resonance frequency is given by

$$\omega_c = \frac{eH}{m_c^*}; \qquad (16.13)$$

resonances are observed at multiple frequencies $\omega_c n$, where n is an integer. Here the mass m_c^*, which measures the inertia connected with the carrier motion along the cyclotron path, is

$$m_c^* = \frac{\hbar^2}{2\pi}\left[\frac{\partial A(k)_z}{\partial \mathcal{E}}\right]_{\mathcal{E}_F}. \qquad (16.14)$$

Inserting $\mathcal{E} = \hbar^2 k^2/2m$ valid for free electrons, one finds $m_c^* = m$. Here $A(k_z)$ is the area of that part of the plane $k_z = $ const which is cut off by the Fermi surface. Values of m_c^*/m are represented in Table 16.4.

Table 16.4. Ratio of cyclotron mass to free-electron mass (m_c^*/m)

	Na (K)	Cu		
Exper. 1.24(1.21 ± 0.02)		1.38 ± 0.01	exper.	belly orbit $H \parallel [100]$
		1.1 ± 0.1	theor.	
Theor. 1.00(1.035 – 1.092)		1.23 ± 0.01	exper.	dog's-bone orbit of holes around a deficiency of electrons
		1.12 ± 0.06	theor.	
		0.6	exper.	neck orbit
		0.41 ± 0.02	theor.	

The differences of the cyclotron masses of different orbits in Cu have been linked with s-d interactions.[16]

It should be mentioned in passing that cyclotron resonance, like electrical conductivity in general, measures also the relaxation times of electrons. The problems relating to relaxation times are not yet solved quantitatively from first principles, because of their complexity. An instructive representation of the Fermi surface of Cu, in which six techniques are indicated that provide experimental information, has been given by Mackintosh.[17]

Cohesion

Cohesion, although a structure-sensitive property of metals, can be calculated quite accurately in simple

[16] M. H. Cohen, *The Fermi Surface*, W. A. Harrison and M. B. Webb, Eds., John Wiley and Sons, New York, 1960, p. 176.

[17] A. R. Mackintosh, *Sci. Am.* **209**, 110 (1963).

cases. Wigner and Seitz[18] give a qualitative analysis for alkali metals in terms of three contributions: The first is the so-called boundary correction, which is related to the deformation that the atomic wave functions suffer in bonding to Bloch functions for $k = 0$. These functions must be continuous in the crystal and have a continuous radial derivative on the boundary of the atomic polyhedra in real space. A crystal wave function $k = 0$, constructed from atomic s functions, causes an expansion of the atomic wave function (which far from the nucleus has ψ and $\partial\psi/\partial r$ of opposite sign) into real space (Fig. 16.12).[10] A $k = 0$ function constructed

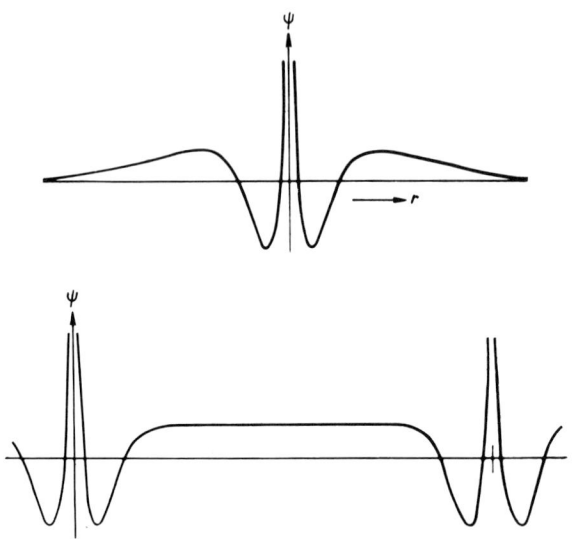

Fig. 16.12. Part of valence electron wave function of a Na atom as compared with $k = 0$ wave function of Na conduction band in crystal. (After Pippard.[10])

from atomic p functions (which change sign at the nuclei), however, may require a compression of the atomic wave function in order to produce a continuous ψ and $\partial\psi/\partial r$ at the boundary of the atomic polyhedron in the solid. Thus the $k = 0$ function derived from atomic s functions gives a boundary correction producing cohesion, while a crystal wave function originating from atomic p functions does not (or much less).

A second contribution, which can be considered separately for the alkali metals, is the sum of the kinetic energies related to the electron motion through the lattice as a whole. And a third contribution, small for alkali metals, stems from electron correlation effects.

Table 16.5 gives cohesive energy and equilibrium lattice distances as calculated and measured for Na and Cu.

[18] E. Wigner and F. Seitz, "Qualitative Analysis of the Cohesion in Metals," *Solid State Physics*, Vol. 1, F. Seitz and D. Turnbull, Eds., Academic Press, New York, 1955, p. 97.

Table 16.5. Cohesive energy and equilibrium lattice distances

Na	Cu		
26.0	81.2	exper.	cohesive energy (kcal/mole)
26.3	61.7	theor.	
4.22	3.6	exper.	lattice constant (Å)
4.26	4.2	theor.	

Transition metals

In the transition metals, some of which have already been mentioned in connection with band calculations, more than one band is partly empty. In some of these metals cooperative phenomena such as ferromagnetism (Fe, Co, Ni) and antiferromagnetism (Cr, Mn) occur. This is true not only for the $3d$ transition elements but for many rare-earth metals of the $4f$ group. Strong magnetic phenomena presuppose that the magnetic interaction between the moments on neighboring lattice sites is not too small in comparison to the bandwidth. For a calculation or even a semiquantitative treatment of such magnetic phenomena the one-electron approach is insufficient. Special models must be developed.

Magnetic interaction through conduction electrons

Conduction electrons in metals present a means of indirect magnetic interaction over large distances between atoms with incomplete $3d$ shells. Zener[19] showed that a strong exchange interaction between the d electrons and the s conduction electrons, irrespective of sign, would lead to ferromagnetism. Strong, in the sense of Zener's model, means: as compared to d-d interaction between adjacent atoms. Moreover, the d-s-d interaction was assumed to be homogeneous, with no change of sign occurring with distance between atoms with partly empty d shells.

Such d interactions via conduction electrons have been substantiated in a revised form;[20] the sign may now depend on distance. These interactions explain cooperative magnetic and other phenomena encountered in the temperature region of $10°$ K, when

[19] C. Zener, *Phys. Rev.* **81**, 440 (1951); C. Zener and R. R. Heikes, *Revs. Modern Phys.* **25**, 191 (1953).
[20] M. A. Ruderman and C. Kittel, *Phys. Rev.* **96**, 99 (1954); T. Kasuya, *Progr. Theoret. Phys.* (Kyoto) **16**, 65 (1956); K. Yosida, *Phys. Rev.* **106**, 893 (1957); A. Blandin and J. Friedel, *J. phys. radium* **20**, 140 (1959).

transition metal ions such as Mn are incorporated in small concentrations (e.g., ca. 0.1 atomic %) in a neutral metallic matrix such as Cu.[21]

In passing we might mention that just as the d electrons of distant Mn atoms are coupled through interaction with conduction electrons, distant nuclear spins in metals are coupled through the hyperfine **I S** interaction (**I** the nuclear spin, **S** the conduction electron spin). It has been found, e.g.,[22] that a scalar coupling $A\mathbf{I}_1\mathbf{I}_2$ exists with a dependence of A on distance r between nuclei 1 and 2:

$$A \cong (2k_F r)^{-3} \cos 2k_F r - (2k_F r)^{-4} \sin 2k_F r, \quad (16.15)$$

with k_F the wave vector at the Fermi level. The oscillations with half the wavelength of the electron waves at the Fermi surface π/k_F encountered in this magnetic interaction are due to the diffraction of electron waves by the magnetic discontinuity caused by the nuclear spins.

Transition metal solute atoms in alloys

In the discussion of the magnetic interaction between Mn atoms in Cu through conduction electrons, the existence of magnetic moments in the presence of the conduction electrons of Cu remains to be explained. Friedel and his school[23] have advanced a treatment of a transition metal impurity imbedded in a conductor like Cu, based on a local concentration of "virtual bound levels" through resonance of the d electron levels, with the s electron levels within the continuous band (Fig. 16.13). This corresponds to a local rearrangement of electrons which screens the difference in charge Z' between the impurity atom and the matrix atoms.

The impurity also causes a long-distance diffraction phenomenon of the conduction electrons like that encountered in the magnetic spin coupling of nuclei. At large distances from the impurity characteristic, phase shifts δ_l of the wave functions are introduced. They are connected with oscillations of the charge density with half the wavelength of the electron waves at the Fermi surface π/k_F. The index l is the number of the

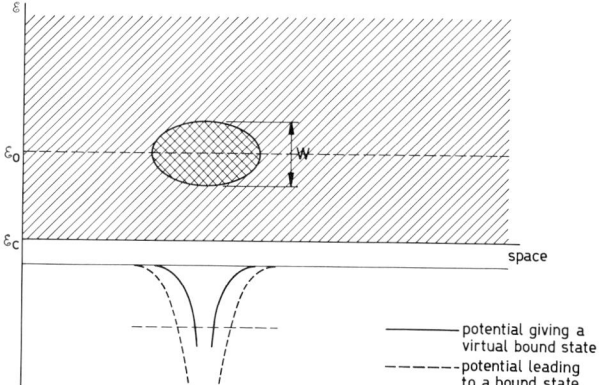

Fig. 16.13. Schematic representation of a virtual bound state due to an impurity atom. The mean energy ε_0 of this state lies within the continuum of conduction-band states. Its energy width is W. Bottom of the conduction band at ε_c. The virtual bound state represents a wave package centered upon the impurity. (After Blandin and Friedel.[23])

spherical harmonic in the development of the scattered conduction-electron wave functions. If the impurity is a $3d$ transition element atom, the phase shift δ_2 will be prominent, because the quantum number l for d electrons equals 2. The phase shifts δ_l are connected with the charge difference Z'. The amplitude of the phenomenon decreases as $1/r^3$, where r is the distance from the impurity. The situation is illustrated in Fig. 16.14.[24] The phenomenon has been measured, for example, by nuclear magnetic resonance in dilute alloys for a matrix atom that has a quadrupole moment and is consequently sensitive to electric field gradients.[25]

After this excursion into alloy structures let us ask how we get a splitting of the virtual bound levels, separating those for spin up from those of spin down, as required for a permanent moment. Blandin and Friedel[23] discuss a model in which the splitting is analogous to that known as Hund's rule for isolated atoms. It is found that splitting occurs when the exchange energy, gained by filling one subband (↑) more than the other (↓), exceeds the kinetic (Fermi) energy which must be provided for this excitation (Fig. 16.15). Narrow virtual bound levels near the Fermi surface and large exchange energy appear required. The theory seems supported by experiments.

Previously we found localized magnetic moments in the neighborhood of, e.g., Mn atoms dissolved in Cu. In the rare-earth metals and in the $4d$-element Pd, the

---

[21] G. J. van den Berg and J. de Nobel, *J. phys. radium* **23**, 665 (1962); F. J. Blatt, *ibid.* **23**, 597 (1962); A. J. Dekker, *ibid.* **23**, 702 (1962).

[22] M. A. Ruderman and C. Kittel, *Phys. Rev.* **96**, 99 (1954); N. Bloembergen, *J. phys. radium* **23**, 658 (1962).

[23] J. Friedel, *Phil. Mag.* [7] **43**, 153 (1952); A. Blandin and J. Friedel, *J. phys. radium* **20**, 160 (1959); J. Friedel, *ibid.* **23**, 593 (1962); E. Daniel, *ibid.* **23**, 602 (1962); J. S. Langer and S. H. Vosko, *J. Phys. Chem. Solids* **12**, 196 (1959); P. W. Anderson, *Phys. Rev.* **124**, 44 (1961); P. A. Wolff, *ibid.* **124**, 1030 (1961); A. M. Clogston, *ibid.* **125**, 439 (1962); S. Alexander and P. W. Anderson, *ibid.* **133** (6A), 1594 (1964).

[24] T. J. Rowland, *Phys. Rev.* **119**, 900 (1960); W. Kohn and S. H. Vosko, *ibid.* **119**, 912 (1960).

[25] R. M. Bozorth, D. D. Davis, and J. H. Wernick, *J. Phys. Soc. Japan* **17**, Suppl. B-I, 112 (1962); A. M. Clogston *et al.*, *ibid.* **17**, Suppl. B-I, 115 (1962); A. M. Clogston *et al.*, *Phys. Rev.* **125**, 541 (1962); R. M. Bozorth *et al.*, *ibid.* **122**, 1157 (1961).

presence of localized moments can also be observed. Additions of as little as 0.1 atomic % of Co produce ferromagnetism. It seems that the 12 nearest-neighbor Pd atoms surrounding a Co atom are polarized, adding $12 \times 0.6 \mathbf{m}_B$ to the $1.7\mathbf{m}_B$ of the Co atom. The total moment should then be about $9\mathbf{m}_B$; the experimental value found is 9 to $10\mathbf{m}_B$ per Co atom.[24]

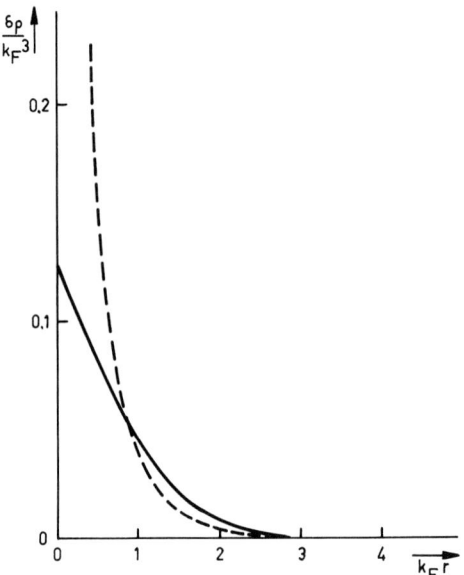

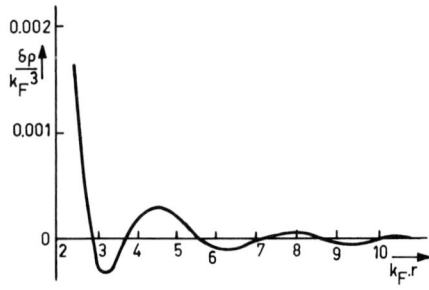

Fig. 16.14. Screening and charge density oscillations in free-electron gas around localized unit charge. (Electron density in real space corresponds to that of Ag or Au.) (After Kohn and Vosko.[24])

In contrast, a model with nonlocalized moments works well for many $3d$-element, notably Ni, alloys. In ferromagnetic iron and iron alloys the magnetic moment per atom appears to increase with applied magnetic field. If both subbands ($\downarrow$ and $\uparrow$) are partly empty, the relative shift by the field causes a noticeable increase in the number of electrons in the lower band.[26]

[26] R. Gersdorf, *Phys. Rev. Letters* **10**, 155 (1963); *J. phys. radium* **23**, 726 (1962).

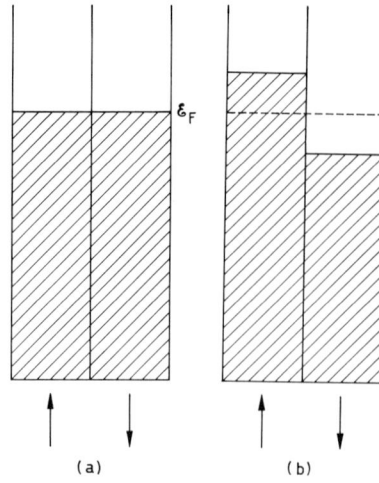

Fig. 16.15. Splitting of a virtual bound state (a) requires kinetic (Fermi) energy (b). (After Friedel.[23])

Exchange fields and Knight shift

In transition metals, say of the $3d$ type, both the $4s$ and the $3d$ band are polarized by external and internal magnetic fields. The exchange interactions created by the fields are not confined to the d electrons; the d-s exchange contributes to an important polarization of the s electrons. This is of importance for neutron diffraction, nuclear magnetic resonance (NMR), and Mössbauer measurements. The neutron spins probe magnetic polarization throughout the unit cell. From the magnetic form factors for neutron diffraction[27] from metallic iron at low temperatures, one derives as contribution to the magnetization per atom:

$3d$ shell:	$2.39\mathbf{m}_B$	composed of spin magnetization 2.31, and orbital magnetization $0.08\mathbf{m}_B$
$4s$ shell:	$-0.21\mathbf{m}_B$.	
Total	$2.18\mathbf{m}_B$	

In Mössbauer and NMR measurements the nuclei feel only the presence of electrons with s character; the $3d$ electron wave functions vanish at a nucleus. Let us first consider a metal with only conduction electrons and no other unfilled shell. In an applied magnetic field there is a surplus of conduction electrons with spin up, due to Pauli paramagnetism, leading to the Pauli susceptibility χ_p (Eq. 16.8). The nuclear spins feel the external field plus the magnetization of the conduction electrons with wave function amplitude $\psi(0)_F$ at the nucleus (F stands for Fermi surface). Nuclear magnetic resonance occurs at larger frequencies than

[27] C. G. Shull and Y. Yamada, *J. Phys. Soc. Japan* **17**, Suppl. B-III, 1 (1962).

for the same nuclei in nonconducting material;[28] the relative frequency shift (Knight shift) amounts to

$$\frac{\Delta \nu}{\nu} \cong \frac{8\pi}{3} \chi_p M \langle |\psi(0)_F|^2 \rangle_{Av}, \quad (16.16)$$

where M is the atomic mass. The shift is of the order 10^{-3}. The change with respect to the atomic wave function at the nuclei is represented by a parameter

$$\xi = \frac{\langle |\psi(0)_F|^2 \rangle_{Av}}{|\psi_A(0)|^2} \quad (16.17)$$

(for Na, $\xi = 0.72$; for Cu, 0.53; for V, 2.7).

If two electron shells are partly empty (e.g., $3d$ and $4s$ in ferromagnetic Fe), the $3d$ electrons are cooperatively magnetized but the nuclei react only to electrons with s character. The exchange interaction between $3d$ and $4s$ is negative; hence, the nuclei see a negative field due to $4s$ electrons. Moreover—and more important—exchange interactions between $3d$ and the core s electrons in completed shells ($3s$, $2s$, $1s$) deforms these s orbitals.[29] The essential point is that $ns \uparrow$ electron orbits are affected differently from $ns \downarrow$ orbits; hence, generally

$$\rho_{ns} = |\psi_{ns\uparrow}(0)|^2 - |\psi_{ns\downarrow}(0)|^2 \neq 0. \quad (16.18)$$

The closed as well as the open s shells contribute to the effective magnetic field at the nuclei.

NMR and Mössbauer experiments show that the effective magnetic field acting upon the Fe^{57} nucleus in ferromagnetic iron is -333 kGauss opposed to the external field. Theory can reasonably account for it. Also Pt metal,[30] transition metal alloys, and intermetallic compounds such as XAl_2, show a negative Knight shift (X stands for a rare-earth metal[31]).

Electron correlation in plasma oscillations

Thus far the wave function of the system of many electrons was obtained in principle by multiplying one-electron wave functions and taking linear combinations of such products in the form of a Slater determinant. As stressed already, in this treatment the spatial correlation of electrons is not correctly taken into account.

Such correlation effects become dominant in plasma oscillations,[32] excited when a charged particle of large momentum causes a density fluctuation of the valence electrons in metals. The restoring force is due to the Coulomb interaction and the inertia to the electron mass m (for the nearly free electrons). Plasma frequencies will be excited,

$$\omega = \sqrt{\frac{4\pi e^2 N}{m}}; \quad (16.19)$$

N is the average density of electrons in real space and $\omega \sim 10^{16}$. A detailed theoretical treatment of the phenomenon is difficult; also, experiments are not yet fully satisfactory.

Superconductivity

Another effect where electron correlations are all-important is superconductivity. Superconducting metals may be "soft" (Type I) superconductors, such as Pb, Sn, In, and Ga, to which we shall limit our attention, or "hard" (Type II) superconductors, like Nb and such intermetallic compounds as Nb_3Sn (transition temperature $T_C = 18°$ K) and carbides (MoC).

Two properties are typical for superconducting materials under static conditions: complete absence of resistivity (electric field strength $\mathbf{E} = 0$) and not only

$$\frac{d\mathbf{B}}{dt} = 0,$$

(a consequence of $\mathbf{E} = 0$) but also (16.20)

$$\mathbf{B} \equiv 0.$$

When a superconductor (e.g., a sphere) is cooled in an

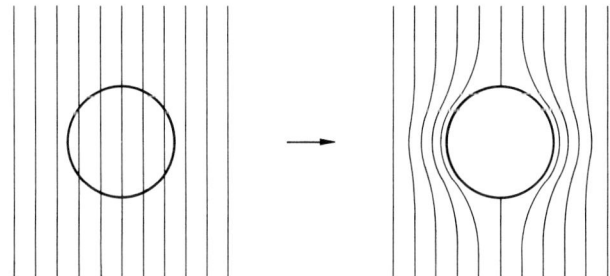

Fig. 16.16. Ejection of magnetic induction lines from a sphere cooled through the transition temperature in a magnetic field. (After London.[33])

external magnetic field, the induction will be ejected at the transition temperature to the superconducting

---

[28] W. D. Knight, "Electron Paramagnetism and Nuclear Magnetic Resonance in Metals," *Solid State Physics*, Vol. 2, F. Seitz and D. Turnbull, Eds., Academic Press, New York, 1956, p. 93.

[29] G. W. Pratt, Jr., *Phys. Rev. 102*, 1303 (1956); J. H. Wood and G. W. Pratt, Jr., *ibid. 107*, 995 (1957); V. Heine, *ibid. 107*, 1002 (1957); D. A. Goodings and V. Heine, *Phys. Rev. Letters 5*, 370 (1960).

[30] R. E. Walstedt, M. W. Dowley, E. L. Hahn, and C. Froidevaux, *Phys. Rev. Letters 8*, 406 (1962); J. Butterworth, *ibid. 8*, 423 (1962); L. E. Drain, *J. Phys. Chem. Solids 24*, 379 (1963).

[31] V. Jaccarino, *J. Appl. Phys.*, Suppl. *32*, 102S (1961).

[32] D. Pines, "Electron Interaction in Metals," *Solid State Physics*, Vol. 1, F. Seitz and D. Turnbull, Eds., Academic Press, New York, 1955, p. 368; S. Raimes, *Repts. Progr. Phys. 20*, 1 (1957).

state (Fig. 16.16).[33] The induction drops to zero within a small boundary layer, a penetration depth of about 500 Å in an ideal superconductor (Meissner effect). The magnetic energy representing this ejection explains that applied magnetic fields counteract the superconducting state.

We will discuss here only certain aspects of the microscopic theory of superconductivity, as developed chiefly by Bardeen, Cooper, and Schrieffer (BCS), and a few verifications.[34] Since superconductivity sets in at very low temperatures, the energy difference between the normal and the superconducting state is very small, and a refined model approach for this property of metallic systems is required.

Important for the development of the theory was the theoretical finding that the filling of k space up to, say, a spherical surface with radius k_{max} need not represent the state of lowest energy.[35] When a net attractive interaction, however weak, exists between electrons, if they are excited out of the Fermi sea or between holes below the Fermi level, "between quasi-particle excitations" in modern language, bound pairs may form and lower the energy of the system with respect to that ground state. The binding energy of such pairs is found of appreciable value only when the sum of the particle momenta is negligible:

$$\mathbf{k}_1 + \mathbf{k}_2 \cong 0. \qquad (16.21)$$

A state in which a large number of such bound pairs occurs can be shown to have superconducting properties.

Pairs may build up wave packages extending over 10^{-4} cm in real space. Such large dimensions had been anticipated from uncertainty arguments:

$$\Delta \mathbf{k}\, \Delta \mathbf{x} \sim 1;$$
$$2\frac{\Delta \mathbf{k}}{\mathbf{k}} \sim \frac{kT_C}{\mathcal{E}_F}; \qquad (16.22)$$
$$\Delta \mathbf{x} \sim \left(\frac{\mathcal{E}_F}{\mathbf{k}}\right)\left(\frac{1}{kT_C}\right) \sim 10^{-4}.$$

They also had been predicted[36] as correlation distance

---

[33] F. London, *Superfluids*, Vol. I, John Wiley and Sons, New York, 1950.

[34] J. Bardeen, L. N. Cooper, and J. R. Schrieffer, *Phys. Rev.* **106**, 163 (1957); *ibid.* **108**, 1175 (1957); J. Bardeen and J. R. Schrieffer, "Recent Developments in Superconductivity," *Progress in Low-Temperature Physics*, C. J. Gorter, Ed., North-Holland Publishing Co., Amsterdam, 1961, p. 153; J. Bardeen, "Superconductivity," *Advances in Materials Research in the NATO Nations*, A Pergamon Press Book (The Macmillan Co., New York), 1963, p. 281; M. Tinkham, "Superconductivity," *Low-Temperature Physics*, C. DeWitt et al., Eds., Gordon and Breach, New York, 1962, p. 149.

[35] L. N. Cooper, *Phys. Rev.* **104**, 1189 (1956).

[36] A. B. Pippard, *Proc. Roy. Soc. (London)* **A216**, 547 (1953).

on experimental evidence, e.g., from the influence of impurity centers on penetration depth (3% In in Sn). This depth increased sharply when the mean free path of the electrons decreased to the order of 10^{-4} cm, but the transition temperature was hardly affected.

How can we explain an attractive interaction between electrons? One such interaction is provided by the quantized lattice vibrations, the phonons. If the velocity of an electron is not too large, it can locally deform the lattice by attracting positive ions. The concentration of positive charge, on the other hand, makes it advantageous for a second electron to join its colleague. The matrix element V_{ph} for the electron-electron interaction via phonon-electron interaction gives rise to attraction if $\hbar\omega_q$, the energy of the phonons with wave vector q, is large compared to the one-electron excitation energy, the Bloch energy ϵ_k of electrons measured from the Fermi level. In this case

$$V_{ph} \cong -\frac{2|M_q|^2}{\hbar\omega_q}, \qquad (16.23)$$

where M_q stands for the electron-phonon interaction.

It is not sufficient that the electron interaction through phonons be attractive: The sum of electron interaction via phonons and the Coulomb repulsion between the electrons, which will be largely screened in the metal, must also be attractive so that bound pairs can be formed. The average over the Fermi surface of $V_{ph} + V_{coul}$ must be negative in an interaction zone, as will be discussed later.

Model for soft superconductors

A reasonable model for the superconductor, based on pair interaction and amenable to calculation, has been postulated by BCS.

The wave function ψ_s for the superconducting ground state (which does not carry current) is a linear combination of normal state configurations, with the proviso that all quasi-particle states are occupied in pairs of opposite momentum and spin. If $\mathbf{k}\uparrow$ is occupied, $-\mathbf{k}\downarrow$ must be as well; if $\mathbf{k}'\uparrow$ is vacant, $-\mathbf{k}'\downarrow$ is likewise. An interaction zone of width $2\hbar\omega_c$, with its center at the Fermi surface $\mathcal{E} = 0$, is defined. In this zone the resultant pair attraction is taken as constant: $-V$, and zero outside. With a variational method the linear combination giving lowest energy is then selected.

The cutoff frequency ω_c that determines the width of the interaction zone is assumed to be about half the Debye frequency. This frequency is representative for the phonon spectrum which is related to pair interaction. Within the interaction zone the condition for attractive interaction: $|\mathcal{E}_k| < \hbar\omega_c$ is fulfilled; quasiparticle excitations are considered well defined relative to energy uncertainty (their characteristic lifetime τ obeys $1/\tau < \omega_c$).

Occupation function and gap parameter

The probability $h(\mathbf{k})$ that a state $(\mathbf{k}\uparrow, -\mathbf{k}\downarrow)$ is occupied decreases from 1 far below the Fermi level to 0 far above it as

$$h(\mathbf{k}) = \frac{1}{2}\left\{\frac{1 - \mathcal{E}(\mathbf{k})}{[\mathcal{E}^2(\mathbf{k}) + \Delta^2]^{\frac{1}{2}}}\right\}, \qquad (16.24)$$

with $|\mathcal{E}(\mathbf{k})| < \hbar\omega_c$. Here $\mathcal{E}(\mathbf{k})$ is the energy (plus or minus) of the electrons (or holes) belonging to a pair, counted from the Fermi level. Figure 16.17 shows

Fig. 16.17. Pair-occupation function in the BCS superconducting ground state (cf. Fig. 16.3).

$h(\mathbf{k})$ resembling the Fermi occupation function f (cf. Fig. 16.3) for the normal electrons at temperature $T \cong \Delta/k$, k here being Boltzmann's constant. It will be seen that in order to gain pair-interaction energy (related to Δ), quasiparticle excitations of the normal state out of its Fermi sea occur even at 0° K.

The denominator

$$[\mathcal{E}^2(\mathbf{k}) + \Delta^2]^{\frac{1}{2}} \simeq \mathcal{E}(\mathbf{k}) \qquad (16.25)$$

in the expression for $h(\mathbf{k})$ is the energy of a quasi particle excited from the superconducting ground state. An isolated electron—originally part of a pair—has an energy at least equal to the dissociation energy Δ. Since all excitations from the superconducting ground state occur in pairs, one expects a minimum excitation energy (energy gap) of 2Δ. The gap parameter Δ of course depends on temperature. It decreases from a maximum at 0° K to 0 at T_C, the transition temperature of the superconducting state. In the following we write Δ_0 for its value 0° K.

Δ_0 can be expressed through the normal state density at the Fermi level $N(\mathcal{E}_F)$ and V, the attractive pair potential:

$$\Delta_0 \simeq 2\hbar\omega_c \exp\left[-\frac{1}{N(\mathcal{E}_F)V}\right]. \qquad (16.26)$$

The exponential function is assumed to be much smaller than unity. The net condensation energy on transition from normal to superconducting state becomes

$$W_{s,o} - W_n \cong -\tfrac{1}{2}N(\mathcal{E}_F)\Delta_0^2. \qquad (16.27)$$

The expression for Δ_0 (Eq. 16.26) contains the isotope effect if one assumes that

$$\hbar\omega_c \sim k\theta_{\text{Debye}}, \qquad (16.28)$$

while $N(\mathcal{E}_F)$ and V are independent of isotopic mass. Since

$$\Delta_0 \sim kT_C, \qquad (16.29)$$

it follows that

$$T_C \sim M^{-\frac{1}{2}}, \qquad (16.30)$$

where M is the isotopic mass.

This relation has been confirmed for Sn, Tl, Pb, and Hg; for Mo the exponent is smaller. Almost no isotope effect has been found for Os and none for Ru. Thus there are possibly other kinds of attractive quasiparticle interactions in transition metals. The interaction between quasi particles up to now is deduced phenomenologically from the measured transition temperature T_C.

Let us examine further the gap parameter and its consequences. While the occupation number n of pair states in k space was not too different from the occupation number of electrons in a normal metal, the excitation spectrum (density of excited states of a superconductor) differs greatly. According to Eq. 16.25, a finite energy must be paid for the excitation of one quasi particle because of the dissociation of pairs, whereas an infinitesimal energy is necessary to excite an electron from the Fermi surface of normal metals. The density of excited states for a superconductor is

$$\frac{dX(\mathcal{E})}{d\mathcal{E}} = \frac{N(\mathcal{E}_F)}{d\mathcal{E}/d\epsilon} = \frac{N(\mathcal{E}_F)\mathcal{E}}{(\mathcal{E}^2 - \Delta^2)^{\frac{1}{2}}} \qquad (16.31)$$

(Fig. 16.18). Far from the Fermi surface 0 the density

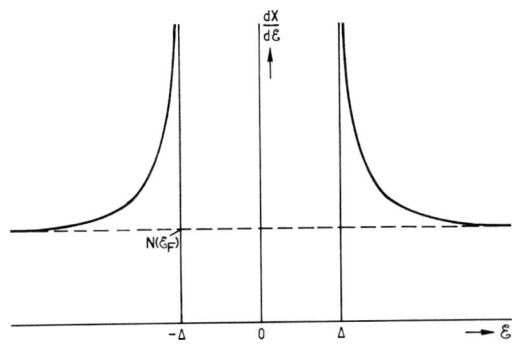

Fig. 16.18. Density of states excited from the superconducting ground state of the BCS theory as a function of $\mathcal{E}$.

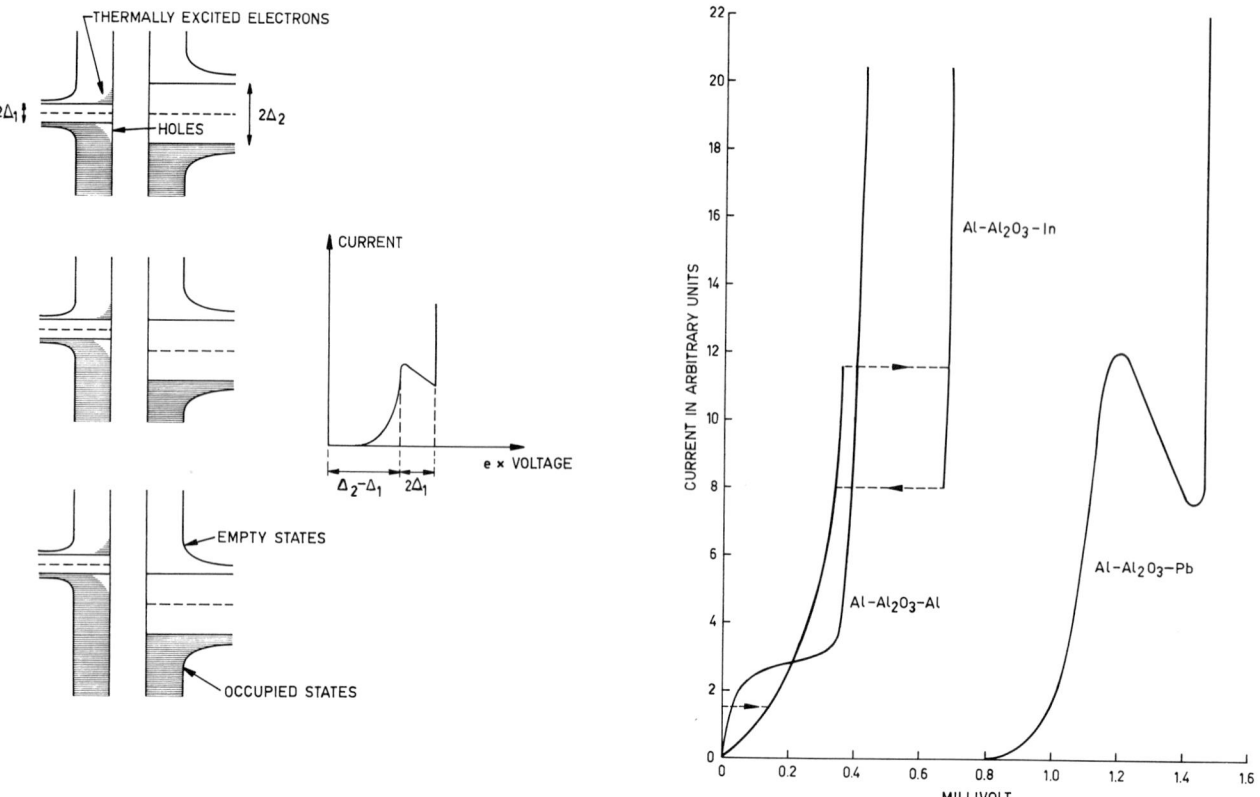

Fig. 16.19. Tunneling experiments with superconducting films separated by a thin oxide layer. (After Giaever.[37]) *Left:* analysis of current voltage characteristic. *Right:* curves for tunneling current, showing agreement with analysis.

of excited states equals that in the normal metal; singularities, however, occur at the gap edges.

This band picture has been confirmed experimentally. One experiment[37] is represented in Fig. 16.19. A sandwich was formed from two different superconducting metals separated by a thin oxide layer (15 to 20 Å). On application of a potential difference between the superconductors, the electrons in the one with the smaller gap, $2\Delta_1$, are raised in energy relative to those in the other one. Current will increase with voltage because the thermally excited electrons above the smaller gap can tunnel through the oxide layer into the region above the gap $2\Delta_2$. A maximum current is found for $\Delta_2 - \Delta_1 = eV$. When the voltage V is increased further, the current decreases because to the right a lower density of states is met. If $\Delta_1 + \Delta_2 \leq eV$, the electrons below the gap $2\Delta_1$ can also tunnel.

Gap values can be measured accurately by such experiments. The BCS theory gives $2\Delta_0 = 3.52\,kT_C$; experimentally $4.33\,kT_C$ was found for Pb, 3.63 for In, and 2.7 for Al.

Space does not permit going into details of other confirmations of the model by measurements of nuclear magnetic-spin relaxation, surface impedance, infrared absorption and transmission, and of magnetic flux quantization. Yet a few words should be said about the main characteristics of the superconducting state, although perhaps this aspect has been deduced least rigorously from theory:

1. By displacing the whole electron distribution in k space one gets a current. The pairs have actually a net momentum, different from zero, of

$$\mathbf{p} = 2\hbar\mathbf{q}, \qquad (16.32)$$

since the configuration is $\mathbf{k}+\mathbf{q}\uparrow$, $-\mathbf{k}+\mathbf{q}\downarrow$. At finite temperature the superconductor has a thermal distribution of excited quasiparticles corresponding to $\mathbf{p}$. The current remains persistent, because the interaction energy, the gap, sees to it that the long-range correlation of average momentum is preserved. After scattering of individual electrons by phonons, an energy of $\sim 2\Delta$ can be recovered by recondensation into the current-carrying state.

2. The Meissner effect, $\mathbf{B} = 0$, is also connected with the energy gap. It gives stiffness to the wave function of the superconducting state and prevents it from

[37] I. Giaever, *Phys. Rev. Letters* **5**, 464 (1960); J. Nicol, S. Shapiro, and P. H. Smith, *ibid.* **5**, 461 (1960).

changing in a magnetic field in a way analogous to the stiffness of atomic orbitals in filled electronic shells, which exhibit diamagnetism. Calculations have been given by Rickayzen[38] and others.

No particular attention has been paid here to Type II superconductors, which are of great importance also because of their large current-carrying capacities. The reader is referred to the proceedings of a recent international conference concerning the status of the field of superconductivity[39] and to summarizing reviews of that conference.[40]

[38] G. Rickayzen, *Phys. Rev. 115*, 795 (1959).

[39] *International Conference on the Science of Superconductivity, Revs. Modern Phys. 36*, No. 1 (1964).

[40] H. R. Hart and R. W. Schmitt, *Science 143*, 57 (1964); *Physics Today*, February 1965, p. 31.

17 · AVALANCHES AND GAS BREAKDOWN

Heinz Raether

The formation of avalanches — Experimental study of single avalanches — Breakdown conditions — Many-generation and streamer mechanisms — The spark chamber — Verifications of the preceding concepts — Extension to large gap distances and to low or high pressures

The formation of avalanches

The elementary process of a discharge is the formation of electron avalanches: n_0 primary electrons leaving the cathode are amplified to

$$n = n_0 \exp(\alpha v_- t) \qquad (17.1)$$

after they had run a time t with the electron drift velocity v_-; α is the Townsend coefficient, i.e., the number of electron-ion pairs created by one electron. In addition, photons are produced by the inelastic impact of the electrons; hence, a photon avalanche is built up simultaneously with the electron avalanche.

The photons regenerate electrons at the cathode, i.e., new avalanches. The production of electrons by positive ions hitting the cathode can, in general, be neglected. These new avalanches start after the transit time of the preceding avalanche through the gap

$$T_- = \frac{d}{v_-}, \qquad (17.2)$$

where d is the gap distance. When this process is repeated, a series of avalanche generations arises, each lasting a time T_-.

If the number of avalanches per generation is n_0 and if this number is reproduced by secondary processes, then we speak of a self-sustaining discharge:

$$n_0 \gamma e^{\alpha d} = n_0, \quad \text{or} \quad \gamma e^{\alpha d} = 1, \qquad (17.3)$$

where $\gamma e^{\alpha d}$ is the number of electrons produced at the cathode by the photons of one avalanche. In the case $\gamma e^{\alpha d} < 1$, the number of starting avalanches per generation decreases and after some time the current reduces to zero.[1,2]

These considerations have been developed further theoretically[3] and were quantitatively verified by the experiments described later (cf. also Ref. 2).

Experimental study of single avalanches

Single avalanches can be investigated by various methods:

1. *Cloud chamber.* The ions produced by the ionizing collisions of the electron within the gap are used as condensation nuclei in a cloud chamber. If a voltage pulse of a length $T < T_-$ is applied to the gap, a single avalanche and the spatial distribution of the ionization processes can be observed[2] (Fig. 17.1).

---

[1] Cf. J. M. Meek and J. D. Craggs, *Electrical Breakdown of Gases*, Clarendon Press, Oxford, 1953; L. B. Loeb, *Basic Processes of Gaseous Electronics*, University of California, Berkeley, 1961.
[2] Cf. H. Raether, *Avalanches and Breakdown of Gases*, Butterworths Scientific Publications, London, 1964.
[3] P. L. Auer, *Phys. Rev.* **111**, 671 (1958).

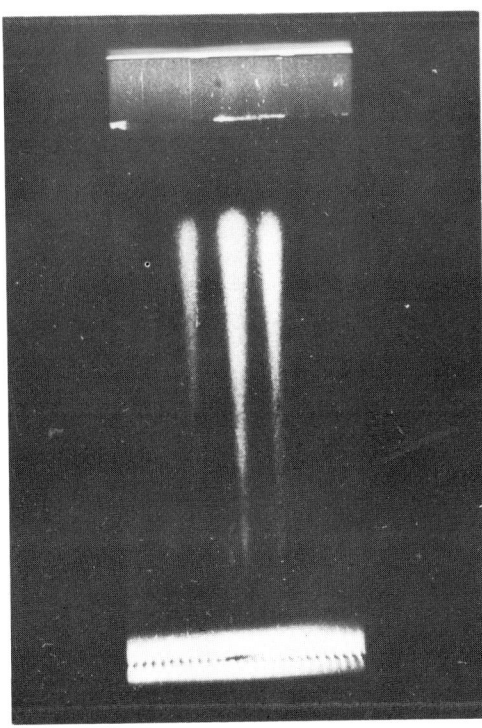

Fig. 17.1. Cloud-chamber photograph of single-electron avalanches in N_2 at 280 torr.

2. *High-gain image intensifier.* To detect the photon avalanche, one scrutinizes the avalanche with an image intensifier.[4] The sensitivity of this method is less than that of the cloud chamber; we see therefore especially the head of the avalanche which radiates very intensively (Fig. 17.2).

With both these methods the drift velocity

$$v_- = \frac{l}{T} \qquad (17.4)$$

(order of 10^7 cm/sec) can be deduced from the length l of the avalanche and the pulse length. From the profile of the avalanche one derives the electron temperature (some eV), in agreement with that calculated from the drift velocity.

3. Instead of imagining the *spatial* distribution of ions and photons, we can concentrate on the *temporal* development of the amplification process by measuring the current produced by the drifting electrons and ions (electrical method) and/or observing the radiation emitted from an avalanche with a photomultiplier (optical method). Each avalanche produces a current pulse on the oscilloscope during the transit of the electrons through the gap; hence,

[4] K. H. Wagner and H. Raether, *Z. Physik* **170**, 540 (1962); K. H. Wagner, *ibid.* **178**, 64 (1964).

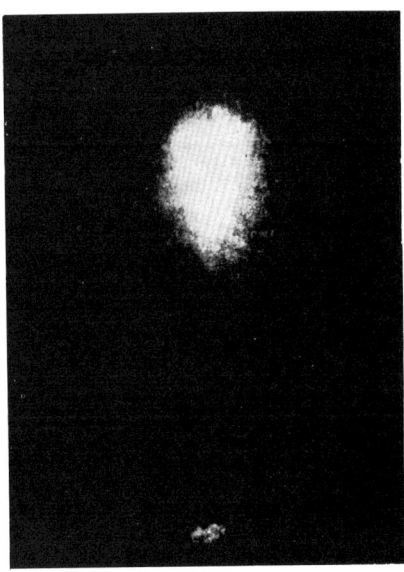

Fig. 17.2. Electron avalanche in N_2 photographed by high-gain image intensifier.

single avalanches and their successors can be observed (Fig. 17.3).

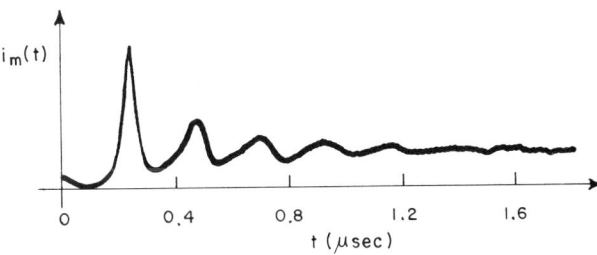

Fig. 17.3. Avalanche generations in nitrogen at $\gamma e^{\alpha d} = 1$. The oscillations represent generations started by photoeffect at the cathode (optical method).

These experiments provide the possibility of studying the growth of the electron and also of the photon avalanches quantitatively. The time constant measured proves to be in good agreement with $(1/\alpha)v_-$, indicating the validity of the exponential law of Townsend (Eq. 17.1). (At electron amplification $> 10^6$, deviations occur because of the effect of space charge of the positive ions.) The sequence of generations was also studied as function of time and was found to agree with theory.[2]

Breakdown conditions

Condition 17.3 for a self-sustaining current is certainly no breakdown criterion. It states that a discharge started, for example, with some 10^4 electrons continues with 10^4 avalanches per generation *ad infinitum*. Breakdown, on the other hand, implies current growths to values of amperes, so that the voltage across the gap drops to small values. Still, Condition 17.3 normally leads to breakdown by the following effect: The electrons of each generation leave the gap after a time T_-; the ions, however, drift much more slowly and need a time $T_+ \gg T_-$ to reach the cathode. The positive ions of many avalanche generations therefore are accumulated in the gap and modify the electrical field by its space charge, so that the ionization efficiency $\exp \int_0^d \alpha \, dx$ of the subsequent electron avalanches is changed. Under ordinary conditions of p and d, the value of $\int_0^d \alpha \, dx$ exceeds that of αd without space charge. Hence, the discharge current grows to very high values on account of the increasing ionization effi-

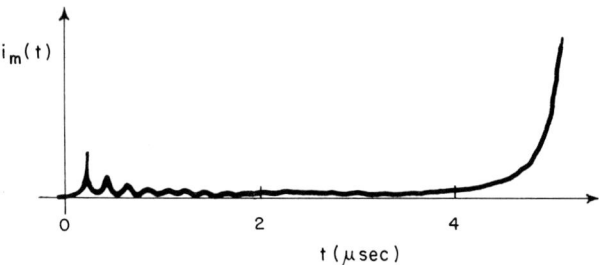

Fig. 17.4. Electron avalanche with photosuccessors in N_2 + 2.5% CH_4 at $\gamma e^{\alpha d} = 1$, leading to breakdown on account of accumulated (positive) space charge (optical method).

ciency $\int_0^d \alpha \, dx$, and breakdown results.[2,5] Figure 17.4 shows an example.[6] We call this breakdown mechanism the Townsend or "many-generation" mechanism.

In addition to this many-generation mechanism, the "streamer mechanism" (also called the "one-generation mechanism") is known: If one avalanche, in general the first one, reaches values of amplification so high that the eigenspace charge field of its ions changes the electrostatic field E_0, the classical Townsend avalanche is transformed. This is the case when the space-charge field

$$E_{sp} = [\epsilon \exp(\alpha x)] \frac{1}{r_D^2}, \qquad (17.5)$$

where r_D is the radius of the avalanche head, is comparable with the applied field E_0. Experiments in the cloud chamber have shown that this takes place at about

---

[5] A. von Hippel and J. Franck, *Z. Physik* **57**, 696 (1929); A. von Hippel, *ibid.* **80**, 19 (1933); A. von Engel and M. Steenbeck, *Elektrische Gasentladungen*, Springer-Verlag, Berlin, 1934.

[6] H. Tholl, unpublished.

$$n = \exp(\alpha x) \sim 10^8$$

or (17.6)

$$\alpha x \sim 20 \quad \text{(critical amplification)}$$

(d some cm, p some 100 torr). At the critical amplification the avalanche transforms into a plasma streamer proceeding to the anode with a velocity greater than v_- (anode-directed plasma streamer). Simultaneously a new phenomenon takes place: A cathode-directed streamer develops with still higher velocity ($\sim 10 \cdot v_-$). In the channel bridging anode and cathode the current grows to breakdown values (Fig. 17.5).

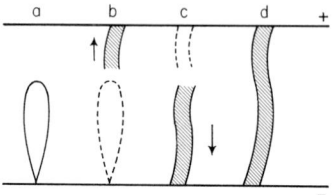

Fig. 17.5. Stages of breakdown in the one-generation (streamer) mechanism: (*a*) avalanche; (*b*) anode-directed streamer, continuing Townsend avalanche; (*c*) cathode-directed streamer; (*d*) plasma channel bridges between anode and cathode.

For the one-generation breakdown it can be observed oscillographically, using the electrical or optical method, that the first avalanche pulse is not followed by further generations; instead, a steep current rise becomes visible which we identify with the streamer formation. This phenomenon can be understood qualitatively by taking into account the gas-ionizing radiation; a critical amplification of $n \sim 10^8$ starts this mechanism.

Many-generation and streamer mechanisms

The criterion of Eq. 17.6 allows to predict which mechanism will be realized under experimental conditions: If in a gas with a given cathode the value of γ is 10^{-5}, the amplification of every avalanche must be 10^5 in order to fulfill Eq. 17.3. Under these conditions, as found in gases such as N_2 or H_2, the many-generation mechanism goes on if the voltage is applied for >5 μsec at least, as in Fig. 17.4. As the cloud chamber testifies in such situations, the one-generation mechanism can be produced by applying a voltage exceeding that for which $\gamma e^{\alpha d} = 1$ is fulfilled (overvoltage). If this overvoltage is so high that the first starting avalanche reaches amplification equal to or greater than the critical one (Eq. 17.6), it starts streamers. The streamer breakdown is much more rapid than the Townsend mechanism: In the case of the many-generation breakdown there is a time lapse (usually about 10^{-5} sec) until the ions become effective by the accumulation of positive space charge; in the one-generation mechanism only the first avalanche has to reach the critical amplification value, then rapid streamer development sets in. This accounts for overvolted gaps breaking down so rapidly. This breakdown can be speeded up further by using very high overvoltages and small electrode distances. In this way the development time to critical avalanche height becomes reduced and the development of the streamer accelerated.

The spark chamber

This quality of breakdown has allowed construction of the spark chamber so frequently used in nuclear physics: It is a characteristic of the one-generation mechanism that the spark channel develops just in the first avalanche. Thus the spark allows the localization of the electron that has triggered the first avalanche. By placing together a great number of such gaps of small distance, a nucleon which crosses the different gaps nearly perpendicularly to the electrodes (or nearly parallel to the electric field) releases in each of them some electrons which initiate sparks just at the place where they are produced. These sparks indicate the path to the nucleon in the different gaps. Figure 17.6 shows such a spark track of a nucleon coming from the right. The upper half of the photo is taken from the same track by a mirror.[7]

Verifications of the preceding concepts

In discharge gaps filled with vapors (ether, alcohol, etc.) the static breakdown proceeds as a one-generation mechanism: Here the value of γ is very small (ca. 10^{-9}). The electron avalanche can therefore reach the critical amplification of $e^{\alpha d} \sim 10^8$ and start streamers while $\gamma e^{\alpha d} < 1$ (Fig. 17.7). If γ is increased ($>10^{-8}$) by

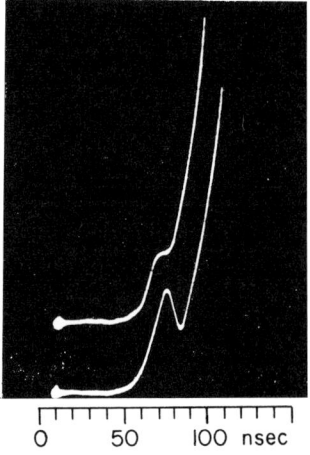

Fig. 17.7. Two oscillograms of a one-generation breakdown in methyl alcohol. Just after the first avalanche (peak at time T_-), a steep current growth indicates breakdown in plasma channel. (Two subsequent events are photographed, indicating fluctuation in start of streamer.)

special preparation of the cathode, the breakdown goes over into the many-generation mechanism.[8]

In discharge gaps filled with normal diatomic gases, γ is about $10^{-4} \to 10^{-6}$, and we can expect a many-generation mechanism. However, even in this situation we can produce the critical amplification in the avalanche without applying overvoltages, by starting a very large number of primary electrons in a spatial region of the cathode comparable with the diffusion radius of the head of the avalanche. For example, if we start $n_0 \sim 10^4$ primary electrons by shooting an α particle into the gap parallel to the electric field and each reaches an amplification of 10^4, then the whole avalanche reaches the critical amplification of 10^8 inside the diameter of the avalanche, and streamers are developed. If we therefore start a discharge with an α

Fig. 17.6. A 90° stereo view of polarized protons scattered from 1-in.-thick graphite plates of an argon-filled spark chamber. The spacing between plates is ¼ in.

[7] A. Roberts, Rev. Sci. Instr. 32, 480 (1961).
[8] J. Pfaue, Z. angew. Phys. 16, 15, 1963.

particle in an ordinary gas, the breakdown goes on not as a many-generation but as a one-generation process at static breakdown voltage.[9]

These two types of breakdown mechanism are prototypes; naturally there are transitions between them. For example, a discharge may begin as Townsend discharge with many generations, but after sufficient accumulation of positive charges one avalanche in the increased electric field will reach the critical amplification, and the Townsend discharge suddenly becomes a streamer mechanism. All such variations can be understood from the two typical processes outlined.

Extension to large gap distances and to low or high pressures

The discussion given dealt with the development of the avalanche process for usual gap distances ($\sim$1 cm) and pressures ($\sim 10^2$ torr). The problem arises whether these concepts are still valid if we go at a constant (e.g., atmospheric) pressure to very large distances, fix the electrode distance, and reduce the pressure appreciably below 1 torr, or else increase the pressure to some 10 atm or more. In the first case the problem approaches that of the mechanism of a lightning discharge. Experiments with long gaps at atmospheric pressure in air in homogeneous fields show that the breakdown mechanism changes its type beyond ca. 6 cm from the many-generation mechanism into a type of streamer mechanism.[10] In such a case the formation of negative ions plays an important role on account of the low E/p value ($\sim$36). At low pressures (in the limit vacuum breakdown) and at high pressures (condensed system[11]) the investigation with oscilloscopic methods promises still better insight into the mechanism of these phenomena.

[9] H. Schlumbohm, *Z. Physik 170*, 233, 1962.
[10] G. A. Schröder, *Z. angew. Phys. 13*, 296 (1961).
[11] Cf. Chap. 20 of this book.

18 · ELECTRIC DISCHARGES AND EXCITED SPECIES

Alfred von Engel

Ionization in gases — Excited species in discharges — Excitation cross sections and coefficients — Application to discharges — Outlook

Ionization in gases

The story of the passage of electricity through gases starts with an inquiry into the nature of the disappearance of an electroscope charge. Well-designed experiments proved convincingly that the carriers of charge actually moved through the gas and not across the surface of insulators. The next step concerned their speed of motion, which brought us the concept of ion mobility and drift velocity (i.e., the swarm's speed) in the direction of the electric field, in analogy with ions in electrolytes.[1] Around 1900 it became known that a potential difference of less than 100 volts applied to two electrodes in a gas at low pressure may give rise to a multiplication of electric charge in the gas; the initial free charges consisted of photoelectrons liberated at the cathode by illumination with ultraviolet light. The explanation was given by Townsend, who introduced the concept of the electron gas. It is "heated" by the electric field, leading to a random motion of electrons by collisions with gas molecules. A small number of fast electrons in the distribution is able to ionize gas molecules, thus adding new electrons and positive ions to the swarm. The electron gas can acquire a high temperature which is sometimes several hundred times larger than that of the neutral gas with which it is mixed. However, because of the large mass difference between electrons and molecules, the electron gas loses little energy and thus is in poor thermal contact with the neutral gas. In addition to charge multiplication in the gas, so-called secondary effects at the cathode have been studied later. Here positive ions, metastable species, light quanta, and other agencies may release electrons.[2]

Excited species in discharges

During the past thirty years one of the main objects of discharge physics was to investigate the mechanisms operating in the various parts of a discharge. The final aim is, of course, to calculate numerically all the properties of a discharge from known atomic data. Today this object has been achieved to a certain extent. However, there are still a large number of well-known phenomena whose mechanism remains obscure.

Since the early days of discharge studies it appeared[3] that one type of process previously not included in the elementary theory of ionization by collision may be of major importance: the ionization by collisions of the second kind (sometimes called superelastic collisions) caused by the presence of excited particles, such as electronically excited atoms and molecules. There were many reasons for neglecting these processes up to now, such as scarcity of reliable numerical data of excitation cross sections and of the life of excited species

---

[1] A. von Engel, *Ionized Gases*, 2nd ed., Oxford University Press, London, 1965.

[2] P. F. Little, "Secondary Effects," *Handbuch der Physik*, Vol. 21, Springer-Verlag, Berlin-Göttingen-Heidelberg, 1956, p. 574.

[3] A. von Engel, *Appl. Sci. Research* B5, 34 (1955).

as well as a lack of knowledge of other decaying mechanisms.[4] Also, for many years theoretical values have been in circulation which were in error, sometimes by several orders of magnitude. It might be asked in what respect this has delayed the development in the understanding of the ionization processes, if we remember that the main contribution was ascribed earlier to collisions between electron and gas molecules. However, where should the energy required to ionize molecules come from if not from the electric field via the faster electrons? The answer is: It is drawn from carriers of potential energy, as for example the excitation energy of gas molecules, which have been excited by slower electrons. As to the question about the incorporation of new ideas, it might be argued for purely experimental reasons that the time was not ripe.

A few examples will illustrate the point discussed earlier. Take a vessel filled with mercury vapor at low pressure and irradiated with the resonance light emitted by an ordinary mercury quartz lamp. If two electrodes are arranged in the vessel so that they are not illuminated, and a potential of a few volts is applied between them, a current is observed in the circuit. This is because instead of photoionization of mercury atoms requiring quanta of more than 10.4 volts, photoexcitation of atoms occurs, followed by thermal collisions between excited atoms. The 5-volt resonance light produces excited mercury atoms ($^3p_1$ and $^3p_0$) which collide in pairs, thus forming positive molecular mercury ions (Hg_2^+) and electrons.[5] In brief, ionization energy of the order of 10 volts is supplied in two steps. This process of "ionization in stages" is of great practical importance, as will be shown later. It may be added that photoionization of other metal vapors, such as cesium, takes place in a similar manner.

Another example of using the potential energy of atoms to produce ionization in the gas is the Penning effect.[6] It was first discovered in spectroscopically pure neon, which usually contains a very small amount of argon. If such neon is electrically excited, a fairly large population of excited neon atoms (metastables) with a long life is set up. These atoms move randomly through the gas without losing their excitation energy unless they hit the wall. However, if a metastable atom hits an argon atom, it transfers its potential energy to argon and ionizes it, since neon carries slightly more than 16 volts, whereas the ionization energy of argon is < 16 volts. As a result, ions and electrons are formed in the neon gas. Similarly, helium metastables ionize neon, argon metastables ionize krypton, etc.

Excitation cross sections and coefficients

It is necessary to discuss first the quantities used to describe the rate with which excited species are produced in the gas. We shall confine ourselves to electron-atom or electron-molecule collisions as far as the primary excitation is concerned.

When an encounter of electrons of uniform energy with gas molecules produces positive ions, the efficiency of this process is expressed either in terms of a cross section in cm² or of an ionization coefficient in ion pairs per cm path in the field direction per electron per unit gas density, usually given in mm Hg. The ionization efficiency is found to be a function of the electron energy and depends on the nature of the gas. For not too large energies, singly charged positive ions are produced. When, however, the electrons have an energy distribution as in a swarm, the rate of production of ions and electrons is proportional to the drift velocity of the electrons and to the ionization coefficient, which depends on E/p (the field per unit gas density being equivalent to the energy an electron has picked up along a mean free path in the direction of the field).

Excitation is a more complex process than ionization, since excitation from the ground state can take place to a variety of electronic states and—for the case of molecules—to various vibrational and rotational states. Fortunately the situation is not as hopeless as it appears, because the probabilities of excitation to different levels often vary by orders of magnitude; also states having a common property such as metastable states can be lumped together. This will now be illustrated in greater detail.

Consider for example inelastic collisions of electrons in helium. The dependence of the cross section of excitation to singlet and triplet states for electrons of uniform energy is shown in Fig. 18.1.[7] The results are based on measurements and theoretical calculations. The sum of the specific inelastic cross sections is equal to the measured total inelastic cross section which for

---

[4] R. G. Fowler, "Radiation from Low Pressure Discharges," *Handbuch der Physik*, Vol. 22, Springer-Verlag, Berlin-Göttingen-Heidelberg, 1956, p. 209.

[5] Cf. p. 64 of Ref. 1, 1st ed.; also G. F. Rouse and G. W. Giddings, *Proc. Natl. Acad. Sci. U. S.* **12**, 447, 1926.

[6] W. de Groot and F. M. Penning, "Anregung von Quantumsprüngen durch Stoss," *Handbuch der Physik*, 2nd ed., Vol. 23, Pt. 1, Springer-Verlag, Berlin, 1933, p. 23; A. von Engel, "Ionization in Gases by Electrons in Electric Fields," *Handbuch der Physik*, Vol. 21, Springer-Verlag, Berlin-Göttingen-Heidelberg, 1956, p. 504.

[7] S. J. B. Corrigan and A. von Engel, *Proc. Phys. Soc.* (London) **72**, 786 (1958).

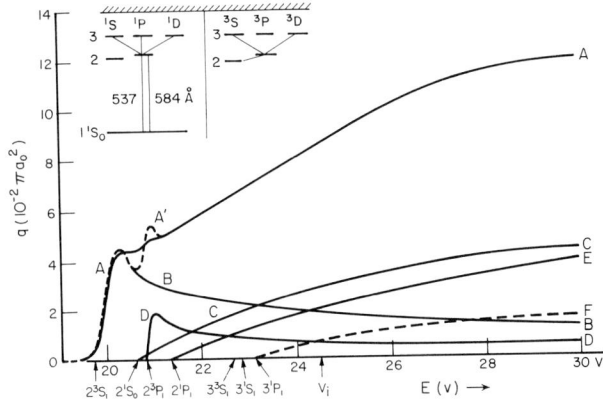

Fig. 18.1. Excitation cross sections q in units of $10^{-2} \cdot \pi a_0^2$ ($= 8.8 \cdot 10^{-19}$ cm^2) of the helium atom for various transitions, as function of the electron energy E (in volts). Insert: Simplified level diagram. A, A', total excitation cross section, low- and high-energy resolution, respectively; B, excitation to 2^3S; C to $2'S$; D to 2^3P; E to $2'P$; F to $3'P$. V_i, ionization potential.

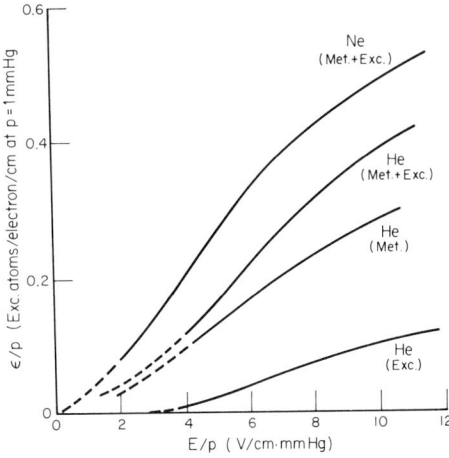

Fig. 18.2. Excitation coefficient ϵ/p (excited atoms per electron per mm Hg per cm in the field direction) to metastable and radiating levels in helium and neon, as function of electric field reduced to unit gas density.

electron energies above 24.6 volts includes ionization.[8] Some of the theoretical curves have in the meantime been experimentally confirmed.[9]

Several points are striking; at lower electron energies excitation to metastable triplet states takes place chiefly. The corresponding cross sections rise steeply with energy potentials. In contrast, the curves representing transitions to the singlet states rise slowly and exhibit maxima far above the critical potentials. Moreover, it can be seen that the maximum values of all cross sections shown are about the same or smaller than 10^{-18} cm^2; their value is always a small fraction of that of an elastic collision cross section. We conclude that in the case of an electron swarm, and thus in ordinary discharges, we should expect large numbers of metastables to exist relative to the rate of emission of ultraviolet or visible light. Only electrons of larger energy should produce quanta, and it seems that the population of metastables has been greatly underestimated in earlier work.

Figure 18.2 shows the excitation coefficient ϵ/p as a function of the reduced field E/p for helium and other rare gases. These curves can either be measured or obtained by calculation from known cross sections and the energy distribution. The outcome is shown in two separate curves: metastables and ultraviolet quanta emitted per cm of path per electron. (The ionization coefficient α/p is also given for comparison, showing that excitation greatly exceeds ionization.) The excitation coefficient of metastables is larger than that of ultraviolet quanta. The relation between excitation

[8] H. Maier-Leibnitz, Z. Physik **95**, 499 (1935).
[9] R. M. St. John, C. L. Bronco, and R. G. Fowler, J. Opt. Soc. Am. **50**, 28, 1960; cf. also Phys. Rev. **122**, 1813, 1961.

and ionization must also hold, of course, at low values of E/p and is simply a consequence of the fact that excitation sets in at lower electron energy than does ionization. At present it is not possible to discriminate between metastables and ultraviolet quanta in the other rare gases shown, since the corresponding cross sections are not known. However, production of metastables seems to exceed light emission in all cases considered.

An example of the electronic excitation in a molecular gas is given in Fig. 18.3. It shows the potential

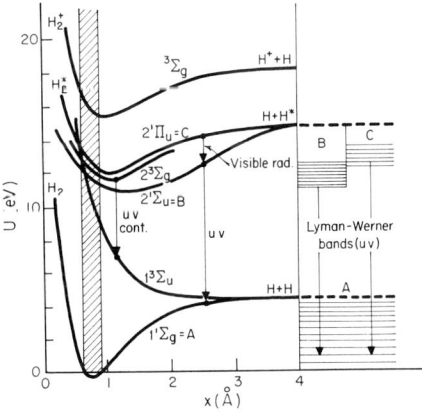

Fig. 18.3. Potential energy (U) curves of H$_2$ for several electronic states and the system of vibrational levels (Lyman-Werner bands).

energy as a function of the interatomic distance for some of the more important lower electronic states of molecular hydrogen. The shaded area indicates the range in which excitation by collisions with electrons would occur if the Franck-Condon principle would strictly hold, an assumption that has been borne out by experience.

Thus, an electron colliding with a molecule in the ground state A produces an electronically excited hydrogen molecule H_2^* that can be either an attractive state as curves B, C or a repulsive state as curve A. The repulsive triplet state a can be produced by electrons of at least 8.5 eV, whereupon the excited molecule H_2^*, because of the falling potential curve, dissociates into two hydrogen atoms in the ground state, each carrying away about 2 eV of kinetic energy. The amount of kinetic energy follows simply from taking the difference between the energy that the primary electron transferred to the molecule and the dissociation energy—better called the "thermal dissociation energy"; the latter corresponds to the critical energy necessary to separate the atoms of a molecule by heating the gas and raising the vibrational energy until the bond is broken.

This demonstrates clearly that dissociation of a hydrogen molecule by electron collision requires approximately twice the energy necessary for thermal dissociation of the hydrogen, a result that holds similarly with other gases. For example, the thermal dissociation energy of nitrogen is 9.7 eV, whereas electrons of about 24 eV are required to dissociate molecular nitrogen into N and $N^+ + e$ by electron collision.

Excitation of hydrogen into the second (bound) triplet state b will, after a time of the order 10^{-8} sec, be accompanied by a transition down to the lower triplet state a. As a consequence of this, quanta are emitted with energies corresponding to the difference in energy between vibrational levels of the b and a state. This gives rise to the well-known continuum of hydrogen which starts somewhere at 3500 Å and extends down to below 2000 Å. The low-pressure hydrogen discharge is a copious source of this continuum radiation.

Electrons of higher energy can give rise to excitation of hydrogen into the B and C states, which are both singlets. There are two groups of bands emitted from these two states which are called Werner and Lyman bands, producing light in the 1000-Å region. The maximum excitation cross sections are probably of the order of 10^{-17} cm^2, but details are not yet known. Light measurements however, have been carried out[7] to determine the lumped excitation coefficient ϵ/p as a function of a reduced field E/p (Fig. 18.4). In the same diagram there is a clear indication that ionization occurs less frequently than light emission in the far ultraviolet or the near ultraviolet.

Finally, a remark on dissociation by electron collisions: Fig. 18.5 shows the measured dissociation

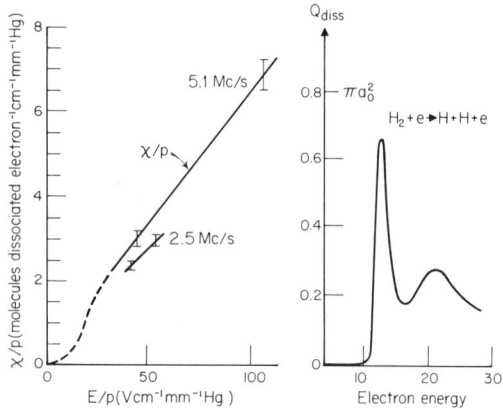

Fig. 18.5. Dissociation coefficient E/p (molecules dissociated per electron per cm per mm Hg) as function of reduced field E/p and calculated dissociation cross section $Q_{d\,ss}$ (in πa_0^2) for H_2.

coefficient χ/p (in dissociations per cm, etc.) as a function of E/p, and the dissociation cross section $q_{diss} = f(\epsilon)$, which is in good agreement with experiment. Approximate selected data are now available for N_2 and O_2.[1]

Application to discharges

Our interest in the role of excited particles in ionized gas has enabled us to look anew at certain questions, such as the mechanism of starting of electrodeless discharges in uniform electric fields.[10] From observations of Ne in the frequency range 10 to 10^7 cps and

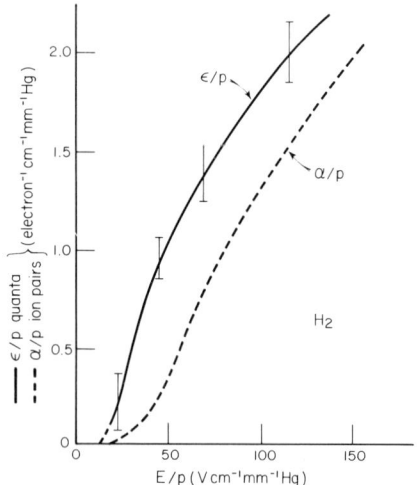

Fig. 18.4. Excitation yield ϵ/p (total number of quanta emitted per electron per mm Hg per cm in the field direction) and ionization yield α/p (ion pairs/cm mm Hg electron), measured in H_2 as $f(E/p)$.

[10] W. L. Harries and A. von Engel, *Proc. Roy. Soc. (London)* A**222**, 490 (1954).

pressure range 2 to 200 mm Hg, it was found that resonance quanta emitted from the gas release secondary electrons from the glass. This process, which leads to the multiplication of electrons and quanta seems to set up a population in the resonance states which can be ionized by slow electrons in the distribution. In the hollow-cathode glow discharge[11] field measurements with a deflected electron beam have shown that ultraviolet quanta often seem to contribute to the secondary emission from the cathode.

The low cathode fall and other parameters of so-called cold arcs, such as arcs with Hg and Cu cathodes, indicate that excited particles and resonance radiation must play a major role, since direct ionization by collision is ruled out.[12] The new theory of these vapor arcs assumes that a very dense and thin layer of vapor exists at the cathode from which excited particles fall on the cathode, thereby releasing electrons. The transition from a thermionic arc cathode (with carbon electrodes in nitrogen) into a vapor arc has been explained by a de-excitation process, namely the quenching of excited species by collisions.[13] Finally, dissociation and ultraviolet excitations in low-pressure hydrogen ring discharges used as sources for fast particles have been studied quantitatively.[14]

Outlook

There are a number of other problems to which the foregoing may be usefully applied. First of all, the role of excited species in gases must have a significant bearing on the chemical changes that occur in ionized gases or gas mixtures. This is of course of practical importance and has therefore been studied in great haste, particularly in connection with the production of ozone, nitric oxide, and acetylene; the lack of data has so far prevented workers in this field from testing the suggested mechanisms quantitatively.

Another problem to which little attention has been paid is the ablation of solids under the impact of excited species.[15] It appears that uncharged atoms are issued from solid surfaces by a step process whereby the energy for moving the atoms from the original positions in the crystal into positions of lower binding energy is supplied in the form of the potential energy of the excited particles. Thus the absolute value of the excitation energy of a particle contributing to ablation can be a fraction of the heat of vaporization.

Good use of the general ideas developed earlier has been made in our investigations on the calorelectric effect: When a torch flame passes through a hot (500° C) and cold (20° C) electrode, a potential difference of about 2 eV is produced between them. Here the energy source is the combustion reaction that sets up a large population of vibrationally excited molecules from which the electrons receive their energy by collisions of the second kind. In this way the electron temperature is raised above the gas temperature, and the difference in electron temperature at the hot and the cold wall gives rise to a difference in wall potential.

With the knowledge now at our disposal it may be possible to make further advances in the field of optical "absorption" in excited gases by means of intense beams of excited molecules and to study the refractive index and polarizability as well as the corresponding magnetic properties of excited species.

[11] P. F. Little and A. von Engel, Proc. Roy. Soc. (London) A224, 209 (1954).
[12] A. von Engel and A. E. Robson, ibid. A242, 217, 1957.
[13] A. von Engel and K. W. Arnold, Proc. Phys. Soc. (London) 79, 1098 (1962).
[14] C. C. Goodyear and A. von Engel, ibid. 79, 732 (1962).
[15] A. von Engel and K. W. Arnold, Nature 187, 1101 (1960).

19 · SPACE-CHARGE-LIMITED CURRENTS

Albert Rose

Vacuum capacitor versus solid-state capacitor — Shallow traps and trap densities — Transient currents — Space-charge-limited currents in semiconductors — Performance of photoconductors and solid-state triodes — Double injection

Vacuum capacitor versus solid-state capacitor

When a voltage V is applied to a plate condenser in vacuum, a negative charge

$$Q = CV \tag{19.1}$$

lies at the inner surface of the cathode, and its countercharge at the positive plate. An electrostatic force, proportional to the square of the electric field strength E, tends to force the charges together. The negative charge of electrons would move into the anode-cathode space were it not restrained by the work function of the metal.

If the cathode is heated sufficiently, a cloud of electrons appears in front of it. Without a field it adheres closely to the metal surface because of the electrostatic attraction of the positive countercharge (image force), but when a voltage is applied, the negative charge is pulled toward the anode, and a significant transient current as well as a steady-state current, a "space-charge-limited" current, is observed.

All field lines from the anode end on electron charges in space (Fig. 19.1). If they were uniformly distributed between cathode and anode, the capacitance of the condenser would read twice its "cold" value, since the mean distance of the negative charge from the anode is now half the plate spacing d. Continuity of current demands that the charge density in space be higher near the cathode, where the field is weaker and the electrons move more slowly, than near the anode. The result is that the total capacitor charge Q for a hot cathode is greater than that in the cold capacitor but less than twice that value.

To compute the current density J, we divide the charge by its transit time T from cathode to anode. In a first approximation we assume that the average electron velocity will be one half that prevailing in vacuum without space charge:

$$T = \frac{2d}{v} \cong \frac{2d}{(2\,\mathrm{eV}/m)^{1/2}}. \tag{19.2}$$

(The actual transit time is somewhat longer, because the field near the cathode is weakened by the negative

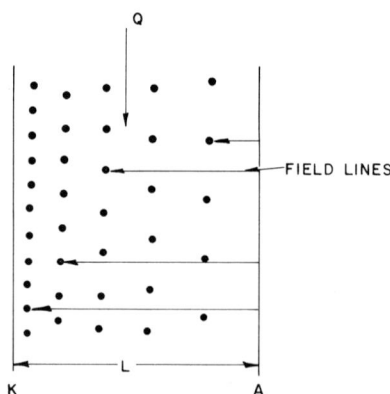

Fig. 19.1. Space charge injected in vacuum diode.

charge distributed throughout the cathode-anode space.) Hence, since the capacitance of a plate condenser of area A and separation distance d in vacuum (dielectric constant ϵ_0) is

$$C = \frac{A}{d}\epsilon_0 \quad [\text{farad}], \quad (19.3)$$

$$I = \frac{Q}{T} = \left(\frac{e}{2m}\right)^{1/2} \frac{V^{3/2}}{d^2} \epsilon_0. \quad (19.4)$$

Equation 19.4 matches the rigorous result within about 10 percent, because the increase of Q by space charge approximately balances the underestimate in transit time.

In vacuum, the cathode must be heated to provide a reservoir of free electrons in space. The work function of metals "looking into vacuum" is generally 2 to 5 eV, a value high compared with the thermal energy at room temperature ($kT = 1/40$ eV). For metals "looking into a solid" the work function may be of the order ~ 0.1 eV; i.e., already at room temperature a reservoir of electrons can exist just outside the metal surface in the conduction band. Application of a voltage across such a condenser injects a charge into the cathode-anode space just as before (Fig. 19.2). If

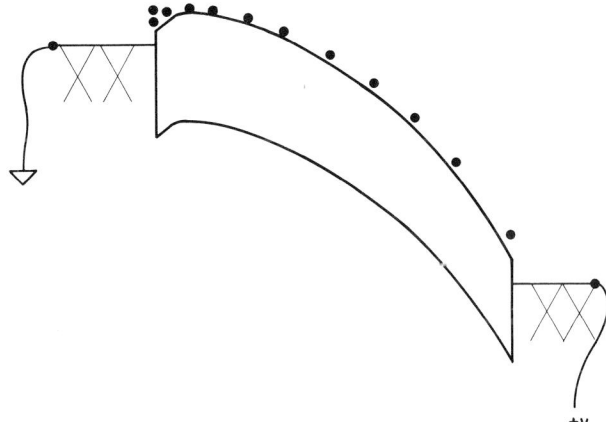

Fig. 19.2. Space-charge-limited current in solid.

the insulator is free of traps (i.e., of localized states in the forbidden zone), the injected charge remains free and migrates according to Ohm's law with a transit time in the undistorted field

$$T = \frac{d}{v} = \frac{d}{b(V/d)}. \quad (19.5)$$

Hence,

$$I = \frac{Q}{T} = \frac{CV}{L^2/Vb} = \frac{\epsilon' b V^2}{d^3}, \quad (19.6)$$

where ϵ' is the dielectric constant of the medium and b the electron mobility. For an insulator of $d = 10^{-3}$ cm, $\kappa' = \epsilon'/\epsilon_0 = 10$, and $b = 10^2$ cm^2/V sec, a voltage of 10 V would yield a current of 10 amp/cm^2. Fortunately, such high currents are not common for dielectrics, else we would have difficulties making acceptable condensers. The work function of a metal-insulator contact is normally ≥ 1 eV, limiting the emission to the order of 10^{-12} amp/cm^2 at room temperature.

Unfortunately, information on contact potentials is too sparse to claim that high metal → solid work functions are the major reason that insulators insulate. A second reason is the enormous attenuation of space-charge-limited currents by traps; for this we have ample evidence.[1-3]

Shallow traps and trap densities

The concepts just outlined allow an immediate qualitative assessment of the effect of traps and a quantitative one with minor refinement. The charge drawn into the cathode-anode space, as given by the simple condenser relation (Eq. 19.1), is independent of whether that charge is free or trapped. The injected charge density is not likely to exceed $\sim 10^{13}$ electrons/cm^3. Since trap densities well in excess of this number are the rule, they could easily immobilize most of the injected charge. For a quantitative estimate of the fraction of injected charge that remains free in the conduction band, we must refer to the Fermi level.

In a solid in thermal equilibrium the allowed energy states are substantially filled below the Fermi energy and above it, mostly empty. The probable occupancy of states located ε electron volts above the Fermi level is about $\exp(-\varepsilon/kT)$; at room temperature it decreases by about 10^{-2} per 0.1 eV.

Consider an insulating solid in thermal equilibrium with a single set of traps located above the Fermi level but $\Delta\varepsilon$ eV below the conduction band. When a space charge of electrons is injected by an applied field, the ratio of free to trapped electrons becomes

$$\frac{n_f}{n_t} = \frac{N_c}{N_t} e^{-\Delta\varepsilon/kT}, \quad (19.7)$$

with N_c the effective state density in the conduction band ($\approx 10^{19}$/cm^3) and N_t the density of traps. The presence of shallow traps thus modifies Eq. 19.6 by a multiplication factor $\theta \equiv n_f/n_t$:

$$J = \theta \frac{\epsilon' b V^2}{d^3}. \quad (19.8)$$

[1] R. W. Smith and A. Rose, *Phys. Rev.* **97**, 1525 (1955).
[2] N. F. Mott and R. W. Gurney, *Electronic Processes in Ionic Crystals*, Clarendon Press, Oxford, 1940, p. 172.
[3] A. Rose, *Phys. Rev.* **97**, 1538 (1955).

For reasonable values of N_t ($10^{16}/cm^3$) and $\Delta\mathcal{E}$ (0.8 eV), θ becomes about 10^{-12}. It is clear that Eq. 19.8 can account for the exceedingly small space-charge-limited currents normally encountered in insulators and that it can also be used to obtain the density and depth of shallow trapping states[1,3,4] ("shallow" means lying above the Fermi level).

When the voltage is increased to the point where the injected space charge of electrons just suffices to fill the shallow traps, an abrupt current rise by a factor θ^{-1} will occur. This "trap-filled-limit voltage"[4] directly measures the density of trapping states.

If the trapping states are distributed more or less continuously between Fermi level and conduction band,[1,3,5,6] the injected space charge Q_0 will first fill the trapping states lying in an energy interval $\Delta\mathcal{E}$ just above the Fermi level,

$$Q_i \simeq N_t e \frac{d}{2} \Delta\mathcal{E}. \tag{19.9}$$

The fact that the Fermi level has been raised by $\Delta\mathcal{E}$ volts signifies that the free-carrier density increases by the factor $\exp(\Delta\mathcal{E}/kT)$; the space-charge-limited current

$$I_{scl} = I_{thermal}\, e^{\Delta\mathcal{E}/kT}. \tag{19.10}$$

Since the density of free carriers is many orders of magnitude smaller than that of the trapped carriers, the correction to the condensed charge of Eq. 19.9 is negligible, but the space-charge-limited current increases exponentially with applied voltage.

In general, for trap densities that vary continuously as a function of energy, the space-charge-limited current will increase as some higher power of the voltage.

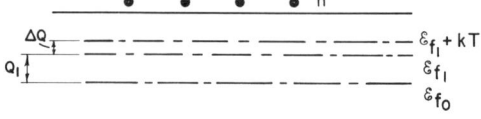

Fig. 19.3. Measurement of trap density and energy position by injected space charge (voltage V_l raises Fermi level from $\mathcal{E}_{f0}$ to $\mathcal{E}_{f1}$; the additional voltage ΔV raises it further to $\mathcal{E}_{f1} + kT$).

Assume that the applied voltage is raised further to increase $\Delta\mathcal{E}$ by kT. This is accomplished by injecting enough charge to fill the traps in the energy slice $(\Delta\mathcal{E} + kT) - \Delta\mathcal{E}$ (Fig. 19.3). The current will increase

[4] M. A. Lampert, *Phys. Rev.* **103**, 1648 (1956).
[5] R. W. Smith, *RCA Rev.* **20**, 69 (1959).
[6] M. A. Lampert, A. Rose, and R. W. Smith, *J. Phys. Chem. Solids* **8**, 464 (1959).

according to Eq. 19.10 by the Napierian base factor $e \equiv 2.718$. Hence, the operation of raising the voltage so that the current is little more than doubled gives, according to Eq. 19.9, an immediate measure of the trap density near the Fermi level. The Fermi level itself is known at the outset from the starting current or conductivity.

The energy distribution of traps has been measured in this way by Smith[5] in CdS, DeVore[7] in CdSe, Mark and Helfrich[8] in anthracene, and Lanyon[9] in amorphous selenium. In the case of selenium good correlation was found between the trap densities so measured and the optical absorption near the band edge. Trap-density distributions often range from about $10^{17}/cm^3\text{-}kT$ near the conduction band to about $10^{13}/cm^3\text{-}kT$ a volt below it.

Transient currents

The currents discussed thus far are steady-state currents after the injected carriers have settled into traps. Since the carriers are initially injected into the conduction band, the current must decrease from a large trap-free value to its steady trap-controlled state. The rate of decay is proportional to the density of traps and their capture cross sections for free carriers. Since we already have a means for measuring trap density, we can compute capture cross sections from the decay time of the transient current. Such cross sections were measured by Smith and Rose[1] in CdS and Mark and Helfrich[8] in anthracene.

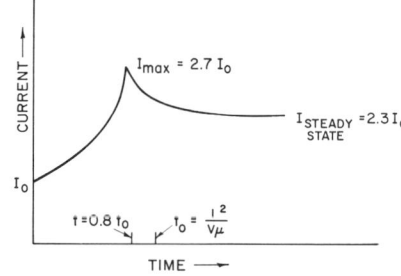

Fig. 19.4. Initial transient of space-charge-limited current flow.[8,10]

If the capture time for a free carrier lies below its transit time from cathode to anode, the current will decay smoothly after voltage is applied, but if it is longer, there will be an initial current rise followed by a decay, as first analyzed by Many et al.[10] and Mark

[7] H. B. DeVore, *RCA Rev.* **20**, 79 (1959).
[8] P. Mark and W. Helfrich, *J. Appl. Phys.* **33**, 205 (1962).
[9] H. P. D. Lanyon, *Phys. Rev.* **130**, 134 (1963).
[10] A. Many, S. Z. Weisz, and M. Simhony, *ibid.* **126**, 1989 (1962).

and Helfrich[8] (Fig. 19.4). The distinct cusp indicating the time when the first sharp front of space charge arrives at the anode allows computing the drift mobility of carriers (cf. Many et al.[10] for iodine and Mark and Helfrich[8] for anthracene). The reason for the cusp is roughly that more charge is injected into the solid during first transit than can be accommodated in the steady state. Fast trapping prevents this overshooting.

Space-charge-limited currents in semiconductors

For the injection of space-charge-limited currents into insulators the thermal density of free carriers can be neglected. In semiconductors we will, of course, initially observe Ohm's law at low voltages. At sufficiently high voltages the thermal density of carriers can again be neglected in comparison with the injected density, and the resulting space-charge-limited currents follow the arguments previously advanced.

The crossover from ohmic to space-charge-limited flow occurs at that voltage at which the injected space-charge density equals the thermal density.[1,3] At lower voltages the injected space charge is dissipated by the conducting solid before it reaches across the sample. The transit time of the injected carriers must be shorter than the thermal-equilibrium relaxation time of the solid before space-charge currents become significant. This relaxation time is defined as

$$\tau_r = \frac{\kappa' \times 10^{-12}}{4\pi n_0 eb} \quad [\text{sec}], \quad (19.11)$$

where κ' is the relative dielectric constant, n_0 the density of thermal carriers, e the electronic charge in coulombs, and b the mobility of free carriers in cm²/V sec. The transit time of a free carrier is, as before,

$$T = \frac{d^2}{Vb}. \quad (19.5)$$

By equating these two times, we obtain the condition

$$Q = VC = n_0 ed \quad (19.12)$$

for the onset of space-charge-limited current flow.

The relaxation time τ_r remains constant in the Ohm's law range since the density of carriers n_0 remains constant. At the onset of space-charge-limited current flow the carrier density increases linearly with increase in voltage; hence, τ_r decreases as $1/V$. In the range of space-charge-limited current flow, τ_r remains equal to the transit time for a trap-free solid.

Performance of photoconductors and solid-state triodes

The fact that ohmic currents give way to space-charge-limited currents as the applied voltage is increased leads to a far-reaching relation governing the performance of photoconductors.[11,12] In general, the photoconductive gain (ratio of photo to photon current) is given by

$$G = \frac{\tau_l}{T}, \quad (19.13)$$

where τ_l is the lifetime of a free carrier and T its transit time between electrodes. The meaning of this relation is that during the lifetime of an extra free carrier, τ_l/T extra electron charges can pass through the photoconductor and external measuring circuit. As already noted, the transit time becomes equal to the relaxation time τ_r in the space-charge-current regime for a trap-free solid. Also, for a trap-free solid the lifetime of a free carrier is equal to the response time τ_0 of the photoconductor to changes in light intensity. Hence, we can rewrite Eq. 19.13 for the space-charge-current regime as

$$G \times \frac{1}{\tau_0} = \frac{1}{\tau_r}. \quad (19.14)$$

Equation 19.14 holds also in the presence of shallow traps, since their effect is to increase gain at the onset of space-charge flow and response time by the same factor, i.e., by the ratio of trapped to free carriers.

The left-hand side of Eq. 19.14 is the familiar gain-bandwidth product. The significance of this relation is that high performance, i.e., high gain-bandwidth products, can be achieved only in highly conducting materials for which τ_r becomes small. Conversely, if constrained to a relatively insulating material, one must sacrifice either sensitivity (gain) or speed of response.[13] Insulating materials are required for detecting small amounts of light, for constructing the storage-type television camera tube (Vidicon) and the electrostatic-type photographic layers (Xerography and Electrofax).

While Eq. 19.14 is a valid description of the performance of most photoconductors, special distributions of traps and recombination centers can lead to improved performance. The more general form[11] is

---

[11] A. Rose and M. A. Lampert, *RCA Rev.* **20**, 57 (1959).
[12] R. W. Redington, *J. Appl. Phys.* **29**, 189 (1958).
[13] The electric field is assumed to be below that needed for multiplication by impact ionization.

$$G \times \frac{1}{\tau_0} = \frac{1}{\tau_r} \times M, \qquad (19.15)$$

where M is the ratio of anode charge to trapped charge in thermal contact with the conduction band. "Thermal contact" means that free carriers undergo one or more trapping events before final recombination with the centers from which they were excited optically.

The physics of the gain process in photoconductors is substantially the same as that for solid-state and vacuum triodes. Hence, the same generalization on the performance of photoconductors can be applied to such triodes.[14,15] The gain process is the introduction of an extra free carrier in the cathode-anode space for a time greater than the transit time of a carrier. The gain is in fact equal to the ratio of lifetime of extra carrier to transit time.

[14] E. O. Johnson and A. Rose, *Proc. IRE 47*, 407 (1959).
[15] A. Rose, *RCA Rev. 24*, 627 (1963).

Double injection

While the flow of one sign of carrier in a solid follows essentially the corresponding flow in vacuum, the flow of both signs of carriers has an element of novelty.[16,17] In the case of vacuum, the space-charge-limited current of electrons is increased only by a factor of about 2 when positive charges are emitted from the anode. In the solid, the simultaneous injection of electrons at the cathode and holes at the anode can give rise to currents many times larger than those due to either carrier alone. The difference is that the two signs of carriers can neutralize each other over most of the cathode-anode space in a solid, while in vacuum the neutralization is confined to only a plane in space. The difference arises from the fact that carriers in a solid may have nearly constant drift velocity, whereas those in vacuum are continuously accelerated between the electrodes.

[16] M. A. Lampert, *RCA Rev. 20*, 689 (1959).
[17] M. A. Lampert, *Proc. IRE 50*, 1781 (1962).

20 · CONDUCTION AND BREAKDOWN

Arthur von Hippel

Electric strength and external electron supply — Dynamic field distortion and plasma formation — Failure of Paschen's law; emergence of new breakdown conditions — Feature changes in condensed systems — Prebreakdown currents in liquid hexane — Breakdown phenomena in liquids and solids

Internal electric fields create the molecular organization of matter, external fields tend to unbalance it in actions ranging from reversible polarization and carrier motion to irreversible transformation and destruction. Preceding chapters have explored various aspects of this theme: fluorescence and phosphorescence that reveal the intricate flow of excitation energy and the variety of obstacles encountered by electrons on their return to lower energy states (Chaps. 13 and 14); conduction in metals and semiconductors with its information on internal fields, energy surfaces, and occupation density of electron bands (Chaps. 15 and 16); impact ionization in gases that leads to avalanche breakdown (Chap. 17); the role of excited atoms and molecules in triggering and sustaining gas discharges (Chap. 18); the modification of current flow through solid dielectrics by trapping phenomena and space-charge fields (Chap. 19). The present chapter aims to round out this survey by inquiring how phenomena of unbalance develop in strong unidirectional fields as one progresses from gases to liquids and solids.

Electric strength and external electron supply

Electrons are the most effective transducers between external field energy and the internal energy of a molecular system. Their small mass allows rapid energy accumulation with negligible loss in elastic collisions, and their strong coupling to bound electron clouds provides efficient energy transfer by inelastic collisions. Hence electrons must be kept out or rendered harmless if high electric strength is desired.

Electrons cannot be kept out completely, as the conduction of gases testifies. There is always some ionizing background radiation, stemming from radioactivity and cosmic rays. It averages at atmospheric pressure only a few electrons and ions per cm^3 and sec, but in high fields such electrons can remain free and multiply catastrophically. The random creation of starting electrons in gases is reflected in the statistical scatter of the impulse strength. Overvoltages are reached and statistical waiting times encountered (Fig. 20.1).[1] This "statistical time lag" can be eliminated systematically by increasing the electron supply: in our example by boosting the illumination intensity of the cathode from I_0 successively to $2000\,I_0$ and thus reducing the delay for the appearance of photoelectrons to the vanishing point.

The breakdown voltage V_{max} of the various gases can be displayed in "Paschen characteristics" (Fig. 20.2).[2]

---

[1] R. C. Fletcher, *Phys. Rev.* **76**, 1501 (1949).

[2] Data taken from B. Gänger, *Der elektrische Durchschlag von Gasen*, Springer-Verlag, Berlin, 1953, pp. 176–177.

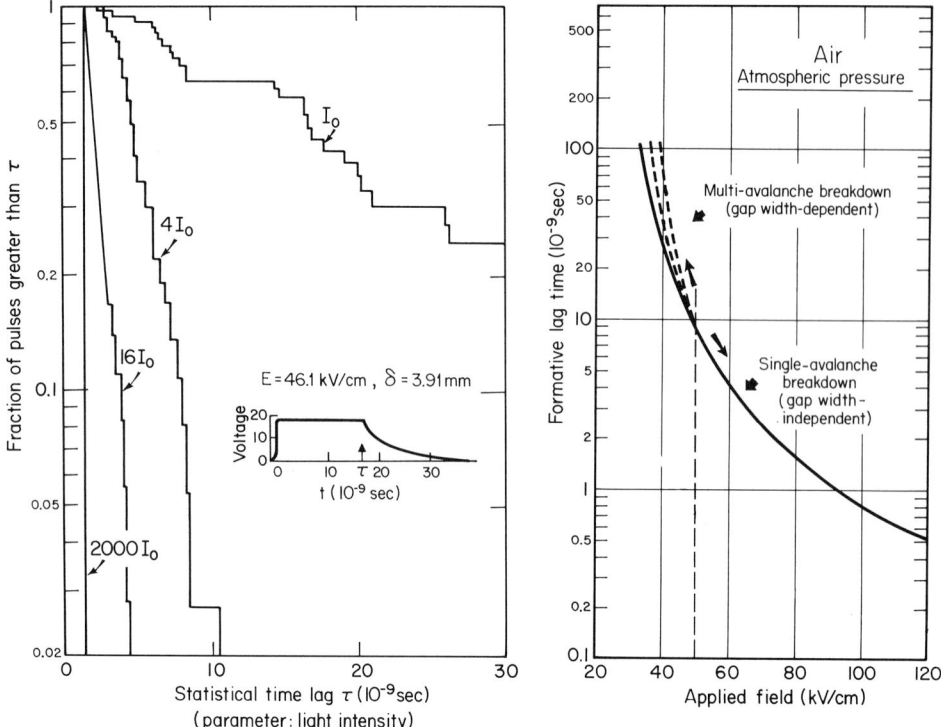

Fig. 20.1. Statistical and formative time lag in gas breakdown.

Paschen found empirically at an early date[3] that the breakdown strength of a plane gap did not depend on gas pressure p and electrode distance d separately but only on their product pd. Gas pressure (more accurately, gas density ρ) is inversely proportional to the free path l of the electrons; the product $pd \approx d/l$ therefore characterizes the average number of electron collisions across the gap.

This similarity law became understandable in conjunction with Townsend's concept[4] that breakdown occurs at a critical avalanche height (cf. Eq. 17.3):

$$H = e^{\alpha d} = \frac{1}{\gamma}. \qquad (20.1)$$

Here α is the probability coefficient of ionization (number of ionizing impacts per electron and unit distance in field direction) and γ the probability coefficient of regeneration (number of new starting electrons released at the cathode per positive ion or photon). Townsend's condition (Eq. 20.1) predicts breakdown when a starting electron re-creates its successor with certainty $(1 \to 1)$. In the pd diagram this V_{max} curve corresponds to the Paschen curve. It is the boundary link between voltage regions of under-$(1 \to <1)$ and over-$(1 \to >1)$ electron supply (cf. Fig. 20.6).[5]

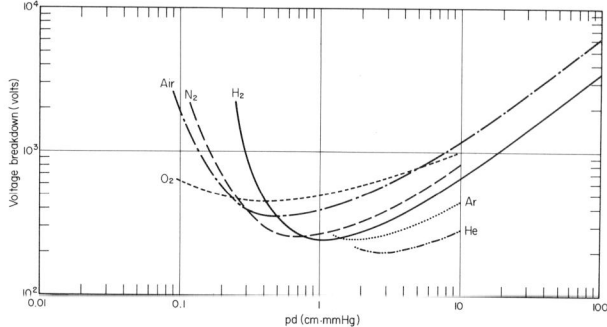

Fig. 20.2. Paschen characteristics of some gases.

The shape of the Paschen curve reflects the effectiveness of translating external field energy into impact ionization. At the left of the minimum the number of impacts is too small; the electrons must be speeded up appreciably beyond the ionization limit to make each impact count with sufficient yield. At the right of the minimum the number of inelastic collisions is too large; in this region of high electron-excitation probability

---

[3] F. Paschen, *Ann. Phys. 37*, 69 (1889).

[4] J. S. Townsend, *Electricity in Gases*, Clarendon Press, Oxford, 1915.

[5] Cf. A. von Hippel, *Molecular Science and Molecular Engineering*, The M.I.T. Press and John Wiley and Sons, New York, 1959, Chap. 3.

much energy is squandered through secondary processes, e.g., by light emission. Differently expressed: In this region the barrier of electron excitation states protects the gas against an early onset of breakdown.

The statistical time lag disappears and Paschen's law becomes invalid at high gas densities (Fig. 20.3).[6]

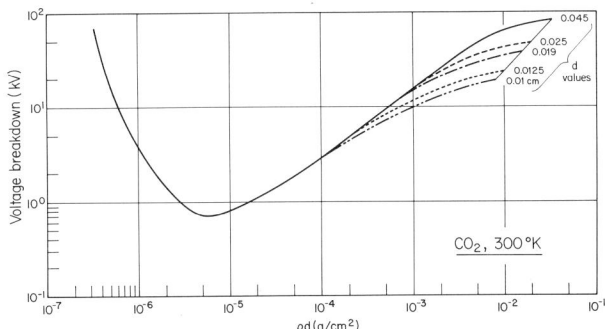

Fig. 20.3. Paschen law for CO_2 and its invalidation at high densities.

Now, the illumination of the cathode begins to prove superfluous, because—as current measurements show—a more copious primary electron supply supersedes the photoelectric one: ejection of electrons by field emission.

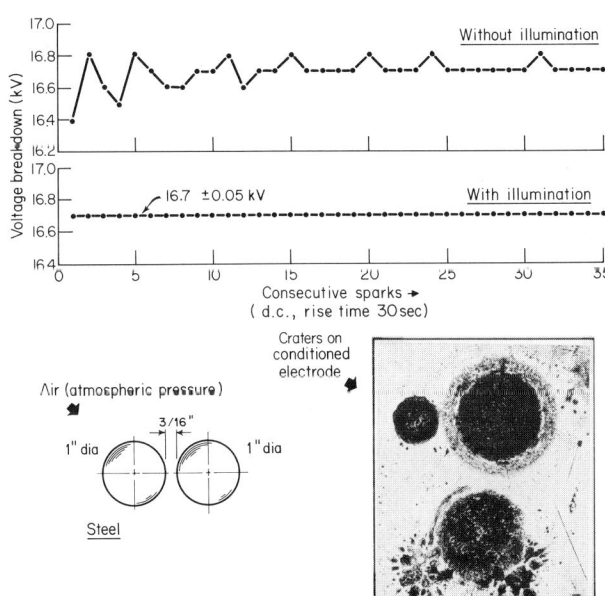

Fig. 20.4. Scatter of d-c breakdown strength of air and effect of conditioning.

The spurious onset of this phenomenon begins to be noted disturbingly already at lower pressures, since considerable scatter is observed in the d-c breakdown voltage of a spark gap when the cathode is not illu-

[6] D. R. Young, *J. Appl. Phys.* **21**, 222 (1950).

minated; it diminishes as the number of sparks drawn increases (Fig. 20.4). The engineer, therefore, has resorted to "electrode conditioning" by presparking, a somewhat brutal scorching treatment that leaves the previously polished metal surfaces badly scarred but better behaved. The procedure burns off sensitive spots of incipient field emission. They can be seen in the self-photographs of Lichtenberg figures (Fig. 20.5).

Fig. 20.5. Field-emission spots on cathode, as shown by Lichtenberg figures.

Without countermeasures, reliable d-c data cannot be obtained in this range[7] until field emission takes over with certainty.

Dynamic field distortion and plasma formation

Townsend's condition of perpetual regeneration (Eq. 20.1) can be equated to breakdown, because overshooting of this critical yield leads to instability. Positive space charge, left behind in the gas by fast emigrating electron avalanches, may rapidly contract the applied field into a steep cathode fall.[8] At low pressures and voltages, this spontaneous field distortion comes to a halt when the Townsend condition is reapproached at the left-hand branch of the Paschen curve (Fig. 20.6). Here the collapse of the gas insulation terminates in a stable glow discharge. If, however, the surging cathode fall triggers field emission, Townsend's regeneration criterion loses its stabilizing influence: A negative spark develops from the cathode

[7] Cf., e.g., N. L. Allen, *Tech. Rep. 107*, Lab. Ins. Res., Mass. Inst. Tech., April, 1956.
[8] A. von Hippel and J. Franck, *Z. Physik* **57**, 696 (1929); A. von Hippel, *ibid.* **80**, 19 (1933).

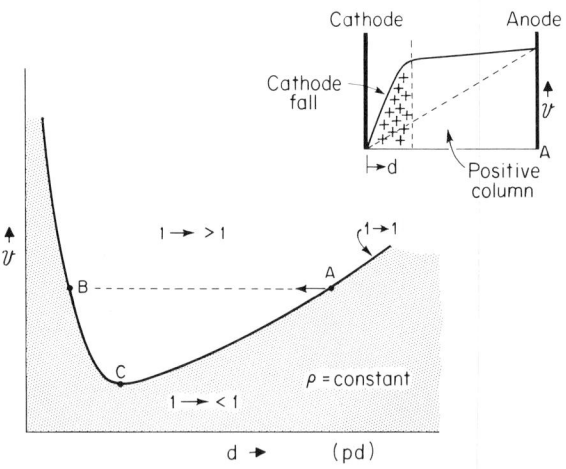

Fig. 20.6. Paschen curve and instability.

Fig. 20.7. Negative spark from cathode by plasma formation.

by plasma formation[9] (Fig. 20.7). "Plasma" here designates a concentration of free electrons and counterbalancing positive ions of such density that it practically acts as a metallic short circuit across the affected gap section. The high field strength at the free ends of this quasi wire and the photon stream emitted from its luminous interior—transformed into photoelectrons—extend the spark channel rapidly across the whole gap if sufficient voltage is maintained. Raether[10]

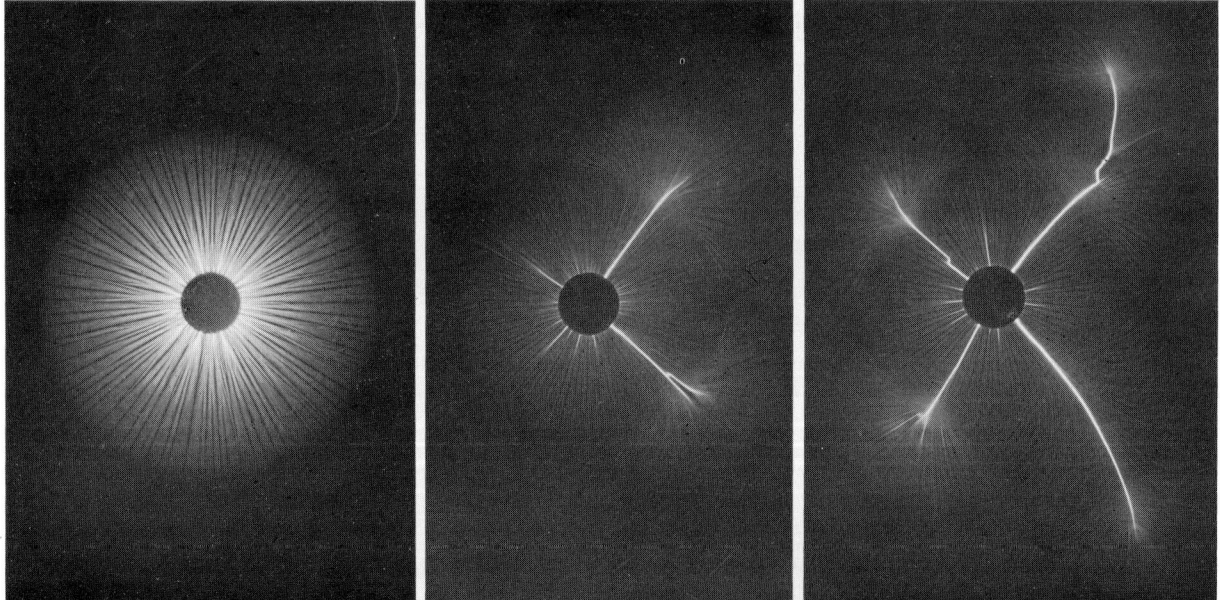

Fig. 20.8. From space charge to plasma as function of time.

has clarified how this phenomenon may set on in space at a critical avalanche height (multiplication factor $\geq 10^8$) and cause single-avalanche breakdown by propagating toward cathode and anode in the "streamer mechanism" (cf. Chap. 17).

The critical avalanche height can be reached in various ways: by application of overvoltage, which leads to single-avalanche breakdown after a "formative time lag" (cf. Fig. 20.1); by using long spark gaps, which generate their own overvoltage dynamically by

[9] F. H. Merrill and A. von Hippel, J. Appl. Phys. 10, 873 (1939).
[10] Cf. Chap. 17; also H. Raether, Ergeb. exakt. Naturw. 33, 175 (1961).

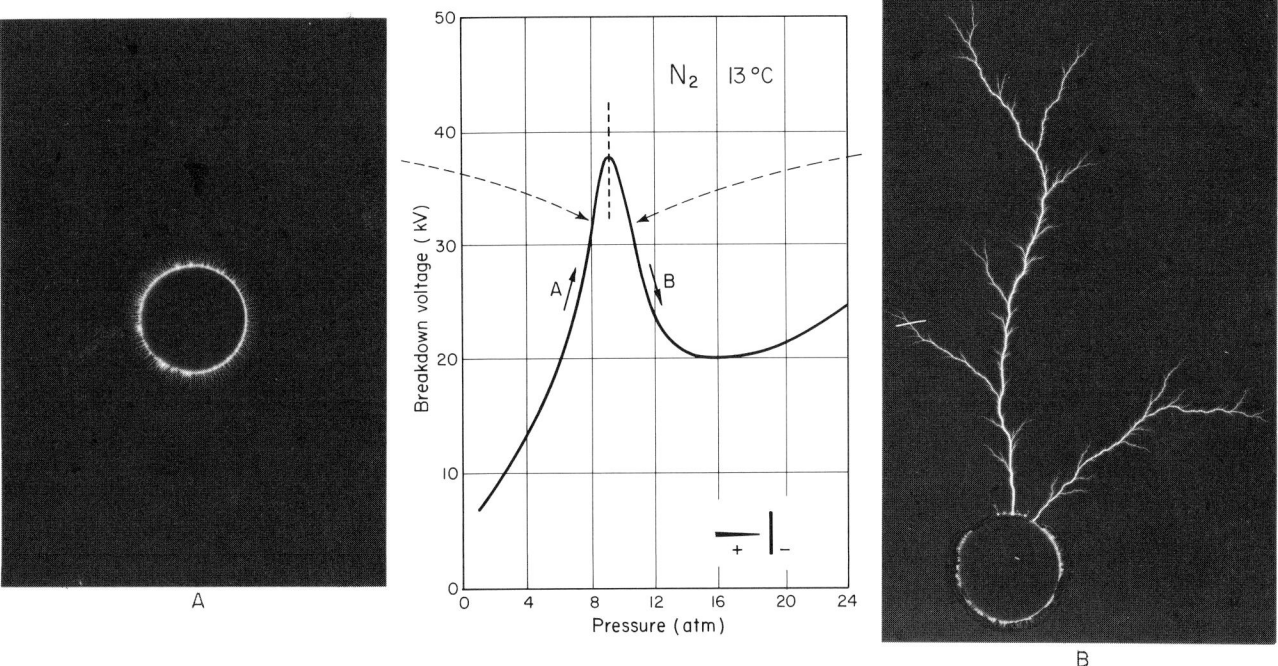

Fig. 20.9. Anomalous breakdown-voltage dependence caused by onset of positive spark.

field distortion; by poisoning the cathode with organic vapors against electron regeneration ($\gamma < 10^{-8}$),[10] a remedy well known as a preventive against repetitive discharges in Geiger counters; and—an amusing paradox—by just doing the opposite, i.e., promoting the electron generation at the cathode, in our example by field emission (cf. Fig. 20.7) and in Raether's research by generating ca. 10^4 electrons simultaneously by an α ray, so that their avalanches act as one.

Two additional plasma-promoting parameters must be added to this list. Increased time of voltage application allows the transformation of a space-charge-guided "primary Lichtenberg figure" into a plasma-guided "secondary figure" (Fig. 20.8). And high gas pressure p (i.e., density ρ) increases the space-charge density in the avalanche path and with it the dynamic field distortion to critical values. This pressure effect explains an old observation by Goldman and Vul:[11] an anomalous maximum in the breakdown voltage of the positive point-to-plane gap in nitrogen at about 8 atm near room temperature. Lichtenberg figures demonstrate that here the charge density in the positive space-charge branches reaches the critical magnitude for plasma onset at the anode. The previously harmless corona discharge turns thus into a spark that rapidly bridges the gap from anode to cathode (Fig. 20.9).

Summarizing: Dynamic field distortion by predischarges and the electron supply determine the onset of plasma and the development of "positive and negative sparks," from gas discharges to thunderstorms.[12]

Failure of Paschen's law; emergence of new breakdown conditions

Townsend's multiavalanche breakdown embodies Paschen's similarity law. The buildup of a regenerating avalanche from cathode to anode requires a critical voltage drop per free path in field direction and a critical number of such free paths ($\approx pd$) across the gap; hence, $V_{\max} = f(pd)$. Differently expressed: As long as the total gap length d dominates the avalanche build-up, the exponential rise of ionization with d permits reducing $E_{\max}$ as distance increases.

The influence of the cathode material (γ) is buried insensitively in the exponential growth term of the avalanche. A slight change in d or α can vary the calculated regeneration yield γ by orders of magnitude. Still, this uncertainty cannot explain the tremendous change of γ for air of atmospheric pressure when the gap is lengthened from 10^{-2} to 10 cm (Table 20.1).[13] The data must be read as indication that here the total gap length d loses control over the breakdown strength beyond the distance of millimeters.

---

[11] I. M. Goldman and B. M. Vul, *Tech. Phys. (U.S.S.R.)* **1**, 497 (1935).

[12] A. von Hippel, *Naturwissenschaften* **22**, 701 (1934); *Ergeb. exakt. Naturw.* **14**, 79 (1935).

[13] Cf. Ref. 2., p. 163.

Table 20.1. Breakdown in air of atmospheric pressure

d cm	E_{max} kV/cm	αd	$\gamma = 1/eH$
10^{-2}	95.6	8.9	1.4×10^{-4}
10^{-1}	45	8.8	1.5×10^{-4}
1	31.7	17	4.1×10^{-8}
10	26.6	42	5.7×10^{-19}

A corresponding invalidation of Townsend's law occurs at constant gap distance with rising pressure. When instability sets in before distance runs out, the voltage drop per free path becomes the decisive yardstick. In consequence, E_{max} assumes independence of d and rises proportionally to gas density as the liquid state is approached (Fig. 20.10).

Single-avalanche breakdown does not require the whole gap length and can start instability, as shown, in various ways. The space charge left by one primary electron in front of the cathode may trigger field emission, plasma formation, and a negative spark; an elec-

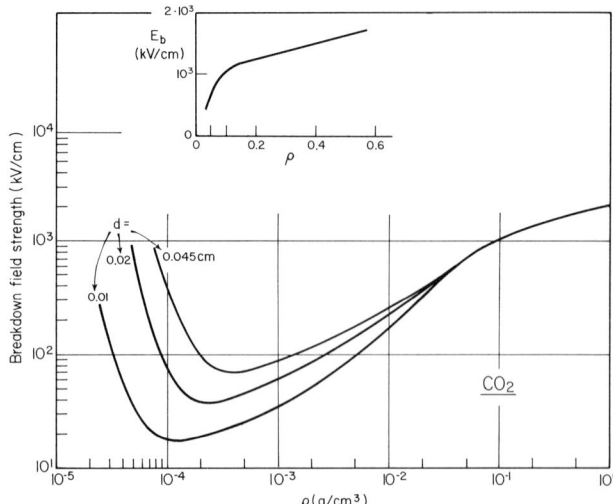

Fig. 20.10. E_{max} as $f(\rho)$ and $f(d)$ for CO_2 inside and beyond Paschen's law.

mining the electric strength. A field-emission current[14]

$$J_f = aE_c^2 e^{-b/E_c} \tag{20.2}$$

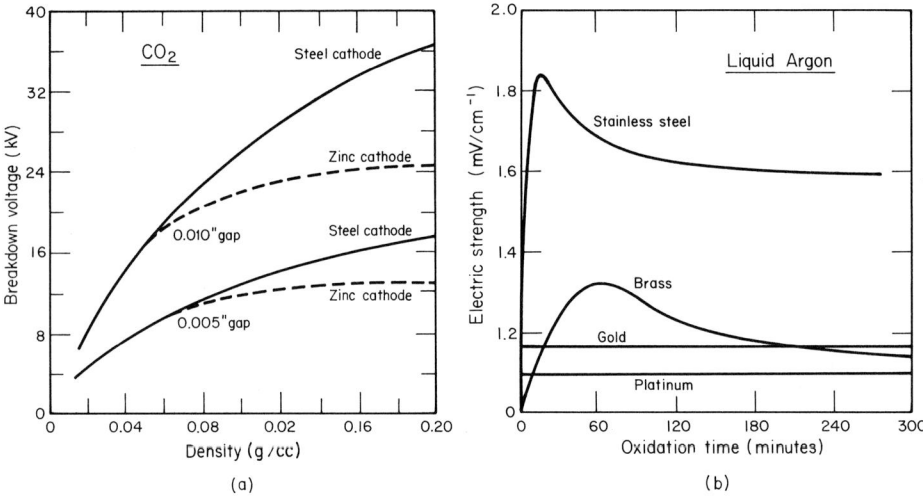

Fig. 20.11. Dependence of breakdown strength on cathode material: (a) in CO_2 gas at high density; (b) in liquid argon as function of preoxidation.

tron track traversing a fraction of the gap can produce plasma onset in space, bridging toward both electrodes; or plasma formation may occur at the anode and send a positive spark backward toward the cathode (cf. Fig. 20.9).

Thus the rules change subtly with increasing gap distance and/or gas pressure. The onset of impact ionization still precedes breakdown, but field emission or plasma formation may be the decisive event deter-

should, in our experiment, appear amplified by impact ionization in the gas to

$$J = J_f e^{\alpha d} \tag{20.3}$$

as long as the field is relatively uniform and the total gap length counts. This was confirmed by Young.[6]

[14] R. H. Fowler and L. Nordheim, *Proc. Roy. Soc. (London)* A119, 173 (1928).

At higher pressures the factor a—according to theory dependent solely on Fermi level and work function—began to increase exponentially with gas density, probably reflecting an increase in cathode fall E_c and in emitting area. (Lichtenberg figures show the emission concentrated in spots and the increase of erupting spots with pressure.)

The factor b in the simple Fowler-Nordheim equation (Eq. 20.2) depends only on the work function of the metal. If the breakdown is triggered by field emission, V_{max} should be strongly reduced when a stainless steel cathode of high work function is replaced by a zinc cathode of low work function. This dependence was observed in CO_2 at high densities,[6] where Paschen's law loses its validity (Fig. 20.11a).

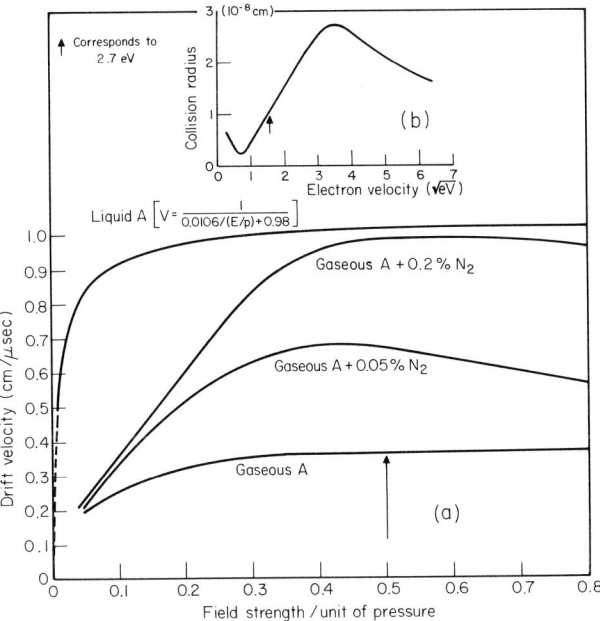

Fig. 20.12. Electron mobility and Ramsauer effect in argon. (Gaseous argon after Bortner et al.; liquid argon after Williams.[16])

When stainless steel is subjected to progressive oxidation, its field-emission properties vary drastically with surface-layer formation. Such changes again are graphically reflected in the breakdown strength, as recent measurements on liquid argon testify (Fig. 20.11b).[15]

Feature changes in condensed systems

The testimony of liquid argon could be introduced in the preceding discussion of gas breakdown, because it is the only inert liquid obtained pure enough to keep

[15] D. W. Swan and T. J. Lewis, *J. Electrochem. Soc.* **107**, 180 (1960).

electrons free at low fields. Electronic mobilities, onset of the Ramsauer effect,[16] impact ionization, and avalanche formation have been observed, as expected for an inert gas of that density (Fig. 20.12); and the avalanche build-up extends across the gap, as the decrease of E_{max} with distance d certifies (Fig. 20.13).[15,17]

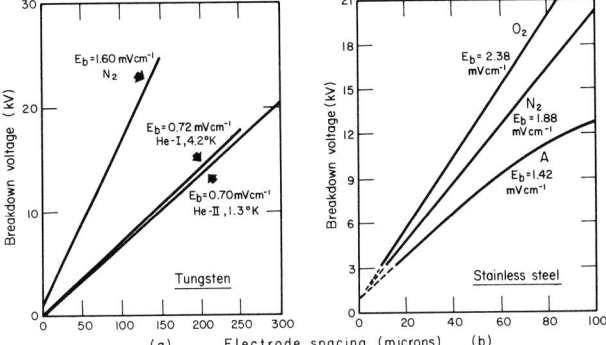

Fig. 20.13. Breakdown strength of liquefied gases: (a) after Blaisse et al.;[17] (b) after Swan and Lewis.[15]

However, this freedom of electrons is a precarious one: The pulse height in liquid-argon counters becomes rapidly quenched by small amounts of impurities, especially those of high electronegativity (Fig. 20.14).[18]

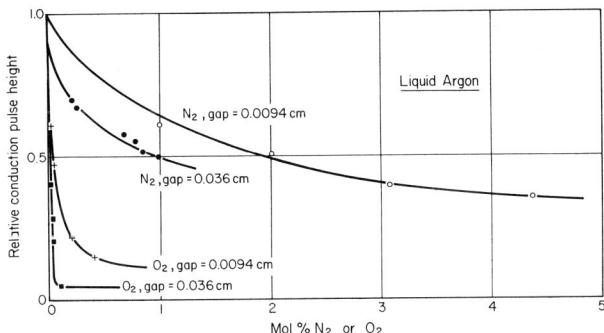

Fig. 20.14. Quenching of pulses in liquid-argon counter.

The electrons are trapped, forming negative ions, the breakdown strength rises and tends to become independent of gap width as for other liquefied gases (cf. Figs. 20.10 and 20.13).

Changes in the laws of electron motion occur grad-

[16] T. E. Bortner, G. S. Hurst, and W. G. Stone, *Rev. Sci. Instr.* **28**, 103 (1957); R. L. Williams, *Can. J. Phys.* **35**, 134 (1957); C. Ramsauer, *Ann. Phys.* [4] **64**, 513 (1921).

[17] B. S. Blaisse, A. van den Boogaart, and F. Erne, "Electrical Breakdown in Liquid Helium and Liquid Nitrogen," *Problems in Low Temperature Physics and Thermodynamics*, Proc. Meeting Comm. I Intern. Inst. Refrig., Delft, Netherlands, 1958, published in 1959, p. 333.

[18] N. Davidson and A. E. Larsh, *Phys. Rev.* **77**, 707 (1950).

ually with increasing gas density. At low pressures the electrons are essentially in free flight, interrupted by short-time collisions. The external field near breakdown amounts to electron volts per free path in field direction, and only the barrier of electronic excitation states shields effectively against a premature collapse of the insulation. As liquid densities are approached, the free path, as an average collision distance punctuating free flight, vanishes. It has become an atomic distance: the voltage drop across it has shrunk to $\sim 10^{-2}$ eV; simultaneously a system of phonon modes has developed for the condensed state. The electron finds itself in a state of permanent coupling to the surroundings; the excitation of vibrations is the main obstacle against electron acceleration and the main promoter of electron trapping.[19]

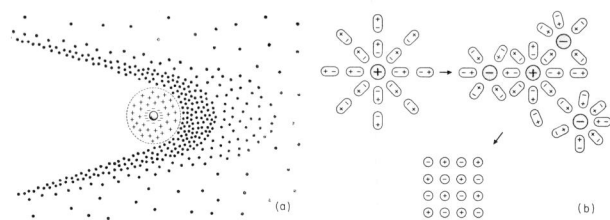

Fig. 20.15. Atmospheres of countercharges: (a) for satellite; (b) from dipolar atmosphere in liquid to ionic atmosphere in electrolyte and crystal.

In gases at normal pressures simple carriers could not survive in field-free space, because the attractive energy of several electron volts between opposite charges enforces recombination. This argument loses validity for complex ions and condensed systems, because here carriers can decrease their free field energy by polarization of the surroundings. In macroscopic theory, this energy decrease is inversely proportional to the dielectric constant κ' of the embedding medium, and the mutual attraction falls—in first approximation—with $1/\kappa'^2$. Molecular analysis must distinguish between electronic, ionic, and dipolar atmospheres that may form around a reference carrier, lowering its field energy and increasing its mass.[20] Such atmospheres of countercharges—changing in magnitude, composition, and shape with particle speed and surroundings—are even felt in the motion of satellites,[21] dominate the theory of electrolytes,[22] organize the structure of polar crystals (cf. Chap. 2), and strongly alter the migration of electrons as "polarons" through dielectrics[20,23] (Fig. 20.15).

Since the restrictions on electronic freedom enter gradually, no discontinuity in electric strength marks the transition from gas to liquid for CO_2[5] or hexane[24] near the critical point. However, such insensitivity must not be expected for transitions replacing the short-range order of an amorphous with the long-range order of a crystalline system.

In perfect crystals, quantized electron states traversing the whole system promise a new freedom of motion, but electrons placed in such conduction bands are "free" in a different sense as before. In gases, an average travel *distance* between obstacles, a "free path," determined the lifetime. In a crystal, when disordered only by thermal agitation, the life expectancy is figured as in an insurance policy. After a statistical *time* τ a terminating event is bound to occur through a disturbing incident, be it a collision with a phonon or an automobile. If, however, the perfection of the crystal lattice is marred by localized disturbances (defects and dislocations) overshadowing the thermal disorder, the life expectancy becomes again linked to a collision distance—insurance prudently excludes such risks as "acts of God."

The influence of long-range order on electron motion is more pronounced for insulators and semiconductors than for metals (e.g., the conductivity of metals is in general not strongly affected by melting). This difference can be traced to the composition and response time of the electron atmosphere. Electrons in metals stay free, because they move fast and are shielded from the interaction with the localized nuclei by an electronic counteratmosphere that adjusts itself about instantaneously to the environment. In dielectrics the charge carriers are shielded mainly by ionic and dipolar atmospheres of much longer relaxation time and strong localization. Their drag slows electrons down by multiphonon excitation and tends to trap them in amorphous systems.

Prebreakdown currents in liquid hexane

Liquid *n*-hexane has served as the guinea pig for many dielectric studies, because it is chemically inert; melting ($-94.3°$ C) and boiling points ($+69°$ C) lie convenient for purification by crystallization, refluxing, and other methods; and a conductivity as low as 10^{-19}

---

[19] A. von Hippel, *Z. Physik* 75, 145 (1932); *J. Appl. Phys.* 8, 815 (1937).
[20] A. von Hippel, *J. Chem. Phys.* 8, 605 (1940).
[21] R. Jastrow, *Sci. Am.* 201, 37 (1959).
[22] P. Debye and E. Hückel, *Physik. Z.* 24, 185 (1923); cf. H. Falkenhagen, *Elektrolyte*, Hirzel, Leipzig, 1953.

[23] I. Pekar, *Untersuchungen über die Elektronentheorie der Kristalle*, Akademie-Verlag, Berlin, 1954.
[24] A. H. Sharbaugh and P. K. Watson, Annual Report, Conference on Electrical Insulation, National Research Council, 1961, p. 95.

to 10^{-20} ohm^{-1} cm^{-1} has been claimed.[25] Still, when the transconduction was measured as function of field strength, the situation became extremely confused: The cathode seemed to enter as a current emitter,[26-29] impact ionization with a gap-length dependence as in gases was proclaimed[30] and denied,[26] ionic currents and polarization effects were observed,[31] dissociation of the hexane in high fields proposed,[32] a time lag of breakdown ascribed to the mobility of positive ions,[29] and so on. Investigations carried through in this laboratory,[33,34] interpreted with the concepts developed in the preceding discussion, probably clarify the major issues.

dielectric constant, contains ions stemming from impurities and generated in the material proper by radioactivity, cosmic rays, catalytic agents, light, etc. Such ions, probably partly of complex structure (polyelectrolytes), disappear from sight by loose association but are successively called into play as the applied field increases. The ease of their discharge depends on ion structure, electrode material, and surface layers. As positive charge accumulates in front of the cathode, electron emission sets in, as described for gases and demonstrated quantitatively also in condensed systems, e.g., on the example of the alkali halides.[35] Thus, in hexane an ionic current triggers a field-emission current (Fig. 20.17). The electrons, trapped outside the

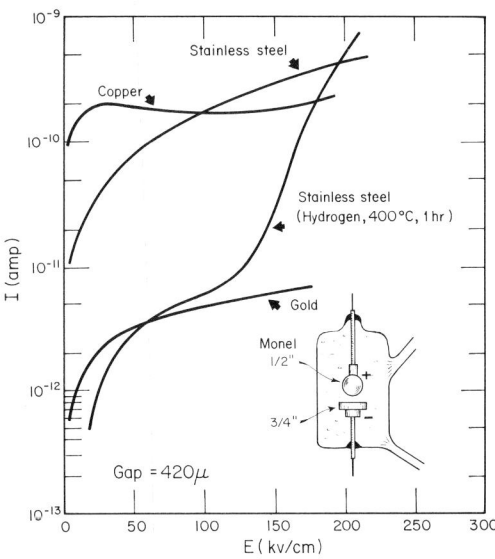

Fig. 20.16. Various current-voltage characteristics in liquid n-hexane.

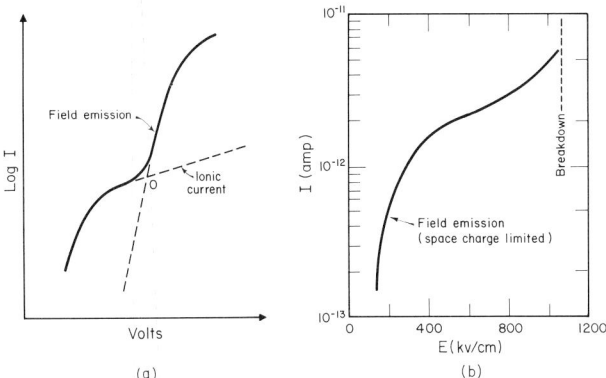

Fig. 20.17. (a) Ionic current preceding and triggering field-emission current; (b) d-c current-voltage characteristic up to breakdown.

The d-c current-voltage characteristics in highly purified hexane may assume a bewildering variety of shapes, depending on cathode material and pretreatment (Fig. 20.16). This versatility can be traced to the interplay of two fundamental current components. Hexane, in spite of its nonpolar character and low

steep cathode fall, limit the promoting field gradient; the electronic current component levels off by negative space-charge formation, until the applied voltage becomes high enough for impact ionization. At this point single-avalanche breakdown causes collapse of the insulation.

When the cations promote field-emission but their supply is at a premium, the ionic current can be raised and an earlier emission onset enforced by drawing on a larger volume of hexane, i.e., by increasing the gap width at equal field strength. Figure 20.18 proves the existence of this effect and the fallacy of equating a gap-length dependence of the current *a priori* with impact ionization. The magnitude of the influence depends on the cathode fall build-up, hence on geometry, material, and surface layers of the cathode, which determine the ease of positive ion discharge and of electron emission. Figure 20.19 certifies that in a cylindrical arrangement, for example, emission can be precipitated by raising the field at the cathode surface through

---

[25] A. Nikuradse, *Das flüssige Dielektrikum*, Springer-Verlag, Berlin, 1934.

[26] L. D. Inge and A. Walther, *Tech. Phys.* (U.S.S.R.) *1*, 539 (1934).

[27] E. H. Baker and H. A. Boltz, *Phys. Rev.* *51*, 275 (1937).

[28] W. R. LePage and L. A. DuBridge, *ibid.* *58*, 61 (1940).

[29] D. W. Goodwin and K. A. Macfadyen, *Proc. Phys. Soc.* (London), *B66*, 85 (1953).

[30] A. Nikuradse, *Z. Physik* *77*, 216 (1932).

[31] G. Jaffe and C. Z. LeMay, *J. Chem. Phys.* *21*, 920 (1953).

[32] H. J. Plumley, *Phys. Rev.* *59*, 200 (1941); C. S. Pao, *ibid.* *64*, 60 (1943).

[33] W. B. Green, *J. Appl. Phys.* *26*, 1257 (1955); *27*, 921 (1956).

[34] R. Coelho and M. Bono, *J. Electrochem. Soc.* *107*, 94 (1960).

[35] A. von Hippel, E. P. Gross, J. G. Jelatis, and M. Geller, *Phys. Rev.* *91*, 568 (1953).

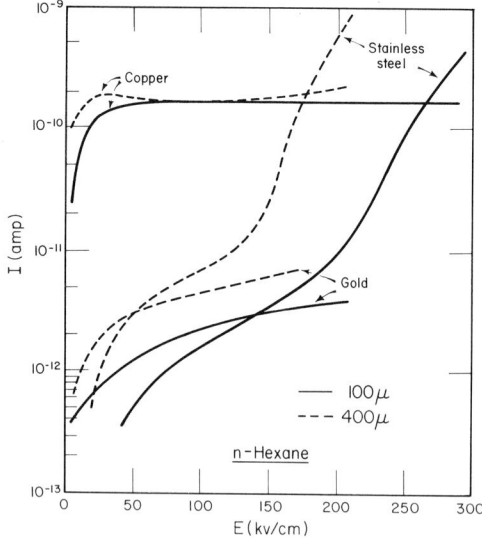

Fig. 20.18. Gap-width effect in hexane.

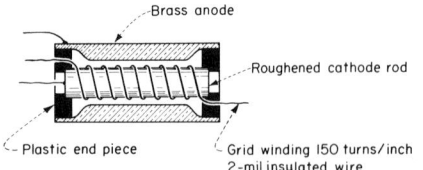

Fig. 20.20. Field-emission triode in hexane.

initially the wire as cathode produced the greater current, but after some weeks the opposite result prevailed (Fig. 20.22a). Furthermore, the temperature dependence after such conditioning shows opposite trends: For wire positive, it falls; for wire negative, it rises with temperature increase (Fig. 20.22b). Also the polariza-

geometrical changes—increase of electrode diameter or surface roughening. An auxiliary grid winding can transform such a coaxial diode into a field-emission triode (Fig. 20.20); Green[33] succeeded in modulating the current through hexane in that way by 30 percent, with a grid bias of 50 volts.

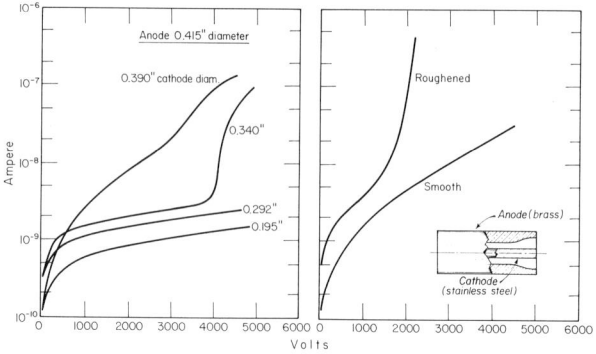

Fig. 20.19. Change of field-emission onset in cylindrical arrangement.

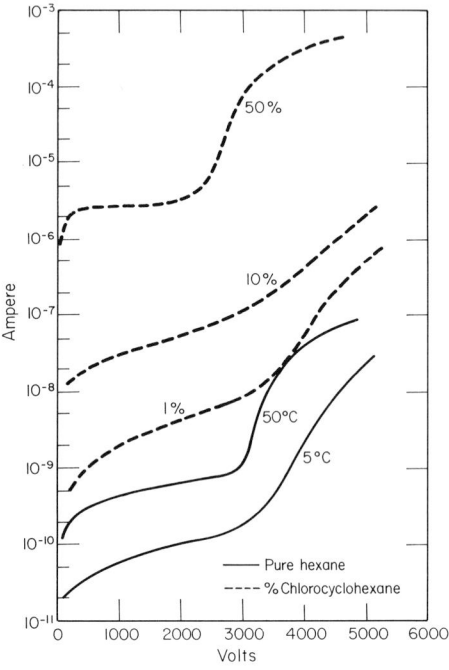

Fig. 20.21. Current variation by temperature or addition agent in liquid hexane.

If the positive ions discharge freely at the cathode, the ionic current can be varied in wide limits by temperature change or addition agents without affecting field emission (Fig. 20.21). Clearly, experimental results which at first sight seem quite contradictory actually need not be in conflict.

Since space-charge formation and field-enhanced dissociation appeared to play a decisive role, a closer study of these effects was made in a highly inhomogeneous field configuration (thin wire in coaxial cylinder).[34] Here the conditioning as function of time became very pronounced and had the seemingly strange effect that

tion proves quite polarity dependent (Fig. 20.22c): For wire negative, equilibrium is established faster under voltage, and the subsequent short-circuit current reverses quickly and disappears in simple decay. For wire positive, the approach to a stationary state (the "conditioning") continues much longer, and, on short-circuit, the reverse current sluggishly traverses a maximum before decaying.

In the case of wire positive, Coelho and Bono[34] proved that the field-strength dependence of the current flow obeys Onsager's theory of weak electrolytes. For wire negative, the field-emission component is bound to hamper field-induced dissociation. In addition, the geometrical distribution of the negative space

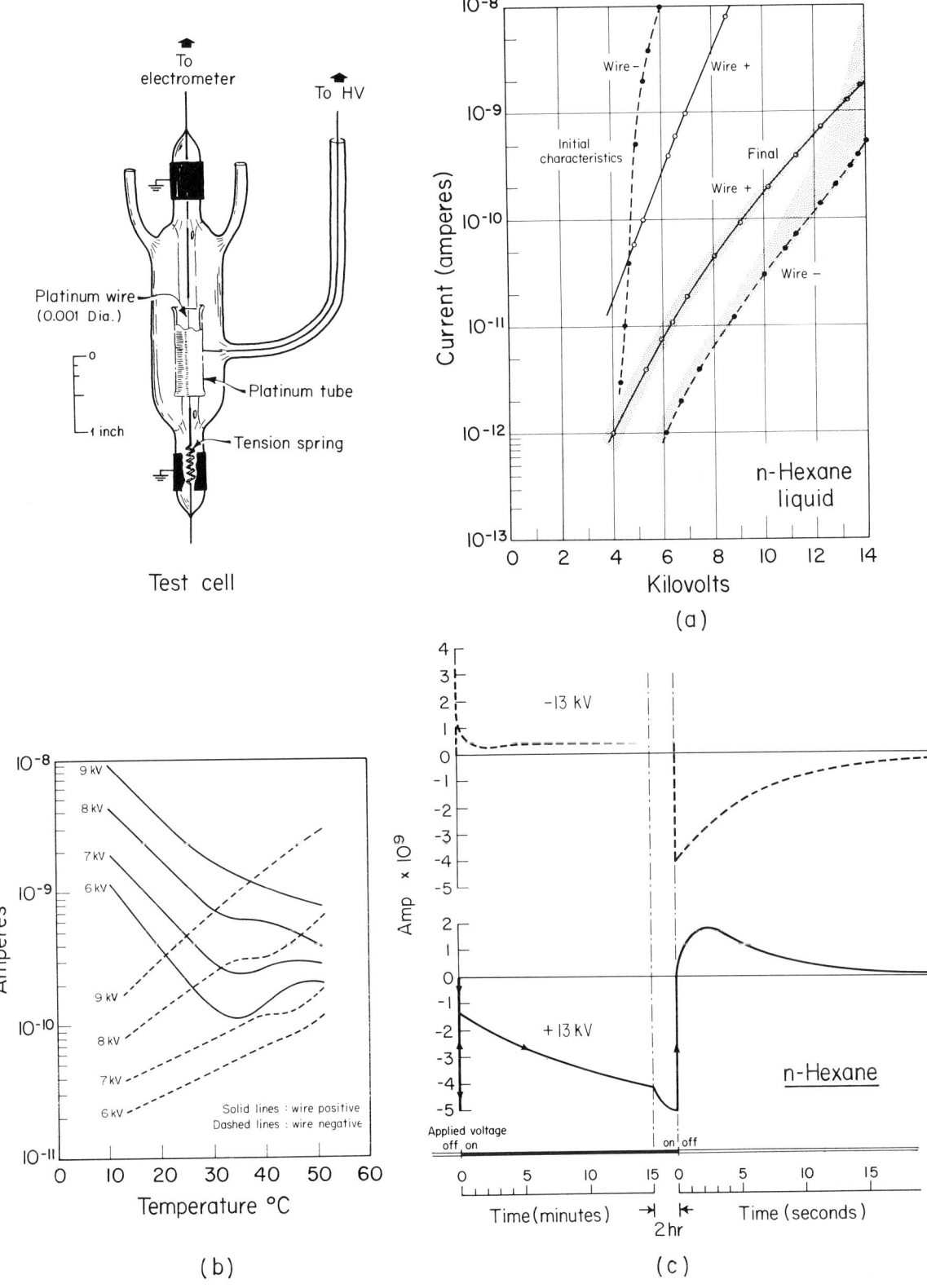

Fig. 20.22. Polarization effects in liquid hexane: (a) current before and after several weeks; (b) temperature dependence; (c) short-circuit currents after formation.

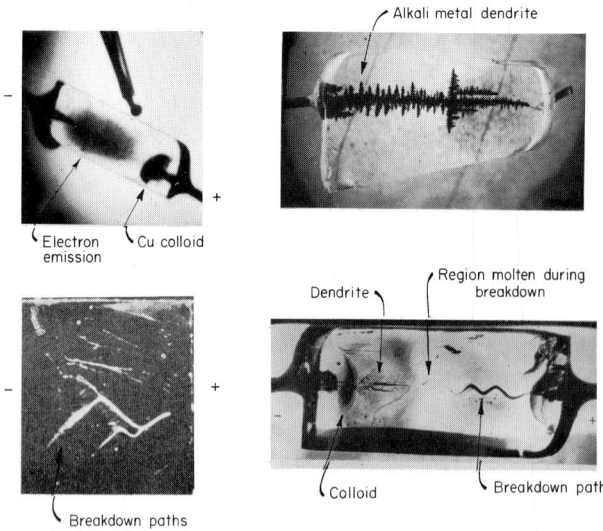

Fig. 20.23. Succession of events in NaCl crystal at high fields and elevated temperature.

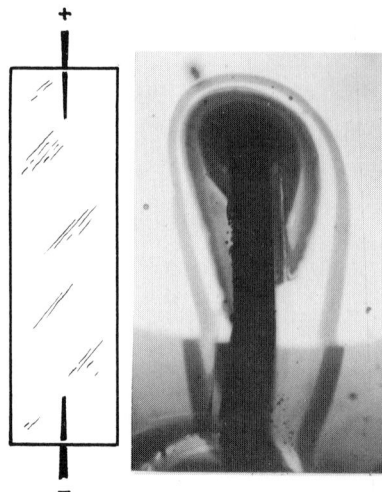

Fig. 20.24. Shattering of glass by field emission and electron trapping.

charge around the positive wire is unlike the two-layer structure of positive and negative space charge around the negative wire. The effect of temperature and time on dissociation, the types of ions liberated, the field distribution and cathode conditioning resulting—all such influences conspire to make the macroscopic responses more baffling as the number and interplay of molecular parameters increase.

Breakdown phenomena in liquids and solids

The scientist interested in isolating fundamental phenomena is frequently worlds apart from the practical engineer, whose yardstick is technical performance under service conditions. In the case of liquid dielectrics, commercially available oils normally provide the insulating ambient for transformers, capacitors, and switchgear housed in not too clean containers; and phenomena dismissed in the laboratory as "dirt effects" prevail. In such situations, a microanalysis that can point the way to optimum performance under economically profitable conditions is an often neglected task.

Any scientist or engineer observing liquids microscopically in electric fields will have watched with amused exasperation the antics of fibers and suspensions clinging to the electrodes or bouncing between them, lining up in bridge formation and collapsing such bridges by microdischarges. The aligning of moist cellulose fibers in oil[36] or of fat globules in milk[37] are early described examples. Kok[38] in recent years has made the "impurity breakdown of oil" the subject of

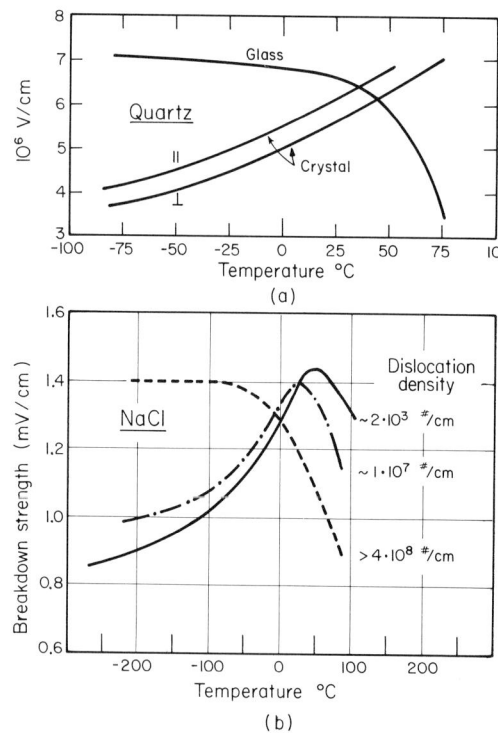

Fig. 20.25. Temperature dependence of breakdown strength: (a) crystal versus glass; (b) effect of dislocation density.

a fascinating study, with special emphasis on colloidal chemistry.

[36] T. Hirobe, W. Ogawa, and S. Kubo, *Electrician* 78, 656 (1917).
[37] E. Muth, *Kolloid-Z.* 41, 97 (1927).
[38] J. A. Kok, *Electric Breakdown of Insulating Liquids*, Philips Technical Library, 1961.

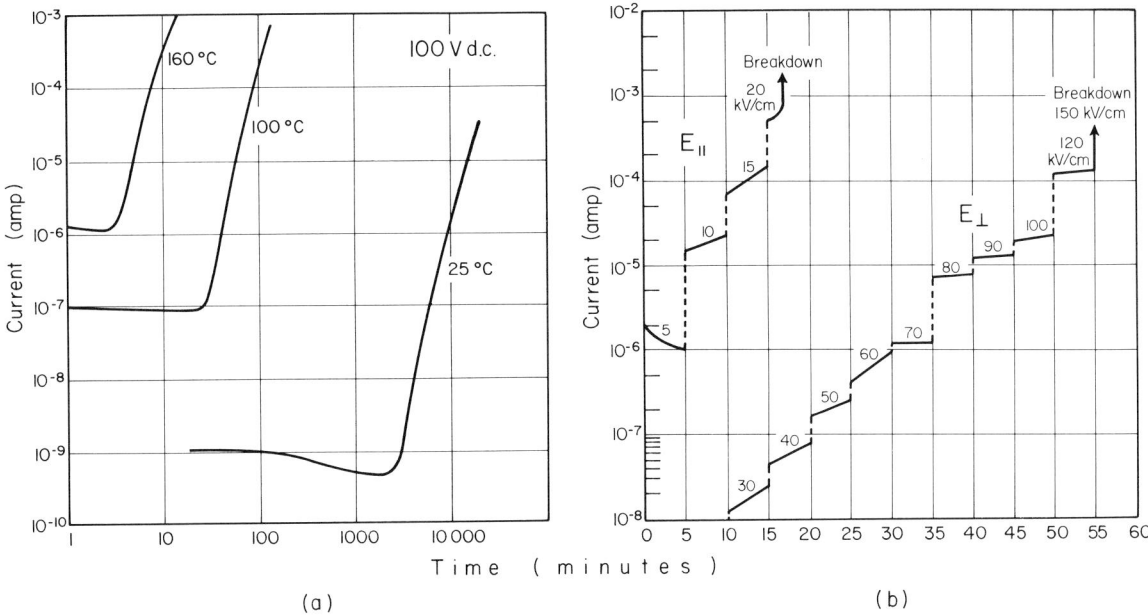

Fig. 20.26. Electron injection in rutile: (a) weak primary current triggers runaway current; (b) strength of rutile at direct current parallel and perpendicular to the c axis.

These types of destruction of the insulation, triggered by the alignment or coagulation of suspended material, by bubble formation and also cavitation, can be avoided by operating with sufficiently short single voltage pulses or increasing the viscosity of the medium. The latter remedy leads to the solid state, to polymers, glasses, and crystals. Space permits only a short glance at some situations of special interest.

Alkali halide crystals with pointed copper electrodes, viewed at elevated temperatures in d-c fields, reveal a succession of phenomena—each of them potentially destructive (Fig. 20.23):[39] electron injection from the cathode as a color-center cloud; electron trapping near the anode by immigrating copper ions resulting in copper colloid formation; growth of alkali metal dendrites building a bridge from cathode to anode; and finally electron impact ionization driving a crystallographically oriented "direction breakdown" path from anode to cathode. In glasses, electron injection alone may prove deadly (Fig. 20.24): The glass shatters under the mechanical strain caused by neutralization of cations and space-charge formation.[40]

A clear distinction between short-range and long-range ordered systems appears in the temperature dependence of breakdown exemplified by the d-c characteristics of quartz glass versus quartz crystal (Fig. 20.25a).[41] Both curves fall at high temperatures, but a good crystal shows a rising trend at low temperatures, while the breakdown field strength for glass proves here practically constant. Our explanation that electron scattering by phonons causes the positive slope in crystals could now be confirmed in a direct experiment: Introduction of an excessive amount of dislocations into a NaCl crystal transformed its breakdown characteristic from that of a crystal to that of a glass (Fig. 20.25b).[42]

The falling part of the breakdown curves depends on the time of voltage application and disappears for impulse breakdown of sufficiently short rise time.[43] The suspected cause—field distortion by ionic conduction—has been verified by current measurements.[42] The corresponding situation in four glasses of different sodium content has been investigated by Vermeer in beautiful detail.[44]

Excessive conduction—electronic or ionic—can produce the thermal destruction of a dielectric by Joule heat without recourse to impact ionization. This "thermal breakdown," well known since the early work of Wagner[45] and Inge, Walther, and Semenoff,[26,46] has

---

[39] A. von Hippel, *Z. Physik* **98**, 580 (1936).
[40] A. von Hippel, *J. Appl. Phys.* **8**, 815 (1937).
[41] A. von Hippel and R. J. Maurer, *Phys. Rev.* **59**, 820 (1941).
[42] R. Nevald. *Tech. Rep. 193*, Lab. Ins. Res., Mass. Inst. Tech., Oct., 1964.
[43] A. von Hippel and R. S. Alger, *Phys. Rev.* **76**, 127 (1949).
[44] J. Vermeer, *Physica* **20**, 313 (1954); *ibid.* **22**, 1247, 1257, 1269 (1956).
[45] K. W. Wagner, *J. Am. Inst. Elec. Engrs.* **41**, 1034 (1922).
[46] Cf. N. Semenoff and A. Walther, *Die physikalischen Grundlagen der elektrischen Festigkeitslehre*, Springer-Verlag, Berlin, 1928.

reappeared in new disguise in recent work: The conductivity may not exist in the dielectric *a priori*, but may be created by injection after a preparation period. Rutile single crystals, for example, conduct much better parallel than perpendicular to the *c* axis.[47] If a low voltage is applied, hundreds of hours may pass at room temperature before suddenly the current begins to rise catastrophically (Fig. 20.26*a*).[48] The cause seems to be the onset of hole injection under the prodding of an anode fall, which in turn increases—by space-charge compensation—the electron injection. D-c voltage will destroy the crystal parallel to the axis at about 20 kV/cm, and normal to the axis at about 150 kV/cm by thermal breakdown at room temperature in a 10-minute experiment (Fig. 20.26*b*); in impulse tests with 10^{-6}-sec rise time we reached ca. 600 kV/cm for both orientations.[49]

Here this short sketch must end.[50] It may not have created more than curiosity in a large field of research, but it hopes to have inspired confidence that molecular analysis is now able to link the various states of aggregation successfully.

---

[47] A. von Hippel, J. Kalnajs, and W. B. Westphal, *J. Phys. Chem. Solids 23*, 779 (1962).

[48] J. A van Raalte, *Tech. Rep. 195*, Lab. Ins. Res., November, 1964.

[49] Measurements by D. A. Powers, Lab. Ins. Res., Mass. Inst. Tech.

[50] A monograph on the electric strength of materials is in preparation.

21 · THE LIVING CELL*

Murray D. Rosenberg

Introduction — Cytology — Dimensions of cells — Molecular constituents and their reproduction — Nucleus and cytoplasm — The cytoplasmic matrix — Dynamic activities — The cell surface — Packing and cell dynamics — Cells on mono- and multilayer substrates — Experiments on liquid-liquid interfaces — Possible causes of cellular motion and surface movements — Interfacial dynamics — Outlook

Introduction

The reader may question the inclusion of a chapter on the living cell in a book entitled *The Molecular Designing of Materials and Devices*. Certainly a living cell is not in the usual sense of the word a material or device. It is the smallest organization of matter meaningly designated as being alive. It is a composition of molecular and macromolecular complexes which interact in a manner that allows it to grow, develop, and reproduce its own likeness and to maintain itself within definite limits. It is a thermodynamically open system, capable of exchanging matter and energy with the exterior to sustain the processes necessary for life.

The physicist and engineer foresee the time when they will be able to synthesize materials and devices as needed, and they have a well-structured scaffolding to support such ambitions. The biologist must fall back primarily on qualitative observations. It would be delusion to assert that he can explain how one designs living things. In more modest terms, however, he can say something about their design and place their complexities in perspective.

This paper attempts two brief tasks: to review some of the structural components within cells, with special emphasis on the dynamic properties of membranous elements; and to discuss cell surfaces and observations on their interactions with synthetic interfaces. It is hoped that this presentation will provide some understanding of the complex interdependent operations that make up the biological organization designated as "the living cell."

Cytology

Cytology, the branch of biology treating the structure and functions of cells, is a young science. In 1665 Robert Hooke used the term "cell" to describe the "great many little boxes" in a piece of cork, but these observations were not fully appreciated until the early part of the nineteenth century. Mirbel, Lamarck, Turpin, Meyen, von Mohl, Schleiden, Schwann, and others then established the theory[1] that animals as

---

* Based on research supported by the National Institutes of Health under Grant CA-06375 to Dr. Paul Weiss as principal investigator. The author is currently a Career Scientist of the Health Research Council of the City of New York.

[1] E. D. De Robertis, W. W. Nowinski, and F. A. Saez, *General Cytology*, W. B. Saunders Co., Philadelphia, 1960.

well as plants are aggregates of cells. An extension was provided by Virchow (1858) with the statement that all cells are derived from pre-existing ones. During the latter half of the nineteenth and early part of the twentieth century progress was extensive but patchy. Anatomic and taxonomic studies were soon supplemented by research on biochemical pathways within cells and cell fractions. Electron microscopy made it possible to determine fine structures within cells and to designate regions where specific biochemical processes are localized. This correlation between function and structure of macromolecular groupings represents one of the great achievements of modern cytology. A hundred years ago Schwann recognized the connection between metabolic and morphologic phenomena; in this sense he can be looked upon as the father of modern cytology.

Successes in associating the form and compositions of macromolecular complexes with their roles in cellular operations are now setting the stage for the attack on more complex problems: the energetics of organization within cells; transfer among chemical, mechanical, and electrical energy; mechanisms of regulation and control; exchange of matter and energy with the external milieu and their effect on the organization and function of the internal environment; and the circulation paths of molecules throughout the cell. These types of problems must be clarified before one can begin to alter living cells in premeditated fashion.

Dimensions of cells

The vast majority of cells are small (1 to 20 μ); the extremes of size and shape, however, are great. Smallest among living cells are some bacteria (0.25 μ in diameter); human red blood cells measure approximately 7.5 μ across; many epithelial cells range from 15 to 40 μ; and ostrich-egg cells (without shell) are 24 mm in diameter. These cell types are essentially round (flat disks or spherical; the human red blood cell is a biconcave disk). In addition there are elongate muscle and nerve cells; the latter can be several feet long with a central body only 100 μ in diameter (Fig. 21.1).[2] The ratio of volumes between largest and smallest cells is roughly $5 \times 10^5 : 1$.

Marked differences of size and shape occur not only between different species but also (to a lesser extent) within a single one. They appear to mirror a complex set of interactions demonstrated by factors such as ratio of cell surface to cell volume, nuclear-cytoplasmic

[2] G. G. Simpson, C. S. Pittendrigh, and L. H. Tiffany, *Life: An Introduction to Biology*, Harcourt, Brace and World, Inc., New York, 1957, Chap. 3, p. 41.

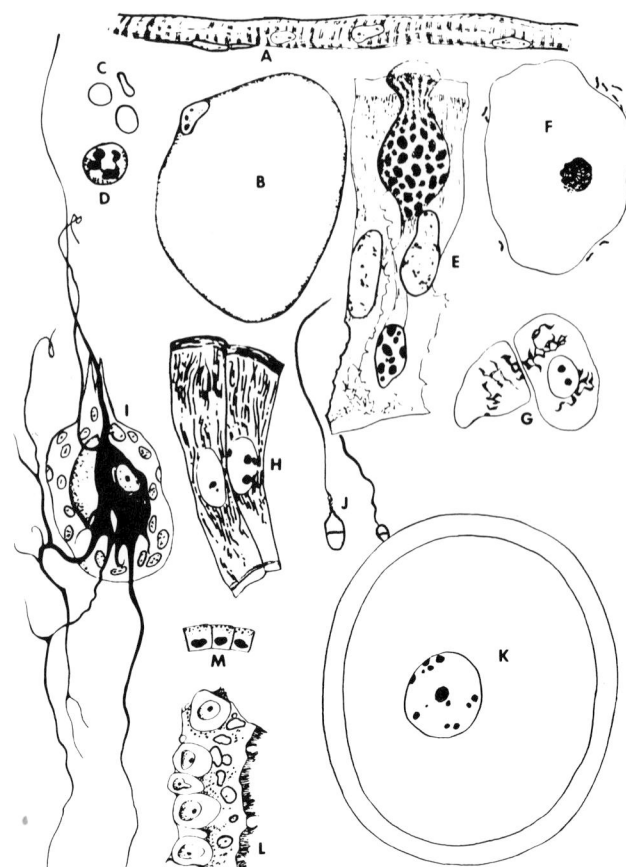

Fig. 21.1. A diversity of cells. *A*. Human muscle cells. *B*. A fat-storage cell from human connective tissue. *C*. Human red blood cells. *D*. Human white blood cells. *E*. Three epithelial cells from the intestine of an axolotl, a larval salamander. *F*. An epithelial cell from the lining of the human vagina. *G*. Two cells from the liver of a mouse. *H*. Two epithelial cells from the rat's intestine. *I*. A human nerve cell. *J*. Human sperm cell. *K*. Human egg cell, with a sperm entering it. *L*. Part of the human placenta. *M*. Three cells from the human eye (retina), containing pigment granules. (After Simpson *et al.*[2])

ratio, rate of metabolism, cell-environment and cell-cell interactions. To complicate the issue, some cells can assume variable shapes in response to environmental pressures, others retain relatively stable shapes. With respect to functional activities, we even find marked variations among cells of the same tissue, even though their shapes and sizes appear similar.

Molecular constituents and their reproduction

Evolutionary pressures have led to features and structures common to all cells and to principal constituents that are similar. The most common constituent is water. In many cells, water accounts for about 85 percent of the fresh weight, protein macro-

Nucleus and cytoplasm

Fundamentally a cell consists of two parts: a central nucleus and the surrounding cytoplasm. The nuclear components control heredity and development of the organism, the cytoplasmic ones are responsible for respiration, growth, secretion, contraction, etc. These assignments are carried in close cooperation between nucleus and cytoplasm.

Membranes, organelles, particles, fibrils, tubules, etc., abound and make, in a sense, the cell an ever-changing lattice of surfaces and fibrils imbedded in an ill-defined matrix. The interfacial regions separate colloid-rich from colloid-richer phases.

The nuclei of all plant and animal cells are well delineated. The primary carrier of genetic information is the molecule deoxyribonucleic acid (DNA),[4] associated with the hereditary material of cells, the chromosomes. It is a stable macromolecule generally complexed with the basic protein, histone (Fig. 21.2). The stability of this genetic material is obviously necessary for the maintenance of the species.

Within the nucleus are one or more condensed structures called nucleoli,[5] aggregations of particles rich in ribonucleic acid (Fig. 21.3). These particles are

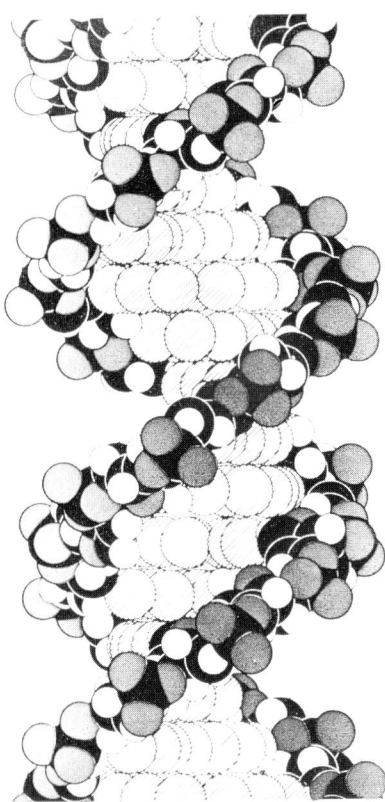

Fig. 21.2. Model of the DNA molecule: double-helical chain. (Courtesy of L. D. Hamilton, Brookhaven Laboratory.)[4]

molecules for 10 percent, fats for approximately 2 percent; the remaining 3 percent are other organic and inorganic molecules. Differently stated: For an average protein macromolecule there may be 18,000 water molecules, 10 molecules of fat, 20 other small organic molecules, and 100 inorganic molecules; or, in a cell of 2×10^{-9} g, protein macromolecules account for approximately 4×10^{-10} g.

If such a cell were to divide with a generation time of 24 hours, it must synthesize roughly 30,000 protein molecules (average molecular weight 10^5) per second. In other words, the cell must assemble each second roughly 30×10^6 amino acids in proper sequence and relevant locations. At the same time all molecular species, such as phospholipids, lipids, polysaccharides, etc., must not only be duplicated but also assembled in proper location into formed elements, such as membranes, tubules, and fibrils. It is interesting to note that a similar calculation for the smallest cells (0.25 μ) indicates that there is probably only one or two of each required macromolecule for metabolic reactions,[3] a far cry from the crystalline or random statistical assemblies of molecules with which we are most familiar.

[3] A. G. Loewy and P. Siekevitz, *Cell Structure and Functions*, Holt, Rinehart, and Winston, New York, 1963.

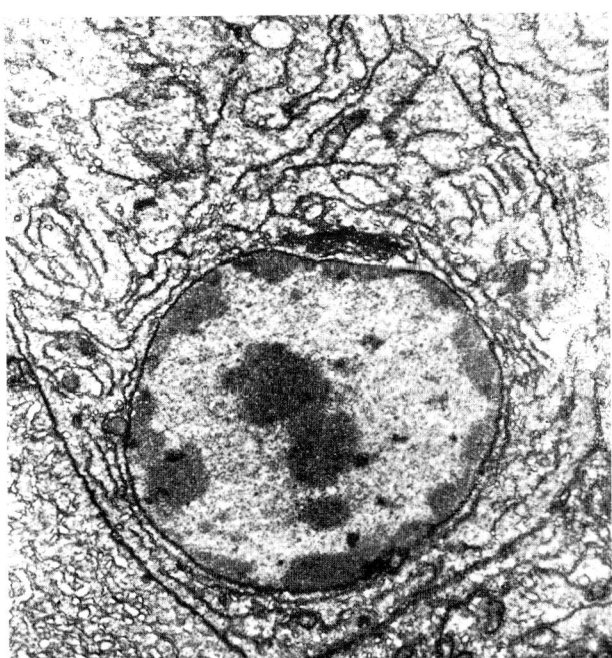

Fig. 21.3. Cell nucleus with prominent nucleoli. (From Brachet.[5])

roughly similar in size and shape to ribosomes commonly associated with protein synthesis within the

[4] *The Cell*, The Upjohn Company, Kalamazoo, Mich., 1962.
[5] J. Brachet, *Sci. Am.* **205**, 50 (1961).

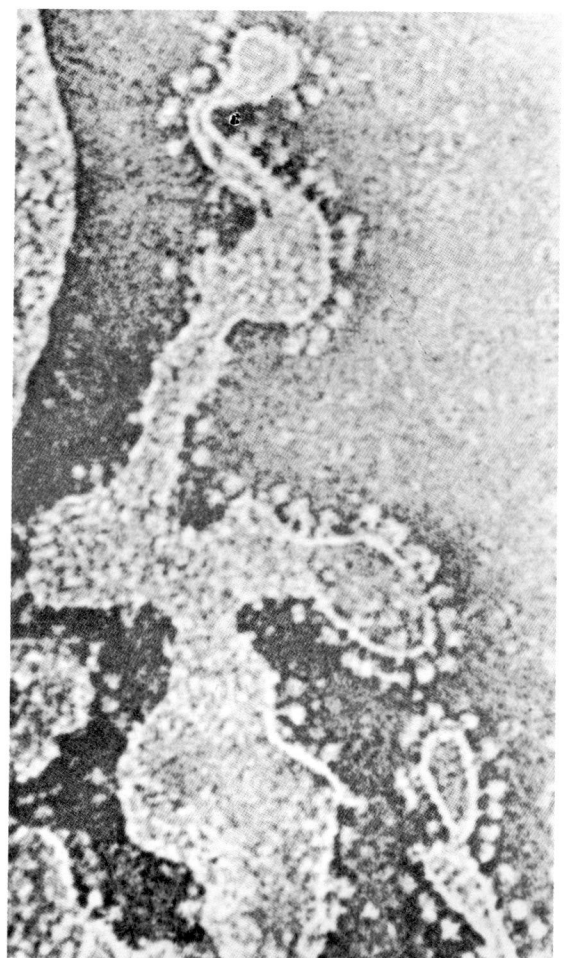

Fig. 21.4. Electron transfer particles of mitochondria (elementary particles). (From Green.[6])

cytoplasm. The hereditary materials of the cell are closely packed in the nuclear ground substance (matrix, sap, or karyoplasm): Little is known of the fine structure of this matrix. Encircling the nucleus is the specialized nuclear membrane, in some cells perforated by pores.

The cytoplasm is rich in organelles and inclusions, among them mitochondria, smooth and granular endoplasmic reticulum, ribosomes, the Golgi complex, lysosomes, centrioles, lipid droplets, secretion droplets, microtubules, microfilaments and microfibrils.

The mitochondria serve in animal cells as power plants extracting energy from food by oxidation; in plant cells similar structures, called chloroplasts, extract energy from sunlight by photosynthesis. There is evidence that much of the energy conversion in mitochondria takes place in specialized regions or elementary particles,[6] roughly 80 Å in diameter and located

[6] D. E. Green, *Sci. Am. 210*, 63 (1964).

in ordered arrays on these membranous organelles (Fig. 21.4).

The granular lysosomes contain hydrolytic "digestive enzymes," including much of the acid phosphatase contained within cells. The endoplasmic reticulum is a network of membranes, a "cytoskeleton." It ranges from smooth to granular as seen in the electron microscope, and the granules (called ribosomes) contain large amounts of ribonucleic acid (RNA), active in the synthesis of protein (Fig. 21.5).[5] It has been suggested

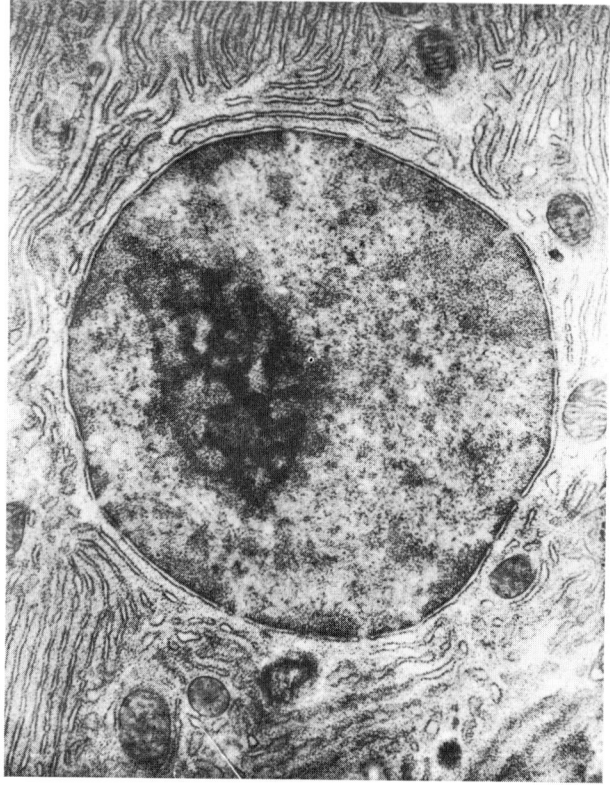

Fig. 21.5. Cell nucleus with prominent nucleolus. Membraneous endoplasmic structure in cytoplasm. (From Brachet.[5])

that the endoplasmic reticulum might serve as canaliculi for the transport of materials from the external medium into the cytoplasm and even into the nucleus.

The centrosome contains a pair of structures called centrioles, which in animal cells form the poles of the apparatus that separates two duplicate sets of chromosomes during cell division (cf. Chap. 22). Centrioles are not present in plant cells. The Golgi apparatus (or complex) is often located near the centrosome and represents an accumulation of smooth membranes and vesicles invariably implicated in the synthesis of secretions and the formation of new membranes.

Other cytoplasmic inclusions are variously described

as vacuoles, fat bodies, secretion granules, pigment granules, crystals, etc. Microtubules 200 to 270 Å in diameter have been stated as being present in many animal and plant cells subjacent and tangential to the outer cell membrane. These "tubular" structures (up to several microns in length) appear to have an inner core of different consistency from their outer walls.

The outer cell surface (plasmalemma, cell- or plasmamembrane) encloses the entire cell. It is universally present and takes part in the regulation of the exchange of energy and movement of material between cell and outside. Although most membranous structures appear similar in electron microscopy, their thickness varies approximately between 70 and 130 Å, and they differ enormously in terms of function.

The cytoplasmic matrix

Many structures undoubtedly have not yet been seen by light or electron microscopy, detected by microspectrophotometry, or isolated by ultracentrifugation or other techniques. Among the least understood is the cytoplasmic matrix (ground substance) which is in intimate relationship and balance with the cytoplasmic organelles and inclusions. This matrix has been pictured as a jellylike non-Newtonian liquid, as a reversible sol-gel system, or as a fluid matrix with protein macromolecules linked to one another and to lipids and polysaccharides. These complexes are assumed to be capable of aggregation into fibrils (diameter from 150 to 1000 Å), into tubules (100 to 300 Å across), and into broad membranes (40 to 150 Å thick) in the colloid-richer regions of the cytoplasm. The aggregations of molecules and their associations are assumed to be variable in space and time as the cell responds to external influences and internal energy generation and transduction.

Bretschneider[7] has suggested that 4 to 10 helical chains of protein macromolecules are interwoven in a lattice of units called "leptons." End-to-end linkages together with transverse cross-linkages are supposed to impose a strict regularity on the lepton matrix. Although evidence for such matrices is strong in muscle cells that contain ordered arrays of myofilaments and myofibrils, there is no strong evidence for the presence of such a periodic system within most animal or plant cells.

Frey-Wyssling[8] considered proteins as the structural elements of cytoplasm. These proteins are assumed to be loosely interconnected by junctions of a dynamic nature; i.e., the protein molecule may be reversibly joined or released.

Ling[9] prefers to view cytoplasm as a bonded lattice of protein macromolecules (unit cell ~20-Å diameter) that, together with water and salt ions, acts as a quite rigid fixed-charge system with relatively uniform spatial distribution of ionic sites. Ling writes, "Charge fixation increases the average degree of ion association through its effect on energy and entropy of dissociation. The ultimate state of charge fixation is visualized as a three-dimensional lattice-bearing fixed-charge system." Unlike ionic crystals, the counterion in the fixed-charge system retains considerable freedom of motion, and the spacings between charged sites are sufficiently great so that interaction between neighboring sites can be neglected. The protein-water-salt system is fundamentally a labile one, capable of cooperative changes from one metastable state to another, such as the folding of protein and so on. On the basis of this *liquid-crystal* model of cytoplasm, Ling proceeded to explore the properties of proteins, permeability, and diffusion within and into cells, contractile mechanisms leading to cell movement, and antigen-antibody reactions.

Undoubtedly these molecular models are inadequate in many ways. The principal criticism regarding this hypothetical composition of cytoplasm is that such highly ordered lattice systems have not been observed by electron microscopy. Nor do these models explain the separation, concentration, and association of membranous or filamentous units into specialized cellular organelles. The basic question is the degree of statistically ordered orientation and the presence of association colloids[10] in contrast to a more definite bonded organization.

Dynamic activities

To the static description of cells has to be added a picture of their activities. Complicated chemical processes take place continuously in all parts. Materials are transported by direct and indirect mechanisms to various regions, detectable as changes in chemical constituents and electrical activity, flow of cellular material, changes in cell morphology, and movements of cellular organelles or of the cell itself.

---

[7] L. M. Bretschneider, "The Fine Structure of Protoplasm," *Survey of Biological Progress*, Vol. 2, G. S. Avery, Jr., Editor-in-Chief, Academic Press, New York, 1952, p. 223.

[8] A. Frey-Wyssling, *Submicroscopic Morphology of Protoplasm*, Elsevier Publishing Co., New York, 1948.

[9] G. M. Ling, *A Physical Theory of the Living State*, Blaisdell, New York, 1962.

[10] H. L. Booij and H. G. Bungenberg de Jong, "Biocolloids and their Interactions," *Protoplasmatologia, Handbuch der Protoplasmaforschung*, Vol. 1, Springer-Verlag, Wien, 1956, p. 3.

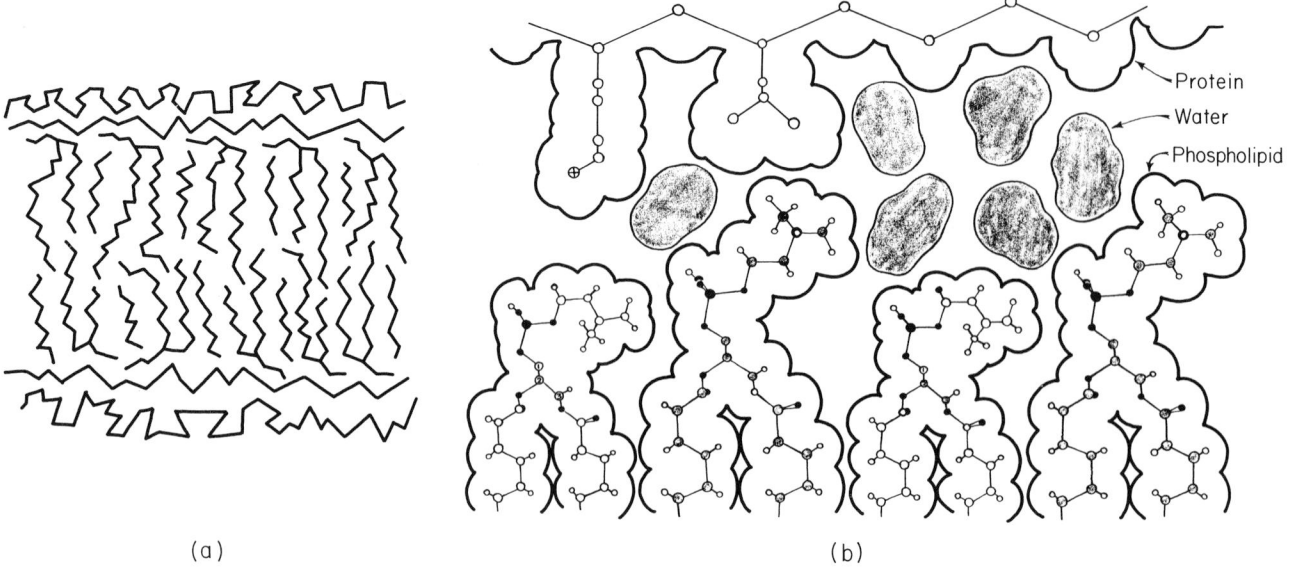

Fig. 21.6. (a) Bimolecular leaflet of phospholipids and associated adsorbed proteins; (b) diagrammatic steric association between lecithin and chain of amino acids. (After Vandenheuvel.[11])

Such dynamic activities are most dramatically illustrated in cells cultivated in tissue culture. Time-lapse photography shows incessant movements: cell locomotion, changes in shape and contour, motions of cell organelles such as mitochondria, etc. These activities point to continuous interaction with the substratum and with surfaces of adjacent cells. The full description must also include a study of intracellular movements under a broad spectrum of conditions, and any theory of cell organization must allow for highly localized changes in ionic content, macromolecular constitution, and energy transfer. It must provide for the reconstitution and dissolution of membranes and fibrils and predict intracellular movements and streaming consistent with observation. No all-inclusive theory of that kind has yet been formulated, but some narrow aspects have been put in theoretical and experimental perspective. For example, fairly precise experiments can be performed to demonstrate adjustments to the environment and a marked sensitivity of cells to the orientation and composition of molecular arrays with which they are in contact.

The remainder of this chapter will deal with the surface of some vertebrate cells for which we have a physical model, consistent experimental data, and considerable speculation. Special attention will be given to the interaction between the cell surface and some well-defined monomolecular or multimolecular interfaces.

The cell surface

The cell surface is an interfacial region separating two phases. This transition region between two phases can be viewed as a separate phase of surfactant molecules concentrated within the interface, which may assume highly ordered configurations. Figure 21.6a is the well-known Danielli-Harvey model of the cell surface: a bimolecular leaflet of phospholipids with primarily adsorbed coats of proteins arranged transversely to the phospholipid components. X-ray diffraction, electron and polarization microscopy, as well as less direct techniques, have confirmed several aspects of this picture, especially the arrangement of phospholipids. Molecular models of the bimolecular leaflet, based on the three-dimensional configuration of the molecules (Fig. 21.6b), suggests steric fitting between lecithin molecules and a protein coat.[11] Depending on steric matching and complementarity there will be a variable number of vacancies that can be occupied by fairly tightly bound water molecules. The degree of dissociation of these water molecules is unknown, but a certain amount should occur by interaction with ionizable groups of the protein and phosphatide. The arrangement of Fig. 21.6b is hypothetical and over-simplified. More complicated models include mixed leaflets of various phospholipids and of phospholipids and sterols. Packing arrangements, the amount of included water, and the flexibility and stability of the structure can then be altered in countless ways. At the same time the interaction between macromolecular coat and leaflet can undergo considerable changes.

Packing and cell dynamics

It is difficult to determine the stability of various

[11] F. A. Vandenheuvel, J. Am. Oil Chemists Soc. 40, 455 (1963).

arrangements of surface molecules. Estimates have been made of the interaction energies between certain molecular groupings. For example, according to Salem[12] the interaction energies between the charged polar groups of the phospholipids are approximately 4.1 kcal/mole at 5-Å spacings; polarization energies are negligible. Between neutral, nonpolar fatty acid chains, the interaction energies from London-van der Waals forces are approximately 0.4 kcal/mole per CH_2 unit at 5-Å spacing. Since these are additive, they can become appreciable between the long-chain fatty acids (thermal energy ≈ 0.62 kcal/mole).

At closest packing, saturated phospholipid molecules will occupy an area of approximately 40 Å^2 per molecule. Evidence for such close packing is given in Fig. 21.7, which illustrates the relation between surface

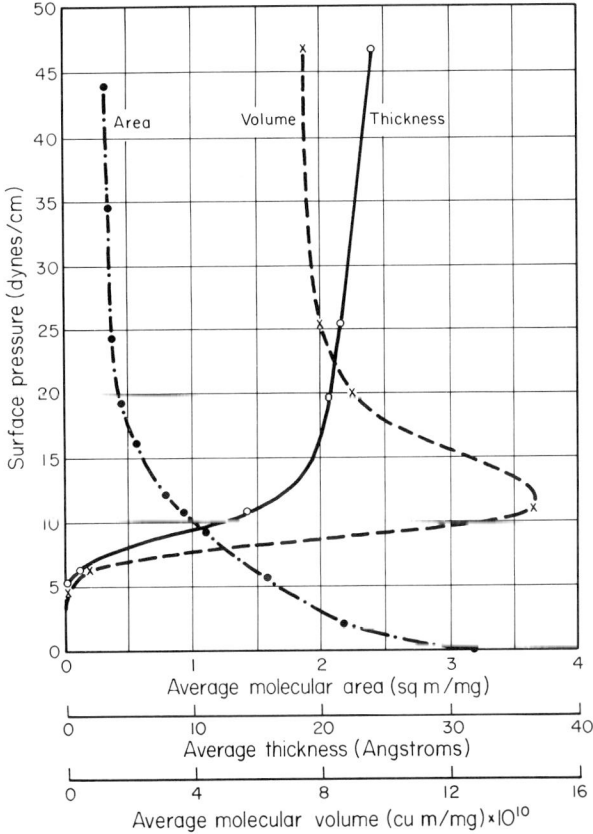

Fig. 21.7. Surface pressure versus average area per molecule and average thickness for synthetic lecithin at air-liquid interface.

pressure versus area, volume, and average thickness for synthetic lecithin molecules placed at an air-water interface. Most natural phospholipids, however, have an unsaturated fatty acid chain in the β position, and the double bonds are in *cis* configuration. Condensed leaflets of natural phospholipids should therefore have

[12] L. Salem, *Can. J. Biochem. and Physiol.* **40**, 1287 (1962).

less cohesion due to steric hindrance between neighboring molecules. This would give them a more fluid state.

Time-lapse observations on the behavior of tissue-cultured cells dramatically impress one with this surface fluidity. Cell surfaces in contact with an appropriate substratum expand, plicate, invaginate, and form ruffles, microextensions and intermittent contacts. The difficulties in analyzing the dynamic behavior of these surfaces lie not only in developing pertinent experimental methods but also in formulating the problem with sufficient precision. In the subsequent discussion some experimental procedures employed by the author are used as examples of a few techniques developed to investigate some characteristics of bimolecular leaflets and their macromolecular coats.

Cells on mono- and multilayer substrates

One method explores the interaction between cell surfaces and monomolecular films. In brief, one constructs a film of known molecular composition, orientation, and structure, places living cells in contact with this synthetic film, and observes their behavior and activities during a controlled exposure. After removal of the cells the synthetic surface is examined for alterations in structure and orientation. The film thus mirrors characteristics of cell-surface and substrate operation during interaction.

Methods for working with monomolecular films were developed by Langmuir. With a Langmuir trough it is possible to form at air-water or oil-water interfaces monolayers of molecules that have hydrophobic and hydrophilic segments. The hydrophilic (polar) end will be directed toward the water phase; the hydrophobic (nonpolar) end toward the air or hydrocarbon phase. Orientation and packing of these molecules are adjustable by compression of monolayers with a mechanical barrier.

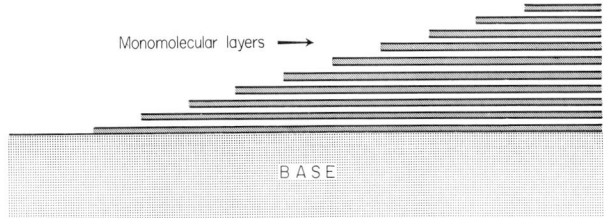

Fig. 21.8. Terracing of fatty acid monolayers on base slide.

At sufficiently high compression (especially at air-water surfaces) a sequence of monolayers can be transferred to solid surfaces by successive immersion and emersion. Figure 21.8 illustrates such a build-up of fatty acid layers on solid surfaces. The initial monolayer is 25 to 30 Å thick, depending on the length of the fatty acid molecule.

If various strains of cells are maintained in contact with such substrata, one observes in a medium free of large adsorbable molecules the gradual deposition on the solid surface of a material ("microexudate") approximately 40 Å thick.[13] Figure 21.9 gives deposition-

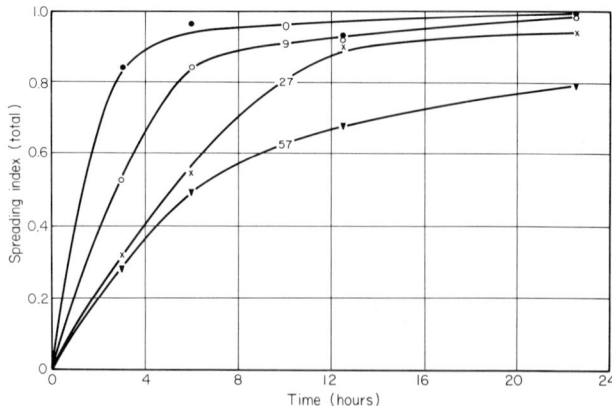

Fig. 21.10. Time course of cell spreading as a function of number of subjacent monomolecular layers of stearic acid and barium stearate.

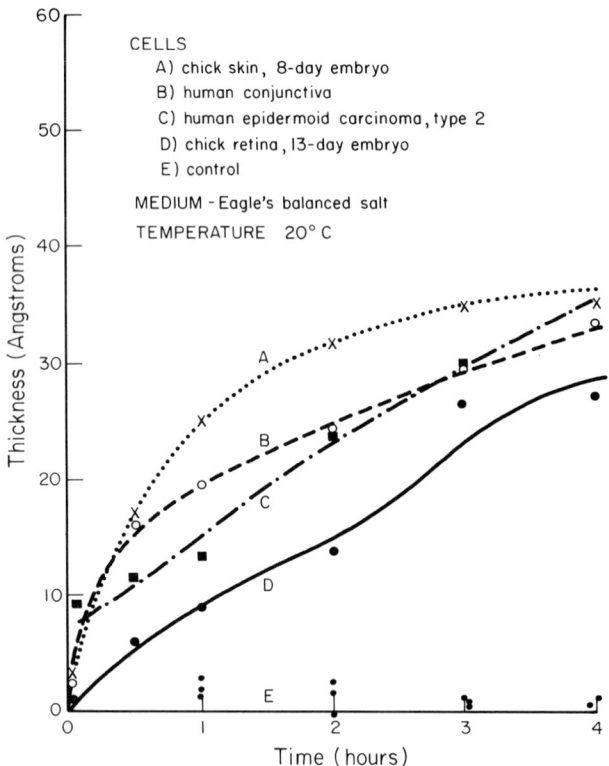

Fig. 21.9. Time course for formation of microexudate in vicinity of cells on a chromed-glass substratum.

is increased (Fig. 21.10).[15] The reason for this behavior is not fully clear, but the cell surface proves highly sensitive in its activities to the molecular structure, thickness, and orientation of contact surfaces. Figure 21.11 demonstrates that the degree of cellular attach-

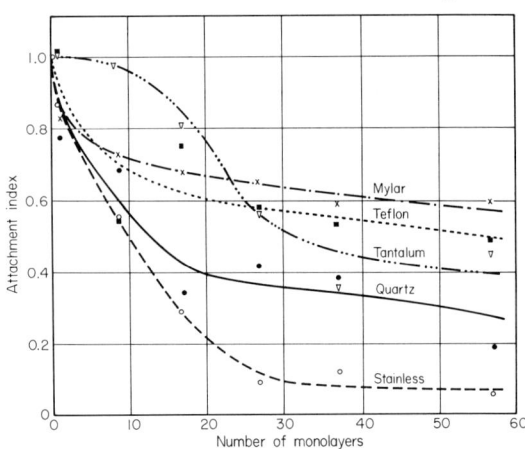

Fig. 21.11. Cell attachment as a function of number of subjacent monomolecular layers and type of base slide.

time characteristics. The material, not yet identified, appears to stem from three sources: cellular secretions, detached portions of the cell surface, and portions of the outer coat of cells transferred to the solid surface. The last-named material most likely accounts for most of the deposit. If adsorbable molecules are added to the culture medium, little microexudate is detectable.

It is possible to observe the behavior of cells in contact with multilayers of fatty acids by measuring the rate of attachment or spreading of cells.[14] The rate of cellular spreading is reduced on stearic acid when the number of monolayers (i.e., the substratum thickness)

ment depends not only on the number of monomolecular films subjacent to the cell surface but also on the base to which the film was transferred.

These observations led to the prediction that if the molecular carpet of fatty acids were in some regions thinner than in others, the cells might be entrapped in the thinner regions as a result of increased adhesivity. In this manner surface activity can indeed be oriented, as Fig. 21.12 certifies. A random population of cells was maintained for four hours on a substratum of behenic acid. In the region of the cross the substratum

[13] M. D. Rosenberg, *Biophys. J.* **1**, 137 (1960).

[14] The attachment index is the percentage of a given cell population in contact with a substratum that is not detached by a gravitational or shearing force. The spreading index is a measure of the flattening and spreading of individual cells on the substratum.

[15] M. D. Rosenberg, *Proc. Natl. Acad. Sci. U.S.* **48**, 1342 (1962).

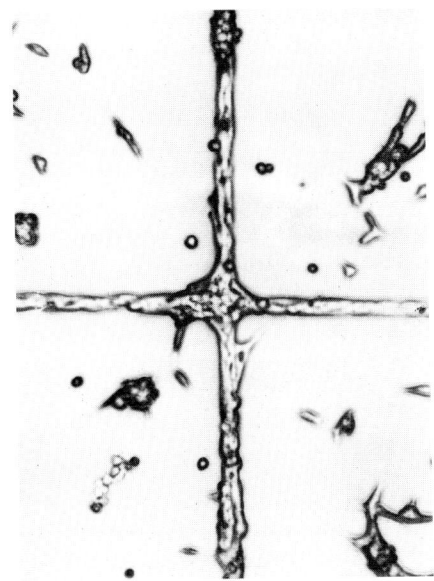

Fig. 21.12. Alignment of tissue culture cells within a trough cut in multilayers of behenic acid.

is 180 Å thinner. The cells, which are roughly 200,000 Å in diameter, have outlined this region and oriented along the trough.

The interactions between cell surfaces and monomolecular films at liquid-solid interfaces can serve in other ways. For example, it is possible to determine the activity of enzymes adsorbed to cell surfaces. If a proteolytic enzyme such as trypsin is adsorbed to the surfaces of tissue-cultured cells and the cells are then placed in contact with a film of protein, the digestion of the protein film is comparable with that which occurs when the proteolytic enzyme is applied in an equal amount directly to the protein film (Fig. 21.13).[13]

Experiments on liquid-liquid interfaces

Additional information can be obtained by observing cells and their surface activities *in vitro* at liquid-liquid rather than liquid-solid interfaces. The use of liquid interfaces presumably avoids molecular inhomogeneities generally present at surfaces of solids, and should simulate a situation closer to reality.

Strains of human conjunctival cells, cultured *in vitro*, or suspensions of freshly dissociated embryonic chick-liver cells were maintained in contact with interfaces separating two liquid phases. The upper liquid was hydrophilic and consisted of a synthetic balanced salt—amino acid solution with or without a supplement of 10 percent fetal calf serum. The lower liquid phase, hydrophobic or lyophilic, consisted of either siliconated or fluorinated hydrocarbons. The experimental setup (Fig. 21.14) consisted of a high-density oil phase in a shallow trough whose inner surfaces were hydrophobic. This trough was placed

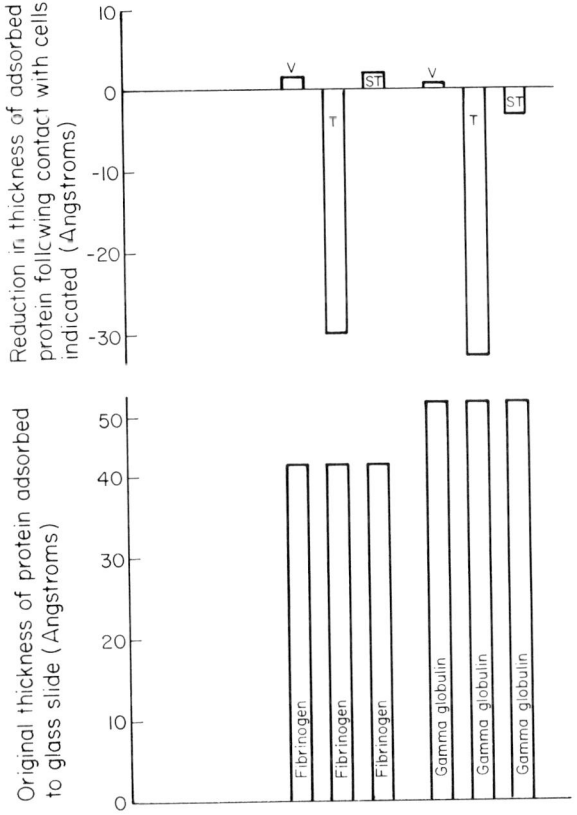

Fig. 21.13. Change in average thickness of various adsorbed substrates due to presence of trypsin on surfaces of cells.

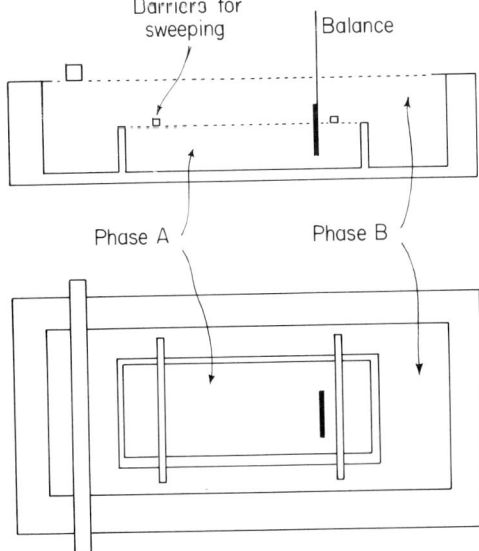

Fig. 21.14. Trough-within-trough technique for the study of liquid-liquid interfaces.

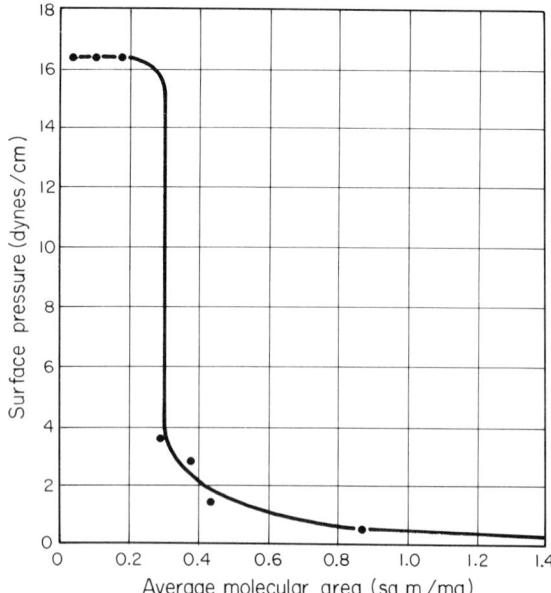

Fig. 21.15. Interfacial tension versus average area per molecule of synthetic lecithin at liquid-fluorocarbon interface.

in an outer one filled with saline solution whose inner surfaces were hydrophilic. By means of barriers one could sweep and clean the saline-oil interface and measure its surface pressure. The atmosphere was saturated with water vapor and the temperature kept constant.

Many types of molecules can be injected into the interface and their pressure-area relationship measured. The interactions between cells and soluble or insoluble surfactants of interfacial films are observable as functions of the packing, composition, and orientation of the interfacial region. Figure 21.15 gives the pressure-area curve for synthetic DL-dipalmitoyl lecithin at this liquid-liquid interface. The shape of the curve indicates that at low packing density the lecithin molecules are highly compressible, possibly associated with one another in the form of islands or pockets. With increased packing the surface pressure rises rapidly, suggestive of a highly condensed solid film. At this point the experimental average area per molecule is 38 $Å^2$; theoretical calculation gives a value close to 40 $Å^2$. In a sense, these phosphatide monolayers together with adsorbed proteins and other components of the medium simulate one half of a cell surface (cf. Fig. 21.6). The advantage of the technique is that the molecular structure, orientation, and composition of this synthetic surface are controllable.

If cells are allowed to contact the liquid-liquid interface where no lecithin is present, the cells spread as well as they would on a clean solid surface (Fig. 21.16a). This striking behavioral pattern raises interesting questions regarding the cell surface. On an isotropic interface the principal cause of surface movement should be intrinsic metabolic processes; one must look toward the possibility that cytoplasmic structures subjacent to cell membranes provide reversible rigidities for the extension and retraction of pseudopods.

When the same cells contact phospholipids (which seem to be associated in small islands), the behavioral pattern of the cells is intermediate between aggregation and spreading (Fig. 21.16b). A few flattened cells can be found, while the majority interact poorly with the interface and remain spherical in shape. If the same cells are placed in contact with a highly condensed monolayer of lecithin molecules at the interface, the interaction between cell and substratum has been so

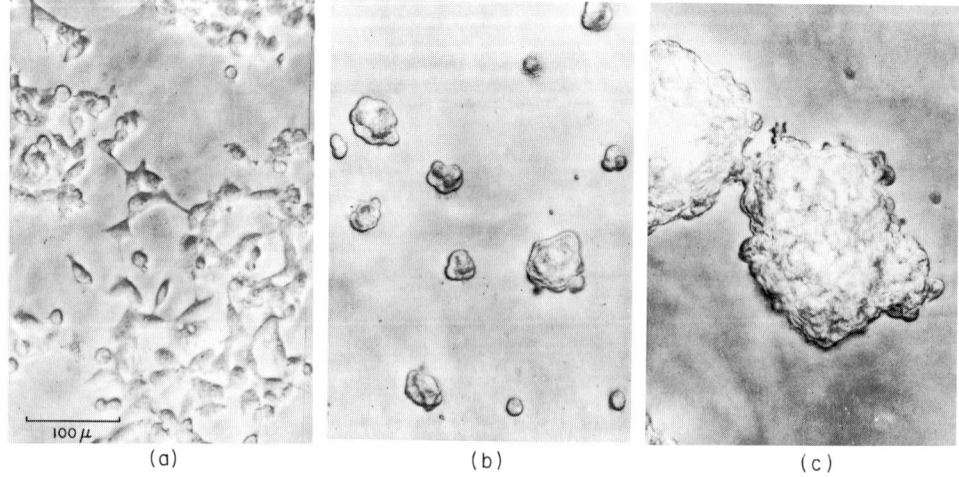

Fig. 21.16. Spreading of tissue culture cells at (a) liquid-fluorocarbon interface; (b) liquid-fluorocarbon interface in presence of partially condensed monolayer of synthetic lecithin; (c) liquid-fluorocarbon interface in presence of fully condensed monolayer of synthetic lecithin.

greatly reduced relative to cell-cell interactions that the cells aggregate into large clusters rather than to spread on the interface (Fig. 21.16c). Thus the cell surface performs a bioassay of the molecular structure of the artificial membrane with which it is in contact.

A few other observations obtained with this technique might be summarized: Lecithin in solution (in equal quantities) does not affect the behavior of cells. Phosphatidyl-l-ethanolamine at the interface has an effect similar to lecithin. Phosphatidyl-l-serine with its additional carboxyl group acts similarly to loosely packed lecithin; stearic acid has no noticeable influence. Transfer experiments indicate that the lecithin monolayer is stable and does not alter the cell surface *per se*. The adsorption of macromolecules from the nutrient medium to the monolayer seems markedly influenced by the molecule type and its charge. Alterations in the charge and type of phospholipids in cell surfaces appear to lead to alterations in the orientation and foldings of macromolecular coats and vice versa.

To test this hypothesis further, the following experiments have been tried: Crystalline ovalbumin was injected into the liquid-liquid interface in amounts sufficient to form a monolayer. Compression of the protein led to no alteration in interfacial pressure, showing that the protein remained globular and went into solution. Adsorption studies were then carried out by adding fixed amounts of protein to the saline phase. For each aliquot of protein added to the bulk phase the interfacial pressure rapidly increased (in a matter of minutes) to an equilibrium value, consistent with Gibbs adsorption theory. The kinetics of adsorption were studied as a function of concentration. Figure 21.17 gives the equilibrium value of the interfacial pressure as function of bulk concentration. The dependence starts linear, but at higher concentrations there is a region of inflection. Mechanical compression of molecules adsorbed from a solution of low concentration produced no change in interfacial pressure, while molecules adsorbed from a solution of relatively high concentration produced marked changes in interfacial pressure. It appears that at these bulk concentrations some of the protein is unfolded (surface-denatured) or in a special phase at the interface.

That the protein monolayer has undergone changes at the surface can be shown by replacing all of the bulk saline phase with fresh saline free of protein. The presence of noninsoluble surface protein is demonstrable by mechanical compression and measurement of the concomitant changes in interfacial pressure. In the presence of a condensed layer of synthetic lecithin, increased concentrations of ovalbumin in the saline bulk phase result in no changes in interfacial pressure.

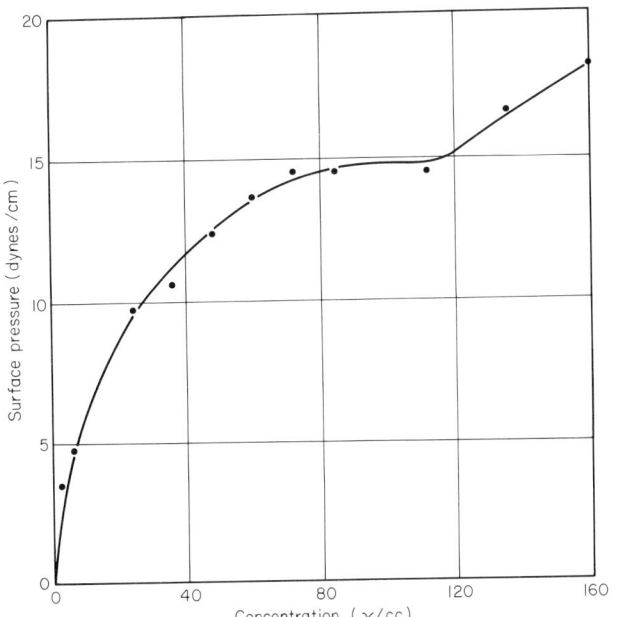

Fig. 21.17. Changes in interfacial tension due to adsorption of ovalbumin from solutions of different concentrations.

This indicates that the interaction between the condensed monolayer of synthetic lecithin and the ovalbumin is not very strong.

These and other results allow the following tentative conclusions: Cell surfaces have a capacity for great variations in shape and configuration. They vary physically and most likely structurally from one region to another and as a function of time. Changes in type of orientation or packing of membrane components may alter the structure of adjacent macromolecules. Surfaces can expand and contract through the formation of new surfaces or the emergence of surfaces previously invaginated within cytoplasm. The surface is markedly sensitive to the molecular configurations of overlying and contiguous structures and reacts through alterations in the outer macromolecular coat. Cell surfaces must be looked upon as consisting of a bimolecular membrane, with outer macromolecular coats both primarily and secondarily adsorbed, and an inner ectoplasmic structure of considerable thickness.

Possible causes of cellular motion and surface movements

Within the ectoplasm (subjacent and contiguous with the cell membrane) several physical changes can cause contractions. Perhaps the most familiar is a sol-gel transformation, a mechanism Marsland[16] has favored as the cause of protoplasmic streaming. The

---

[16] D. Marsland, *Intern. Rev. Cytol.* 5, 199 (1956).

ectoplasmic matrix is pictured as a complex intermesh of fibrillar units of macromolecules in a gel-like structure that is potentially contractile. Changes in volume of the gel result from forcible foldings of the proteins with maintenance of intermolecular linkages. Contraction of a gel of this type is accompanied by the desorption of heat, unlike the reverse process that occurs in most gels. It is assumed that the movement of the surface region is caused by contraction of ectoplasm, an ATP-dependent process (i.e., dependent on the presence of high-energy phosphate bonds).

Other investigators have described formed organelles in the cortical ectoplasm and elsewhere in the cytoplasm that may provide the motile power for movement of the ectoplasm as well as cytoplasmic flow. Leadbetter and Porter[17] have ascribed to cortical microtubules the development and orientation of displacement forces. Whether flow of ectoplasm as part of cytoplasmic flow is a product of undulating motions of the tubules or a result of interactions at the surface of stationary tubules could not be decided. Their proposal certainly merits attention, especially since similar structures are now being observed by electron microscopy in numerous cell types and are frequently subjacent to and in tangential arrays with the plasmalemma (Fig. 21.18). As described earlier, these microtubules are 230 to 270 Å in diameter, of undetermined length, with walls about 70 Å thick and a lumen 100 Å in diameter. In a sense the word tubule may be misleading because the inner core may be as structured as the walls appear to be.

In numerous other instances cytoplasmic tubules are found. Among these are the spindle fibers of the mitotic apparatus in the giant amoeba (*Pelomyxa carolinensis*), the spindle fibers of dividing cells of sea-urchin embryos, the filamenture of cilium and flagellum, the marginal band of nucleated red blood cells, the developing spermatid, the cones of microspikes, certain neurotubular structures, and so forth. Jarosch[18] has described some fibrillar elements in extruded protoplasm of plant cells, capable of wavelike motion, and postulated that such structures are composed of macromolecules in multiple-coiled arrangements like twisted springs or helices wound parallel to one another. A torsional force applied to either end can cause considerable shortening and lengthening, or contractions and expansions, accompanied by corresponding movement of cytoplasmic or ectoplasmic matrix. For plant cells,

[17] M. C. Leadbetter and K. R. Porter, *J. Cell Biol.* **19**, 239 (1963).
[18] R. Jarosch, "Screw-Mechanical Basis of Protoplasmic Movement," *Primitive Motile Systems in Cell Biology*, R. D. Allen and N. Kamiya, Eds., Academic Press, New York, 1964, p. 599.

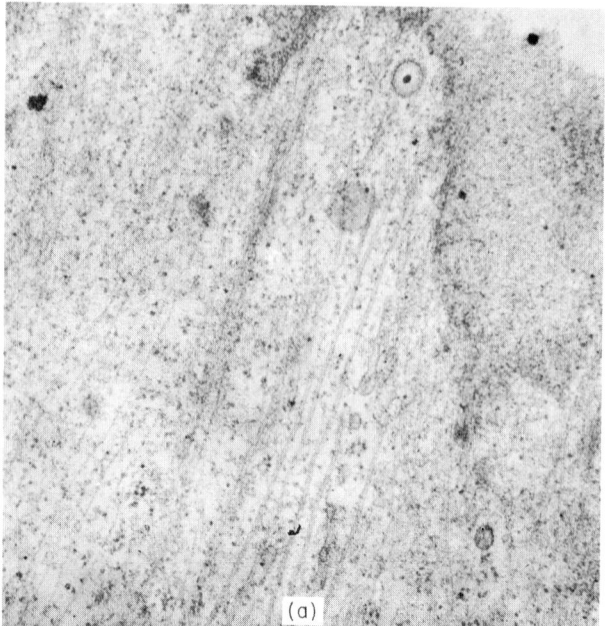

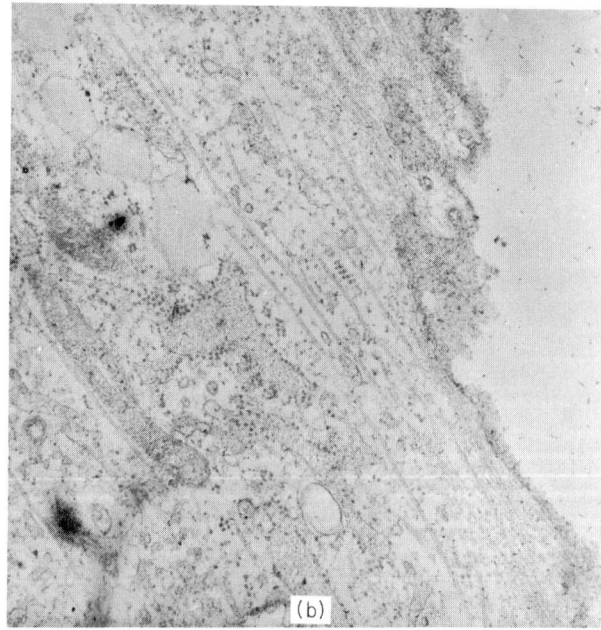

Fig. 21.18. (*a*) Microtubular and (*b*) microfilamentous elements in cytoplasm. (Courtesy of Dr. A. C. Taylor, the Rockefeller Institute.)

Kamiya[19] has presented evidence that displacement forces are generated by an ATP-sensitive system of fibrillar units at the interface between endoplasm and cortical gel.

About the development of motive forces by these tubules little is known. The contraction or expansion

[19] N. Kamiya, "The Motive Force of Endoplasmic Streaming in the Ameba," *ibid.*, p. 257.

of some elements depends on the presence of adenosine triphosphates (ATP) which supply high-energy chemical bonds. The work of Katchalsky[20] on the mechanochemistry of fibers may be pertinent; he demonstrated that the stretching or contraction of collagen fiber can be measured as changes in the chemical potential of the system. The conversion between mechanical and chemical energy was found to pass through extremely sharp maxima, depending on the concentration of the salt solution surrounding the fiber. These maxima occurred at the melting point of the collagen, as though the collagen were undergoing maximum changes in length by melting and recrystallization.

This phenomenon of reversible melting and recrystallization has not been adequately studied, especially in relation to biological systems. In a liquid polymer, crystallization consists of the alignment of fibrils in orderly array. Apparently the transition between the crystalline and noncrystalline phases can be made quite rapidly (in less than 10 sec in the case of collagen) and as a cooperative process could lead to considerable movement of surrounding ground substance. Surface loci at the inner portion of the cell membrane could provide large numbers of nucleation sites to instigate the formation of lamellar micelles; strain imperfections within the ectoplasmic matrix would encourage the formation of cylindrical micelles.

There is a small amount of evidence that crystallization of some macromolecular units may be associated with surface movements. Taylor and Robbins[21] have described microspikes extending from the surfaces of cells. These projections are roughly 1000 Å in diameter, form very rapidly, may become several microns long and have very sharp bends, fold upwardly to merge once again with the cell surface, and at times can be shown by electron microscopy to have a crystallinelike core. Thus there may be semicrystalline, transient structures within cells, particularly in the ectoplasmic region, that can cause movements of cell surface and of cytoplasm.

Interfacial dynamics

Turning to the cell-surface region in general, especially the cell membrane and immediately adjacent macromolecular coats, Kavanau[22] has recently made the attractive suggestion that metaphase changes in the ultrastructure of the membranes lead to the forceful removal of water molecules in certain regions. The

[20] A. Katchalsky, Weitzmann Inst., Israel, private communication.
[21] A. C. Taylor and E. Robbins, *Develop. Biol.* 7, 660 (1963).
[22] J. L. Kavanau, *ibid.* 7, 22 (1963).

pumping out of low-molecular-weight ions is assumed to provide a propulsive force that can bring about movement of the membrane. Whether such transitions are significant is not known, but the proposal warrants investigation.

It is well known that changes in chemical potential at certain regions of an interface can result in mechanical movement of the interface. Such effects were first described in 1855 by the British scientist James Thomson (older brother of Lord Kelvin) in an article "On Certain Curious Motions Observable at the Surfaces of Wine and Other Alcoholic Liquors." Thomson inferred that surface-tension-driven flow was responsible for these motions. Considerable interest in these phenomena has re-emerged in recent years, especially among chemical engineers. The phenomena are designated as Plateau-Gibbs-Marangoni effects or "interfacial dynamics" and consist of two related surface effects: a movement *in* a fluid interface due to local variations of interfacial tension caused by differences in composition, temperature, electric charge or pressure, and a movement *of* the interface by differential deformation, extension, or contraction (Fig. 21.19).

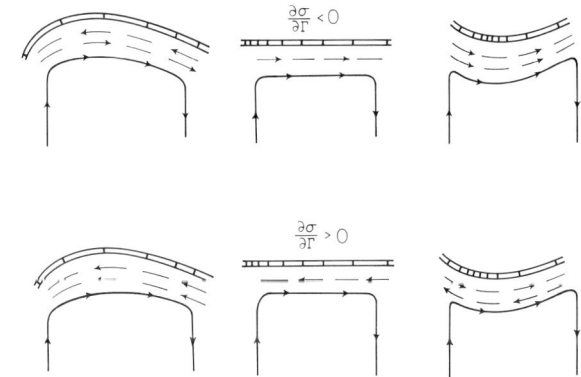

Fig. 21.19. Vector fields resulting from surface tension gradients at an interface.

The model system is a mobile interfacial region separating two liquid phases. Several interdependent and interacting processes are assumed to take place (e.g., chemical reactions, transformations, selective adsorption, or desorption in localized areas). These heterogeneous reactions produce differences in interfacial tensions which in turn lead to tangential stresses and movements of materials within the interface. The rates of these processes are controlled by the relative diffusion rates of reactants to the interfacial region, chemical and physical reaction rates, and the rate of movement within the interface. The direction of movement is determined by the sign of the dependence of interfacial tension on concentration. In addition, changes in shape of the interface are associated with the development of normal stress components (capillary pressure) and concentration gradients of the surface-active materials as interfacial regions are compacted and expanded. Concurrent alterations in the tangential and normal stress components result. The direction of these stresses will depend on whether the surface-active materials tend to decrease or increase interfacial tension.

The movements of materials to the interfacial region will be governed by molecular diffusion due to concentration differences and convective transport by the bulk solution. Excellent studies and reviews have been provided by Scriven[23] and Levich.[24] These effects can be viewed as a form of mechanochemistry similar to that described by Katchalsky for fibers, where an alteration in chemical potential produces mechanical motion and vice versa.

Whether similar processes take place at biological interfaces is not known. However, protein components of biological surfaces can provide localized sites for chemical reactions or transformations, or—in their interactions with the polar end groups of the phospholipids and other molecular constituents of the membranous leaflet—can be selectively adsorbed and desorbed. Local variations in interfacial tension will arise from such changes in composition or concentration, but many other factors can cause similar disturbances (e.g., changes in electric charge distribution, variation in surface temperature and pH, nonuniform flow of the bulk fluid, etc.).

Interdependent transductions in the interfacial structure may provide an inherent degree of self-regulation. For example, an excessive change in interfacial tension in one region may be reduced by corrective forces that reduce the diffusion of reactants to that region. To reach a deeper understanding, an analysis in terms of macroscopic parameters such as equilibrium interfacial tension, surface dilatational viscosity, surface density, and surface shear viscosity must be complemented with a study of the molecular mechanics of interfaces.

[23] L. E. Scriven, *Chem. Eng. Sci. 12*, 98 (1960).
[24] V. G. Levich, *Physicochemical Hydrodynamics*, Prentice-Hall, Englewood Cliffs, New Jersey, 1962.

The interactions between the phospholipid-sterol complex of the bimolecular leaflet and the protein and other macromolecular constituents in contact with this membrane may be a starting point, since they depend not only on the individual molecular species involved but also on their packing arrangements, orientations, charge distributions, ionic atmospheres, and so forth. The stability of the surface is possibly related to a steric complementarity between the protein coats and the phospholipids of the leaflet. A perturbation of the order of 10 percent in the packing arrangement of phospholipids in some regions could lead to rarefactions elsewhere and the propagation of two-dimensional hypersonic waves. The maintenance of such waves would require energetic priming sources, but dissipation might be low. Associated foldings and unfoldings of contiguous macromolecular coats would occur. The net effect would be reversible and coordinated transformations in molecular packing and orientation *in* the surface and movements *of* the surface. Suggestions such as these have recently been discussed by Weiss.[25] Further complexities can be introduced, e.g., permeability changes associated with changes in molecular groupings, turnover rates of portions of the phospholipids, introduction into the bimolecular leaflet of additional molecules from an ectoplasmic molecular pool, and so on.

Outlook

Whether investigators shall ever be able to synthesize the living cell from its component parts is doubtful. Weiss[26] in describing the "cell as a unit" has succinctly stated the problem as follows:

We have arrived at last at a point which comes rather close to what might be defined as "molecular control of cellular activity," only to discover that the "controlling" molecules have themselves acquired their specific configurations, which are the key to their power of control by virtue of their membership in the population of an organized cell, hence under "cellular control."

[25] P. Weiss, Rockefeller Institute, private communication.
[26] P. Weiss, *J. Theoret. Biol. 5*, 389 (1963).

22 · STUDIES OF CELL DIVISION WITH AN IMPROVED POLARIZING MICROSCOPE *†

Shinya Inoué

Spindle fibers and mitosis — Improvement of the polarizing microscope — Verification of spindle-fiber formation

Spindle fibers and mitosis

The mechanisms of cell division and the mitotic movement of chromosomes have puzzled biologists for a long time. By the turn of the century the structures of the gene-bearing chromosomes and of the mitotic spindle, allegedly responsible for the orderly segregation of the chromosomes, were clearly described following Abbe's perfection of the compound microscope and the introduction of techniques for tissue fixation, sectioning and staining.[1]

The fibrous organization of the spindle, so clearly visible (Fig. 22.1), was soon to be challenged, however, as representing artifacts of the chemical treatment employed. The spindle fibers generally could not be seen in healthy living cells (Fig. 22.2), and destruction of the spindle fibers by microsurgery did not prevent subsequent movement of the chromosomes to the spindle poles. As elegantly reviewed by Schrader,[2] several decades of controversy ensued regarding the nature of the spindle fibers and the mechanism of chromosome movement.

In 1937, Schmidt[3] discovered on sea-urchin eggs under a sensitive polarizing microscope that the whole spindle in these living cells was birefringent, with the higher index of refraction parallel to the spindle axis. He also noted a drop of the birefringence during anaphase separation of the chromosomes. These findings opened the possibility of visualizing the behavior of the spindle-fiber components during mitosis, provided the fibers do in fact exist, and probing into the molecular mechanism of mitosis. The change in birefringence could indicate, as Schmidt[4] pointed out, a folding of the polypeptide chains of the protein making up the spindle fibers.

The regular petrographic microscopes available at that time lacked the resolution and sensitivity required for such studies; attempts to improve upon Schmidt's observations were not successful.[5–7] We therefore

---

* Dedicated to Professor W. J. Schmidt, father of biological polarization microscopy, on his 80th birthday, February 12, 1964.

† Supported in part by grants from National Science Foundation (G19487) and National Cancer Institute, U.S. Public Health Service (CA 04552).

[1] E. B. Wilson, *The Cell in Development and Heredity*, 3rd ed., Macmillan Co., New York, 1928.

[2] F. Schrader, *Mitosis. The Movements of Chromosomes in Cell Division*, 2nd. ed., Columbia University Press, New York, 1953.

[3] W. J. Schmidt, *Die Doppelbrechung von Karyoplasma, Zytoplasma und Metaplasma*, Protoplasma-Monographien, Vol. 11, Verlag von Gebrüder Borntraeger, Berlin, 1937.

[4] W. J. Schmidt, *Chromosoma 1*, 253 (1939); *Ergeb. Physiol. 44*, 27 (1941).

[5] A. F. Hughes and M. M. Swann, *J. Exptl. Biol. 25*, 45 (1948).

[6] S. Inoué and K. Dan, *J. Morphol. 89*, 423 (1951).

[7] M. M. Swann, *J. Exptl. Biol. 28*, 434 (1951).

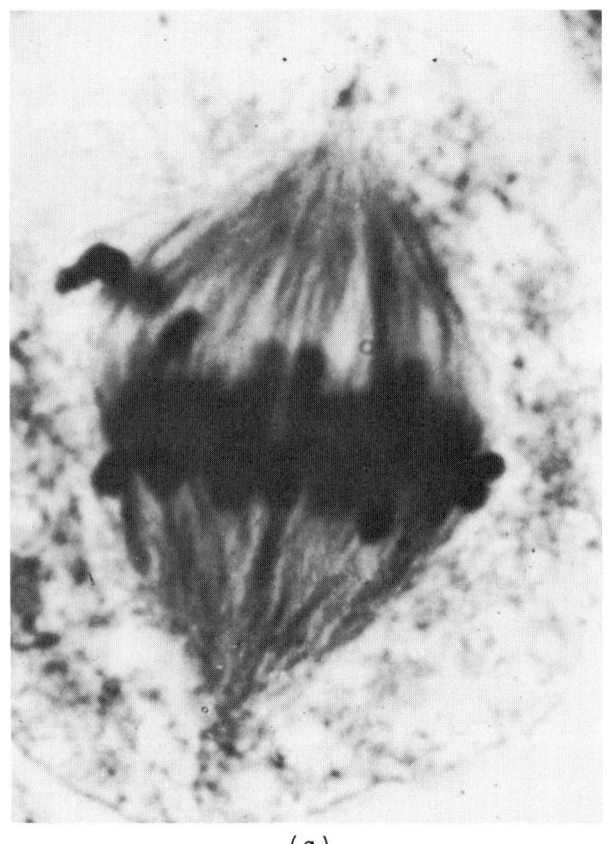

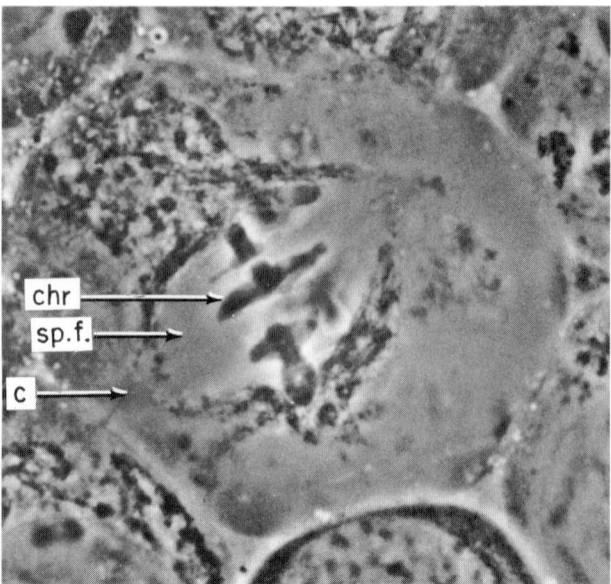

Fig. 22.2. Living grasshopper (*Chloealis genicularibus*) spermatocyte, showing chromosomes (*chr*) and spindle outline with phase-contrast microscope. Half-spindles appear empty and continuous, and spindle fibers (*sp. f.*) are not visible. Expected position of a centriole (*c*). (Courtesy, Prof. K. Shimakura, Faculty of Agriculture, Hokkaido University, Sapporo, Japan.)

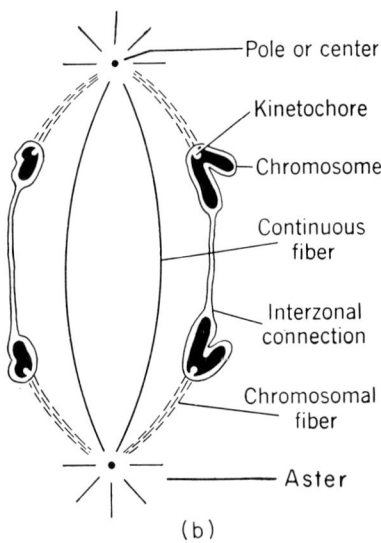

Fig. 22.1. (a) Division figure (in side view) from first meiotic metaphase of a pollen mother cell of wheat (*Triticum æstivum* × *turgidum*). Fixation: Acetic alcohol 2 minutes, followed by Müntzing's modification of Navashin's fixative for 2 years. Microtome section stained with combination of acetocarmine and crystal violet. (Courtesy, Dr. G. Östergren, University of Lund, Sweden.) (b) Schematic diagram of an animal-cell mitotic spindle with centrioles. Plant cells generally lack centrioles and asters. Continuous and chromosomal spindle fibers are indicated (modified from Schrader[2]).

undertook a study to improve the polarizing microscope.

Improvement of the polarizing microscope

The ultimate limitation of conventional polarizing microscopes was found to lie in their low extinction at higher numerical apertures, a factor which made high resolution and high sensitivity incompatible.[8] This low extinction was caused by depolarization of light at lens surfaces, multiple reflection, and rotation of the polarization plane. The rotation arose from the differential reflection losses of the parallel and perpendicular vectors at each plane of incidence, except where the polarization azimuth was 0 or 90 degrees. This circumstance gave rise to a dark cross at the objective back aperture, resembling that from a uniaxial crystal between crossed polarizers.

The multiple reflection, and to some degree the rotation, could be reduced by low-reflection coatings. The residual rotation of up to 7 to 8 degrees for high-aperture lenses was removed by the introduction of a polarization rectifier consisting of (1) a half-wave birefringent plate oriented parallel to the polarizer axis, which reverses the sense of the rotation introduced by the lenses, and (2) a zero-power meniscus which introduces appropriate additional rotation to cancel the

[8] S. Inoué, *Exptl. Cell Research* 3, 199 (1952).

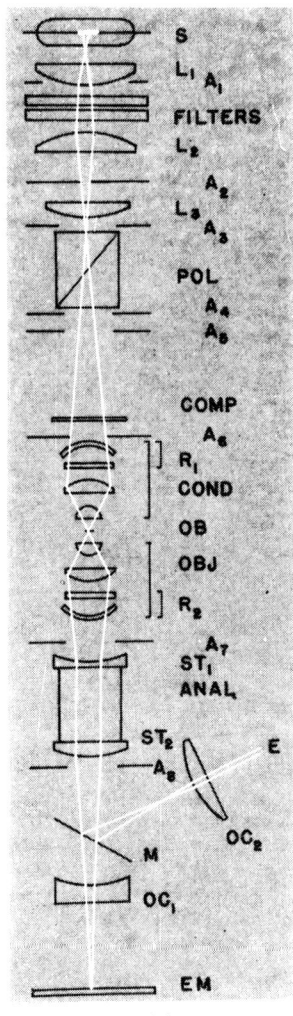

(a) (b)

Fig. 22.3. (a) Transilluminating polarizing microscope designed for maximum sensitivity and image quality. Inverted system with light source (S) on top and detectors (EM and E) at the bottom. Light from high-pressure mercury arc is filtered to provide monochromatic illumination 546 mµ and focused by L_1 and L_2 on pinhole aperture A_2. This restricts size of source image; after projection by L_3 it just covers the condenser aperture diaphragm A_6. The polarizing Glan-Thompson prism POL is placed behind stop A_4 away from condenser COND to prevent light scattered by polarizer from entering condenser. Half-shade plates are placed at level A_5, and compensators COMP above condenser. Condenser and objective OBJ lenses are rectified by R_1 and R_2 and coated for low reflection. The image of field diaphragm (A_3 or A_5) is focused on object plane OB by condenser, whose NA can be made equal to that of the objective. Stigmatizing lenses St_1 and St_2, low-reflection coated on their exterior face, are cemented directly onto the analyzing Glan-Thompson prism ANAL to protect surfaces of prism. Aperture stops A_1–A_8, placed at critical points, minimize scattered light from entering the image-forming system. Final image is directed by OC_1 on a photographic or photoelectric sensor EM or to the eye E via mirror M and ocular OC_2.[11] (b) Rectified polarizing microscope designed by author with help of the staff of the Research Center, American Optical Co., and the Institute of Optics, University of Rochester.[13]

reversed rotation over the whole lens aperture.[9] Such a rectifier not only provides high extinction at high numerical apertures but also eliminates spurious diffraction of weakly birefringent specimens, since the vectorial transmission through the objective-lens aperture is no longer inhomogeneous.[10]

Figure 22.3 shows an advanced polarizing microscope we have built which incorporates rectified optics and other general improvements described elsewhere.[11] This instrument has been used to measure retardations down to 0.1 Å with a lateral image resolution of 0.15 µ.

[9] S. Inoué and W. L. Hyde, *J. Biophys. Biochem. Cytol. 3*, 831 (1957).
[10] S. Inoué and H. Kubota, *Nature 182*, 1725 (1958).

[11] S. Inoué, "Polarizing Microscope Design for Maximum Sensitivity," *Encyclopedia of Microscopy*, G. Clark, Ed., Reinhold Publishing Corp., New York, 1961, p. 480.

Verification of spindle-fiber formation

With this improved instrument, the dividing cells of many plants and animals have been studied, and the general presence of birefringent spindle fibers in living cells (Figs. 22.4 and 22.5) has been established,[12-14] similar to those seen after careful fixation and staining (cf. Fig. 22.1). Time-lapse motion pictures of dividing cells have been taken, and changes in spindle-fiber birefringence during normal mitosis and upon experimental alteration of mitosis have been analyzed.

During the natural course of mitosis the birefringent fibers appear in one region and disappear in another which, in the time-lapse movies, sometimes resemble flickering northern lights. When the cell is cooled to approximately 4° C, the birefringent spindle fibers disappear rapidly but reappear in a few minutes after the cell is returned to room temperature. Chromosome movement is arrested by cold and is restored soon after spindle birefringence returns. The region of a spindle fiber locally irradiated with a microbeam of ultraviolet light loses its birefringence but recovers it in a few minutes, with little effect on subsequent chromosome movements.[13]

These observations, coupled with thermodynamic analysis of the spindle birefringence, suggest that the protein molecules making up the spindle fibers[15] are in metastable equilibrium between oriented and non-oriented states, the former being birefringent.[13]

After appropriate fixation, 200-Å thick filaments have now also been observed under the electron microscope, not only as components of spindle fibers but also in regions of the cell associated with other types of localized cell motility.[16] It is probable that localized motility in cells requires a temporary building up of "contractile" fibers upon demand, which are resolved quickly after completion of the cells' mechanical function. The dynamic behavior of such metastable, transient fibers, which can now be followed in active

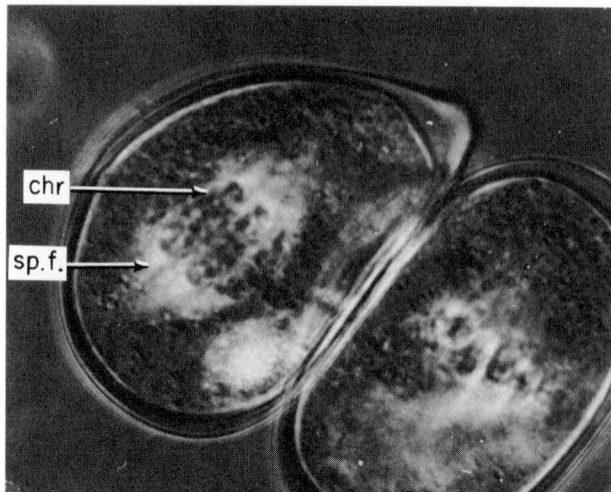

Fig. 22.4. Living pollen mother cell of Easter lily (*Lilium longiflorum*). Spindle fibers (*sp. f.*) show strong birefringence adjacent to helical chromosomes (*chr*).[12]

living cells with the improved polarizing microscope, may ultimately lead to an understanding of the molecular mechanism underlying the orderly movement of

Fig. 22.5. Birefringent spindle fibers in living oöcyte of marine worm (*Chaetopterus pergamentaceus*). Egg centrifuged to move away the birefringent yolk granules.[12]

chromosomes, cell division, and other types of localized cell movements.

---

[12] S. Inoué, *Chromosoma* 5, 487 (1953).

[13] S. Inoué, "Organization and Function of the Mitotic Spindle," *Primitive Motile Systems in Cell Biology*, R. D. Allen and N. Kamiya, Eds., Academic Press, New York, 1964, p. 549.

[14] S. Inoué and A. Bajer, *Chromosoma* 12, 48 (1961).

[15] D. Mazia, "Mitosis and the Physiology of Cell Division," *The Cell*, J. Brachet and A. E. Mirsky, Eds., Vol. 3, Academic Press, New York, 1961, p. 77.

[16] H. Nakajima, "The Mechanochemical System behind Streaming in Physarum," *Primitive Motile Systems in Cell Biology*, R. D. Allen and N. Kamiya, Eds., Academic Press, New York, 1964, p. 111; L. E. Roth, *ibid.*, p. 527; K. E. Wohlfarth-Bottermann, *ibid.*, p. 79.

23 · MUTUAL SUBSTITUTION OF NETWORKS AND MATERIALS*

Heinz M. Schlicke

Simulation of material properties by network approaches — Specific targets — Distributed parameter effects in high-dielectric-constant ceramics — Suppression of resonances in ceramic interference filters — Transformation properties of simple elements — Quasi materials created by series-parallel feedback — Generalized verters — Design procedures — Parametric transmutation — Conclusions

Simulation of material properties by network approaches

Despite all the ingenuity of physicists and chemists, the molecular designing of materials and devices is still often inadequate for engineering purposes. By necessity, then, the electronics engineer becomes a synthesizer himself in developing an unconventional network approach that allows the utilization of available materials. In other words, by incorporating such materials into certain circuits they are made to act like materials we wish to have but do not have.

As scientists build with molecules, we shall manipulate simple basic structural units such as the two building blocks of Fig. 23.1: (a) a cylindrical body of high-dielectric-constant ceramic (e.g., barium titanate) and (b) an amplifier with series and parallel feedback impedance.

We shall use the dielectric cylinder unconventionally as a passive network by changing its boundary conditions.

The amplifier, as an active network component, provides a basic structural unit of much greater flexibility. Such amplifiers can be arranged in groups and modified, as a molecular engineer would handle a macroscopic "supermolecule."

Specific targets

We shall limit our efforts of substitution to two industrial demand areas where satisfactory solutions are

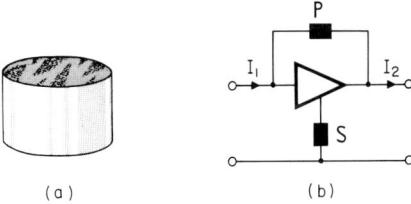

Fig. 23.1. The basic building blocks: (a) high-dielectric-constant cylinder; (b) amplifier with series-parallel feedback.

---

* The author wishes to thank Mati Tiitus for his patient help in mathematical tabulations, which this paper can impart only fractionally and in simplified form.

not yet at hand: interference elimination filters and true thin-film microelectronics. Our specific targets (designated by **T**) will be numbered consecutively for reference later, when applicable solutions come into sight.

Uncontrolled interaction within or between systems deteriorates the information handled, but such interference can be eliminated by low-pass filters. At high frequencies, barium titanate might serve for this purpose; however, detrimental electromagnetic resonances make such dielectric filters worthless unless properly modified.

We ask (**T1**) how resonances of dielectric bodies may be utilized, and (**T2**) how such resonances can be suppressed.

At low frequencies and when large d-c or a-c power must be carried by the conductors, conventional LC filters become prohibitively bulky. Active filters, using for example the great enhancement of a capacitance by a transistor in the common base arrangement (cf. Fig. 23.7e), have been built, but the power has to pass through the transistor. Hence, our next goal is (**T3**) to create new types of active filters where the power bypasses the active elements.

Turning to microminiaturization: size reduction, increased reliability, fast response, and flexibility are desired, but the greatest impact should result from cheap mass production. Complete thin-film microelectronics, furthermore, depends on the availability of planar triodes. Such active elements might be based on space-charge-limited currents[1] (cf. Chap. 19), field-effect triodes,[2] or hot-electron devices.[3,4] (For brevity, we omit two other classes of active elements: negative resistors[5] and nonlinear reactances used in parametric devices,[6-8] although we shall refer on occasion to the concept of negative resistance.)

Thin-film techniques readily allow the deposit of conductors, resistors, and capacitors. Once a technique of depositing planar active elements on substrates is perfected, mass production becomes possible, and we can afford to be generous with active elements without fear of excessive cost or severely reduced reliability. As a tentative list of additional targets we name

(**T4**) Substitutes for large inductances and conventional transformers.

(**T5**) Temperature stability.

(**T6**) Nonreciprocity (a mandatory requirement for stability of densely packaged systems).

(**T7**) Duality (a network responds to current as its dual does to voltage).

(**T8**) Possible compensation of nonlinearity.

(**T9**) Exclusion of detrimental stray effects.

(**T10**) Finally and most importantly, a new way of thinking about circuitry. Instead of only arranging components into networks, we must operate on a higher plane of composition:

 a. by combining networks to transform them into seemingly incommensurable ones (permutative transposition); and

 b. by external switching of one parameter (e.g., by light) to change a network so fundamentally as if it were completely rewired or used basically different components (parametric transmutation).

Distributed parameter effects in high-dielectric-constant ceramics

At high frequencies, bodies of high-dielectric-constant ceramics exhibit three phenomena of special interest:

1. Their electromagnetic resonance frequencies lie $\sqrt{\kappa'}$ times lower than those of empty microwave cavities. Since temperature-stable barium titanate mixtures can be made with $\kappa' \simeq 2500$, a size or frequency reduction of 1:50 is common. A microwave cavity becomes a VHF cavity, but its Q is relatively low.

2. In dielectric cavities, metallic walls can be replaced (totally or in part) by air-dielectric interfaces approximating reflecting boundaries of nearly infinite impedance. This leads to new modes not found in standard microwave theory. In particular, quasi-degenerated *TE* modes are possible, while normal microwave cavities exhibit only degenerated *TM* modes.[9]

3. Certain nonmetallized surfaces of the dielectric bodies leak magnetically, thereby permitting convenient coupling to associated circuits (**T1**).

Figure 23.2 shows three cylindrical high-dielectric-constant cavity modes of the quasi-degenerated type[10] and their magnetic flux patterns. For $\kappa' = 2500$ and a

---

[1] G. T. Wright, *J. Brit. Radio Engrs.* **20**, 337 (1960).

[2] H. Borkan and P. K. Weimer, *RCA Rev.* **24**, 153 (1963).

[3] C. A. Mead, *J. Appl. Phys.* **32**, 646 (1961).

[4] M. M. Atalla, *NEREM Record 4*, 1962, p. 162.

[5] G. D. Sims and I. M. Stephenson, *Electronics and Control* **9**, 349 (1960).

[6] R. Maurer and K. H. Loecherer, *Archiv der elektrischen Übertragung* **15**, No. 2, 71 (1961).

[7] V. W. Vodicka and R. Zuleeg, *Electronics* **33**, August, 56 (1960).

[8] U. L. Rohde, *ibid.* **35**, June, 46 (1962).

[9] H. M. Schlicke, *Essentials of Dielectromagnetic Engineering*, John Wiley and Sons, New York, 1961, p. 23.

[10] Here the first superscript designates the boundary impedance of the face of the cylinder, and the second that of the periphery of the cylinder; the subscripts follow the normal convention.

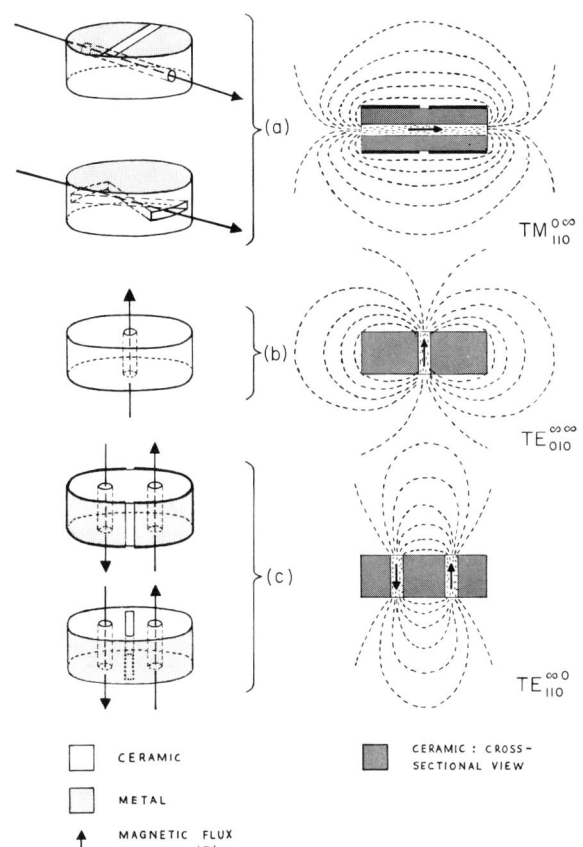

Fig. 23.2. High-dielectric-constant cavity modes: (a) faces metallized; (b) base ceramic cylinder; (c) periphery metallized.

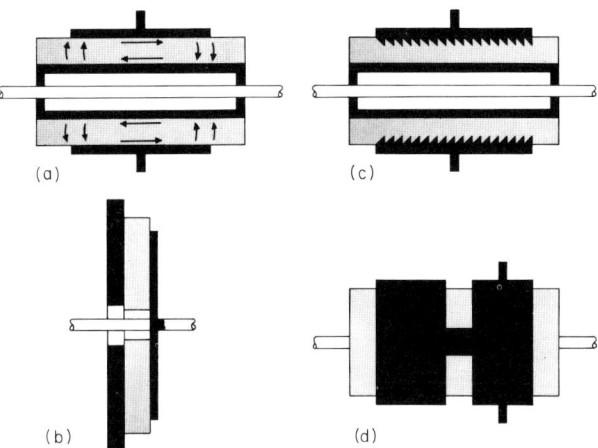

Fig. 23.3. Deresonating of ceramic feed-through capacitors, using nonuniform transmission lines.

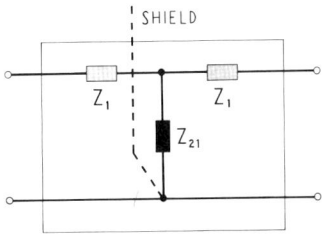

Fig. 23.4. Four-pole equivalent of feed through capacitor.

Suppression of resonances in ceramic interference filters

Effective elimination of conducted noise or interference requires low-pass filters. Ceramics of high dielectric constant may serve for this purpose, provided their resonances can be suppressed (T2).

The simplest form of a ceramic filter is the tubular ceramic feed-through capacitor (Fig. 23.3), supposedly representing a short circuit for very high frequencies. In contradistinction to conventional filter theory, we encounter here filters that are completely mismatched, made of material of very high propagation constant, and confined to the simplest of geometries. For a mechanical length of 1 cm and a κ' of 2500, the $\lambda/2$ frequency is 300 MHz.

With a diameter of 2.5 cm, the resonances are near 150 to 300 MHz. The field in cylinder (b) has axial symmetry; its polarization cannot be influenced. The two other cavities in their natural state have no preferred polarization; here the polarization depends on the direction of the exciting magnetic field. For practical circuit work it can be directed by a slot in the metallized surfaces (a) or by a polarization slot cut into the ceramic (a) and (c).

The T network of Fig. 23.4 is an equivalent of the ceramic transmission line with a transfer impedance of

$$Z_{21} = -\frac{jZ_0}{\sqrt{\kappa'}\sin b}, \qquad (23.1)$$

where $b = \omega\sqrt{\kappa'}l/3 \times 10^{10}$ (rad), and l is the mechanical length of the line (in cm). For $b = n\ 180$ degrees, Z_{21} becomes ∞ for a lossless dielectric. For actual, loss-afflicted dielectrics, the undesirable peaks shown in Fig. 23.5a occur.

The objective to deresonate such ceramic filters can be met by either of two generic approaches:

1. By introducing nonuniformities, thus diffusing the resonances of the essentially open-circuited transmission line or at least shifting them to much higher frequencies.

2. By introducing losses selectively for very high frequencies only.

In the first category the problem is to eliminate the parallel resonances of the transfer impedance. Solutions to this peculiar broadband problem are sketched

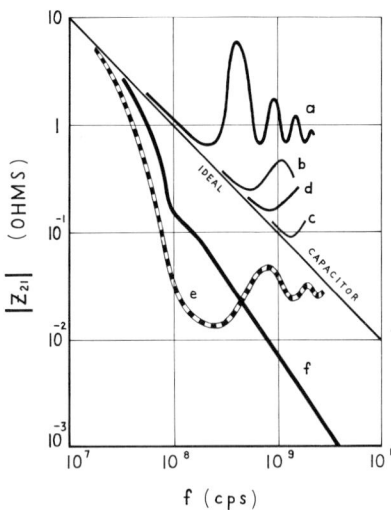

Fig. 23.5. Transfer impedance Z_2 of various ceramic feed-through capacitors.

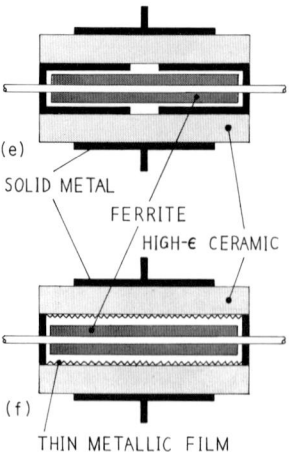

Fig. 23.6. Ceramic filters: (a) cascaded; (b) simulated-skin effect.

in Figs. 23.3b, c, d, and the corresponding performance characteristics in Figs. 23.5b, c, d. Solution b is the simplest one: The nonuniformity of a disk line causes so much internal reflection that the first detrimental resonance occurs for the $TM_{010}^{0\infty}$ mode, determined by the first zero of the Bessel function

$$J_0'(b) = 0. \quad (23.2)$$

In the tubular case, the first critical resonance occurs at the frequency for which the length of the tube represents a phase measure of π. In the discoidal configuration, the radius (not the diameter) has to have the phase measure 3.81 (cf. Fig. 23.5b).

The same general approach of diffusing resonances is used for solution c. The sawteeth represent miniature exponential lines which are highpass filters. Below the cutoff frequency (chosen very high—hence the very short teeth), such filters exhibit no phase shift and therefore cannot resonate.

The cross-slotted solution (Figs. 23.3d and 23.5d) suppresses the unwanted resonance by a radial interruption in the outer electrode with a connecting metallic bridge of rather critical dimensions. By proper dimensioning of this bridge,[11] $f_2 = 0$ is made to coincide with $f_1 = 0$ for the same frequency. A pole is canceled by a zero:

$$Z_{21} = \frac{f_1(f)}{f_2(f)} \neq \infty \quad \text{for} \quad f_2 = 0. \quad (23.3)$$

In Category 2, pertaining to deresonating by admitting frequency-dependent losses, solutions e and f of Figs. 23.5 and 23.6 are the cascaded filter and the simulated-skin-effect filter, respectively.

[11] See Ref. 9, p. 131.

For cascaded filters[12] the inner (or outer) electrode of the tubular capacitor is split in order to introduce ferrite beads. At very high frequencies these act approximately as frequency-independent resistors ($\kappa_m'' \approx 1/f$). The domain walls in the ferrites are exploited to introduce frequency-determined losses that suppress resonances in the open-circuited transmission line. Excellent RC filters result for which the parallel resonances of the sectional transmission lines are shifted to quite low impedance levels.

But an even better performance is achieved by configuration e, which is essentially a simulated skin effect.[13] For the large cross sections and the extremely short dimension of the inner conductor, we cannot utilize the real skin effect. It would be much too weak. Thus, we have to create it artificially. We assume that we have a film which is thin enough to be resistive and work backward. Since the film is thinner than the skin depth, we would shunt it out if we put a plain heavy conductor through the center. This short circuit is prevented by the insertion of the ferrite.

Transformation properties of simple elements

Before manipulating whole circuits we recall some transformation properties for simple elements.

The ideal transformer (Fig. 23.7a) (secondary to primary turn ratio n, input impedance Z_i, and load impedance Z_l) is described in its current-voltage behavior by

$$\left. \begin{array}{l} V_2 = nV_1, \\ I_2 = \dfrac{I_1}{n}, \end{array} \right\} Z_i \equiv \frac{V_1}{I_1} = \frac{Z_l}{n^2}. \quad (23.4)$$

[12] Ibid., p. 169.
[13] H. M. Schlicke, IEEE Trans. Electromagnetic Compatibility EMC-6, No. 1, 47 (1964).

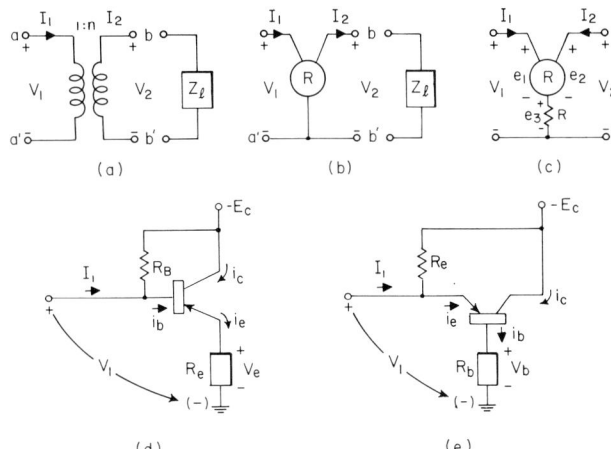

Fig. 23.7. Transformation properties of simple elements: (a) transformer; (b) gyrator; (c) isolator; (d) common-emitter-type impedance multiplier; (e) common-base-type admittance multiplier.

The gyrator (Fig. 23.7b), realizable for example by Hall effect in a semiconductor,[14,15] Faraday effect in a ferrite,[16] or a nonreciprocal combination of electromechanical transducers,[17] gives

$$\left.\begin{array}{l}V_1 = RI_2,\\ V_2 = RI_1,\end{array}\right\} Z_i \equiv \frac{V_1}{I_1} = \frac{R^2}{Z_l}; \quad (23.5)$$

i.e., an inverted relation between input and output impedance results. The isolator (Fig. 23.7c), composed for example of a Hall effect gyrator and a resistor of R ohms to absorb reverse energy, passes energy in one direction only:

$$V_1 = e_1 + e_3 = -RI_2 + (I_1 + I_2)R = RI_1,\\ V_2 = e_2 + e_3 = RI_1 + (I_1 + I_2)R = 2RI_1 + RI_2; \quad (23.6)$$

i.e., the input is not affected by changes in the output.

Active elements, such as the transistor, add another dimension of flexibility. For an ideal transistor (Fig. 23.7d) β_0 is the ratio of collector current to base current i_c/i_b. Input current and voltage

$$I_1 \cong i_b = i_e - i_c = \frac{i_e}{\beta_0 + 1},\\ V_1 = v_e = i_e R_e, \quad (23.7)$$

result in an input impedance

$$Z_i = \frac{V_1}{I_1} = \frac{i_e R_e}{i_e/(\beta_0 + 1)} = (\beta_0 + 1)R_e. \quad (23.8)$$

The resistor R_e appears transformed into a much larger

[14] R. F. Wick, J. Appl. Phys. 25, 962 (1954).
[15] W. P. Mason, W. H. Hewitt, and R. F. Wick, ibid. 24, 166 (1953).
[16] C. L. Hogan, Bell System Tech. J. 31, 1 (1952).
[17] M. Onoe and M. Sawabe, Proc. IRE 50, 1967 (1962).

resistor $(\beta_0 + 1)R_e$, where β_0 may be 50 to 200. We could have similarly chosen to multiply an inductor by suitably placing the inductor in the emitter circuit. For multiplication of admittances, or specifically capacitances, we have to replace the common emitter circuit by the common base circuit (Fig. 23.7e). Hence, large capacitors and inductors now become available, with the multiplier being a function of the transistor characteristics.

Another method of utilizing an active device to obtain corresponding results is to deposit a field-effect triode, which behaves much like a vacuum-tube pentode, and then utilize the Miller[18] effect which multiplies the gate-to-drain capacitance. Cancellation of circuit losses[19] is also possible, but often only with sacrifice of stability.

Much more complex results can be achieved when a larger number of circuit elements are introduced.

Quasi materials created by series-parallel feedback

The language most suitable for operating with whole circuits instead of individual components is matrix algebra.[20] For instance, instead of writing the relations of input (subscript 1) and output (subscript 2) voltages and currents

$$V_1 = a_{11}V_2 + a_{12}I_2,\\ I_1 = a_{21}V_2 + a_{22}I_2, \quad (23.9)$$

we write in "shorthand"

$$\begin{bmatrix}V_1\\ I_1\end{bmatrix} = \begin{bmatrix}a_{11} & a_{12}\\ a_{21} & a_{22}\end{bmatrix} \times \begin{bmatrix}V_2\\ I_2\end{bmatrix}. \quad (23.10)$$

This $[A]$ matrix is most useful in cascading fourpole networks: The matrices are simply multiplied.

Rearrangement of the input and output relationships, as in row i of Table 23.1, produces the four corresponding additive matrices suitable in the sequence listed for parallel (Y), series (Z), series-parallel (H), or parallel-series (G) connection of fourpoles with others (see also Fig. 23.8).

We turn now to the four principal types of three-electrode active elements. In rows d, e, f, and g of Table 23.1, each type is represented by a characteristic matrix of the additive kind, each based upon one of the four A-matrix terms indicative of a specific activity. For instance, Type I, characterized by the Y matrix,

[18] J. M. Miller, Sci. Papers Natl. Bur. Standards 15, 367 (1920), Sci. Paper No. 351.
[19] J. L. Dalke and R. C. Powell, Electronics 24, 224D (1951).
[20] Ref. 9, p. 124.

Table 23.1. Characteristic matrices of idealized active elements

a	Type	I	II	III	IV
b	Mechanism	$V_1 \to I_2$	$I_1 \to V_2$	$I_1 \to I_2$	$V_1 \to V_2$
c	$\|A\|$	$\begin{Vmatrix} 0 & a_{12} \\ 0 & 0 \end{Vmatrix}$	$\begin{Vmatrix} 0 & 0 \\ a_{21} & 0 \end{Vmatrix}$	$\begin{Vmatrix} 0 & 0 \\ 0 & a_{22} \end{Vmatrix}$	$\begin{Vmatrix} a_{11} & 0 \\ 0 & 0 \end{Vmatrix}$
d	$\|Y\|$	$\begin{Vmatrix} 0 & 0 \\ a_{12}^{-1} & 0 \end{Vmatrix}$	—	—	—
e	$\|Z\|$	—	$\begin{Vmatrix} 0 & 0 \\ a_{21}^{-1} & 0 \end{Vmatrix}$	—	—
f	$\|H\|$	—	—	$\begin{Vmatrix} 0 & 0 \\ a_{22}^{-1} & 0 \end{Vmatrix}$	—
g	$\|G\|$	—	—	—	$\begin{Vmatrix} 0 & 0 \\ a_{11}^{-1} & 0 \end{Vmatrix}$
h	Example	Field-effect transistor	Madistor	Common base transistor	Common collector transistor
i	Characteristic Fourpole eq.	$I_1 = y_{11}V_1 + y_{12}V_2$ $I_2 = y_{21}V_1 + y_{22}V_2$	$V_1 = z_{11}I_1 + z_{12}I_2$ $V_2 = z_{21}I_1 + z_{22}I_2$	$V_1 = h_{11}I_1 + h_{12}V_2$ $I_2 = h_{21}I_1 + h_{22}V_2$	$I_1 = g_{11}V_1 + g_{12}I_2$ $V_2 = g_{21}V_1 + g_{22}I_2$

represents a field-effect transistor where an input voltage controls an output current. The 21 terms in the Z, Y, H, and G matrices indicate transfer from terminals 1 to terminals 2, whereas the terms 12 hold for the reversed direction. Since we postulate ideal, unilateral active devices, the 12 (or feedback) terms of the characteristic matrices are set to zero. In practice this may have to be done by neutralization or by balancing of one feedback factor by another of opposite sign. In the additive matrices, the 11 and 22 terms pertain to the input or output immittances. They are ideally equated to zero for perspicuity of presentation.

In approaching the macroscopic synthesis of quasi materials, we report here only some representative findings without reconverting the effects into equivalent material properties (e.g., the gyrator will be described in its effects and not in terms of the tensor permeability of the ferrites that produces the microwave gyrator).

Generalized verters

To investigate the generalized transformation of a load impedance Z_l into an input impedance Z_i of a fourpole, we introduce the expression "verter" as a generic term encompassing inverters (I) and converters (C) of impedances and distinguish between positive (P) and negative (N) verters. Four types are possible (Table 23.2): PII, a Positive Impedance Inverter (gyrator); PIC, a Positive Impedance Converter (transformer); NII, a Negative Impedance Inverter (negator); and NIC, a Negative Impedance Converter.

Generalized verters are those for which two impedances (Z_3 and Z_4), the significant parameters, can have any value desired. For converters, the ratio $(\pm)Z_4/Z_3$ determines the transformation; for inverters, the square root of their product, $\sqrt{\pm Z_3 \cdot Z_4}$, is the inverting impedance. NII's or NIC's invert or convert positive into negative impedances.

Table 23.2 surveys the characteristics of the generalized verters. The A matrices (second column) are virtually all we have to know, since the other functions

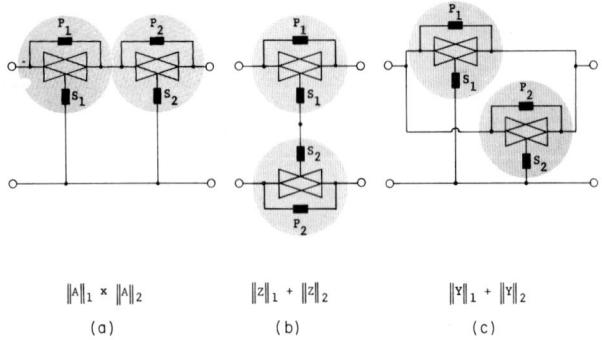

Fig. 23.8. (a) Cascade, (b) series, and (c) parallel combinations of the building block, Fig. 23.1b.

Table 23.2. Properties of verters

	Generalized Device		Cascade Matrix $[A]$	$|A|$	Input Impedance Z_i General	For resistive Z_3 and Z_4 and if $Z_L =$ $1/j\omega C$	$j\omega L$	0	∞	Z_3, Z_4 inductive and if $Z_L =$ 0	∞	Z_3, Z_4 and Z_L resistive or reactive	Other Outstanding Characteristics
a	PII (gyrator)		$\begin{matrix} 0 & Z_4 \\ \frac{1}{Z_3} & 0 \end{matrix}$	$-\frac{Z_4}{Z_3}$	$\frac{Z_4 Z_3}{Z_L}$	L'	C'	∞	0	$-\infty$	0	$-R\Omega^2$ $-R/\Omega^2$ $-j\Omega^3 L$ $-1/j\Omega^3 C$	Converts networks into duals*
b		Reciprocal $PIC(R)$	$\begin{matrix} \frac{Z_4}{Z_3} & 0 \\ 0 & \frac{Z_3}{Z_4} \end{matrix}$	1	$\left(\frac{Z_4}{Z_3}\right)^2 Z_L$	C'	L'	—	—	—	—	$R\omega^4$ R/Ω^4 $j\Omega^5 L$ $1/j\Omega^5 C$	Transforms also from d.c. on
c	PIC (trafo)	Nonreciprocal $PIC(U)$	$\begin{matrix} \frac{Z_4}{Z_3} & 0 \\ 0 & \frac{Z_4}{Z_3} \end{matrix}$	$\left(\frac{Z_4}{Z_3}\right)^2$	Z_L	C	L	—	—	—	—	—	Transforms only power level, not impedance
d			$\begin{matrix} 1 & 0 \\ 0 & \left(\frac{Z_4}{Z_3}\right)^2 \end{matrix}$	$\left(\frac{Z_4}{Z_3}\right)^2$	$\left(\frac{Z_3}{Z_4}\right)^2 Z_L$	C'	L'	—	—	—	—	$R\Omega^4$ R/Ω^4 $j\Omega^5 L$ $1/j\Omega^5 C$	Transforms from d.c. on
e	NII (negator)		$\begin{matrix} 0 & Z_4 \\ \frac{-1}{Z_3} & 0 \end{matrix}$	$\frac{Z_4}{Z_3}$	$-\frac{Z_4 Z_3}{Z_L}$	$-L'$	$-C'$	$-\infty$	0	∞	0	$R\Omega^2$ R/Ω^2 $j\Omega^3 L$ $1/j\Omega^3 C$	Negates networks and itself†
f	NIC (−)		$\begin{matrix} 1 & 0 \\ 0 & \frac{-Z_3}{Z_4} \end{matrix}$	$-\frac{Z_4}{Z_3}$	$-\left(\frac{Z_4}{Z_3}\right) Z_L$	$-C'$	$-L'$	0	$-\infty$	0	$-\infty$	$R\Omega^2$ R/Ω^2 $j\Omega^3 L$ $1/j\Omega^3 C$	Negates networks, including itself‡

* The cascade arrangement of a gyrator, a fourpole, and a gyrator results in the dual of the fourpole; e.g., an $N:1$ trafo is transformed into a $1:N$ trafo.

† An NII cascaded between two fourpoles gives an NII; this system cascaded again gives a $1:-1$ trafo.

‡ An NIC cascaded between two fourpoles gives an NIC; this system cascaded again is equivalent to a straight connection (all participating fourpoles, including NIC's, being canceled).

listed are derived from them. But since verters have uncommon, peculiar properties, we shall point out some specificities.

The determinant of the A matrix ($|A|$) is the square root of the ratio of the backward to the forward gain. It is unity for a reciprocal network, and 0 or ∞ if the network is completely unilateral (**T**6).

If Z_3 and Z_4 are resistors, a gyrator changes a capacitor (which is easily deposited in a planar circuit with a controllable temperature coefficient) into an equivalent inductor (**T**4).

The requirement for planar transformers is also contained in (**T**4), but our synthetic transformer (PIC) can surpass decisively a real transformer with magnetic cores and copper windings:

1. It can operate at direct current (corresponding to an infinite permeability of our equivalent iron core), provided the amplifiers employed also operate for direct current.

2. It can be unilaterized [$PIC(U)$, row (d) of Table 23.2, **T**6]. The A matrix of the ideal transformer is

$$\begin{bmatrix} \left.\frac{V_1}{V_2}\right|_{I_2=0} & 0 \\ 0 & \left.\frac{I_1}{I_2}\right|_{V_2=0} \end{bmatrix} = \begin{bmatrix} N & 0 \\ 0 & \frac{1}{N} \end{bmatrix} \quad (23.11)$$

and of the corresponding reciprocal PIC (row b)

$$\begin{bmatrix} \frac{Z_4}{Z_3} & 0 \\ 0 & \frac{Z_3}{Z_4} \end{bmatrix} = [A]_{PIC(R)}. \quad (23.12)$$

In both cases, if the voltage is stepped up $N:1$, the current is reduced $1:N$ and the impedance changes as $N^2 Z_L$ or $(Z_4/Z_3)^2 Z_L$, respectively. However, according to row (c), we can also have another type of transformer, characterized by

$$\begin{bmatrix} \frac{Z_4}{Z_3} & 0 \\ 0 & \frac{Z_4}{Z_3} \end{bmatrix} = [A]_{PIC(U)}. \quad (23.13)$$

Table 23.3. Forward and reversed A, Z, and Y matrices of Types I, II, III, and IV amplifiers with combined series-parallel feedback

Type of Amplifier		I (Y) $V_1 \to I_2$		II (Z) $I_1 \to V_2$		III (H) $I_1 \to I_2$		IV (G) $V_1 \to V_2$	
$\|A\|$	Forward	$\dfrac{Sy_{21}-1}{y_{21}(S-P)-1}$	$\dfrac{P(Sy_{21}-1)}{y_{21}(S-P)-1}$	$\dfrac{S(P-z_{21})}{PS+z_{21}(P-S)}$	$\dfrac{-SPz_{21}}{PS+z_{21}(P-S)}$	$\dfrac{S(1-h_{21})}{Ph_{21}+S(1-h_{21})}$	$\dfrac{PS(1-h_{21})}{Ph_{21}+S(1-h_{21})}$	$\dfrac{P+S(1-g_{21})}{g_{21}P+S(1-g_{21})}$	$\dfrac{PS(1-g_{21})}{g_{21}P+S(1-g_{21})}$
		$\dfrac{Sy_{21}-1}{y_{21}(S-P)-1}$	$\dfrac{Sy_{21}-1}{y_{21}(S-P)-1}$	$\dfrac{P-z_{21}}{PS+z_{21}(P-S)}$	$\dfrac{S(P-z_{21})}{PS+z_{21}(P-S)}$	$\dfrac{1-h_{21}}{Ph_{21}+S(1-h_{21})}$	$\dfrac{P+S(1-h_{21})}{Ph_{21}+S(1-h_{21})}$	$\dfrac{1-g_{21}}{g_{21}P+S(1-g_{21})}$	$\dfrac{S(1-g_{21})}{g_{21}P+S(1-g_{21})}$
	Reverse	1	P	1	$\dfrac{Pz_{21}}{z_{21}-P}$	$1+\dfrac{P}{S(1-h_{21})}$	P	1	P
		$\dfrac{y_{21}}{Sy_{21}-1}$	1	$\dfrac{1}{S}$	1	$\dfrac{1}{S}$	1	$\dfrac{1}{S}$	$1+\dfrac{P}{S(1-g_{21})}$
$\|Z\|$	Forward	$\dfrac{Sy_{21}-1}{y_{21}}$	$\dfrac{1-Sy_{21}}{y_{21}}$	S	$-S$	S	$-S$	$S+\dfrac{P}{1-g_{21}}$	$-S$
		$\dfrac{y_{21}(S-P)-1}{y_{21}}$	$\dfrac{1-Sy_{21}}{y_{21}}$	$S+\dfrac{Pz_{21}}{P-z_{21}}$	$-S$	$S+\dfrac{Ph_{21}}{1-h_{21}}$	$-S-\dfrac{P}{1-h_{21}}$	$S+\dfrac{Pg_{21}}{1-g_{21}}$	$-S$
	Reverse	$\dfrac{Sy_{21}-1}{y_{21}}$	$\dfrac{y_{21}(P-S)+1}{y_{21}}$	S	$-S-\dfrac{Pz_{21}}{P-z_{21}}$	$S+\dfrac{P}{1-h_{21}}$	$-S-\dfrac{Ph_{21}}{1-h_{21}}$	S	$-S$
		$\dfrac{Sy_{21}-1}{y_{21}}$	$\dfrac{1-Sy_{21}}{y_{21}}$	S	$-S$	S	$-S$	S	$\dfrac{Pg_{21}}{g_{21}-1}-S$
$\|Y\|$	Forward	$\dfrac{1}{P}$	$\dfrac{-1}{P}$	$\dfrac{z_{21}-P}{Pz_{21}}$	$\dfrac{P-z_{21}}{Pz_{21}}$	$\dfrac{1}{P}$	$\dfrac{-1}{P}$	$\dfrac{1}{P}+\dfrac{g_{21}}{S(1-g_{21})}$	$\dfrac{-1}{P}$
		$\dfrac{1}{P}+\dfrac{y_{21}}{1-Sy_{21}}$	$\dfrac{y_{21}}{1-Sy_{21}}$	$\dfrac{z_{21}(S-P)-PS}{SPz_{21}}$	$\dfrac{P-z_{21}}{Pz_{21}}$	$\dfrac{1}{P}-\dfrac{1}{S(h_{21}-1)}$	$\dfrac{h_{21}}{S(h_{21}-1)}-\dfrac{1}{P}$	$\dfrac{1}{P}+\dfrac{1}{S(1-g_{21})}$	$\dfrac{-1}{P}$
	Reverse	$\dfrac{1}{P}$	$\dfrac{-1}{P}$	$\dfrac{z_{21}-P}{Pz_{21}}$	$\dfrac{PS+z_{21}(P-S)}{SPz_{21}}$	$\dfrac{1}{P}$	$\dfrac{h_{21}-1}{S(h_{21}-1)}-\dfrac{1}{P}$	$\dfrac{1}{P}$	$\dfrac{-g_{21}}{S(1-g_{21})}-\dfrac{1}{P}$
		$\dfrac{1}{P}$	$\dfrac{-1}{P}$	$\dfrac{z_{21}-P}{Pz_{21}}$	$\dfrac{P-z_{21}}{Pz_{21}}$	$\dfrac{1}{P}$	$\dfrac{-1}{P}$	$\dfrac{1}{P}$	$\dfrac{-1}{P}$

Here both voltage and current are simultaneously stepped up or down by Z_3/Z_4, while the impedance remains unchanged. We have here the case of a real power amplifier in the truest sense of the word (**T6**). The first footnote of Table 23.2 also indicates that a gyrator can fulfill (**T7**).

A negator with inductive Z_3 and Z_4 inverts a short circuit into an open circuit (**T7**). Since negators can be made with direct current passing through Z_3 and Z_4 only while bypassing the active elements, we can realize goal (**T3**) pertaining to active filters (e.g., Fig. 23.9).

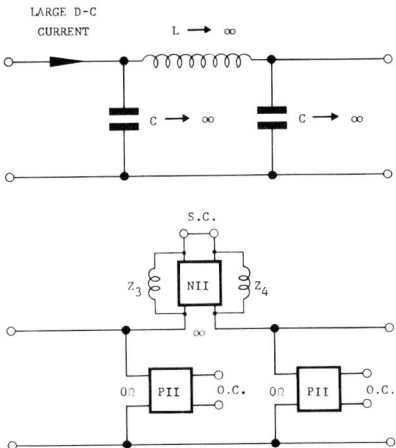

Fig. 23.9. Idealized low-pass filter (top) and its active equivalent (bottom).

For the equivalent shunt capacitors of our π network we need not observe the critical d-c condition. Hence, to convert an open circuit into a short circuit corresponding to an infinitely large capacitor, we can utilize a gyrator with Z_3 and Z_4 being resistive.

The next-to-last column of Table 23.2 lists some possibilities of realizing strongly frequency-dependent positive and negative resistors, capacitors, and inductors, depending on the choice of Z_3, Z_4, Z_L, and the type of verter (**T4**). Thus, with frequency-dependent resistors, filters having no phase shift can be made [stability of active filter (**T3**)] or real, undesired capacitances can be eliminated by means of negative capacitances (**T7**).

The negation of complete networks (cf. footnotes to Table 23.2) by NII's and NIC's can be proved mathematically and gives prospects of overcoming the difficulty outlined as (**T9**). We encounter here a phenomenon where NII's and NIC's not only negate (or virtually abolish) adjoining networks but also themselves ($1:-1$ or $1:1$ transformer, respectively).

Not listed in Table 23.2 is one gyrator property of great importance to (**T8**): A gyrator used as a pre-network permits the inversion of nonlinearity, which can be exploited for linearization.

Design procedures

We shall now operate in the same fashion as molecular designers but on the macroscopic level. Our basic "supermolecule" (cf. Fig. 23.1b) is too large to permit complex "crystal" structures to be built; our bonds are circuit connections. We therefore restrict ourselves to some simple complexes (cf. Fig. 23.8): two of our supermolecules—not necessarily with the same P and S—combined in cascade, series, or parallel connection of the fourpoles. (This synthesis is often simplified by equating P or S to 0 or ∞.)

Table 23.3 refers to the individual supermolecule in the three combinations of Fig. 23.8; the amplifiers may operate either forward or backward (R). They are highly idealized, as in Table 23.1. A comparison of the four columns indicates that the type of amplifier chosen plays a significant role. According to Fig. 23.8, each type of connection pertains to a corresponding matrix; hence, whatever synthesis is desired has to be expressed in the appropriate matrix form.

Table 23.4 represents two major groups: The first

Table 23.4. *A*, *Z*, and *Y* matrices of generalized verters and isolators

	[*A*]	[*Z*]	[*Y*]
PII	$\begin{matrix}0 & Z_4 \\ 1/Z_3 & 0\end{matrix}$	$\begin{matrix}0 & Z_4 \\ Z_3 & 0\end{matrix}$	$\begin{matrix}0 & 1/Z_3 \\ 1/Z_4 & 0\end{matrix}$
PIC	$\begin{matrix}n_1/n_2 & 0 \\ 0 & n_2/n_1\end{matrix}$	∞	∞
NII	$\begin{matrix}0 & Z_4 \\ -1/Z_3 & 0\end{matrix}$	$\begin{matrix}0 & Z_4 \\ -Z_3 & 0\end{matrix}$	$\begin{matrix}0 & -1/Z_3 \\ 1/Z_4 & 0\end{matrix}$
NIC	$\begin{matrix}1 & 0 \\ 0 & -Z_4/Z_3\end{matrix}$	∞	∞
$Is(f)$	$\begin{matrix}1/2 & R/2 \\ 1/2R & 1/2\end{matrix}$	$\begin{matrix}R & 0 \\ 2R & -R\end{matrix}$	$\begin{matrix}1/R & 0 \\ 2/R & -1/R\end{matrix}$
$Is(r)$	∞	$\begin{matrix}R & -2R \\ 0 & -R\end{matrix}$	$\begin{matrix}1/R & -2/R \\ 0 & -1/R\end{matrix}$
Sep	∞	$\begin{matrix}z_{11} & 0 \\ 0 & -z_{22}\end{matrix}$	$\begin{matrix}y_{11} & 0 \\ 0 & -y_{22}\end{matrix}$

four rows are the generalized verters concerned with impedance transformation, already known to us in their effects, while the remaining rows refer to the general class of isolators important for (**T3**) and (**T6**), to be correlated later with verters. Specifically, there are the following:

Is(f), the forward isolator, transmitting energy from terminal pair 1 to terminal pair 2 and blocking in the reversed direction.

Is(r), the reversed isolator, in which terminal pairs 1 and 2 are exchanged.

Sep, the separator, separating input and output terminals completely both ways.

Some simple realizations of verters, based on Tables 23.3, 23.4, and Fig. 23.8, are depicted in Fig. 23.10.

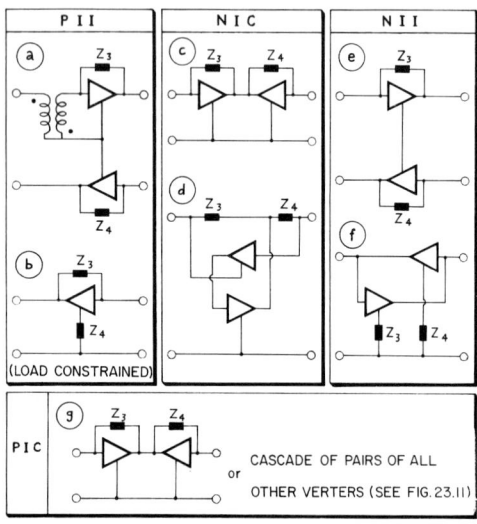

Fig. 23.10. Examples of verter realizations.

One variation in structure is shown in Fig. 23.10d, Sandberg's version[21] of the *NIC*, where a cross-coupling between a pair of building blocks is applied. The *NII* of Fig. 23.10e was designed by Honnell.[22] Its dual (Fig. 23.10f) was derived by our general method. With $P = \infty$, $S_1 = Z_3$, and $S_2 = Z_4$, Table 23.3 yields

$$\begin{bmatrix} 0 & 0 \\ \dfrac{y_{21}'}{1 - Z_3 y_{21}'} & 0 \end{bmatrix} + \begin{bmatrix} 0 & -y_{21}'' \\ 0 & \dfrac{1 - Z_4 y_{21}''}{0} \end{bmatrix}$$

$$= \begin{bmatrix} 0 & \dfrac{1}{Z_4} \\ \dfrac{-1}{Z_3} & 0 \end{bmatrix}. \quad (23.14)$$

This is the *Y* matrix of a negator for $|Z_3 y_{21}'| \gg 1$ and $|Z_4 y_{21}''| \gg 1$. This version of the *NII*, however, is

[21] I. W. Sandberg, *Bell System Tech. J.* **39**, 947 (1960).
[22] P. M. Honnell, *Matrix and Tensor Quarterly* **13**, No. 1, 1 (1962).

inherently unstable unless additional conditions are introduced.

The verters described thus far were of a universal type. For many practical purposes, restrictions are fully permissible or even desirable. For instance, the gyrator of Fig. 23.9 has only to invert an open circuit into a short circuit but not also vice versa, as a general *PII* would have to function. With such constraints, we can forget pairing of our basic building blocks and manipulate simple building blocks as, e.g., those for which the 11 or 22 terms of the additive matrices may not be zero.

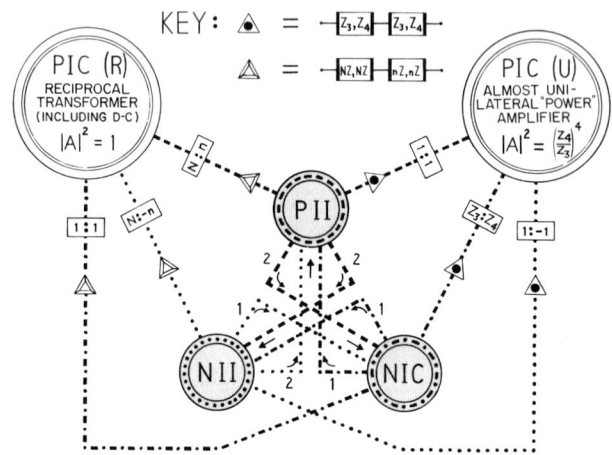

Fig. 23.11. Permutative transposition of verter pairs.

Figure 23.11 presents a different approach: the derivation of one of the verters by cascading two others (**T10a**). The inner triangle encompassing *NII*'s, *PII*'s, and *NIC*'s indicates that the cascading of two verters in the proper sequence yields the third (e.g., *NIC* × *NII* = *PII*). The peripheral area of Fig. 23.11 refers to *PIC*'s generated by cascading any congeneric pair of the inner verters. Nonreciprocal transformers are effectuated by cascading any two identical verters. Reciprocal *PIC*'s result from the tandem connection of two nonidentical verters of the same type (e.g., one *NIC* having $Z_3 = Z_4 = NZ$, the other $Z_3 = Z_4 = nZ$).

In the next section we shall briefly illustrate a simple stability criterion for one of our verters. For more general stability considerations, we refer to Table 23.3, which also makes obvious the reason for the selection of our basic building block, Fig. 23.1b: *P* and *S* are not only dual by representing voltage versus current feedback, they may also generate different signs of feedback, depending on the sign of the 21 matrix terms of the amplifier, thereby greatly facilitating efforts of stabilization.

Parametric transmutation

Target (**T**10*b*) stated the intention to design a network such that the change of one parameter transmutes the original network into another of wholly different properties. In tackling this problem, we find simultaneously elegant solutions for some more of our targets.

In contrast to the previous approach of combining basic building blocks, we now consider only the block itself and change either P or S or even replace the amplifier by another active fourpole. In short, instead of working with a combination of the basic units of type of Fig. 23.1*b*, we modify a single unit (Fig. 23.12).

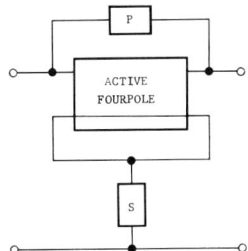

Fig. 23.12. Change of kernel.

Again, representative examples provide a better understanding than general statements.

If the active fourpole is a gyrator and $S = R/2$ and $P = 2R$ (with $R = Z_3 = Z_4$, the gyrating impedance), the total fourpole is an isolator (input and output impedances equal to R) transmitting energy from left to right as though the isolator did not exist, but preventing any energy flow in the opposite direction. This operation can be reversed for reversed premises. If the active fourpole is a readily built isolator and $S = -R/2$ and $P = -2R$, the resulting fourpole is a *PII*.

Table 23.5. Conditions for parametric transmutations of gyrators and isolators

Resulting Fourpole \ Original Fourpole	Gyrator	Isolator
Reversed isolator† + (1:−1) trafo	$(s+1)p = -1$	$(s^* + 2)p^* = -1$
Gyrator	$sp = -1$	$(s^* + 1)p^* = -1$
Isolator	$(s-1)p = -1$	$s^* p^* = -1$

† With negative input and output impedances
Note: $s = S/R$
  $p = P/R$

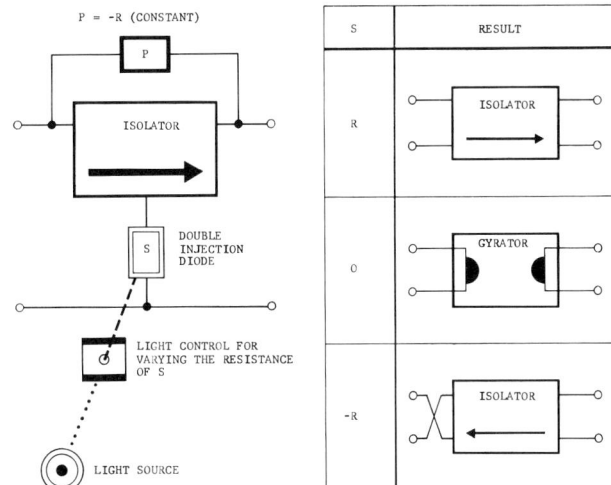

Fig. 23.13. Example of parametric transmutation.

More generally (Table 23.5) by changing either S or P, we can obtain gyrators, isolators, or even reversed isolators from the same structure. Figure 23.13 illustrates such parametric transmutation (**T**10*b*) by means of a light-controlled resistor S. Varying the light intensity seemingly completely reorganizes the network.

Target (**T**5) also enters into these considerations. Assume we have built two gyrators from an isolator, according to Table 23.5. We single out the z_{11} term

$$z_{11} = R + \frac{R^2}{P} + S \qquad (23.15)$$

of the gyrator (the other terms follow correspondingly). In case *a* an isolator with $S = -R$, $P = \infty$,

$$z_{11}' = R - R = 0 \quad \text{(Table 23.4).} \qquad (23.16a)$$

In case *b*, with $P = -2R$ and $S = -R/2$,

$$z_{11}'' = R - \left(\frac{R^2}{2R} + \frac{R}{2}\right) = 0. \qquad (23.16b)$$

These cases seem to be identical under the conditions postulated. Let, however, R remain constant and S and P change 10 percent, say by drift or temperature effects. Then $z_{11}' = -0.1R$ for case *a*, while in case *b* $z_{11}'' = -0.0045R$; hence, this gyrator is much more stable. Thus we have here an excellent means of satisfying (**T**5).

Another set of possibilities related to (**T**4) and based on the reciprocal relationship of P and S in Eqs. 23.16 is an isolator with

$$P = \frac{1}{j\omega 2C} \quad \text{and} \quad S = \frac{2}{j\omega C};$$

$$z_{11} = R + R^2 j\omega 2C + \frac{2}{j\omega C} \qquad (23.17)$$

$$= R - \left(j\omega L' + \frac{1}{j\omega C'}\right).$$

The two capacitors have been commuted into a series resonant circuit. Utilization of such frequency effects for filters in conjunction with gyrators and isolators is obvious.

The foregoing observation brings us back to our active filter problem (**T3**). We can now become more specific about the separator mentioned only briefly before. If, for instance, the fourpole of Fig. 23.12 is a Type IV (*G*-Type amplifier), Table 23.3 yields

$$[Z] = \begin{bmatrix} S - P \dfrac{1}{g_{21} - 1} & -S \\ S - P \dfrac{g_{21}}{g_{21} - 1} & -S \end{bmatrix}. \quad (23.18)$$

For $P = S = 0$, one possible solution is

$$[Z] = \begin{bmatrix} 0 & 0 \\ 0 & 0 \end{bmatrix}. \quad (23.19)$$

For stability, g_{21} must be < 0, since otherwise $z_{11} < 0$, because $z_{11} = S + (P - g_{22})/(Pg_{11} + 1 - g_{21} - g_{11} g_{22})$ in the nonidealized case. Here we have the peculiar situation of disconnection or complete separation between input and output, although on account of P being zero, direct current is passed through without hindrance, provided the amplifier does not operate for direct current.

If in Fig. 23.14 the active fourpole chosen is a negator with $Z_3 = Z_4 = R$, other separators can be obtained, with

$$[Z] = \dfrac{1}{P + 2R} \begin{bmatrix} -R^2 + SP + 2SR & PR + R^2 - SP - 2SR \\ -PR - R^2 + SP + 2SR & R^2 - SP - 2SR \end{bmatrix}. \quad (23.20)$$

Since separation results for z_{12} and z_{21} being zero or

$$P(R - S) = R(2S - R), \quad (23.21)$$

P may again be zero for $R = 2S$. If the active element is blocked by a capacitor for direct current, direct current passes through P, which can be zero. For alternating current, the active element is effective, thus

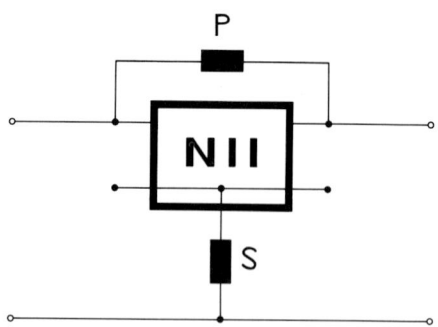

Fig. 23.14. A separator derived from a negator.

separating input and output or equivalently making the transfer impedance zero (**T3**).

Conclusions

A complement to true molecular designing is the mutual substitution of networks and materials. As molecular designers manipulate basic building blocks, we can manipulate basic units on the macroscopic level. By systematic network synthesis we have shown

1. Imperfect materials can be idealized.
2. Materials of not yet existing properties can be simulated.
3. Networks can be parametrically transposed as if completely rewired.

Typical examples are

For 1: Artificial creation of the skin effect.
For 2: A quasi magneticum, permitting the construction of reciprocal and even nonreciprocal transformers, operating from direct current on and in a completely planar structure without magnetic materials or windings.
For 3: The transposition of an isolator into a gyrator by changing light intensity impinging upon an element in the structure.

24 · NEW COMBINATIONS OF PHYSICAL PHENOMENA IN DEVICE DESIGN

Morris Tanenbaum

Introduction — Light beams as information carriers — Modulation of light beams — The magneto-optical modulator — The electro-optical modulator — The zigzag modulator — Piezoelectric semiconductor devices — Magnetoelastic devices — Conclusion

Introduction

Preceding chapters of this book discussed principles and phenomena of solid-state physics. The present chapter will deal with specific applications of such phenomena, selected to illustrate future trends in electronics technology.

The beginning of electronics technology dates from the first decade of this century with the invention of the first amplifying device, the vacuum-tube triode. This soon evolved into more complex devices such as tetrodes and pentodes, and in the 1930's and 1940's new classes of vacuum electronic devices were invented: the klystron, the traveling-wave tube, the magnetron, etc. In all these structures a stream of electrons interacts in vacuum with changing electric and magnetic fields, producing amplification, frequency conversion, and other important electronic functions. For the most part, solids are used only as structural elements, except for the thermionic emitter. Thus, the internal structure of the solids does not often participate in the primary interactions that provide the active electronic function.

The invention of the transistor signaled a strikingly new approach to electronic devices by performing the principal electronic functions inside a solid; thus internal structure becomes exceedingly important. This provides large technological advances by reducing power requirements and greatly improving device reliability. However, in the transistor, as in the vacuum tube, the governing interactions are still between free conduction electrons and externally applied electromagnetic fields.

As attempts are made to extend device capabilities to perform increasingly complex functions, increasingly difficult problems are encountered if consideration is restricted to direct interactions between free electrons and electromagnetic fields. Solutions to these difficulties may lie in the more subtle properties of matter, and it may be advantageous to consider the use of the more complex interactions inherent in the crystal lattice. The examples described in the following pages illustrate how such interactions, involving new combinations of physical phenomena, are being used in device design.

Light beams as information carriers

The properties of coherent light, as generated in lasers, provide many technological opportunities, such as the use of light as a high-capacity communications channel.

One of the principal problems to be solved before the coherent light of a laser can serve as a communications channel at its full potential is modulation at near-optical frequencies. It is common practice to operate well-developed communications carriers at bandwidths of the order of 10 percent of the carrier frequency, because at smaller bandwidths the carrier medium is not efficiently employed. Since optical frequencies are of the order of 10^{14} Hz, full use of the carrier would suggest modulation at frequencies of 10^{12} to 10^{13} Hz. A usable bandwidth of only 0.1 percent of the optical carrier frequency would provide 100 GHz of information capacity. Compared to present microwave carrier systems, with maximum bandwidths of about one gigacycle, such an optical system would present a significant challenge for systems engineers to develop economically other attractive uses for all of this capacity.

Modulation of light beams

Most of the successful light-modulating systems constructed to date are based on phase modulation, because materials are available in which the phase velocity of propagation can be varied at optical frequencies. In contrast, it has proved difficult to find a medium whose optical absorption can be changed with facility at high frequencies. However, since present optical detectors (e.g., photomultipliers or semiconducting photodiodes) are sensitive to energy and not phase, the optical beam must be converted after phase modulation by a suitable phase discriminator into an amplitude-modulated beam for detection.

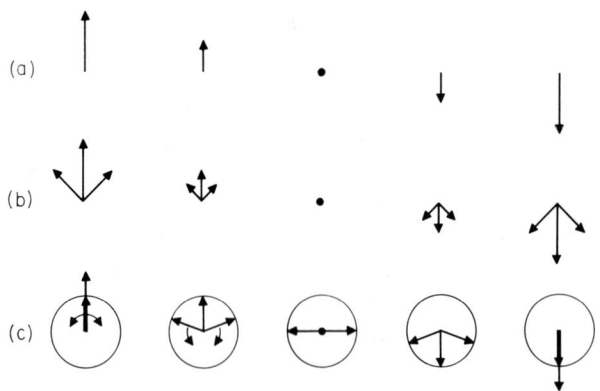

Fig. 24.1. (a) Resolution of plane-polarized light beam; (b) into two orthogonal components; (c) into two circularly polarized components.

Figure 24.1 indicates the electric field of a plane-polarized beam as it varies with time. The field vector, which traverses one half-cycle in (a) can be resolved into two electric vectors vibrating in phase at right angles in (b) or equivalently into two circularly polarized components in (c). Representation (b) will be used to explain phase modulation by the electro-optical effect and representation (c) for the magneto-optical effect.

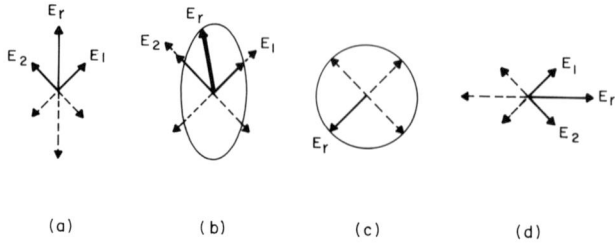

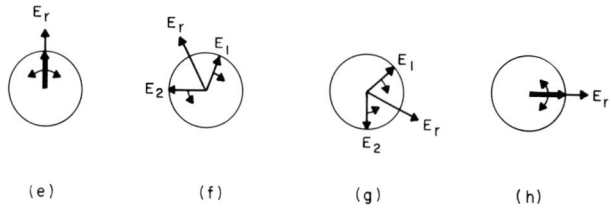

Fig. 24.2. Rotation of the plane of polarization of a light beam. Solid arrows indicate instantaneous value of resultant electric field E_r and its components E_1 and E_2. Dashed arrows indicate locus of these fields as function of time. In (a) E_1 and E_2 are in phase and E_r is plane-polarized. In (b) E_2 lags E_1 and E_r traverses an ellipse (elliptically polarized light). In (c) E_1 and E_2 are 90 degrees out of phase and E_r traverses a circle (circularly polarized light). In (d) E_2 has lagged E_1 by 180 degrees. E_r again traverses a plane now rotated 180 degrees from its original direction. Figures e to h indicate how the phase delay of the circularly polarized component of E_1 relative to E_2 produces the rotation of the resultant polarization E_r.

Figure 24.2 illustrates how the polarization characteristics of the light beam can be altered: In (a) the beam is passed through an optically anisotropic material so that the relative phase of the two orthogonal component vectors is changed. When one of the components is delayed by 180 degrees relative to the other, the plane of polarization of the beam is rotated by 90 degrees.

In (b) the light beam traverses a medium in which the propagation velocity depends upon the direction of rotation of the circularly polarized components. In similar fashion, a relative delay of one of the components produces a rotation of the plane of polarization.

These phase shifts can be converted into amplitude

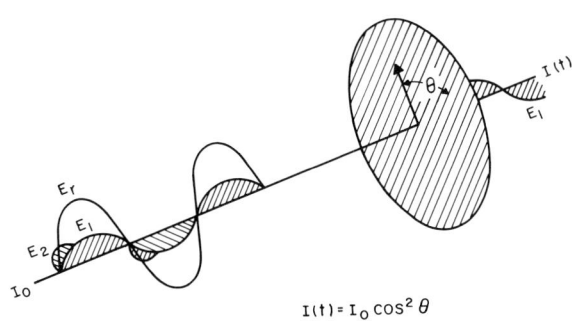

Fig. 24.3. The conversion of phase modulation to amplitude modulation. The analyzer is transparent to component E_1 and opaque to component E_2. Therefore, as the direction of polarization of E_r is varied, the magnitude of component E_1 changes and the amplitude of the component passing the polarizer is varied.

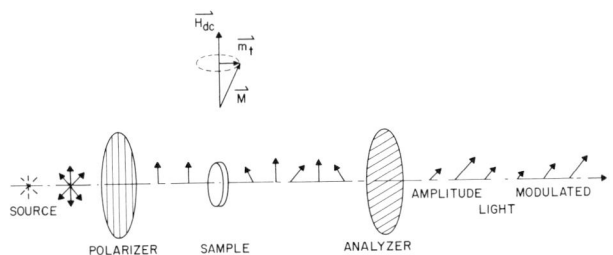

Fig. 24.4. Magneto-optical light modulator. (After Anderson.[1])

variations by use of a polarization analyzer, as illustrated in Fig. 24.3.

The magneto-optical modulator

The magnetic properties of solids arise from the spins of unpaired electrons. When such spins in solids are oriented by a d-c magnetic field and then subjected to the torque of a small a-c field normal to the former, the magnetic dipoles precess around the d-c magnetic-field axis. The frequency and direction of this precessing motion depends upon the magnitude and direction of the applied magnetic fields. This fact endows the materials with a circular magnetic anisotropy; a circularly polarized electromagnetic wave can interact with the precessing electron spin. This is the principle of microwave ferrite devices, e.g., gyrators, isolators, circulators, and switches. Such devices operate near the ferromagnetic resonance range (typically 10^6 to 10^{10} Hz).

It is well known that electronic transitions at optical frequencies reflect the circular anisotropy of the precessing electron spin in the Zeeman effect. In solids, the rotation of the polarization plane of light is known as Faraday rotation; it is especially strong in ferromagnetic materials. Rotations of the order of 5×10^5 deg/cm have been estimated for thin films of iron.

If this rotation arises from the magnetic properties of the material, one should be able to modulate it through the magnetization. On this basis, Anderson[1] has developed a magneto-optical modulator whose operation is indicated in Fig. 24.4. The active material, a thin slice of ferrimagnetic yttrium iron garnet (YIG), is placed in a microwave cavity, and an external d-c magnetic field is applied perpendicular to the propagation direction of the light beam. The sample is excited with an r-f field perpendicular to the applied d-c field. In the presence of the d-c field alone there is no magnetization along the direction of light propagation and hence no rotation of the polarization plane of the light. The r-f field produces an alternating net magnetization along the direction of light propagation and thus an alternating rotation of the polarization plane. As the intensity of the r-f field is varied, the amount of rotation is increased or decreased. If the light is now passed through an analyzer normal to the original direction of polarization, no light is transmitted in the absence of the microwave field. In the presence of a microwave field the intensity of transmitted light varies in synchronism with the microwave field.

With this device, measurable modulation has been achieved at frequencies near 10 GHz. In principle it could be extended to substantially higher frequencies; in practice, however, there are severe limitations.

The modulation sensitivity M of the device is

$$M = \rho l \exp\left(\frac{\alpha l}{2}\right) \sqrt{\frac{P \chi_m}{\omega \mu_0}}. \quad (24.1)$$

The Faraday constant ρ is expressed in units of degrees rotation per unit length and l is the length of the crystal. The product ρl measures the total rotation. The factor under the square-root sign contains the magnetic susceptibility χ_m, the r-f frequency ω, the permeability μ_0 of free space, and the microwave power P, required to give unit magnetization along the direction of propagation of the light beam. The exponential factor measures the optical loss due to the absorption coefficient α of the sample.

Optical absorption is an extremely important factor. Essentially, all materials with large Faraday rotation at room temperature also have large optical absorption. In magnetic metals, the specific rotation is very large, but the absorption is so prohibitive that it would not be possible to use samples more than about 100 Å thick. Even magnetic insulators, such as the ferrites and the garnets, have large optical absorption.

This limitation due to optical absorption prescribes short samples and hence large specific rotation per unit

---

[1] L. K. Anderson, *Appl. Phys. Letters* 1, 44 (1962); *J. Appl. Phys.* 34, 1230 (1963).

length; this implies large Faraday constants and high power input. The power P is limited by the temperature rise of the specimen. Anderson considered these effects in order to determine what kind of devices can be built with present materials (Table 24.1). The

Table 24.1. Performance of various materials in a Faraday rotation optical modulator

Material	Wavelength	Optimum Thickness	Maximum Percent Modulation at 10 GHz/sec	Modulation Power (mW)
YIG	0.85	92	0.10	65
Ga-YIG	0.7	44	0.06	30
CrBr$_3$	0.48	25	2.3	17

three materials he considered are yttrium iron garnet, a gallium-substituted yttrium iron garnet, and chromium tribromide. The parameters for chromium tribromide refer to 1.5° K; those of the others to room temperature. CrBr$_3$ is one of the most transparent ferromagnets but its Curie temperature is low (around 35° K). The maximum modulation that can be achieved in these materials at 10 GHz is 2 percent for CrBr$_3$ and a small fraction of a percent for the garnets, at relatively low power. Principal limitations are the very small thickness of the material, enforced by optical absorption, and saturation of the r-f susceptibility that occurs for large r-f power.

Intrinsically the magneto-optic modulator is attractive for modulating light at microwave frequencies. Modulation frequencies of many tens of gigacycles are conceivable, but new materials, with high specific rotation in a frequency region of good optical transparency, are required.

The electro-optical modulator

At present the most successful optical modulators make use of the electro-optical effect in crystals. Figure 24.2 shows schematically that a material with different refractive index in two mutually perpendicular directions can change the ellipticity of a light beam. Modulation requires change of the refractive index by electrical means.

Crystals can be divided into three optical classes (Fig. 24.5): In the cubic class the index of refraction is identical in the three principal directions; the crystal is optically isotropic. In the biaxial crystal the indices

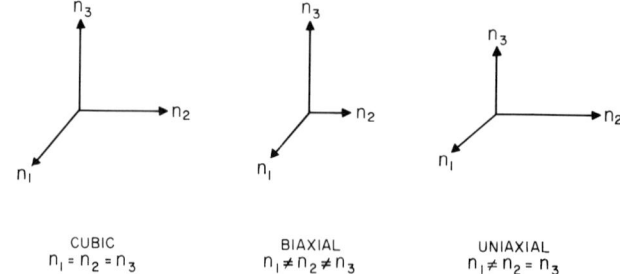

Fig. 24.5. Refractive properties of cubic, biaxial, and uniaxial crystal.

of refraction differ in all three principal directions; and in the uniaxial crystal the index of refraction has one value in a given direction (the optic axis) and a different but constant value perpendicular to that direction (tetragonal and hexagonal crystals are generally of this type). Light propagating along the optic axis will behave identically irrespective of its plane of polarization, because the index of refraction is the same in any direction perpendicular to the optic axis. In any other direction, differing indices of refraction are encountered, depending on the direction of polarization.

To build a modulator it is necessary to vary the index of refraction in one or more directions at high frequency. Here, nature has provided a way. It is well known that certain crystals, when placed in an electric field, respond mechanically with a change in physical dimensions. This is known as the "piezoelectric" effect. Pockels[2] observed that many piezoelectric materials are also electro-optically active. An electric field produces not only a change in mechanical dimensions but also in the refractive indices. The change is generally directly proportional to the applied electric field.

In the past fifty years a great many piezoelectric and electro-optically active crystals have been found. Indeed, any crystal that does not have a center of symmetry may be piezoelectric. Centrosymmetric crystals can also be electromechanically and electro-optically active,[3] but here the effect arises from an induced polarization that varies with the square of the applied field.

These effects are especially large in ferroelectrics just above their Curie temperature T_C.[3] (Ferroelectrics are the electric dipole analogue of ferromagnets; below a critical temperature T_c, the electric dipoles order and give a net spontaneous polarization.)

Figure 24.6 shows the electro-optical effect as a function of temperature in potassium dihydrogen phosphate

[2] F. Pockels, *Abhandl. Ges. Wiss. Göttingen* **39**, 1 (1893).
[3] For a review, cf. W. Känzig, "Ferroelectrics and Antiferroelectrics," *Solid State Physics*, Vol. 4, F. Seitz and D. Turnbull, Eds., Academic Press, New York, 1957, p. 5.

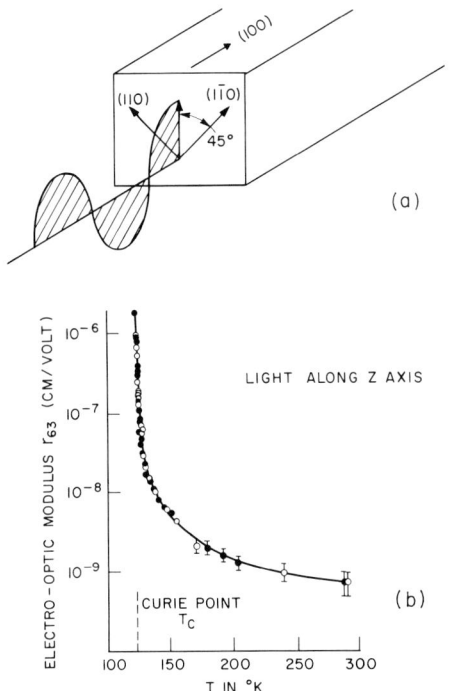

Fig. 24.6. The electro-optical properties of potassium dihydrogen phosphate (KDP).

(KDP),[4] a material which is tetragonal and piezoelectric at room temperature but becomes ferroelectric below 120° K. Just above this temperature there is a large anomalous increase in the electro-optical coefficient, covering about three orders of magnitude and obeying the Curie-Weiss law.

The electro-optical effect is closely related to the electromechanical effects and to the change in polarization with electric field. For example, in KDP both the piezoelectric coefficients and the polarization obey a Curie-Weiss law similar to that exhibited by the electro-optical coefficient, r_{63}. The bulk of these effects stems from small atomic displacements in the crystal lattice. Since the response of atomic movements extends to infrared frequencies, there is reason to expect electro-optical response to very high frequencies.

Figure 24.6a indicates how the electro-optical effect is used in the uniaxial KDP crystal. The optic axis lies along the [100] direction. A beam of plane-polarized light propagated along the [100] direction will emerge unaffected; however, if an electric field E_z is applied along the [100] direction, the crystal becomes biaxial and the refractive indices in the [110] and [1$\bar{1}$0] directions differ as

$$n_{[110]} - n_{[1\bar{1}0]} = n_1^3 r_{63} E_z, \qquad (24.2)$$

where $r_{63} \sim C/(T - T_c)$. The difference in refractive

[4] B. Zwicker and P. Scherrer, *Helv. Phys. Acta* 17, 346 (1944).

index is directly proportional to the electric field and relatively small because of the smallness of the electro-optical coefficient r_{63} (Fig. 24.6b). However, because the wavelength of light is short, even a small effect per unit wavelength can result in substantial over-all phase retardation in a thin crystal.

One of the earliest devices for light modulation at microwave frequencies was built by Kaminow,[5] who placed a KDP crystal in a microwave cavity excited by a magnetron at a frequency of 9.25 GHz (Fig. 24.7).

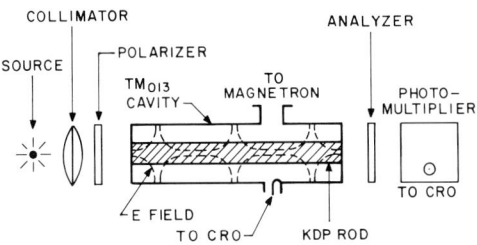

Fig. 24.7. Electro-optical modulator. (After Kaminow.[5]) (The polarized beam of light travels along a KDP crystal located in a TM_{013} microwave cavity.)

If the polarized beam travels through the crystal at exactly the same phase velocity as the traveling microwave excitation, the components of the light beam will experience a constant electric field, producing a relative phase retardation. If both the field and length of optical path are properly chosen, the plane of polarized light may be rotated by 90 degrees for certain parts of the wave train; the other portions of the light beam travel in phase with another part of the microwave excitation and experience a different phase retardation; and that light which travels in phase with a node in the microwave excitation will have no differential delay. Thus, the light beam is modulated at the microwave frequency.

In this discussion it has been assumed that the phase velocity of the light wave equals that of the microwave radiation. This condition is not easily met, since in most electro-optical materials the microwave dielectric constant is substantially larger than that at optical frequencies. The necessity to select materials and geometries so that the phase-velocity matching condition is approximated imposes an important restriction.

Another significant selection criterion is the magnitude of the electro-optical coefficients, since it determines the voltage required to produce a given phase retardation. For example, in the case of KDP several thousand volts are needed to produce a phase retardation of $\lambda/2$ at 7000 Å. Such large voltages lead to large power requirements; these can be reduced by in-

[5] I. P. Kaminow, *Phys. Rev. Letters* 6, 528 (1961).

Table 24.2. Data for a zigzag modulator (bandwidth 10 GHz)

	Electro-optical Coefficient	Index of Refraction	Dielectric Constant κ'	θ_1 (degrees)	D_b (mm)	d (mm)	l (cm)	P (for Modulation 50 Percent) ($M = \frac{1}{2}$)
KDP	7×10^{-10}	1.47	20	19	1	1.5	3.3	25 W
SrTiO$_3$	$2 \times 10^{-14} E$	2.41	300	8	1	3.6	1.4	5 W
CuCl	6×10^{-10}	1.93	8	42	1	0.75	6.7	1 W

creasing the Q of the cavity, but this results in restrictions on bandwidth. Kaminow's first KDP device required 760 watts of microwave power to obtain 50 percent modulation in a cavity whose Q limited the bandwidth to about 60 MHz.

The zigzag modulator

These first experiments represented a very primitive and not very practical form of the electro-optical modulator. Later investigators introduced refinements with greater practical promise.[6] Recently, DiDomenico and Anderson[7] analyzed a device called the zigzag modulator which had been proposed by Rigrod and Kaminow[8] (Fig. 24.8). The principal difference between the zig-

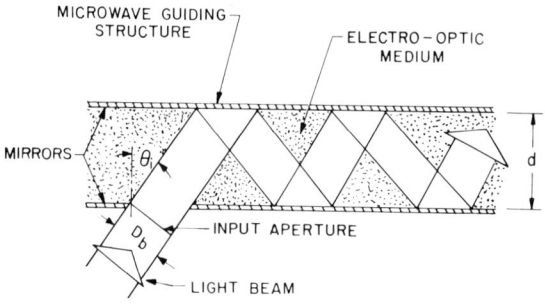

Fig. 24.8. Schematic view of the zigzag modulator.

zag modulator and Kaminow's first model is that the light beam bounces back and forth through the electro-optical medium between two reflecting surfaces which also act as the microwave waveguide. This geometric change has two principal advantages: a more effective use of the electro-optical material due to the longer optical path length and a new degree of freedom in achieving the critical matching condition between the phase velocities of light and microwave. The path length of the light beam can now be adjusted by chang-

[6] C. J. Peters, *Proc. IEEE* 51, 147 (1963).
[7] M. DiDomenico and L. K. Anderson, *Bell System Tech. J.*, in press.
[8] W. W. Rigrod and I. P. Kaminow, *Proc. IEEE* 51, 137 (1963).

ing the angle of incidence θ_i. A small θ_i results in a longer path length and hence slower effective traversal of the structure by the light.

The geometric parameters in this structure are θ_i (angle of incidence), D_b (diameter of light beam), d (diameter of waveguide), w (width of guide), and l (length of cavity).

Table 24.2 summarizes design calculations for the zigzag modulator for three electro-optical materials and for a modulation bandwith of 10 GHz.

Strontium titanate at room temperature is a cubic, centrosymmetric material, hence not piezoelectric. Unlike the piezoelectrics KDP and cuprous chloride, with their linear electro-optical effect, SrTiO$_3$ displays a quadratic response to the applied electric field unless a d-c bias is applied. A field of approximately 30,000 volts/cm will raise the electro-optical coefficient into the range of KDP. The refractive index of SrTiO$_3$ is 2.41, but its microwave dielectric constant of 300 is more than 10 times that of KDP. Hence, the microwave phase velocity is significantly slower than in KDP, and the optical beam must be slowed down by use of a relatively small θ_i (cf. Table 24.2).

Cuprous chloride has an electro-optical coefficient comparable to that of KDP and is cubic, noncentrosymmetric and optically isotropic in the absence of an electric field. In the zigzag structure the light beam cannot be constrained to a single crystallographic direction because of its bouncing pattern. Hence, in a material like KDP it is not possible to have the light beam travel always along the optic axis, and even in zero field there is a large net phase delay. Not only is this fact disturbing but the effect has a large temperature coefficient, thus increasing the compensation problem. The advantage of CuCl is the absence of this anistropy.

The small microwave dielectric constant of CuCl requires a large θ_i, causing an increase in the optimum length l of the modulator (cf. Table 24.2). The large angle of incidence provides an additional advantage, since it is within the critical angle for total reflection. Therefore, mirrors are not needed, because the light beam is trapped within the crystal by virtue of its refractive index.

The last column of Table 24.2 shows the calculated power P required for 50 percent modulation, and is especially encouraging. With any one of these materials it is theoretically possible to get large modulation at large bandwidth with relatively small power requirements. It must be emphasized, however, that the calculations assume perfect crystals with uniform optical and microwave properties. Unfortunately the best electro-optical materials are especially susceptible to strain-induced variations in refractive index. Random strains produce randomly distributed optical anisotropies that degrade modulator performance.

In summary: Materials are available which, through use of the electro-optical effect, have the potential for efficient modulation of coherent light at frequencies in the range of several tens of gigacycles. (Materials with even larger electro-optical coefficients would be desirable.) One of the most important present needs is large single crystals of high perfection with respect to internal strains. Of particular interest are crystals, such as CuCl, whose symmetry is cubic.

This discussion has been restricted to modulation frequencies four orders of magnitude below the modulation capacity of coherent light. There is much room for future invention. Efficient light modulation presupposes a deepening understanding of the fundamental factors that give rise to the optical response of materials.

Piezoelectric semiconductor devices

In the preceding section a relatively old phenomenon, the action of an electric field on the optical properties of a solid, was considered in combination with a new device discovery, the optical maser, in order to achieve a new system of potential technological values. Here we will illustrate how the discovery of a new class of interactions in solids has led to the invention of new types of devices. Until recently the piezoelectric effect in solids has been studied principally in insulators, which found their way into technology as electromechanical transducers, etc. Indeed, one of the principal requirements placed on these materials was very high resistivity, because electrical conductivity degrades performance. However, there is no fundamental reason for an electrical conductor not having interesting piezoelectric properties as well.

No one seemed inclined to look into these possibilities until Hutson[9] pointed out that the anomalous behavior of the electron mobility in semiconducting zinc oxide

[9] A. R. Hutson, *Phys. Rev. Letters* *4*, 505 (1960).

might be explained by its piezoelectricity. He found that the piezoelectric coefficients of zinc oxide were substantially larger than those of many of the commonly used piezoelectric materials, such as quartz. He also observed that other noncentrosymmetric semiconductors, such as gallium arsenide and cadmium sulfide were strongly piezoelectric.

Following these observations, White[10] suggested the possibility of using the interaction between conduction electrons and the piezoelectric properties of semiconductors for the direct amplification of acoustic waves in solids. This suggestion constitutes the invention of the piezoelectric semiconductor ultrasonic amplifier. Its principle is shown in Fig. 24.9.

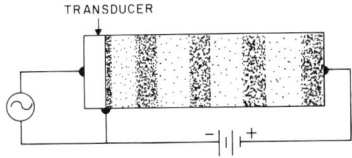

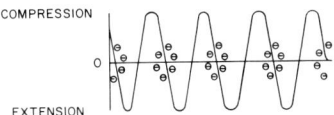

Fig. 24.9. Electron bunching in piezoelectric semiconductor.

If an acoustic wave is generated at one end of a piezoelectric semiconductor and travels through the material, the mechanical strain produced by the acoustic wave will create local electric fields that will interact with the conduction electrons and cause bunching. The electron bunches, in turn, tend to follow the moving field distribution. Therefore, as the acoustic wave travels through the material, it tends to drag the electron bunches with it. Since there is resistance to the movement of the electrons, there is also dissipation of energy in the form of resistive heating. The electrons extract energy from the acoustic wave and thereby add to its attenuation.

If now an electric field is applied to the sample moving the electrons in the direction in which the acoustic wave is attempting to drag them, the field can provide part of the energy that the electrons dissipate in the resistive medium. The attenuation of the acoustic wave is thus decreased. White pointed out that if the field is so great that the electron velocity exceeds the velocity of the acoustic wave, the electrons can deliver energy to the acoustic wave: The acoustic wave is amplified.

[10] D. L. White, *J. Appl. Phys.* *33*, 2547 (1962).

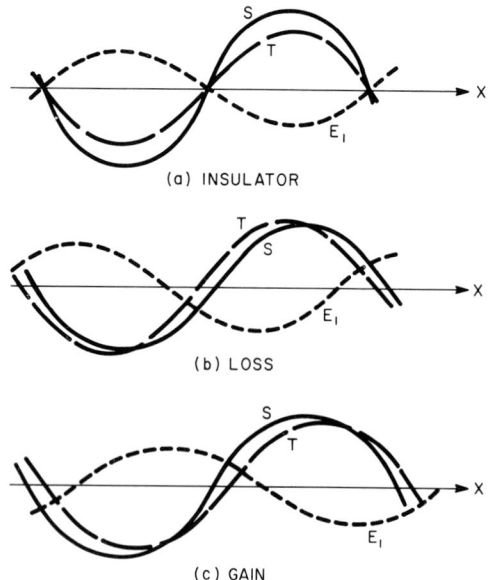

Fig. 24.10. Stress (T), strain (S), and a-c electric field (E_1) in an acoustically driven piezoelectric semiconductor; (a) without free carriers, (b) without d-c field, (c) d-c field applied.

Figure 24.10 shows stress (T), strain (S), and field (E_1), plotted for these various cases. Without free carriers (a) the electromechanical stress and strain are in phase and the field 180 degrees out of phase. If electrons are present (b) in the absence of a d-c electric field, the flow of current causes the electric field to lag the strain, as does the stress; therefore, a mechanical hysteresis causes loss of acoustic energy. When a d-c electric field is applied (c), the electrons move along faster, thereby forcing the piezoelectric field to lead the strain. This also forces the stress to lead the strain, causing the acoustic wave to be amplified.

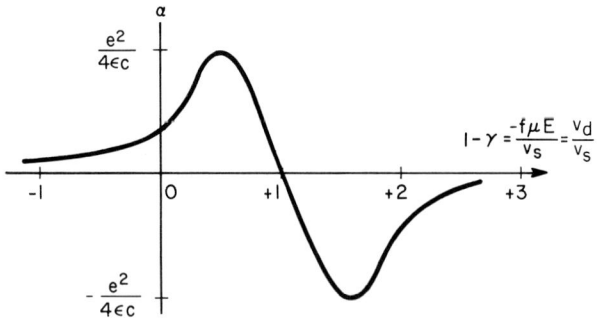

Fig. 24.11. Attenuation α as a function of $(1 - \gamma)$ in a piezoelectric semiconductor.

Figure 24.11 shows the conditions for actual amplification. The attenuation α is plotted as function of a dimensionless parameter $(1 - \gamma)$, equal to the electron drift velocity divided by the acoustic velocity. When the electron velocity is exactly equal to the acoustic velocity, the attenuation is zero. For greater electron velocity the attenuation coefficient becomes negative, indicating acoustic gain.

There is an interesting similarity between this device and the electro-optical modulators discussed earlier. The modulation of the light wave was greatest when it and the microwave excitation traveled with the same phase velocity along the traveling-wave structure. The piezoelectric semiconductor amplifier is also a traveling-wave device; amplification begins when the velocity of the electron clusters exceeds the velocity of the traveling acoustic wave.

Both systems have analogies to other electronic traveling-wave amplifiers. In the traveling-wave electron tube an interaction between a traveling electromagnetic wave and a beam of electrons is produced, and again the problem exists of matching the phase velocity of the electromagnetic wave with the particle velocity of the electron beam. This can be accomplished by making the electromagnetic wave follow a tortuous path (along a helix of wire for example), while the electron beam travels in a straight line down the axis of the helix. In the zigzag electro-optical modulator the light wave is forced along a tortuous path by reflection between mirrors. In the ultrasonic amplifier the acoustic wave travels at the velocity of sound, and the velocity of the conduction electron bunches is controlled by adjusting the accelerating field. In all three cases strong interactions occur when the velocities of the interacting energy streams are matched.

The interaction between electrons and acoustic waves in a piezoelectric semiconductor is relatively small, principally because the piezoelectric effect is also relatively small compared with interactions between electric fields and electrons in the traveling-wave tube. On the other hand, very strong interactions are not required for significant effects, because at high frequencies the length of the ultrasonic wave is small. For example, at a frequency of 2 GHz in cadmium sulfide, the wavelength of an ultrasonic wave is about 1μ; a sample of 1-cm length will therefore contain 10,000 wavelengths. Thus the over-all effect can be quite large.

Figure 24.12 shows a theoretical curve for the amplification of an ultrasonic wave in cadmium sulfide as function of frequency. Note in particular the very high gains that are theoretically possible and the very high frequencies at which these devices may be expected to operate. These curves show that there is promise for these devices not only as ultrasonic but also as electric amplifiers. Here the electrical signal would be converted into an acoustic wave, amplified, and reconverted to an electrical signal.

At frequencies of a few gigacycles, other kinds of

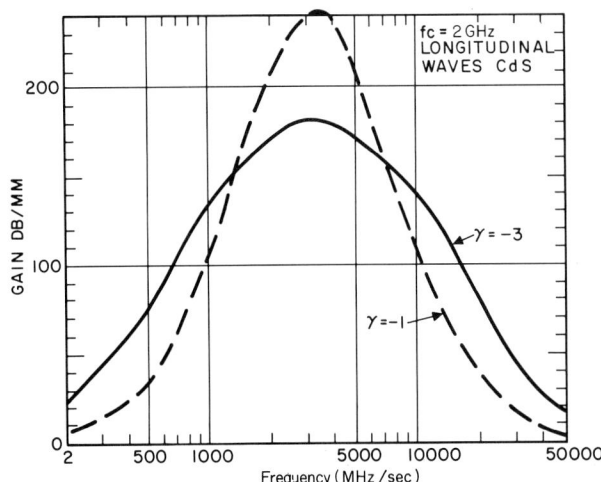

Fig. 24.12. Theoretical plot of acoustic amplification in cadmium sulfide for longitudinal acoustic waves and two values of γ.

solid-state amplifiers, such as transistors, become extremely difficult to fabricate because such devices must be above all small. During the development of vacuum-tube amplifiers a similar evolution occurred. With higher and higher frequencies it became increasingly difficult to make the standard types of triode, tetrode, or pentode amplifier because of the small dimensions required. It was the traveling-wave type of device that opened up the high-frequency spectrum in vacuum electronics. The traveling-wave nature of the ultrasonic amplifier may very well permit the application of solid-state devices at increasingly high frequencies.

It must be emphasized, however, that no piezoelectric semiconductor ultrasonic amplifier that produces usable net gain at microwave frequencies has as yet been announced. Electronic gain has been reported from ~50 MHz up to 1 GHz, and net gain has been reported at 60 MHz.[11] One problem is that present methods for the interconversion of electrical and ultrasonic energy are relatively inefficient. Once the ultrasonic wave is produced in the piezoelectric semiconductor crystal, gains of 50 to 60 db are readily obtained, but the losses of going into the crystal and out again are often greater than this. White[12] proposed a particularly attractive way of producing an efficient high-frequency transducer that may lead to a completely integrated piezoelectric amplifier.

[11] J. H. McFee, private communication; cf. also F. S. Hickernell and N. G. Sakiotis, Motorola, Inc., Paper D-7, 1963 Ultrasonics Symposium, sponsored by the Professional Technical Group on Ultrasonic Engineering of the IEEE.
[12] D. L. White, *IEEE Trans. on Ultrasonic Eng.* UE-9, 21 (1962); cf. also N. F. Foster, *J. Appl. Phys. 34*, 990 (1963).

Figure 24.13 shows a classical piezoelectric transducer in the form of an insulating piezoelectric slab with conducting electrodes attached to opposite faces. The piezoelectric might be a quartz crystal or a thin wafer of a ceramic such as barium titanate. The principal difficulty in this construction is that at higher frequencies the transducers must be made thinner. In general, a transducer should have a thickness of about one half an acoustic wavelength. At 1 GHz this means that the transducer should be about $1\,\mu$ thick; such devices are almost impossible to fabricate. White suggested that the properties of the piezoelectric semiconductor might permit the practical fabrication of structures of the required dimensions.

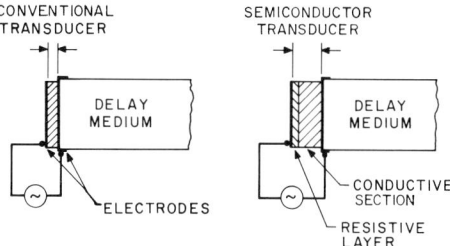

Fig. 24.13. Conventional and semiconducting piezoelectric transducer.

For practical operation of a transducer an insulator is required so that the exciting a-c signal is efficiently converted to mechanical vibrations without loss from conduction processes. White pointed out that it is possible to vary the resistance of a piezoelectric semiconductor over wide ranges by controlling its impurity content. By adding certain impurities (e.g., Cu) to CdS, materials with resistivities $\geq 10^9$ ohm cm result; the crystal becomes a good insulator. It is possible to introduce such impurities by solid-state diffusion; hence, extremely thin layers can be produced. Thus one can conceive of a completely integrated device where all functions are performed in a monolithic piece of crystal.

The effort to produce a practical amplifier is well under way. Obviously one must have control of crystal properties comparable to that required for more conventional devices, such as germanium and silicon transistors.

Magnetoelastic devices

One other group of classical interactions whose potential has greatly expanded during the past few years are magnetoelastic effects in solids. For many decades single crystals of quartz represented the peak of performance in ultrasonic devices; the highest mechanical

Q ever measured ($\sim 10^6$) was obtained on quartz crystals.

In 1961 LeCraw et al.[13] reported measurements of mechanical Q of $\sim 10^7$ in yttrium iron garnet (YIG) single crystals at 9 MHz, indicating mechanical losses at that frequency approximately six times smaller than ever measured in any material.

Yttrium iron garnet is a ferrimagnetic material; hence, there is coupling between its elastic and its magnetic properties. This phenomenon is known as magnetostriction; the magnetostriction of metals has been used in simple acoustic devices for many years. In analogy to the piezoelectric effect, magnetostriction implies that in the presence of a magnetic field the material changes physical dimensions. Conversely, if a mechanical force acts on the material, its magnetization changes. These effects, coupled with the discovery of exceedingly low acoustic losses in magnetic garnets, promise new classes of devices.

As mentioned earlier, an unpaired electron possesses a magnetic moment. In a magnetic field this magnetic moment will precess about the field with a frequency and direction of precession that depend on the magnitude and direction of the magnetic field. If such an electron is placed in a circularly polarized electromagnetic field where the magnetic-field vector rotates in the same direction and with the same frequency as the electron-magnetic dipole, there will be strong interaction between the electromagnetic field and the electron. This is the basis for the Faraday rotation of microwave radiation and permits the construction of many important microwave devices.

Matthews and LeCraw[14] used the same effect to get rotation of the polarization plane of an acoustic wave. The elements of their device are shown in Fig. 24.14. A quartz disk transducer is bonded to one face of a cylindrical single crystal of YIG. A pulse is applied to the face of the quartz disk to generate a 528-MHz acoustic wave. The disk is so cut that it generates an acoustic wave which has a particle displacement transverse to the axis of the YIG cylinder in the direction shown. The wave travels down the YIG rod and is reflected at the opposite end. The reflection travels back and excites the quartz transducer, which now acts as a receiver. It is reflected from this surface also, and a part of the wave begins a second trip down the rod. Because of the high mechanical Q, a given pulse can make many passes through the rod.

The response of the quartz transducer depends upon

[13] R. C. LeCraw, E. G. Spencer, and E. I. Gordon, *Phys. Rev. Letters* **6**, 536 (1961).
[14] H. Matthews and R. C. LeCraw, *ibid.* **8**, 397 (1962).

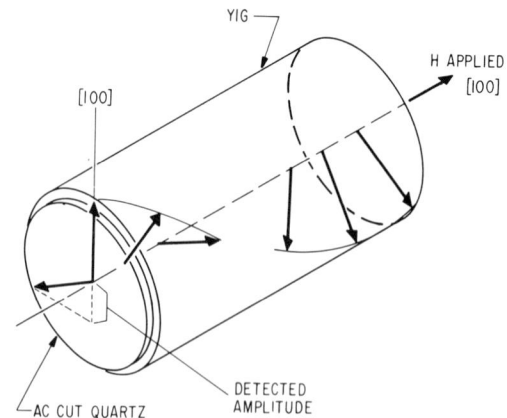

Fig. 24.14. Rotation of a plane-polarized acoustic wave in a magnetoelastic crystal. (An a-c cut-quartz transducer that produces plane-polarized acoustic waves is bonded to the polished end of a cylindrical single crystal of YIG. In the presence of a longitudinal magnetic field the acoustic wave is rotated nonreciprocally.)

the direction of polarization of the acoustic wave exciting it; the maximum response results when the polarization is identical to that of the wave originally launched, and minimum response occurs when it is rotated 90 degrees from the original polarization.

Because of the magnetomechanical coupling, the acoustic wave produces a magnetic field that can interact with the electron spins. The component of the acoustic wave that rotates in the same sense as the precession of the electron spins will interact more strongly and in consequence be retarded or advanced more than the other circular component. This differential retardation gives rise to rotation of the plane of polarization of the total wave. This can be observed by watching the output of the quartz transducers on successive echoes of the acoustic wave.

Figure 24.15 shows the response of the transducer to successive echoes. The amplitude first decreases, then increases, and decreases again. This can be explained

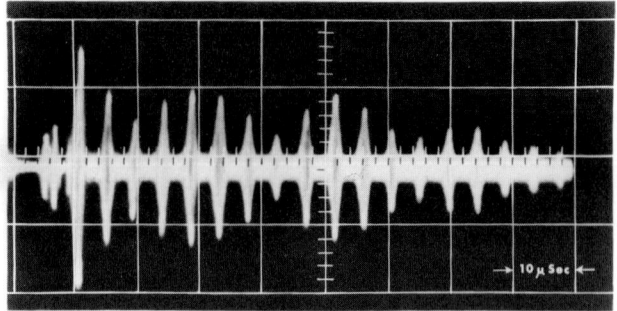

Fig. 24.15. Oscilloscope trace of quartz-transducer response to succeeding echoes of plane-polarized acoustic wave in the device of Fig. 24.14.

by rotation of the elastic wave as it travels down the rod. Furthermore, the picture shows that the rotation is nonreciprocal. If it were reciprocal, an equal and opposite rotation would be produced by the return signal, and no net rotation would result. Thus, this is a demonstration of the Faraday rotation of an acoustic wave in a magnetic medium.

Conclusion

The preceding examples attempt to illustrate how, as electronic science and technology become more sophisticated, new ways are discovered for utilizing interactions between solid-state phenomena, some of which were previously scientific curiosities. The examples chosen were magneto-optical, electro-optical, electromechanical, and magnetomechanical interactions. These are of course but a few of the possibilities. In these examples the fundamental atomic structure of crystals acts as the transducer through which the interactions take place. The ability to construct proper atomic configurations in matter for the synthesis and control of these interactions will determine the future of device design.

25 · PRINCIPLES OF ENERGY CONVERSION

Clarence Zener

Introduction — Free energy of mixing: sea water versus fresh water — Chemical energy: oxygen-concentration cell — Radiant energy: solar cells — Magnetic sieves — Electronic heat engines — Material limitations on thermionic converters — Material limitations on thermoelectric converters — Nernst-Ettingshausen generator

Introduction

The world we live in is, fortunately, a far from equilibrium world. Free energy is being irreversibly dissipated at a great rate in many different ways. Free energy is also stored by nature in many large reservoirs. Those of us who study energy conversion are seeking ways of diverting free energy, now being dissipated, into channels that perform useful work for mankind. We are also seeking new and more economical methods of tapping the available free-energy reservoirs.

Physics forms only a part of the science of energy conversion. It tells us, for example, how much energy is being dissipated by a given natural process, or how much free energy is stored in a given natural reservoir. It also gives us possible methods of diverting the dissipation of free energy and possible new ways of tapping the reservoirs. Technology provides the second necessary part. It guides us to those developments which, if successful, will also be useful. The usefulness will come about because of a favorable, competitive cost if the application is for civilian use, because of a favorable weight advantage if the application is for space application, or because of a more esoteric advantage if the application is for the military. In some cases the bottleneck for a particular energy-conversion process is primarily a problem in physics, such as, for example, in thermoelectric energy conversion. In other cases, all the physics is well understood, and the primary problem is technological. In general, we shall find an investigation of energy conversion more stimulating if we look at both physics and technology.

The tapping of free energy to perform useful work is commonly only half of the general problem of energy conversion, the other half being the complementary problem of using work to replenish a reservoir of free energy. Particular reservoirs of free energy are frequently wanted in particular places. Thus, the conversion of heat into work has the complementary problem of using work to pump heat, i.e., refrigeration. In nearly all cases the physics of the two complementary problems are identical. The technological problems are, however, far from identical. In this chapter, both of these two complementary aspects of energy conversion are discussed.

Free energy of mixing: sea water versus fresh water

Technology has already diverted a sizable portion of the formerly irreversible downward flow of rivers into useful work in the form of hydroelectric power. Technology has not, however, tackled the harnessing of the free energy dissipated by the irreversible mixing of the river waters with the salt water of the oceans. This neglect presumably arises from a negligible value of this free-energy dissipation. We are thus led to ask just how large is this dissipation of free energy.

The osmotic pressure of the salt in the sea water is 25 atm. Now, 25 atm is just the head pressure developed by a column of water 840 feet high. The free energy dissipated by a river through its irreversible mixing with the ocean is thus equal to the dissipation associated with its plunging over a fall 840 feet high.

Our complementary problem, the extraction of fresh water from the ocean, furnishes us with a guide on how to tackle the problem of diverting this free energy of mixing into useful work. The extraction of fresh water from the ocean is being accomplished by two extreme procedures: In the first, one literally plucks molecules of water from the ocean. This procedure is called *distillation*. In the second, one pulls the salt ions away from the water which we wish to freshen. This procedure is called *electrodialysis*.

The distillation procedure is already in large-scale use. Thus, plants generating 1,000,000 gallons a day are in continuous operation. We are therefore led to suspect that for the extraction of the work potential of mixing, the process complementary to distillation would be the most feasible. The basic problem to the distillation of sea water is the use of the heat of condensation of the fresh water for the evaporation of more water from the ocean. Unfortunately, heat can be transferred only by allowing it to flow down a temperature gradient. But such a flow is necessarily accompanied by the generation of entropy and, hence, by the degradation of free energy.

By making use of only known physical constants, by assuming a reasonable cost of heat-exchange surfaces of one dollar per square foot, and by using the standard costs of capital used by many utilities, namely, 2×10^{-3} cent/hr per dollar cost of equipment, one may readily estimate the cost of sea-water conversion by the thin-film distillation process, and also the cost of electrical power generated by mixing fresh water with sea water via the reverse of this distillation process. One obtains about 60 cents per kilogallon for the cost of fresh water from the ocean; about 60 cents for the cost of one kilowatt hour of power generated by mixing fresh water with sea water.

A method designed specifically to minimize this rate of entropy production has been developed at the Westinghouse Research Laboratories under contract with the Office of Saline Water. In this method, the surface of the evaporating sea water is brought ex-

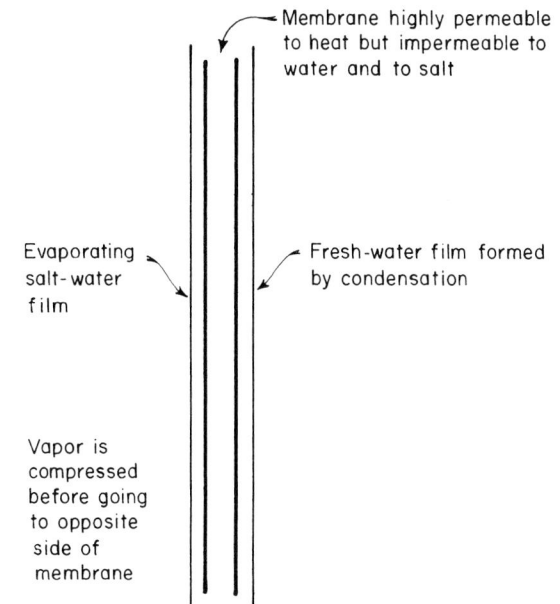

Fig. 25.1. Thin-film distillation process for extraction of fresh water from ocean.

tremely close to the surface of the condensing fresh water, as indicated in Fig. 25.1. In this arrangement, heat need flow through only several mils of water. An arrangement for solving the complementary problem, the extraction of work through mixing fresh water and sea water, is obtained merely by letting the compressor in Fig. 25.1 run backward as a turbine (Fig. 25.2). We have here an example of two complementary processes in which one, the winning of fresh water from the ocean

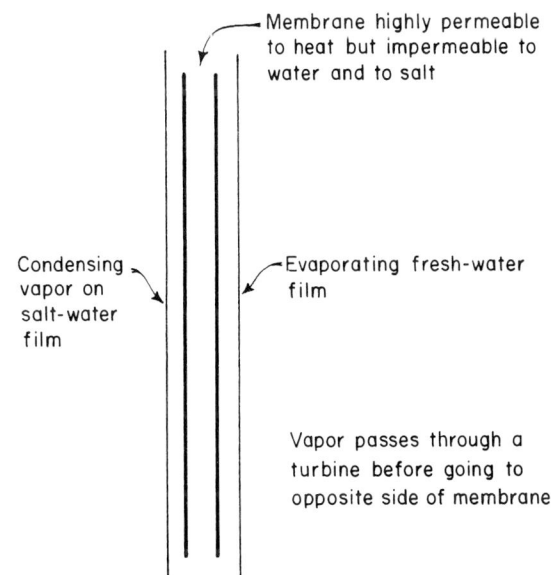

Fig. 25.2. Thin-film distillation process for extracting free energy by mixing fresh water with ocean water.

by a thin-film distillation process, is economically competitive with other processes; whereas the other, the winning of free energy by the mixing of fresh and salt water, is a much more expensive method of obtaining power than is furnished by conventional processes.

Chemical energy: oxygen-concentration cell

Next to nuclear energy, the existence of vast stores of underground hydrocarbons, together with the existence of even still greater amounts of uncombined oxygen above the ground, provides our greatest amount of stored work potential. Currently, this storehouse of work potential is being tapped by irreversibly combusting the hydrocarbons and by using the heat derived therefrom to drive a heat engine. The efficiency of large central station power plants, commonly defined as

$$\text{Efficiency} = \frac{\text{work output}}{\text{heat input}}, \qquad (25.1)$$

is about 40 percent. Since the reversible reaction of high-molecular-weight hydrocarbons with oxygen is an endothermic process, the available work potential is somewhat higher than the heat of combustion. The true efficiency, defined by

$$\text{Efficiency} = \frac{\text{work output}}{\text{potential work}}, \qquad (25.2)$$

is thus a little less, by several percent, than the commonly defined efficiency. One motivation for fuel-cell research is the development of a method for economically tapping this large reservoir of work potential by a method more economical than irreversible combustion.

Fuel cells[1] may be regarded simply as oxygen-concentration cells. Thus, consider two chambers of different oxygen partial pressure separated by a semipermeable membrane permeable to $O^=$ ions but not permeable to free electrons (Fig. 25.3). When the two partial pressures of oxygen are first introduced on the two sides of the semipermeable membrane, $O^=$ ions will start flowing across the membrane from the high- to the low-pressure side. Such a flow will continue until the driving force due to the oxygen partial-pressure difference is exactly counterbalanced by the force arising from the space charge induced on the two membrane faces by the charge transfer. More precisely, the flow will continue until the voltage V generated across the two faces of the membrane is given by

$$4eV = kT \ln\left(\frac{P_H}{P_L}\right). \qquad (25.3)$$

Power is obtained from an oxygen-concentration cell

[1] S. S. L. Chang, *Energy Conversion*, Prentice-Hall, Englewood Cliffs, N. J., 1963, Chap. 7.

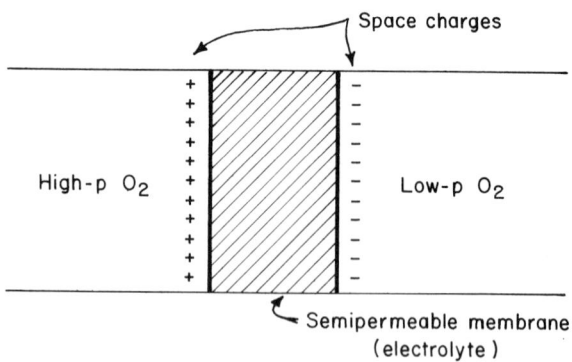

Fig. 25.3. Essential features of an oxygen-concentration cell.

by placing on each surface of the membrane a conductor network porous to O_2 and, finally, by joining these two porous conductors to a load. The identity of a fuel cell with an oxygen-concentration cell is evident when we identify the high-pressure O_2 with the atmospheric partial pressure of oxygen, the low-pressure O_2 with the partial pressure of oxygen in equilibrium with a fuel. The semipermeable membrane is called the electrolyte; the porous conductors attached to each surface of the electrolyte are called the electrodes.

When one burns solid carbon, one can approach the theoretical efficiency of unity with a single cell, provided the current density is sufficiently low. Thus the oxygen pressure will be maintained at a constant unique value determined by the set of equations

$$\frac{(O_2)}{(CO)^2} = K; \qquad (25.4)$$
$$(O_2) + (CO) = 1.$$

If, however, we burn a gas, the ideal efficiency can be approached only by either burning the gas in batches, thereby allowing the voltage to drop as the gaseous fuel becomes diluted with the gaseous end product, or burning the gas continuously in a multiple-staged process. As an example, suppose we wish to complete the combustion of coal by burning CO to CO_2. The oxygen concentration, now determined by the set of equations

$$\frac{(O_2)(CO)^2}{(CO_2)^2} = K'; \qquad (25.5)$$
$$(O_2) + (CO) + (CO_2) = 1,$$

is a function of the burning ratio $(CO_2)/[(CO_2)+(CO)]$. A plot of the voltage, as a function of this burning ratio, is given in Fig. 25.4 for the case of burning of H_2. The ideal work obtained by the combustion of one mole of H_2 is given by the integral of this curve. Such ideal work is approached only by a multiple-staged system, in each stage of which the burning ratio

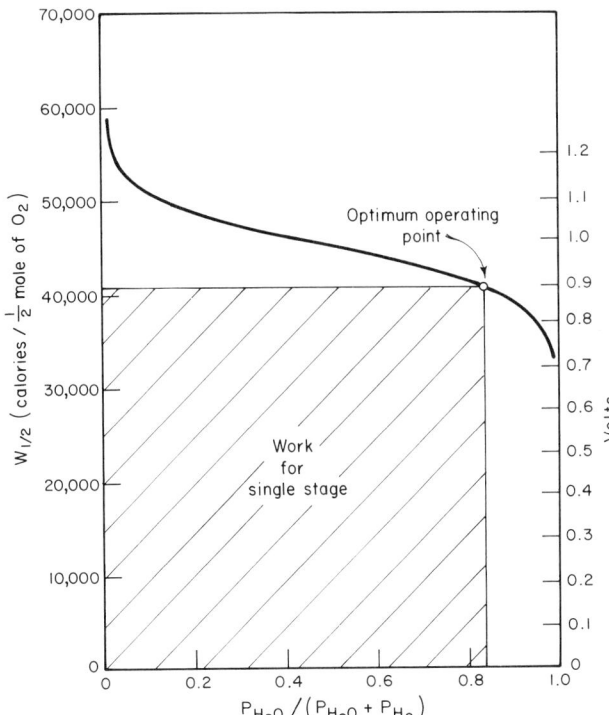

Fig. 25.4. Hydrogen-oxygen fuel cell, 1000° C.

(H_2O)/[(H_2O) + (H_2)] increases by a very small amount. If only a single stage is used at one output voltage, the work obtained is given by a rectangle enclosed by the operating curve. The optimum operating point for a single stage is shown in Fig. 25.4. A single stage introduces a degradation in work output from the ideal work by a factor of 0.78.

A technical problem common to all fuel cells is the rapid combustion of the fuel. One solution to this problem is the use of suitable catalysts; another is to operate at sufficiently high temperatures so that combustion takes place. This second solution promises to allow the use of commercial hydrocarbons as a fuel. It will therefore be discussed in further detail.

In solving the combustion problem by going to a high temperature, we introduce the possibility of using a solid-state electrolyte. The ionic conductivity of any solid must, of course, have a heat of activation; hence, the electrical resistivity of such a solid will decrease rapidly with increasing temperature. A suitable electrolyte material must conduct only through the migration of $O^=$ ions. It must therefore be an oxide in which all the cation sites are filled and in which vacancies are present in the anion ($O^=$) sites. Such a material may thus be fabricated by taking any oxide in which the cations have only complete inner shells, such as ZrO_2, and adding thereto as a doping agent any oxide in which the cation has a lower valency than in the first oxide.

Thus, if we take ZrO_2 as the base oxide and add thereto some CaO, each Ca^{2+} will occupy a site normally occupied by Zr^{4+}, and one vacancy will appear in the $O^=$ sites. Such vacancies will, of course, allow the migration of $O^=$ ions. A challenging problem in solid-state physics is the design and fabrication of an electrolyte with a small heat of activation for the migration of the holes in the $O^=$ sites.

The development of an economical fuel cell or oxygen-concentration cell will, of course, open up the possibility of a new method of separating oxygen from the atmosphere. One would simply run the oxygen-concentration cell in reverse, i.e., by an applied voltage force the oxygen from the partial pressure of the atmosphere to a higher partial pressure.

Radiant energy: solar cells

The most appalling waste of free energy is of course the conversion of the high-grade radiant energy of the sun, corresponding to a temperature of 6000° K, into low-grade heat. The magnitude of this waste is best appreciated by observing that the total triweekly heat falling upon the earth in the form of high-grade radiant energy is equal to all the potential heat stored within the earth in the form of coal.[2,3] The difficulty associated with the direct channeling of the sun's radiation into useful work lies in the capital cost associated with conversion equipment. No method has yet been conceived in which the amortization costs per kilowatt hour of such equipment even approaches the low cost of a kilowatt hour of work generated by conventional equipment. The application of equipment for directly tapping the sun's radiant free energy is therefore in those places where conventional power equipment is not feasible, such as in space satellites. Here we are primarily concerned with a high ratio of power per unit weight, together with reliability and long life under space environment.

Semiconductor solar cells give us one method of directly converting the sun's radiant energy into work.[4,5] The essential features of a solar cell are presented in Fig. 25.5. Photons with an energy greater than E_G create an electron-hole pair when absorbed. This electron-hole pair may be thought of as an extra minority carrier plus an extra majority carrier. If, as

[2] P. C. Putnam, *Energy in the Future*, D. Van Nostrand Co., New York, 1953, p. 107.
[3] M. K. Hubbert, *Energy Resources*, Nat. Acad. Sci., Nat. Res. Council Publication No. 1000-D, 1962.
[4] D. M. Chapin, *Introduction to the Direct Utilization of Solar Energy*, A. M. Zarem and D. D. Erway, Eds., McGraw-Hill Book Co., New York, 1963, Chap. 8.
[5] Ref. 1, Chap. 6.

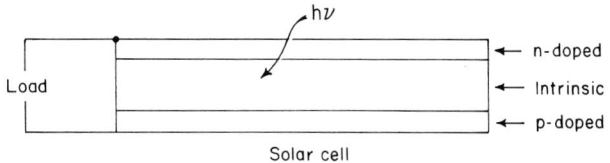

Fig. 25.5. Essential features of a solar cell.

is the case in a well-designed cell, the diffusion length is comparable to or greater than the cell thickness; the minority carrier will migrate to that region where it is a majority carrier. Thus, if the electron-hole pair is created in the n region, the hole will migrate to the p region, and vice versa. This migration, in turn, produces a forward bias across the junction.

Even under ideal conditions the efficiency of a solar cell is considerably less than unity. Thus, of the total radiant energy falling upon a cell, only the fraction

$$\frac{\int_{\nu_G}^{\infty} E_\nu \, d\nu}{\int_0^{\infty} E_\nu \, d\nu} \quad (25.6)$$

can be used in creating electron-hole pairs. Here $E_\nu \, d\nu$ is the radiant energy of frequency ν lying within the frequency range $d\nu$, and $h\nu_G$ is the energy gap of the semiconductor. Under ideal conditions a photon of energy $h\nu$, which creates an electron-hole pair, cannot use all this energy in useful work, but at most $h\nu_G$. The rest is irreversibly lost as heat. The factor by which the efficiency is degraded because of spectrum distribution is therefore

$$\delta_s = \frac{\int_{\nu_G}^{\infty} (\nu_G/\nu) E_\nu \, d\nu}{\int_0^{\infty} E_\nu \, d\nu}. \quad (25.7)$$

In space we may, to a good approximation, regard the spectral distribution function E_ν as that corresponding to a black body at temperature T_s. Such a distribution gives

$$\delta_s = (\tfrac{1}{2}) e^{-x}(x^2 + x), \quad (25.8)$$

with $x = h\nu_G/kT_s$. Upon solving for that value of x which minimizes δ_s, we obtain

$$x^2 - x - 1 = 0 \quad (25.9)$$

and hence,

$$x = 1.6 \quad (25.10)$$

or

$$h\nu_G = 1.6 \, kT_s. \quad (25.11)$$

Substituting the optimum value of x from Eq. 25.10 into Eq. 25.8 gives

$$(\delta_s)_{\max} = 0.42. \quad (25.12)$$

Recalling that k is 86 μeV/°K, and taking 6000°K for T_s, we find

$$h\nu_{G|\text{opt}} = 0.83 \text{ eV}, \quad (25.13)$$

close to the energy gap for germanium.

The next most serious degradation comes from the voltage developed by the cell being less than $h\nu_G$. To estimate this voltage, we investigate the effect of illumination of the p region upon open-circuit conditions. Let J_0 be the "dark" electron current from the p to the n region. This is balanced by the equal "dark" electron current from the n to p region. Under illumination we have an additional induced electron current J_I from the p to the n region arising from the electron-hole pair created by the illumination. Under open-circuit conditions, the total $p \to n$ electron current $J_0 + J_I$ must be balanced by an equal forward $n \to p$ electron current which, however, can be increased above J_0 only by a forward voltage V. This $n \to p$ electron current is just $J_0 \exp(eV/kT)$. We thereby obtain

$$J_0 e^{eV/kT} = J_0 + J_I \quad (25.14)$$

or

$$\frac{eV}{kT} = \ln\left(\frac{J_0 + J_I}{J_0}\right) \simeq \ln \frac{J_I}{J_0}. \quad (25.15)$$

Now

$$J_0 \sim e^{-E_G/kT} \quad (25.16)$$

and, hence,

$$\frac{eV}{kT} = \frac{E_G}{kT} - a, \quad (25.17)$$

where a is a weak function of several parameters, including the illumination intensity and material constants. Our voltage degradation factor may now be defined by

$$\delta_v = \frac{v}{v_G} = 1 - \frac{E_0}{E_G}, \quad (25.18)$$

where

$$E_0 = \frac{a}{kT} \quad (25.19)$$

must be determined from the characteristics of a typical semiconductor and from the assumption of full solar radiation. From a plot of δ_v versus E_G by Wolf,[6] one finds that

$$E = 0.44 \text{ volt}. \quad (25.20)$$

We previously found the E_G that maximizes the single degradation factor δ_s (Eq. 25.13). We shall now find that E_G which maximizes the product

$$\delta_s \delta_v = (\tfrac{1}{2}) e^{-x}(x^2 + x)\left(1 - \frac{0.86}{x}\right). \quad (25.21)$$

[6] M. Wolf, *Proc. IRE* **48**, 1246 (1960), Fig. 6.

In this equation we have utilized the relation

$$\frac{E_0}{kT_s} = 0.86. \qquad (25.22)$$

That value of x which maximizes this product is the appropriate root of

$$x^2 - 1.86x - 1 = 0; \quad x = 2.30, \qquad (25.23)$$

corresponding to

$$E_G = 1.2 \text{ eV}. \qquad (25.24)$$

This is just the energy gap of silicon. Substitution of this value of x into Eq. 25.21 gives

$$(\delta_s \delta_v)_{\max} = 0.24. \qquad (25.25)$$

This value of 24 percent may be regarded as the maximum efficiency to which a solar cell may aspire when exposed to full solar radiation in space.

A solar cell may, at least in principle, be operated in reverse. By passing a forward current across a p-n junction, one introduces an excess of minority carriers. When these recombine with the majority carriers, light is emitted with an $h\nu$ equal to E_G. Electrical energy is thereby converted directly into radiant energy.

In practice, the recombination of electrons and holes takes place more rapidly by nonradiation processes than by radiation. Two approaches have been followed to increase the radiation over the nonradiation recombination. In the first, one depresses the nonradiation processes by removing those imperfections which allow direct recombination. In the other approach, one utilizes the principle of self-stimulation by operating at very high current densities and by appropriate choice of geometry. Reverse solar cells operated in this manner are known as semiconductor lasers.[7]

Magnetic sieves

Semipermeable membranes are exceedingly useful conceptual devices. They allow us, at least in principle, to perform many processes reversibly. Their disadvantage is the difficulty of their practical realization. High magnetic fields give us an opportunity of actually constructing a "membrane" that is selectively permeable, i.e., permeable to heavy positive ions but not to electrons. We denote such a "membrane" as a *magnetic sieve*.

A magnetic field offers an impediment to the motion of a charged particle across the field. Thus, when a charged particle attempts to move across a magnetic field, the field deforms its path into a circle around

[7] A. L. McWhorter, *Solid State Electronics* 6, 417 (1963).

which the particle moves with the angular frequency

$$\omega = \frac{eH}{mc}. \qquad (25.26)$$

Let ω_e, ω_p be the angular frequencies for electrons and positive ions, and let τ_e, τ_p be the times between collisions for these two types of particles. Then, if

$$\begin{aligned}\omega_e \tau_e &\gg 1; \\ \omega_p \tau_p &\ll 1,\end{aligned} \qquad (25.27)$$

the magnetic field essentially blocks the passage of electrons, while allowing positive ions and, of course, neutral atoms to move through essentially unimpeded. In such a case, the magnetic field may truly be said to act as a sieve. The continued flow of positive ions will create a space charge at the two sides of the magnetic sieve that will tend to move the positive ions in the reverse direction. This same space charge will of course tend to move the electrons in the forward direction. By placing electrodes as indicated in Fig. 25.6, we allow the electrons to bypass the magnetic

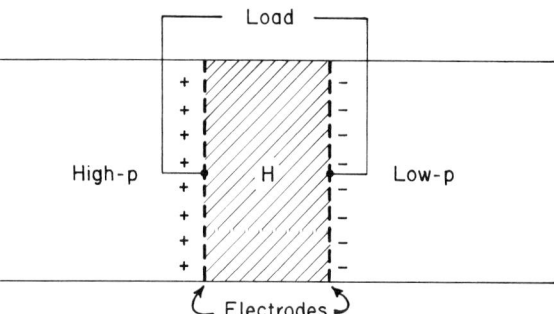

Fig. 25.6. Magnetic sieve.

field. By placing a load in this bypass, we obtain electrical energy.

In the absence of slippage of the neutral gas past the positive ions, flow will cease when the electrical force on the positive ions is precisely balanced by the difference in pressure force; i.e.,

$$N_+ eE = A \, \Delta P, \qquad (25.28)$$

where N_+ is the number of positive ions within the sieve, A the area of the sieve, E the electric field, and ΔP the pressure difference. Upon multiplying Eq. 25.28 by the length of the sieve, we obtain

$$N_+ e \, \Delta V = \text{vol} \times \Delta P. \qquad (25.29)$$

This equation for ΔV is unambiguous only when ΔP is small compared to P_H or P_L. Otherwise, we would have to know just how the pressure varied along the sieve in order to know N_+. In order to handle the case of a finite ΔP, we must consider a large number of magnetic sieves in series and then add their ΔV's and ΔP's. If we consider that the gas flows isothermally through the series sieves, we obtain

$$ze\,\Delta V = kT \ln\left(\frac{P_H}{P_L}\right), \quad (25.30)$$

where z is the fraction of gas molecules that are ionized. This equation is identical to the equation for the voltage of a fuel cell (Eq. 25.3), provided we interpret z as the number of free electrons per molecule.

A characteristic difference between a solid-state sieve (fuel cell) and a magnetic sieve (MHD generator) is that the former works with a medium in which $z \geq 1$, while the latter works with a medium $z \ll 1$. As a consequence, the solid-state sieve is essentially a low-voltage power generator, while a magnetic sieve is a high-voltage power generator.

An efficient magnetic sieve requires the simultaneous satisfaction of the two Conditions 25.27. A prerequisite that these two conditions can be satisfied simultaneously is

$$\omega_e\tau_e \ggg \omega_p\tau_p. \quad (25.31)$$

This inequality can be written as

$$\frac{\tau_e}{m_e} \ggg \frac{\tau_p}{m_p}$$

or

$$Q_e\sqrt{m_e} \lll Q_p\sqrt{m_p}, \quad (25.32)$$

where Q is the collision cross section. Normally, the Q for electrons and for positive ions are of the same order of magnitude, namely, 10^{-14}. However, by taking advantage of the Ramsauer effect, the Q for electrons can be reduced nearly three orders of magnitude. Thus,[8] the Q for 0.1 volt electron in argon is 2×10^{-17} cm², while that for Cs⁺ ions in argon is 1×10^{-14} cm². Upon further observing that the mass of Cs⁺ is 240,000 times that of an electron, we get

$$\frac{(\omega\tau)_e}{(\omega\tau)_p} = 250{,}000 \quad (25.33)$$

for the case of ionized Cs in argon. Thus Condition 25.31 is readily satisfied by a plasma of Cs ions in argon.

We now inquire whether the first of Conditions 25.27 can be satisfied. This condition may indeed be satisfied if again we utilize the Ramsauer effect. Thus, at one atmospheric pressure of argon we obtain

$$\omega_e\tau_e = 8.4\left(\frac{H}{1000}\right). \quad (25.34)$$

Values of 100 for $\omega_e\tau_e$ are thus readily obtained.

[8] *American Institute of Physics Handbook*, McGraw-Hill Book Co., New York, 1957, pp. 7—174, 7—175.

Considerable insight as to the uses of a magnetic sieve may be gained by regarding the fuel cell as a solid-state sieve. The concepts of space-charge induced fields, of collecting electrodes, and of a bypass for the electrons through an electrical load may be transferred unchanged from the solid-state sieve to a magnetic sieve. In particular, the arrangement shown in Fig. 25.7 allows converting compression energy of a gas into electrical energy.

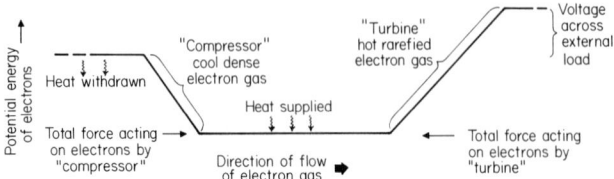

Fig. 25.7. Essential features of electronic heat engines.

A perfect magnetic sieve—one which completely blocks the passage of electrons—would, of course, also completely block that of positive ions, while passage of neutral molecules would not be completely blocked. The latter can, in effect, slip past the positive ions. In order to estimate this rate of slip, we observe that the electric force acting upon the positive ions in a unit volume, namely, eN_pE, is exactly balanced by the friction of the slipping neutral molecules, but the slipping friction acting upon the neutral molecules is in turn balanced by the pressure gradient, dp/dx. We thus have the relation

$$eN_+E = \frac{dp}{dx}. \quad (25.35)$$

If we now use measured values of the mobility b defined by

$$V_s = bE, \quad (25.36)$$

we can estimate the slip velocity V_s. Upon giving b the value of 2.2 (cm/sec)/(volts/cm), corresponding to the measured values of Cs⁺ ions in A at atmospheric pressure, we thus obtain

$$V_s = 5 \times 10^{-2} f^{-1}\left(\frac{dp_\text{atm}}{dx}\right) \quad [\text{cm/sec}], \quad (25.37)$$

where f is the concentration of Cs ions. We conclude that a magnetic sieve one meter long can sustain a pressure drop of one atmosphere in argon gas with a slip velocity of only several cm/sec if a Cs ion concentration of not less than 10^{-4} is present.

Electronic heat engines

An overwhelming fraction of the power used at present is obtained from heat engines, which come in all sizes, shapes, and internal mechanisms, but all operate on precisely the same principle. The working medium is first adiabatically compressed, heat is then pumped into the working medium under conditions of constant pressure, the medium is then allowed to expand adiabatically. In a closed-cycle heat engine, heat is finally extracted from the working medium, bringing it to its original state at the beginning of the cycle.

In a closed-cycle, continuous-flow heat engine, that part of the engine in which the medium is compressed is called the compressor, that part in which it expands, the turbine. This terminology is somewhat misleading, since the compressor and turbine play a symmetrical role in compressing the medium. Thus, while the compressor exerts a force upon the working medium in the direction of its flow, the turbine exerts an equal force upon the working medium in a direction counter to its flow. The working medium is compressed as a result of the balance of these two equal and opposite forces.

The compressors and turbines of all conventional heat engines exert forces upon the bulk of the working medium only indirectly via surface forces, the effects of which are then transmitted through the bulk of the working media via internal stresses. Losses are inevitably associated with these internal stresses.

We are led to investigate the possibility of operating a closed-cycle, continuous-flow heat engine in which the compressor and turbine act directly upon the bulk of the working medium via body forces. Such a heat engine is indeed possible if we employ electrons as the working medium. Here the compressor and turbine need be only suitably placed electric fields. If we cooperate with nature, she will provide us with suitably placed electric fields, such as automatically exist wherever we have an interface between materials. We are therefore led to investigate the possibility of using interfaces as the compressor and turbine elements of an electronic heat engine.

Two distinct types of electronic heat engines have been studied extensively in recent years. In one type, use is made of metal-vacuum interfaces. The anode vacuum-metal interface plays the role of the compressor, the cathode metal-vacuum interface that of the turbine. Such an electronic heat engine is commonly called a thermionic converter.[9] In the second type, use is made of semimetal or semiconductor interfaces. Here the interface, where the electron current leaves the p-type material and enters the n-type material, plays the role of the condenser. The interface where electrons flow from the n-type material into the p-type material plays the role of the turbine. Such an electronic heat engine is commonly called a thermoelectric heat engine.[10-12]

Material limitations on thermionic converters

The cathode of thermionic converters presents a special materials problem arising from the fact that whereas we wish cathodes to emit only electrons, all cathodes emit three type of particles: electrons, photons, and atoms. One would like to operate in such a temperature range that only the electron emission is significant. Unfortunately, nature makes the realization of this wish extremely difficult. In order to avoid unacceptable radiation losses, the rate at which heat is absorbed by the evaporating electrons must be at least five times larger than the rate at which heat is radiated from the cathode. Thus

$$J\mu \geq 5.4 \left(\frac{T}{1000}\right)^4 \quad \text{[watts/cm}^2\text{]}, \quad (25.38)$$

with

$$J = 120T^2 e^{-\mu/kT}. \quad (25.39)$$

Here μ denotes the chemical potential of the electrons in the cathode. The chemical potential and the work function are identical for those cathodes that obey the Richardson equation. We have taken the emissivity of both the cathode and the anode to be 0.4, and have appropriately taken into account back reflection of radiation by the anode. Upon eliminating J between these two equations, we obtain the ratio μ/kT as a very weak function of T. Thus, at $T = 3000°$ K, a threefold change in T introduces only a 6 percent change in the ratio μ/kT. We thereby obtain

$$\frac{\mu}{kT} \leq 16 \quad (25.40)$$

as the condition that radiation losses be not excessive.

The condition for a reasonable life may be formulated by requiring that the cathode evaporate at a rate less than one millimeter per year. Because of the extreme sensitivity of vaporization rate upon temperature, our final conclusions would not be appreciably affected for taking a different rate within two orders of magnitude of one millimeter per year. The rate of vaporization may be written as

$$r = \frac{\gamma P}{(2mkT)^{1/2}} \quad \text{[atoms/cm}^2\text{ sec]}, \quad (25.41)$$

[9] Ref. 1, Chap. 4.

[10] R. Heikes and R. Ure, *Thermoelectricity: Science and Engineering*, Interscience Publishers, New York, 1961.

[11] Ref. 1, Chap. 3.

[12] *Thermoelectricity*, P. H. Egli, Ed., John Wiley and Sons, New York, 1960.

where γ is the sticking coefficient, and the pressure P is given by

$$P = g\left(\frac{kT}{2}\right)\left(\frac{\theta}{T}\right)^3 \left[2\pi m\left(\frac{kT}{h}\right)^2\right]^{3/2} e^{-H/kT}. \quad (25.42)$$

Here g is the statistical weight of the ground state of a free atom, and H the heat of vaporization. Upon requiring that the rate of vaporization be less than 1 mm/yr, we obtain

$$\frac{H}{kT} \geq 42. \quad (25.43)$$

The two Conditions 25.40 and 25.43 are compatible only if

$$\frac{H}{\mu} \geq 2.5. \quad (25.44)$$

Condition 25.44 for low radiation loss simultaneous with long life is not satisfied by any pure metal. Table 25.1 lists the value of the ratio H/μ for various metals.

We conclude that low radiation loss simultaneous with long life must be sought either with appropriate alloys or with operating conditions such that a surface layer of atoms is maintained which lowers the work function μ.

Material limitations on thermoelectric converters

The ability of a material to convert a downhill flow of heat directly into electric power or conversely to convert electric power into an uphill flow of heat is given by a single dimensionless parameter Tz, defined in terms of usual materials constant as

$$Tz = \frac{TS^2}{\rho k}, \quad (25.45)$$

where S is the thermoelectric power, ρ the electric resistivity, and k the thermal conductivity. The physical significance of this parameter may best be appreciated from the relation

$$\frac{k_{V=0} - k_{J=0}}{k_{J=0}} = Tz \quad (25.46)$$

between thermal conductivity measured under conditions of zero potential drop and zero electric current.

The relation of Tz to efficiency of power generation may best be expressed through an intermediate parameter

$$\epsilon(T) = \frac{\sqrt{1+Tz} - 1}{\sqrt{1+Tz} + 1}, \quad (25.47)$$

Table 25.1. **Ratio of heat of vaporization to chemical potential**

Element	H/μ	Heat of Vaporization[13] (kcal/mole)	Work Function[14] (eV)
Nb	2.01	184.5	3.99
Ta	1.95	185	4.13
W	1.93	201.6	4.53
C	1.71	172	4.36
Os	1.66	174	4.55
Re	1.65	189	4.97
Mo	1.60	155.5	4.24
U	1.58	125	3.45
Ir	1.57	165	4.57
Ru	1.54	160	4.52
Pr	1.40	87	2.7
Zr	1.38	125	3.93
Rh	1.29	138	4.65
Ce	1.28	85	2.88
V	1.27	120	4.11
Sm	1.18	87	3.2
Ti	1.16	112	4.16
La	1.16	88	3.3
Nd	1.15	87	3.3
Co	1.07	105	4.25
Si	1.07	88	3.59
Pt	0.99	121.6	5.36
B	0.92	97.2	4.6
Fe	0.91	96.7	4.63
Ni	0.90	101.6	4.91
Be	0.85	76.6	3.91
Pd	0.81	93	4.98
Cr	0.80	80.5	4.45
Cu	0.79	81.5	4.48
Al	0.78	75.0	4.20
Au	0.76	82.3	4.71
Mn	0.75	68.3	3.95
Ge	0.74	78.4	4.62
Ba	0.72	41.96	2.52
Sn	0.71	72	4.39
Ga	0.69	66.0	4.16
Li	0.65	37.1	2.46
Ag	0.64	69.1	4.70
Ca	0.63	46.0	3.20
Sr	0.62	39.2	2.74
Sb	0.58	60.8	4.56
Bi	0.50	49.7	4.34
Pb	0.50	46.3	4.04
Na	0.50	26.0	2.28
Tl	0.46	43.3	4.05
Te	0.44	47.6	4.73
Se	0.43	48.4	4.87
Mg	0.42	35.9	3.76
Cs	0.42	18.8	1.94
Rb	0.42	20.5	2.13
K	0.41	21.5	2.25
Zn	0.32	31.2	4.27
Cd	0.29	27.0	4.04
Hg	0.14	14.5	4.53

[13] F. D. Rossini et al., "Selected Values of Chemical Thermodynamic Properties," Natl. Bureau Stds. Circular No. 500, 1952.

[14] Landolt-Börnstein, *Zahlenwerte und Funktionen*, 6th ed., Vol. 1, Pt. 4, Springer-Verlag, Berlin, 1951, p. 759.

called the differential efficiency coefficient, since for a small temperature interval δT,

$$\text{Efficiency} = \epsilon\left(\frac{\delta T}{T}\right). \quad (25.48)$$

When optimum use is made of cascading, the theoretical efficiency over a finite temperature range from T_L to T_H is

$$\eta = 1 - \epsilon^{\int_{T_L}^{-T_H} \epsilon(T) \left(\frac{dT}{T}\right)}. \quad (25.49)$$

Theoretical efficiency implies no contact resistance losses and no parallel heat leaks.

Within the temperature range from 100° to 1200° K, materials have been developed with Tz lying in the range 0.82 to 1.25, or ϵ in the range from 0.15 to 0.20. Above room temperature,[15] the best materials are semiconductor-doped to nearly the degenerate state, namely, to about 10^{19} charge carriers/cm³. Below room temperature,[16] the best materials are semimetals, having slightly overlapping valence and conducting zones. In the latter case magnetic fields must be applied.

Nernst-Ettingshausen generator

A completely new type of electronic heat engine has recently been studied, based upon the Nernst-Ettingshausen effects.[17] In the utilization of these effects, a magnetic field is established normal to the temperature gradient. Power is then extracted from the emf induced at right angles to both the temperature gradient and the magnetic field.

The expression for the efficiency of a N-E generator is most readily understood if it is derived in a manner parallel to that for a TE (thermoelectric) generator. Toward this end, we write the general equations for the heat current J_h and electric current J_c in terms of the potential drop ΔV and temperature drop ΔT:

$$\begin{aligned} J_c &= L_{11}\,\Delta V + L_{12}\left(\frac{\Delta T}{T}\right), \\ J_h &= L_{21}\,\Delta V + L_{22}\left(\frac{\Delta T}{T}\right). \end{aligned} \quad (25.50)$$

[15] W. G. Evans, *Semiconductor Products and Solid State Technology* **6**, No. 4, 34 (1963).

[16] R. Wolfe and G. E. Smith, *Appl. Phys. Letters* **1**, 5 (1962); *Semiconductor Products and Solid State Technology* **6**, No. 4, 29 (1963).

[17] J. Bardeen, "Conduction, Metals and Semiconductors," *Handbook of Physics*, E. U. Condon and H. Odishaw, Eds., McGraw-Hill Book Co., New York, 1958, Chap. 6, p. 4.

In the TE case, ΔV and ΔT are parallel; in the N-E case, they are normal to one another as well as to the magnetic field H. The coefficients have the following interpretations:

$$\left.\begin{aligned} R &= L_{11}^{-1} \\ K &= \frac{L_{22}}{T}\left\{1 - \frac{L_{12}L_{21}}{L_{11}L_{22}}\right\} \\ S &= \frac{L_{12}}{TL_{11}} \\ &= \begin{cases} \text{thermoelectric power, TE case} \\ H \times \text{Ettingshausen coefficient, N-E case} \end{cases} \\ \theta &= \frac{L_{21}}{L_{12}} \\ &= \begin{cases} 1,\ \text{TE case} \\ -1,\ \text{N-E case} \end{cases} \end{aligned}\right\} (25.51)$$

The efficiency, defined as

$$\eta = \frac{\text{electric power delivered}}{\text{heat input}}, \quad (25.52)$$

is given by

$$\eta = \min_{J_c} \frac{J_c(S\,\Delta T - RJ_c)}{K\,\Delta T + \theta STJ_c}. \quad (25.53)$$

In preparation to giving the results of this minimum problem, we introduce two positive quantities:

$$x \equiv \frac{TS^2}{KR},\quad y \equiv \frac{L_{12}^2}{L_{11}L_{22}}, \quad (25.54)$$

which have the following properties:

$$\left.\begin{aligned} y &= \frac{x}{1+x} \\ 0 &\le x < \infty \\ 0 &\le y < 1 \end{aligned}\right\} \text{TE case} \\ \left.\begin{aligned} x &= \frac{y}{1+y} \\ 0 &\le y < \infty \\ 0 &\le x < 1 \end{aligned}\right\} \text{N-E case} \quad (25.55)$$

The result of the minimum problem is then

$$\eta = \left(\frac{\Delta T}{T}\right)\epsilon, \quad (25.56)$$

where

$$\epsilon = \begin{cases} \dfrac{\sqrt{1+x}-1}{\sqrt{1+x}+1} = \dfrac{1-\sqrt{1-y}}{1+\sqrt{1-y}} & \text{in TE case} \\[2mm] \dfrac{\sqrt{1+y}-1}{\sqrt{1+y}+1} = \dfrac{1-\sqrt{1-x}}{1+\sqrt{1-x}} & \text{in N-E case} \end{cases} \quad (25.57)$$

26 · ENERGY-CONVERSION DEVICES

Edward V. Somers

Introduction — Thermoelectric generators — Thermionic generators — Fuel-cell generators — Magnetohydrodynamic generators

Introduction

Space-age engineering is quick to implement newly discovered concepts of producing power and to improve or discard older techniques. Military and civilian requirements for power, ranging from watts to thousands of megawatts and for time intervals from seconds to many hours, have excited interest in establishing new methods and improving old methods of power generation. In the past, demands for small blocks of power were easily satisfied with batteries and motor generators. Standard apparatus for large blocks of power consisted of steam and gas turbines operating with maximum temperatures of 550° to 850° C. The nuclear reactor, a relative newcomer to the power field, uses steam or gas turbines confined to roughly the same top temperatures. Four means of producing power, unconventional with respect to today's practice, are based on scientific concepts dating back to the early and mid eighteen hundreds: thermoelectricity, thermionics, fuel cells, and magnetohydrodynamics. Since these methods operate without rotating parts, they offer many new design possibilities. Their present state of development is the subject of this discussion.

Thermoelectric generators

A thermoelectric generator produces electric power by flow of heat through a thermocouple system. If a series circuit of thermocouples of materials X and Y is placed between a heat source and a heat sink, a part of the heat flow will be converted to electric power and a part lost to the sink at the cold junctions (Fig. 26.1). To cut this loss, materials must be found that

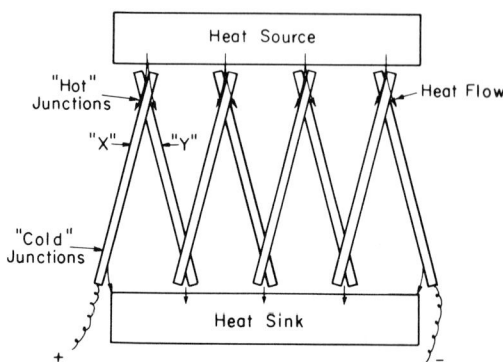

Fig. 26.1. Thermoelectric generator.

will conduct electricity well but not heat. The thermocouple efficiency E, which gives the ratio of the electric power produced to the heat absorbed at the hot junctions (similar to Eqs. 25.45 through 25.48 of the preceding chapter by C. Zener), is

$$E = \frac{T_h - T_c}{T_h} \frac{\left[1 + Z\left(\frac{T_h + T_c}{2}\right)\right]^{1/2} - 1}{\left[1 + Z\left(\frac{T_h - T_c}{2}\right)\right]^{1/2} + \frac{T_h}{T_c}},$$

(26.1)

where $Z = S^2/\rho\lambda$, the thermoelectric materials factor. The exact form is contained in Ioffe.[1] This factor indicates that low thermal conductivity λ, electrical resistivity ρ, and high Seebeck coefficients S will produce high thermal efficiency.

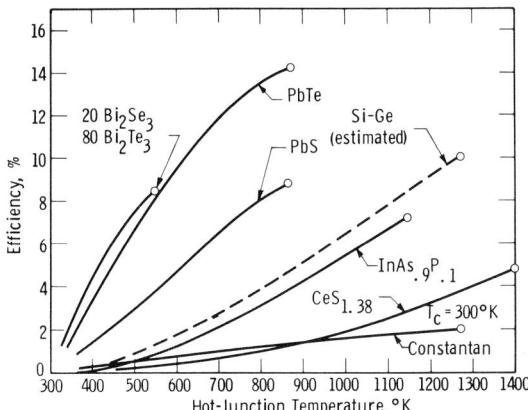

Fig. 26.2. n-type thermoelectric materials.

Years of intensive work aimed at establishing the efficiency limits of thermoelectric materials; the outcome of this research is shown in the data of Figs. 26.2 and 26.3. The efficiency is defined in Eq. 25.49. Of

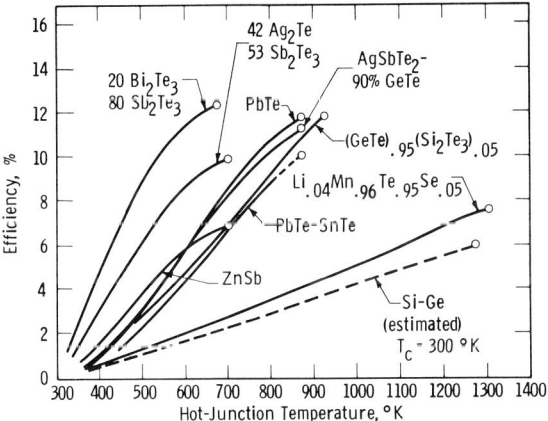

Fig. 26.3. p-type thermoelectric materials.

the n-type materials only the bismuth and lead tellurides are seriously considered for use in ground applications. The germanium-silicon alloy is a contender for space applications because it can serve at high temperatures. The p-type materials exhibit essentially similar characteristics with the exception that the mechanical properties, e.g., of lead telluride and germanium-bismuth telluride, make it difficult to provide reliable electrical contact.

[1] A. F. Ioffe, *Semiconductor Thermoelements and Thermoelectric Cooling*, Inforsearch Ltd., London, 1957, p. 40.

Contact is made by pressure or metallurgical bond. Pressure contacts, generally less desirable, are used when difficulties of joining the thermoelectric material to iron or aluminum conductors are encountered. Either method can provide useful contacts for operation up to 500° C for ground-based generators and up to 700° C for space applications.

Lifetimes of thousands of hours under steady operation with less than 25 percent degradation in the power output have been achieved in isotope-heated generators. Thermoelectric systems must be designed for large thermal gradients. In consequence, high thermal stresses are set up, which frequently cause cracking of materials or contacts. Cycling operation increases the severity of cracking caused by alternating thermal stress. It is expected, however, that within the next few years operating lifetimes of thousands of hours will be realized under cycling conditions.

The theoretical efficiency obtainable is a function of the material parameter Z (Eq. 26.1). After six years of research and many million dollars of expenditure, empirical evidence points to a limit for Z between 1.0×10^{-3} and 1.5×10^{-3}; the resulting theoretical efficiency for a temperature differential of 800° K is about 18 percent (Fig. 26.4).[2] When temperature

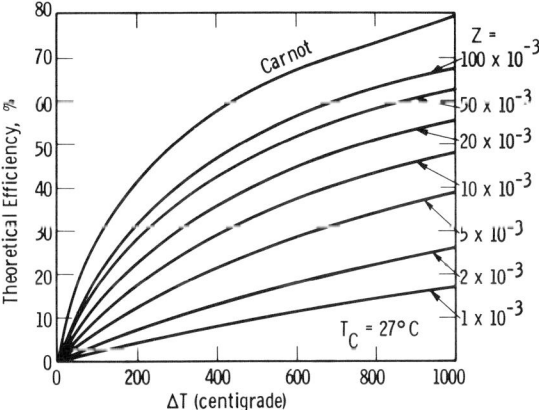

Fig. 26.4. Thermoelectric-generator efficiency as function of temperature differential, with Z as parameter.

drops in the contacts, other heat losses, combustion efficiency, and a reasonable cold-side temperature are considered, practical generator efficiencies between 2 and 10 percent may be expected for useful periods of time. Generators built to date have fallen within this range.

Thermoelectric generators appear practical in remote locations when ratings are limited to a few thousand watts as portable generators of low power rating for

[2] Ref. 1, pp. 39–42.

military use and as auxiliary power units for space applications with power ratings from watts to kilowatts.

Thermionic generators

The conversion of heat to electrical energy by thermionic emission can be described as a "boiling-off" of electrons from a heated cathode surface and condensing of the electrons on a cooled anode surface. The vapor pressure depends on the work functions of the surfaces. The power developed in a thermionic generator is proportional to the difference in work functions (Fig. 26.5). The current increases exponentially with temperature.

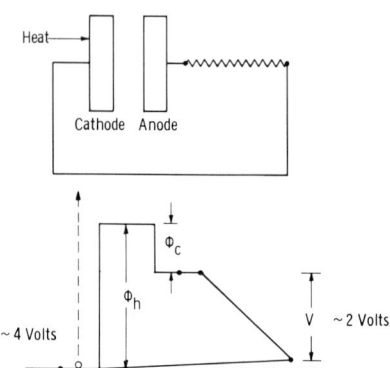

Fig. 26.5. Circuit and potential diagram for a thermionic tube.

The ratio of electric power produced to heat absorbed by the hot cathode is[3]

$$E = \frac{(\phi_h - \phi_c)J}{\phi_h J + Q_l};$$

$$J \sim T_h^2 \exp\left(-\frac{e\phi_h}{kT_h}\right).$$
(26.2)

The factor Q_l sums the heat lost from the cathode, mostly as radiative loss from cathode to anode.

A basic limitation to thermionic energy conversion is space charge. When electrons are emitted, this space charge in front of the cathode creates a retarding field, preventing the emission of more electrons. This space-charge-limited current flow can be partly overcome in two ways: One way is to place cathode and anode so close together that the space-charge cloud cannot develop. While this method works, it requires spacings on the order of 0.0001 inch, which are difficult to construct and maintain. Another method is to neutralize the space charge by an equal number of positive ions in the same region. The best material

[3] K. G. Hernqvist et al., "Thermionic Energy Converter," *Direct Conversion of Heat to Electricity*, J. Kaye and J. A. Welsh, Eds., John Wiley and Sons, New York, 1960, p. 6–1.

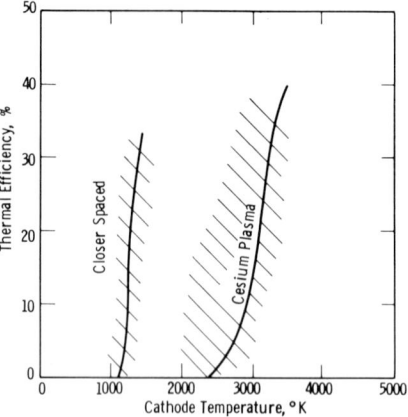

Fig. 26.6. Theoretical thermal efficiency versus cathode temperature.

for creating such ions being cesium, the addition of cesium vapor into the space between the cathode and anode permits the use of wider spacings. However, cesium vapor conducts heat away, interferes with the flow of current, and requires operation at much higher temperatures.

The theoretical efficiency of thermionic energy-conversion devices can be calculated from Eqs. 26.2. This efficiency for the two types of space-charge-eliminating generators is shown in Fig. 26.6. A variety of thermionic generators, some producing up to a few hundred watts, have operated satisfactorily for reasonable periods of time. Published reports on actual efficiencies of such devices have placed them in the range of 5 to 15 percent.

Attractive applications for thermionic generators exist in connection with nuclear reactors. A simple and practical arrangement is to design a cylindrical generator with the nuclear fuel incorporated into the cathode itself. The combined fuel and thermionic generator rod can operate immersed in the reactor coolant (Fig. 26.7). The heat lost to the anode is transferred

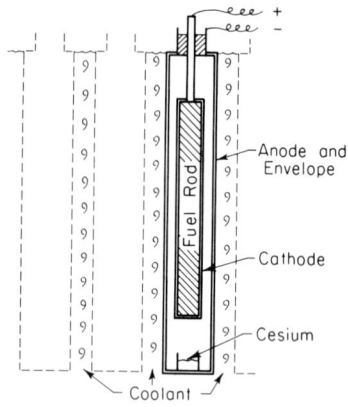

Fig. 26.7. Reactor, incorporating thermionic generators.

to a boiling-water coolant; the steam produced can run a conventional steam turbine. This combination may allow the economical production of commercial power.

A thermionic generator in a nuclear reactor, cooled with a liquid metal at high anode temperatures, may provide an interesting power plant for spacecraft; in devices of small power rating the heat from the anode may be directly radiated into space.

Thermionic generators used in such applications will have efficiencies by themselves of between 10 and 20 percent. When topping a conventional turbine plant, the combined thermal efficiency may exceed 50 percent.

Fuel-cell generators

A fuel-cell generator produces electric power by expanding a gas, either the fuel or the oxidant, through an electrolytic membrane. The operating principle of an oxygen-transport type cell operating at about 1000° C is shown in Fig. 26.8. Oxygen molecules pass

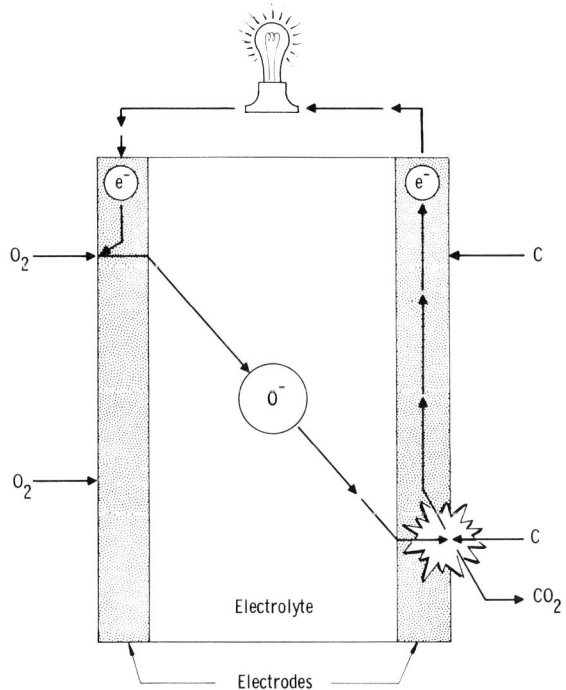

Fig. 26.8. Fuel-cell generator.

through the electrode structure (in this case a solid electrolyte 85% ZrO_2/15% CaO) as negative ions by picking up electrons at the cathode side. At the anode side, the oxygen reacts with a carbon fuel to form carbon dioxide and free electrons. The chemical reaction at the negative electrode reduces the oxygen pressure to an extremely low value and provides the downhill diffusion potential necessary to transport the oxygen ions across the electrolyte. Electrolyte and electrode structures are sufficiently thin to make the process isothermal.

Other types of fuel cells use as electrolytes fused salts, aqueous solutions, or plastic membranes and as the fuel, for example, hydrogen. The principle remains the same, and the thermodynamic voltage $\mathcal{E}_t$ and the volt-ampere relationship for a fuel cell are (cf. Eq. 25.3 and Weissbart and Ruka[4])

$$\mathcal{E}_t = \frac{RT}{n\mathcal{F}} \ln \frac{p_{\text{hi}}}{p_{\text{lo}}}; \qquad (26.3)$$
$$V = \mathcal{E}_t - IR_i.$$

The thermodynamic voltage is effectively the work performed by the gas in isothermal expansion while diffusing through the electrolyte divided by the electrical charge carried during diffusion. Streaming resistance is neglected. This resistance reduces the voltage at the terminals as shown in the second relation.

The choice of the four constituents of a fuel-cell generator depends on the operating temperature (Fig. 26.9). Cells operating above 850° C have several desir-

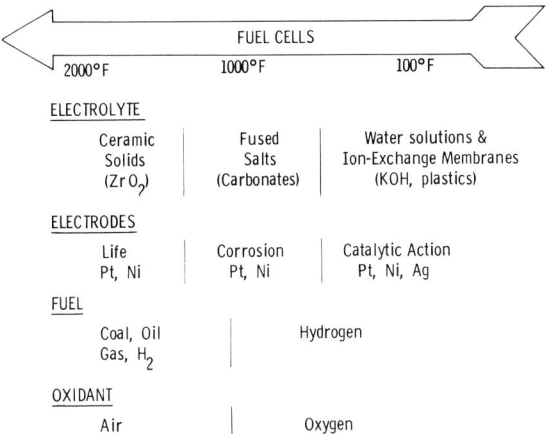

Fig. 26.9. Fuel-cell problems.

able features: a solid electrolyte; an ability to use a variety of cheap fuels (coal, oil, natural gas, hydrogen compounds); and the use of air as the oxidant. Fuel cells of the fused-salt type operating at lower temperatures can still burn carbonaceous fuels and use air as an oxidant but lack the rigid solid structure. Both these cell types have high reaction rates at the electrodes, and impairment of the cell operation due to poor combustion is a negligible factor.

Fuel cells with aqueous and membrane-type electrolytes operate in the low-temperature range—a desir-

[4] J. Weissbart and R. Ruka, "Solid Oxide Electrolyte Fuel Cells," *Fuel Cells*, G. J. Young, Ed., Vol. 2, Reinhold Publishing Corp., New York, 1960, p. 37.

able feature. However, the low temperature limits them in general to hydrogen as fuel and to oxygen as oxidant, because the diffusion rates for other fuels are too low and the reaction between fuel and oxidant must be speeded up catalytically.

Electrodes constitute a major problem for all types of cells: In the high-temperature range, life sets the limit; in the intermediate-temperature range, corrosion by the electrolyte fuel; and in the low-temperature range, poisoning of the catalytic surface.

Fuel cells show promise in many applications. They are being considered for auxiliary power generation in space capsules. Highly efficient portable generators of small power rating for specialized applications in military communications and control appear likely. The introduction of the fuel-cell generator into the commercial sphere, first as special-purpose low-power units, then as larger power units for transportation drives, and eventually as large-power units for central-power station applications, is the long-range goal. To date experimental cells have been built and operated in the power range from watts to kilowatts. Life tests on small cells in the thousands of hours have been accumulated. Systems design and operational studies have shown that fuel-cell power plants for space, military, and commercial uses have attractive characteristics.

Magnetohydrodynamic generators

An MHD power generator essentially combines a turbine and an electrical generator in a single piece of apparatus. Equations 25.28 through 25.30 discuss possible MHD work processes. Figure 26.10 does not illustrate a magnetic sieve but shows a Faraday-type machine as described by Rosa and Kantrowitz.[5] It illustrates how a simulated turbine driving a conducting wire through a magnetic field can be replaced by an ionized-gas conductor (alkali-seeded gas) traversing the magnetic field. The gas of positive ions and electrons moves with velocity v through the magnetic field H, and the magnetic deflection forces an electronic Hall current I from right to left across the gas stream. The electrodes of the generator are the two vertical walls with leads attached; the top and bottom confining walls are electrical insulators.

MHD generators can be multimegawatt machines. Their thermal energy will be derived from carbonaceous fuel for land-based and marine applications; they will use nuclear fuel for military and space applications.

A generator using coal and operating between 2250° and 4000° C is summarized in Fig. 26.11: seed, two

[5] R. J. Rosa and A. Kantrowitz, "Magnetohydrodynamic Energy Conversion Techniques," Ref. 3, p. 12—1.

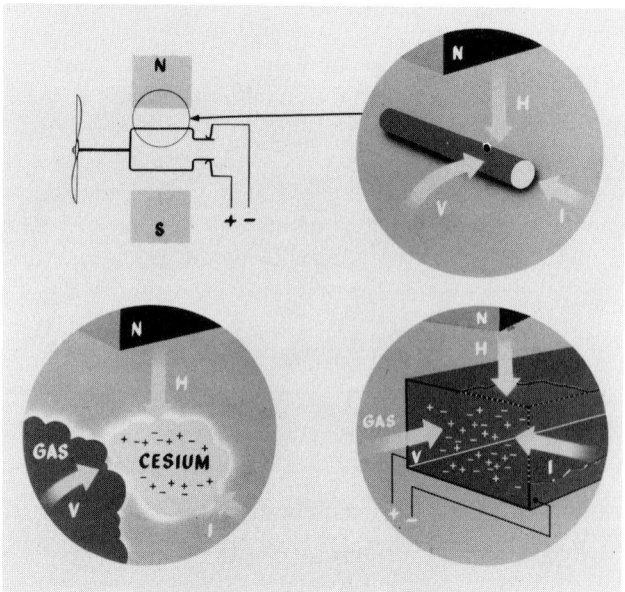

Fig. 26.10. MHD power generator.

types of containing walls (refractory materials and cooled-wall structures), and oxidant. At 4000° C there is no problem with respect to seed; the gas ionizes spontaneously. However, there is no known method of containing the gas within the generator, and pure oxygen must be used to fire the carbonaceous fuel. At 2250° C, seed is expensive, but cheap wall materials are available and preheated air can be used for combustion. The cost of the seed used to produce the electrical conductivity increases as the temperature drops from sodium to potassium and cesium. This increase is offset by a decrease in the cost of containing walls from thoria to zirconia and calcium-alumina.

A wall construction using cooled walls will produce heat losses of the working gas as it passes through the generator; this loss decreases with the temperature of the MHD machine. Since it represents a loss in effi-

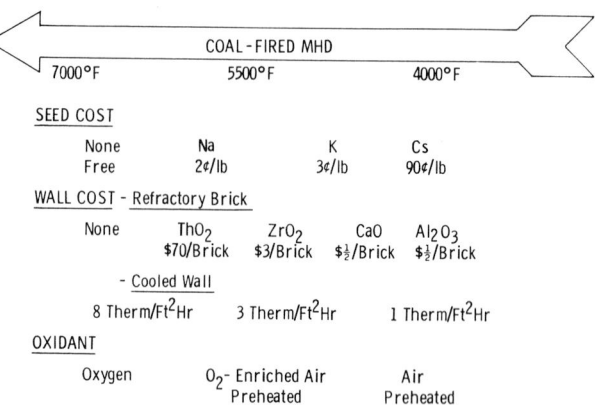

Fig. 26.11. Coal-fired MHD problems.

ciency, hence an increased cost for the power produced, a cooled-wall structure decreases in cost as the temperature decreases. The final problem is the oxidant needed to burn the carbonaceous fuel. To obtain a temperature of 4000° C, oxygen is required for combustion; at 2250° C preheated air is sufficient; the intermediate range of 3000° C requires oxygen-enriched air. Oxygen enrichment results in increased oxidant cost as the temperature rises.

The use of superconducting magnets will greatly reduce the power required to produce the magnetic field. It is necessary only to refrigerate the superconducting coils, a negligibly small power expense. Superconducting magnets of high field strength also compact the generator and thus reduce the ratio of irreversible power loss to power generated. This irreversible loss includes the gas friction and heat losses to the wall. The MHD turbine efficiency, including wall friction, cooling losses, and internal electrical losses, is given as

$$\eta_t = \frac{\eta}{1 + \dfrac{\alpha}{1-\eta}};$$

$$\alpha \sim \frac{1}{B^2},$$

(26.4)

where η designates the generator efficiency including only internal electrical losses. The parameter α, a dimensionless quantity, varies inversely with the magnetic inductance squared and introduces the wall losses. Its effect on the MHD turbine efficiency is shown in Fig. 26.12. For a magnetic induction of $B = 3$

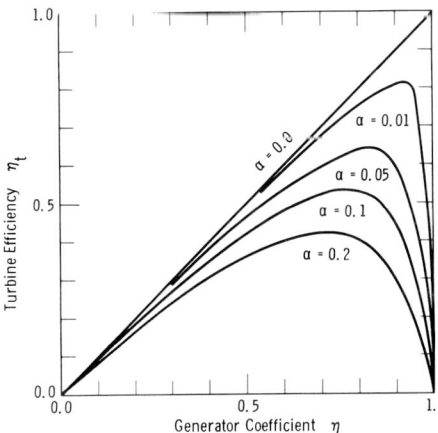

Fig. 26.12. Turbine efficiency with friction and heat transfer versus generator coefficient.

webers/m², the value of α may range from 0.02 to 0.2. For a superconducting magnet of magnetic induction 6 webers/m², the range of α will be reduced to 0.005 to 0.05, with marked improvement of the turbine efficiency. Analysis of a coal-fired MHD plant topping a conventional steam plant has shown that superconducting magnets may be a necessity if the irreversible wall losses are to be reduced to acceptable values.

In planning MHD generators using nuclear heat sources, most of the effort is directed toward better understanding of the electrical conductivity of the working fluid. Such generators will employ a closed loop in which circulates either a noble gas seeded with alkali, for use in large-rating ground-based power stations, or a metallic vapor for generators used in space applications. All working fluids proposed for nuclear-fired MHD generators attain satisfactory thermal ionization (electrical conductivity) between 2000° and 2250° C. Such temperatures lead to severe technical difficulties in the design of nuclear reactors. Accordingly, lower temperatures should be employed and the electrical conductivity enhanced by some nonthermal means, such as ionization by strong internal or external electrical fields. Since this conductivity level is at a high nonthermal value and tends to decay rapidly, much research work is devoted to determining the relaxation times of the nonthermal ions. Should high values of electrical conductivity be maintained in the working fluid as it traverses the MHD duct, lower MHD temperatures can be used in the reactor design. If nonthermal ionization is created by electric fields internal to the MHD duct, a Hall mode of operation is needed for the MHD duct, and a large value of magnetic field is required (cf. Chap. 25). Superconducting magnets appear as a requirement for implementing such a design.

MHD-generator experiments have been under way in many laboratories. The largest experimental generator currently operating has developed about 1500 kilowatts at latest reports. Many generator tests using combustion gases have been completed at various laboratories in the United States and Europe. Several closed loops have been constructed for use with seeded noble gas or with metallic vapor. Experiments aimed at evaluating nonthermal ionization are under way. Satisfactory experimental progress is recorded in practically all problem areas. MHD power systems in both land-based and space applications are undergoing continuous evaluation; the outlook is promising.

LIST OF SYMBOLS

B magnetic induction
e electronic charge
E thermal efficiency of power device
$\mathcal{E}_t$ thermodynamic voltage of fuel cell

$\mathfrak{F}$	Faraday number	T_h	temperature of thermionic cathode
H	magnetic field	T_c	temperature of thermionic anode
I	MHD electric current	V	terminal voltage of thermionic tube
J	electric current per unit area	v	velocity of conductor in MHD machine
k	Boltzmann constant	Z	thermoelectric materials factor
n	number of electrons absorbed per molecule to form ions	α	coefficient for friction and heat loss to walls for MHD generator
Q_l	heat loss rate per unit area from cathode of thermionic tube	η	electrical efficiency of MHD generator
		η_t	turbine efficiency (includes wall friction, heat loss, and internal $I^2 R_i$) of MHD generator
R	international gas constant	ρ	electrical resistivity of thermoelectric material
R_i	internal electrical resistance of fuel-cell or MHD generator	λ	thermal conductivity of thermoelectric material
S	Seebeck coefficient	ϕ_c	anode (cold-side) voltage, thermionic generator
T	temperature of operation in fuel cell	ϕ_h	cathode (hot-side) voltage, thermionic generator

27 · SENSORY CODING IN THE NERVOUS SYSTEM*

Walter A. Rosenblith

Introduction — Some vital statistics and a rudimentary map of a sensory system — Changes in patterns of neuroelectric activity that may lend themselves to stimulus coding — Conclusion

Introduction

The development of formal models for communication processes and coding theory has increased interest in the way in which the nervous system handles and codes sensory messages. There is a quasi-axiomatic belief that, in the course of evolution, genetic and other forms of coding have been optimized. Inquiry into the mechanisms of neural coding is best conducted in relation to the natural information-handling capacities of organisms. There are several methods of assessing these capacities in man. Perhaps for historical reasons, psychophysical methods seem to have exerted a particular appeal to physical scientists.

Psychophysics treats the organism as a black box: Discrete[1] sensory stimuli—measurable by standard physical methods—are the input, and discrete motor (including verbal) responses usually the output. Psychophysical experimentation can be extraordinarily refined; in a controlled, well-intentioned individual, response variability does often not exceed that found in the more established sciences. In spite of this apparent stability for a given experimental paradigm, it is disconcerting to find that estimates of human sensory capacities depend in a rather critical manner upon the way in which sensory tasks are defined.

These findings, superficially paradoxical, are not our major concern here. Instead, we shall deal with those patterns of neuronal activity—primarily neuroelectric—which reflect the ways in which subsystems of the nervous system and even single neurons[2] code the sensory events that impinge upon the organism. In this manner, we may hope to gain a greater understanding of the over-all properties of the nervous system in terms of those of its component parts. Such understanding will hardly permit us to design natural

---

* The preparation of this chapter was supported in part by grants and contracts in the area of Communications Sciences extended to the Research Laboratory of Electronics by the Joint Services Electronics Program under Contract DA36-039-AMC-03200 (E); and in part by the National Science Foundation (Grant GP-2495), the National Institutes of Health (Grant MH-04737-05), and the National Aeronautics and Space Administration (Grant NsG-496).

[1] In contrast, so-called "tracking" experiments make use of continuous stimuli in an attempt to characterize the organism by a transfer function.

[2] The view that neurons are the rather closely coupled "atoms" or "molecules" of the nervous system is not beyond controversy. The classical debate among those who studied the structure of the nervous system in some detail polarized around two doctrines: The adherents of the neuron doctrine were inclined to view each nerve cell as an independent unit that entered merely in contact with other units; the adherents of the somewhat older, reticular theory considered the nervous system as a complex netlike (or reticular) structure in which nerve cells occupied the nodes. In his authoritative, recent monograph, Eccles summarizes this debate in relation to the historical development of ideas on the synapse (cf. Eccles, list of Representative Books and Articles at the end of this chapter).

nervous systems in engineering fashion but might enable us to build and program computers that can be coupled more effectively to brains.

Two types of electrical events shall concern us primarily: (1) Macro- or gross electrode records that give spatial averages of electrical activity from many elements as continuously graded and variable wave forms; these wave forms can also be described by power spectra or correlograms, as has been done for so-called brain waves. When responses evoked by sensory stimuli are recorded, the characteristic deflections in these voltage-versus-time plots are specifiable by amplitude and latency (occurrence with respect to a temporal reference marker). (2) Microelectrodes (made of glass or metal with a tip diameter of the order of a micron), in contrast, record preferentially all-or-none events in a single neuron, i.e., nerve impulses, action potentials, or "spikes." Here the basic data can be described in terms of a firing rate (spikes/sec) or a firing probability for repeated stimuli. In recent years, the time intervals between spikes have been subjected to statistical analysis, and models have been designed to generate distributions of interspike intervals similar to those found experimentally.

Some vital statistics and a rudimentary map of a sensory system

Scientists and engineers are now rather familiar with the estimated number of neurons in the human brain: 10^{10}. Though not quite of the order of Avogadro's number, this number is large. What significance do we attribute to such a number?

Neurochemists have an even more realistic appreciation of this quandary. One of them, Henry McIlwain, recently wrote: "Indeed the chemical explorer now sees among the molecules of one neurone a complexity comparable to that envisaged at a cellular level among the 10^{10} neurones of the brain."[3]

Neurons do not constitute a homogeneous population of elements interacting at random. The organization of brain tissue must be understood in terms of the number, size, configuration, packing density, chemical specificity, and connective pattern of the neurons that compose it. In addition, the roles played by supporting cells other than neurons and by the "fixation of experience" need to be understood.

In a much more modest aim, we shall describe semi-quantitatively certain characteristic features of nervous systems. By getting acquainted with a rough schematic outline of a sensory system, one becomes aware of some of the structural constraints that are operative in the handling of sensory information.

Psychophysical experimentation has given us a reasonably systematic description of human sensory performance. However, the available information about sensory capacities is not of the type that would fit engineering handbooks; we are barely able to state certain general principles.

Higher organisms involved in communication tasks exhibit performance that is statistical in character and depends strongly on the probabilistic structure of the repertory of stimuli and responses. Organisms need more time to handle more information; their capacity to make more subtle discriminations and to react more rapidly depends on stimulus intensity, in the sense that stimuli that are barely detectable are neither well discriminated nor capable of evoking a quick response. The size of the repertory of absolute identifications is rather small and seemingly based upon the ability to make certain crude discriminations simultaneously; that is to say, if one expects to be able to recognize an information-rich display, it is better to encode it grossly into several more or less independent, yes-no dichotomies than to encode it along a single, fine-grained dimension.

This view of sensory performance leads one to look for certain principles of sensory coding which do not predicate that each object in the outside world has a counterpart neuron in the nervous system and that the organism's task is simply to find the "right" neuron whose activity has been affected by the presence of the stimulus object in a unique manner. Rather than discuss these matters *in abstracto*, let us examine a model sensory system about which we have a fair amount of experimental data.

The system chosen is the auditory system in the cat, i.e., those parts of the cat's nervous system that handle primarily acoustic stimuli. At the level of a block diagram, the gross neuroanatomy of the cat's afferent auditory system bears a reasonable resemblance to that of man. This is true in spite of the fact that the feline species carries out its vocal communication with the aid of signals that fall far short of language. Grammar, like other complex symbolic operations of which the human brain is capable, is hardly a property of those afferent mammalian pathways about which we know most.

Figure 27.1 shows a highly schematized and oversimplified outline of the neuroanatomy of a mammalian auditory system.

Such important locations as the cerebellum, the superior olive, the lateral lemniscus, and other regions of the brain from which responses to acoustic stimuli have been recorded are not included;

---

[3] H. McIlwain, *Chemical Exploration of the Brain*, Elsevier Publishing Co., Amsterdam and New York, 1963.

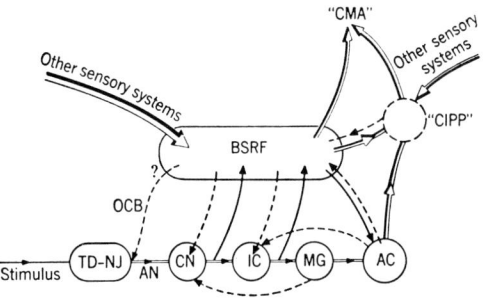

Fig. 27.1. Information flow diagram of a model sensory system.

the still incompletely known and complex organization of the brain-stem reticular formation (BSRF) has not even been suggested. Furthermore, no attempt has been made to indicate connections that correspond to acoustic reflexes of short latency: Middle-ear muscles contract about 10 msec after stimulus delivery, while eye blinks occur about 35 to 40 msec after delivery of a strong click. Such reflexes represent simpler forms of integrated behavior that involve less of the central nervous system than "voluntary" information processing does.

In order to appreciate the number of neural elements present along the afferent pathway, some of Chow's numerical estimates of the auditory system of the rhesus monkey are reproduced here. He found about 10^5 cells in the cochlear nucleus (CN), about 4×10^5 cells each in the inferior colliculus (IC) and the medial geniculate (MG); the sensory projection area labeled auditory cortex (AC) is estimated to have about 10^7 cells. These numbers (presumed to be of the same order of magnitude as in the cat) may be compared with those given further below for the auditory nerve (AN) and for the elements at the transducer-neural junction (TD-NJ). These numerical data, derived from histological investigations, do not permit any direct inference concerning the number of neural units that a given acoustic stimulus either directly excites or inhibits.

As for the remaining symbols: OCB stands for the olivo-cochlear bundle (depicted in dashed lines like the other descending pathways), whose influence upon the activity of the auditory nerve has recently been much investigated; CIPP is the conceptually useful central information-processing pool of neurons whose postulated role is neither neuroanatomically nor neurophysiologically established (the relation of this hypothetical construct to the "sensory-projection" areas and "association" areas of the brain needs to be worked out if a realistic account is to be given of the manner in which organisms handle sensory information); CMA is the postulated control unit of the motor activity that takes place in connection with stimulus-triggered responses.

Once the sound stimulus has been transformed from a disturbance in air into a traveling wave along the cochlear partition of the inner ear, the hydrodynamic events affect thousands[4] of hair cells that lie along the partition, which in cats is 20+ mm long. These perhaps 15×10^3 cells correspond to transducer elements

[4] The numerical estimates given in this chapter should not be assumed to have more than one significant number; to consider them as estimates of orders of magnitude might be even more realistic.

whose mechanical deformation is reflected in changes in cochlear potentials that are not neural in origin.

Beyond the hair-cell-neural junction lies the cat's auditory nerve, whose 5×10^4 parallel fibers funnel information about the pattern of acoustic stimulation brainward. (It is worth noting that the corresponding nerves in man and monkey are currently estimated to have only 3×10^4 fibers.) These fibers of the auditory nerve make synaptic connections with the $\sim 10^5$ cells in the cochlear nucleus, an agglomeration of perhaps a dozen subregions, each of which receives patterns of nerve impulses from the cochlea in some orderly fashion.

Regions like the cochlear nucleus used to be known somewhat metaphorically as "relay" nuclei. Today we are much more impressed with the amount of recoding that occurs in these regions. Here trains of spikes are translated into patterns of graded synaptic potentials which in turn give rise to trains of spikes. The incoming auditory nerve fibers can make various types of connections: one-to-one, one-to-many, and perhaps also many-to-many. Such anatomical variety permits the carrying out of logical operations for the detection of relational properties or abstract distinctive features in the pattern of stimulation.

How incompletely this block diagram of the auditory system describes its function can be illustrated by asking how it can account for the time the organism spends in doing its information processing. To respond to a single acoustic stimulus requires in the alert subject from about 100+ msec when the stimulus is rather loud—to about 400 msec when it is barely detectable. This difference in response times is from one to two orders of magnitude larger than the differences in latencies of evoked electrical responses that are observed along the pathway to the brain over the same range of stimulus intensity. Neural events in the auditory nerve start, depending upon intensity, from 1 to 5 msec after the delivery of the click; at the level of the auditory cortex these onset times of stimulus-evoked responses range from 10 to 20 msec. Thus, changes in the latency of these graded electrical events up to the early components of cortical responses cannot account for changes in response times of several hundred milliseconds. Experiments in which response times to one of a set of possible stimuli are measured support the view that most changes in processing times are imputable to events in CIPP and not to delays in transmission along the afferent pathway.

The top of Fig. 27.2 depicts the reaction time of a human observer in response to clicks and tones of various intensities. The middle shows the latency of evoked cortical responses from the auditory area of a cat to clicks of various intensities. These latency

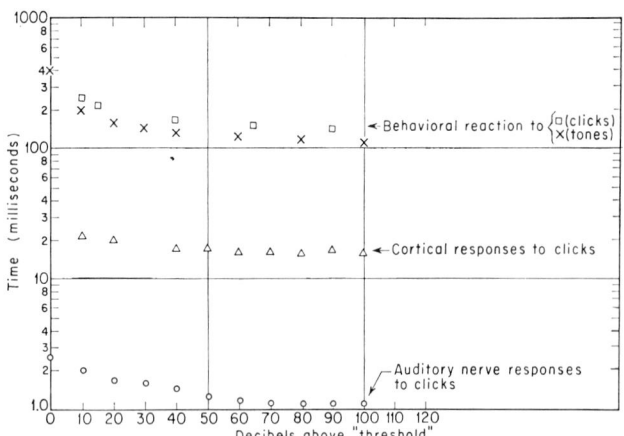

Fig. 27.2. Average response times to auditory stimuli.

values refer to time intervals elapsed between stimulus delivery and the peak of the initial surface-positive component of the evoked response. At the bottom of Fig. 27.2, the latency of neural responses for the cat's auditory nerve is shown as a function of click intensity. Here latency measures the time between delivery of stimulus and occurrence of the negative peak of the earliest neural event (N_1) in the auditory pathway.[5]

Changes in patterns of neuroelectric activity that may lend themselves to stimulus coding

Since it became possible to record electrical events in relation to sensory stimuli, electrophysiologists have looked for measurable aspects of responses that would reflect stimulus parameters. If one divides sensory continua into two classes[6]—prothetic ones (how much?) and metathetic ones (what kind or where?)—one might reasonably assume that a quantitative attribute, such as loudness, would be coded in terms of the number of neural elements that participate in a response. Qualitative attributes, such as pitch, might be coded spatially; i.e., in terms of the location of the neural elements sensitive to a particular stimulus quality. One can try to examine the implications of this model of spatial (or substitutive) versus numerical (or additive) coding somewhat further. Discrimination between two qualitatively different stimuli improves—up to a point—with stimulus intensity (i.e., with the total number of active elements involved). However, how is one going to define the quality of a patch of neural tissue that has appreciable spatial extent, i.e., involves a large number of active elements? Does one appeal to a notion of a center of gravity, does one rely on lateral inhibition as a sharpening process, or does one have to differentiate qualities on the basis of certain differences in spatiotemporal firing patterns or wave forms?

As long as sensory electrophysiologists used mainly gross electrodes and deeply anesthetized brains that exhibit little background activity, it seemed natural to measure the amplitude and latency of certain characteristic deflections in the recorded wave forms. If these wave forms remain reasonably shape-invariant, the measurement is unambiguous, but when new components appear and begin to superpose upon each other, difficulties arise. Without a detailed model of the distribution of generators and elemental wave forms, it becomes difficult to treat the various characteristic deflections as independent of one another. Much ingenuity has been expended in trying to isolate component potentials in the complex wave forms by anatomical or pharmacological procedures. Most of the time, experimenters have still to be satisfied with measuring the amplitude and latency of the least "contaminated" deflections of the evoked responses.

This type of quantitative measurement is based on certain implicit assumptions. The amplitude of a deflection is a measure or at least an estimate of the number of contributing neural elements. Whatever variability exists in the population of contributing neural elements affects mainly the number of units that enter into a given response, and less the instant at which they contribute their elemental wave forms (Figs. 27.3[7] and 27.4). Recently the use of computing techniques has led to widespread emphasis upon averaged evoked responses based upon the assumption that the component potentials are time-locked to delivery of the stimulus. Only relatively rarely are the important concomitant measures of variability and cumulative records also computed.

Before turning from gross electrode records to electrophysiological data on single neurons, it must be noted that data of the type considered permit us to measure reduced responsiveness[8] but not to quantify directly

---

[5] W. A. Rosenblith and E. B. Vidale, "A Quantitative View of Neuroelectric Events in Relation to Sensory Communication," *Psychology: A Study of a Science*, Vol. 4, Sigmund Koch, Ed., McGraw-Hill Book Co., New York, 1962, pp. 334–379.

[6] S. S. Stevens, "The Psychophysics of Sensory Function," *Sensory Communication*, W. A. Rosenblith, Ed., The M.I.T. Press, Cambridge, Mass., 1961, pp. 1–33.

[7] W. T. Peake, "An Analytical Study of Electric Responses at the Periphery of the Auditory System," *Tech. Rep. 365, Research Laboratory of Electronics*, Mass. Inst. Tech., Cambridge, Mass., 1960, pp. 1–62.

[8] Variations in the amplitude of time-locked deflections are interpretable as changes in the number of neural units contributing invariant elemental wave forms to a response.

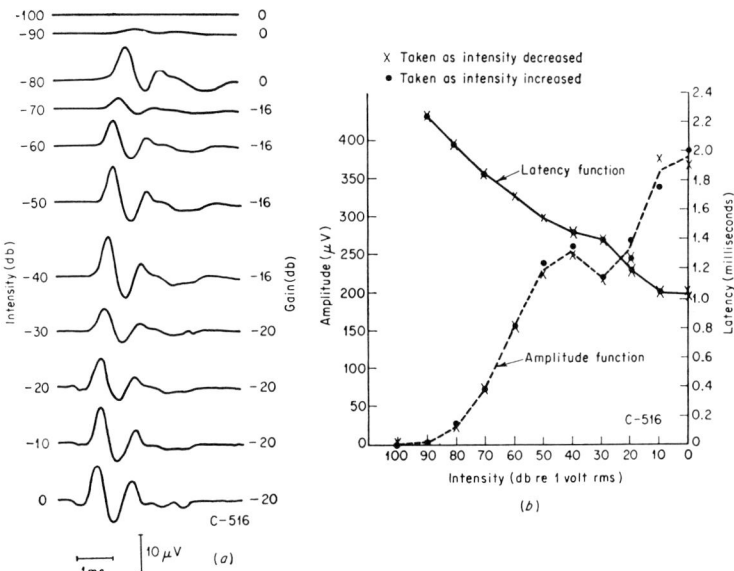

Fig. 27.3. (a) Averaged neural responses to short, repeated noise bursts of various intensities, recorded near the round window of the cochlea of an anesthetized cat.

The averaging operation was carried out by an electronic computing device. Relative positivity at recording electrode indicated by downward deflection. This first major negative-positive deflection in these records corresponds to the so-called "N_1 component"—the earliest and most prominent neural component that can be recorded at this location. It represents the summated action potentials from the auditory nerve. The noise bursts were 0.1 msec in duration and were presented at a rate of 5 per sec. The delivery of the stimulus coincides with the starting point of each trace. Number of responses averaged at the different stimulus intensity levels: 256 responses at −100 and −90 db, 128 at −80 db, and 64 at −70 to 0 db. The 10-μV amplitude calibration marker applies to the 0-db gain setting.

(b) Amplitude and latency of averaged neural responses to noise bursts as a function of stimulus intensity.

The intensity functions are based on the data of Fig. 27.3a. The values of the left-hand ordinate refer to the average peak-to-peak amplitude of the N_1 component while those on the right-hand ordinate represent measurements of the average time interval between stimulus onset and the occurrence of the negative peak of N_1. The latency values include a delay of approximately 0.4 msec that corresponds to the time that it takes for the sound to travel from the earphone diaphragm to the eardrum. (After Peake.[7])

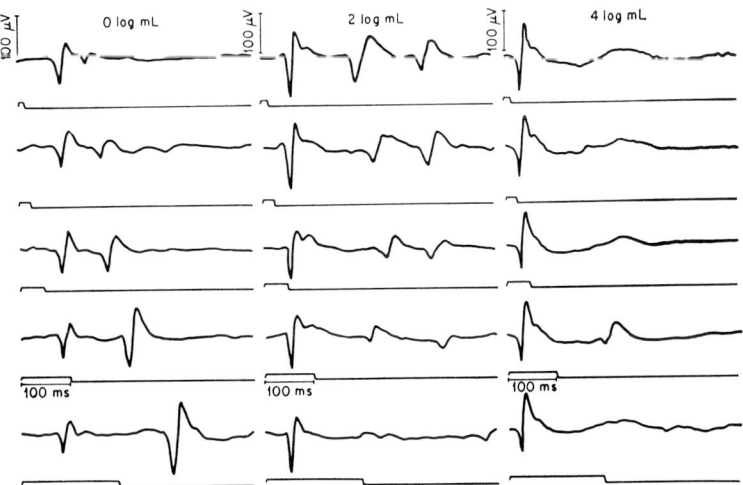

Fig. 27.4. Averaged evoked responses to repetitive photic stimulation, recorded from a location on the visual cortex of an anesthetized cat. (After Rosenblith and Vidale.[5])

The lower of each pair of traces represents the stimulus, and the upper records the corresponding averaged evoked responses. The three columns represent responses at three different luminance levels. For any given luminance, flash duration was varied between the limits of 12.5 and 200 msec; averaged responses to the shortest of the stimuli appear in the topmost row, while succeeding rows depict responses to flashes having durations of 25, 50, 100, and 200 msec, respectively. Repetition rate: one flash every 5 sec. Sixty-four consecutive responses entered into the computation of each average. Surface-positive deflections are plotted downward.

graded electrical events which are combinations of excitatory and inhibitory components. Functional relations for prothetic continua—also named intensity functions—exhibit in general monotonically increasing amplitudes and monotonically decreasing latencies as a function of stimulus intensity. However, the particular shape of an amplitude function depends rather strongly upon the wave form recorded, which in turn depends very much upon the precise electrode location and the state of the organism. It is thus not possible to translate *any* intensity function into an organism's psychophysical responses. The most striking psychophysical-electrophysiological parallelism perhaps resides in the shape of latency and reaction-time functions (cf. Fig. 27.2).

Except perhaps at the most peripheral stations of sensory systems, we do not know how to relate the continuously graded wave forms of evoked responses to statistical data on the discrete events we call action potentials or "spikes" (Fig. 27.5).[9]

Electrophysiologists have tended to record their microelectrode data in an elementary coding scheme: A given unit either *responds* or *does not respond* to a stimulus. In the presence of "spontaneous" activity[10] (occurrence of spikes in the absence of controlled stimulation), responses were said to be excitatory or inhibitory; i.e., the rate of firing increases or decreases within a time interval after stimulus delivery. The PST histogram displays and summarizes these changes for repeated identical stimuli (Fig. 27.6[11]). For some units a given stimulus does not necessarily produce a unidirectional change in firing rate. There may be at first an increase and later a decrease compared to the control level of firing. Other units in the auditory cortex exhibit different PST histograms for different stimuli.

This technique of analysis lends itself also to a description of what happens in the vicinity of a unit's "threshold," where relative frequency of response (response probability), average spike latency, and variability in spike latency all vary substantially. At certain locations in the nervous system strong stimuli give rise to repeated stimulus-locked firings and the PST histogram exhibits sharply defined peaks (cf. Fig. 27.5). Stimulus intensity can thus be coded for a single unit in terms of "extra" spikes within a given

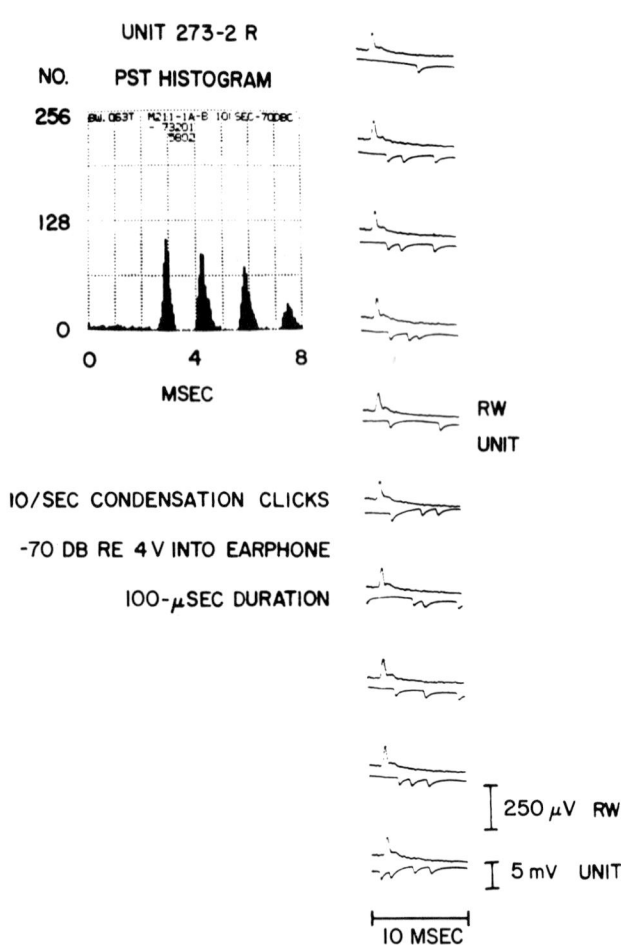

Fig. 27.5. Time pattern of responses of single auditory nerve fiber to repeated clicks, as shown by a poststimulus time (PST) histogram. (After Kiang *et al.*[9])

The column of traces at the right shows individual responses to ten successive clicks; the start of each trace is synchronized with the click presentation. The upper trace of each pair shows the gross potential change recorded by a large electrode near the round window; the large upward deflection is N_1. The visual detection level (VDL) for the N_1 potential is -90 db. The lower trace of each pair shows spike discharges recorded simultaneously with a micro-electrode in the auditory nerve; each downward deflection represents a spike discharge by the auditory nerve fiber. The PST histogram at the left shows the distribution of spikes during a 1-minute recording; the vertical axis represents the number of spikes of a particular latency The horizontal axis gives the time after click presentation.

time interval and/or in temporal relationships between the spikes.

In recent years the distributions of intervals between spikes during spontaneous activity have been examined; the striking stability of some of these distributions with time is illustrated in Fig. 27.7.[12]

[9] N. Y-s. Kiang, T. Watanabe, E. Thomas, and L. Clark, *Ann. Otol. Rhinol. Laryngol.* 71, No. 4, 1009 (1962).

[10] The spontaneous activity must not be too abundant or too irregular; otherwise it becomes difficult to detect the occurrence of a response without the aid of computing techniques.

[11] G. L. Gerstein and N. Y-s. Kiang, *Exptl. Neurology* 10, 1 (1964).

[12] R. W. Rodieck, N. Y-s. Kiang, and G. L. Gerstein, *Biophys. J. 2*, 351 (1962).

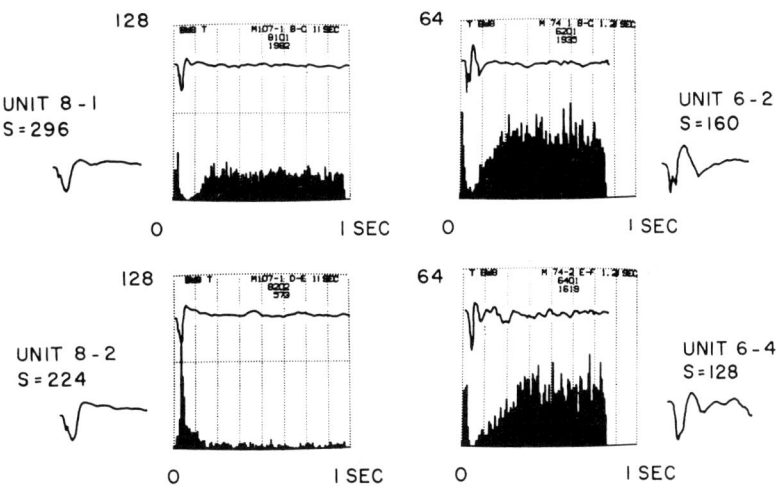

Fig. 27.6. Single-unit response patterns for four units and simultaneously recorded averaged response complex. The "gross" evoked response complex as observed by the microelectrode is shown at twice the time resolution in each insert. (After Gerstein and Kiang.[11])

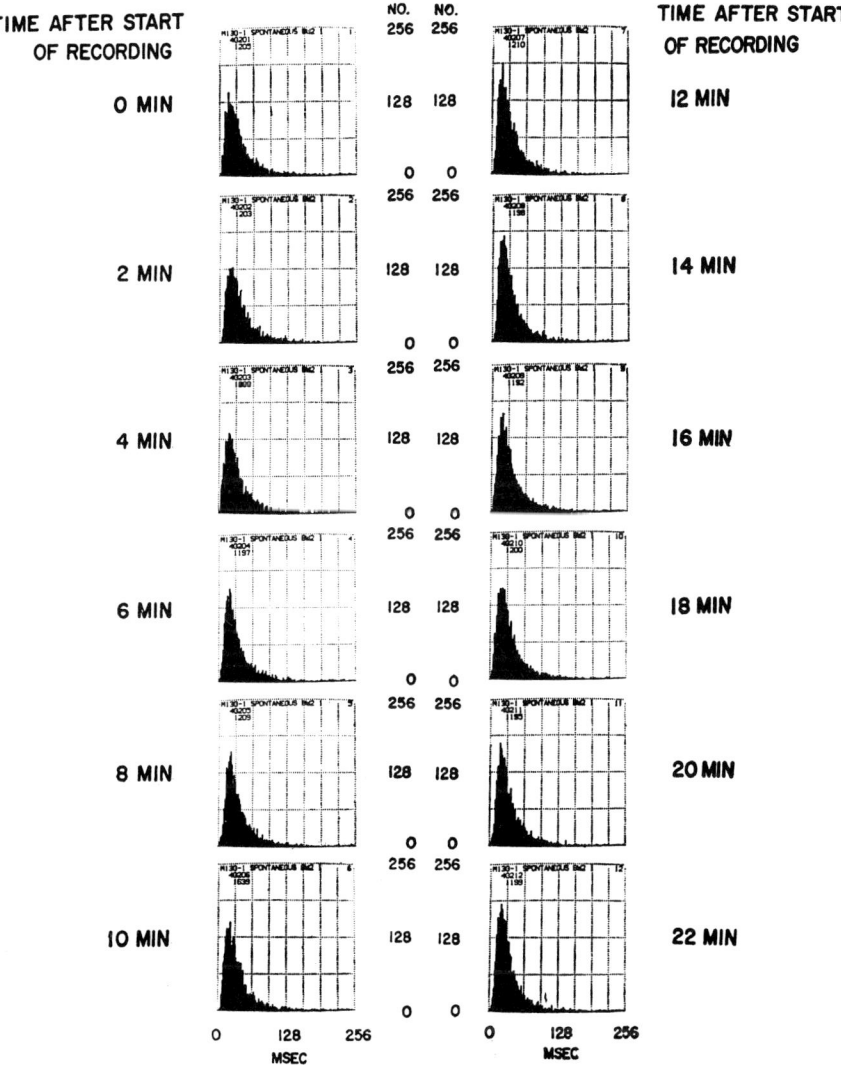

Fig. 27.7. Unit 240-1. Interval histograms for a unit in the cochlear nucleus, successive 2-minute samples from a half-hour length of data. (After Rodieck et al.[12])

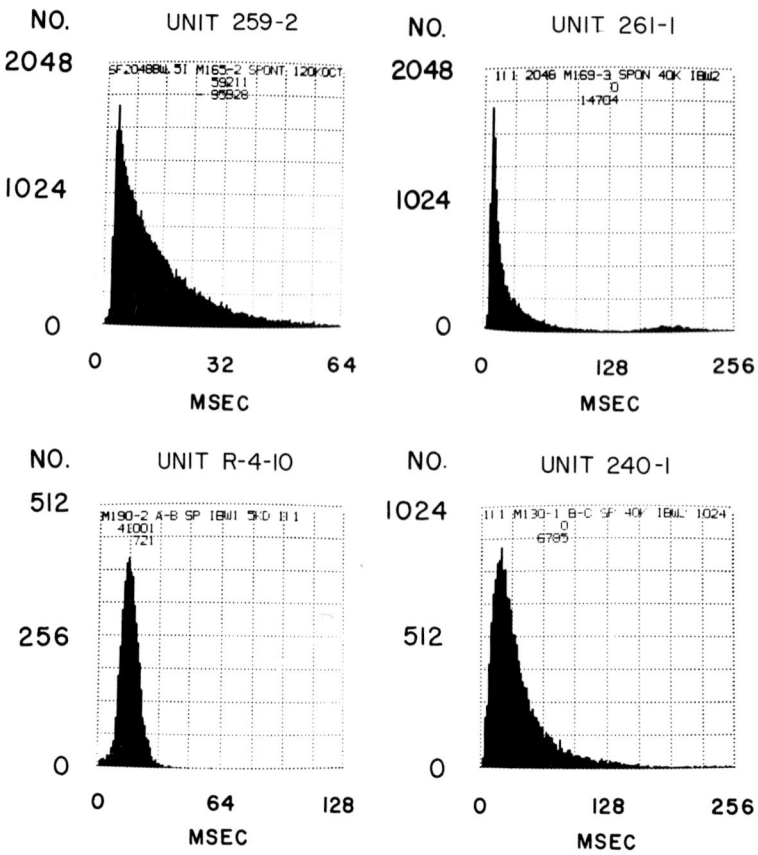

Fig. 27.8. Interval histograms for the spontaneous activity of four selected units in the cochlear nucleus. These histograms are derived from the processing of the following number of interspike intervals; Unit 259-2, $N = 40,960$; Unit R-4-10, $N = 5000$; Unit 261-1, $N = 16,384$; Unit 240-1, $N = 16,384$. (After Rodieck et al.[12])

There is, however, a good deal of variety in the shape of such interval histograms taken from various locations in the nervous system. Even within a given nucleus, e.g., the cochlear nucleus (Fig. 27.8), different distributions can be found for units that lie rather closely together.[12]

Workers have tried to develop mathematical models for these interval distributions and to relate them to functional connections of a given unit. A great deal more work must be done relating detailed morphology to intracellular records of electrical events (both spike and synaptic potentials). Once we can guess what a unit's connectivity pattern is from its interval histogram, we may be able to predict what its PST histogram to a particular stimulus will be.

Some form of coding must involve not only single units but neural populations or "neighborhoods"; systematic and orderly projections from activated receptors have been shown to give rise to spatial patterns such as columns; however, at least in the auditory cortex neither columnar nor laminar organization for types of response patterns has been found. A further complication is introduced by the fact that the response patterns of a given unit may vary substantially for different stimuli. To know the response pattern of a unit in the auditory cortex to a particular stimulus does not enable us to predict what the response patterns to other stimuli will be. This may be a significant difference between the auditory and the visual systems; in the latter, cortical units seem "tuned" to rather specific visual patterns, such as bars of light. The contrast between the two systems makes it tempting to speculate that these features of neural organization furnish some of the biological bases of language and grammar on the one hand and of visual symbolic behavior on the other.

Conclusion

The desire to understand how the nervous system meaningfully codes or represents the organism's environment goes back as far as the Greek notion of eidola. In this conception the sensitive receptors at the organism's boundaries are akin to mirrors. They help focus the image at the "right" place in the brain where it can be held up against previously stored templates or internal standards. Complex stimulus patterns give rise to composite pictures that can be assembled out of simple component elements. Such a hi-fi or perfect mirror view of the function of the nervous system is closer to a caricature than to a realistic model of the brain. As we have seen, sensory systems involve not only repeated energetic and logical transduction but also need- and task-dependent "distortions" or emphases.

In a given species, genetic instructions determine the structural organization (from the chemical to the anatomical level) and its modifiability during the life of the individual organism. Biological needs underlie species-specific information processing. Thus the bullfrog's auditory nerve seems especially well matched to receive the sounds of the male of the species. The frog's visual system seems well adapted to process or abstract those visual features of the environment that, to a frog, are pregnant with information. We are therefore led to consider sensory systems as selective translation devices that take events in physical space and transform or restructure them in a biased manner in a symbolic space.[13] In higher organisms and especially in man such transformations are by no means one-to-one. Many and quite different sound patterns may be interpreted as an "o" depending upon the speaker (man, woman, or child), the language, and even the particular immediate context in which the vowel is imbedded. Signals that are biologically important and effective seem to rely on multiple, perhaps fairly crude, codings throughout the nervous system which have the advantage of being resistant to unavoidable environmental and internal "noise" and interference from the other channels. Bio-logically programmed higher nervous systems are peculiarly sensitive to changes in the environment. As we are told in *Troilus and Cressida*, "Things in motion sooner catch the eye than what stirs not."

What we know about coding is just an extremely modest beginning. We are engaged in a linguistic task singularly more challenging and difficult than those commonly undertaken by anthropologists, ethnographers, linguists, or for that matter of cryptanalysts. To understand the languages and the dialects of nervous systems is a task for which we need more than the utterings of a few neural informants. During the past decade, molecular biology has made dramatic advances in breaking *the* genetic code, i.e., in understanding certain events that occur inside a cell. Nothing comparable has happened in understanding the codes that regulate the flow of intercellular messages.

REPRESENTATIVE BOOKS AND ARTICLES

General References

G. von Békésy, "The Ear," *Sci. Am. 197*, No. 2, 66 (1957).
Biophysical Science, J. L. Oncley et al., Eds., John Wiley and Sons, New York, 1959, p. 1.
M. A. B. Brazier, *The Electrical Activity of the Nervous System*, 2nd ed., The Macmillan Co., New York, 1960, p. 1.
J. C. Eccles, *The Physiology of Synapses*, Academic Press, New York, 1964, p. 1.
IRE Trans. on Information Theory, Special Issue on Sensory Information Processing, *IT-8*, No. 2, 1962, pp. 74–190.
H. W. Magoun, *The Waking Brain*, 2nd ed., Charles C Thomas, Springfield, Ill., 1963, pp. 1–188.
W. A. Rosenblith, "The Quantification of the Electric Activity of the Nervous System," *Quantity and Quality*, D. Lerner, Ed., The Free Press of Glencoe, New York, 1961, pp. 87–102.
Sensory Communication, W. A. Rosenblith, Ed., The M.I.T. Press, Cambridge, Mass., 1961, pp. 1–844.
1962 Symposium on Physiological Optics, *J. Opt. Soc. Am. 53*, 1–201 (1963).

Specialized References

D. Albe-Fessard, "Nouvelles Données sur l'Origine des Composantes des Potentiels Evoqués Somesthésiques," *Actualités Neurophysiologiques*, 3ième série, A.-M. Monnier, Ed., Masson et Cie., Paris, 1961, pp. 23–60.
L. S. Frishkopf and M. H. Goldstein, "Responses to Acoustic Stimuli from Single Units in the Eighth Nerve of the Bullfrog," *J. Acoust. Soc. Amer. 35*, 1219–1228 (1963).
G. L. Gerstein and N. Y-s. Kiang, "Responses of Single Units in the Auditory Cortex," *Exp. Neurol. 10*, 1–18 (1964).
M. H. Goldstein, Jr., "A Statistical Model for Interpreting Neuroelectric Responses," *Information and Control 3*, 1–17 (1960).
J. L. Hall, II, "Binaural Interaction in the Accessory Superior Olivary Nucleus of the Cat—An Electrophysiological Study of Single Neurons," *Tech. Rep. 416, Research Laboratory of Electronics*, Mass. Inst. Tech., Cambridge, Mass., 1964, pp. 1–83.
H. K. Hartline, F. Ratliff, and W. H. Miller, "Inhibitory Interaction in the Retina and its Significance in Vision," *Nervous Inhibition*, E. Florey, Ed., a Pergamon Press Book, The Macmillan Co., New York, 1961, pp. 241–284.
D. H. Hubel, "Integrative Processes in Central Visual Pathways of the Cat," *J. Opt. Soc. Am. 53*, 58–66 (1963).
D. H. Hubel, "The Visual Cortex of the Brain," *Sci. Am. 209*, No. 5, 54 (1963).
D. Kennedy, "Inhibition in Visual Systems," *ibid. 208*, No. 1, 122–130 (1963).
N. Y-s. Kiang, M. H. Goldstein, and W. T. Peake, "Temporal Coding of Neural Responses to Acoustic Stimuli," *IRE Trans. Inform. Theory, IT-8*, No. 2, 113–119 (1962).
N. Y-s. Kiang, T. Watanabe, E. Thomas, and L. Clark, "Stimulus Coding in the Cat's Auditory Nerve," *Ann. Otol. Rhinol. Laryngol. 71*, 1009–1026 (1962).

---

[13] In the foregoing we have used neuroelectric phenomena as a convenient physical substrate of this space.

S. W. Kuffler, "Discharge Patterns and Functional Organization of Mammalian Retina," *J. Neurophysiol.* **16**, 37–68 (1953).

H. R. Maturana, J. Y. Lettvin, W. S. McCulloch, and W. H. Pitts, "Anatomy and Physiology of Vision in the Frog (*Rana pipiens*)," *J. Gen. Physiol.* **43**, Pt. 2, 129–176 (1950).

W. T. Peake, "An Analytical Study of Electric Responses at the Periphery of the Auditory System," *Tech. Rep. 365, Research Laboratory of Electronics*, Mass. Inst. Tech., Cambridge, Mass., March, 1960, pp. 1–62.

T. P. S. Powell and V. B. Mountcastle, "Some Aspects of the Functional Organization of the Cortex of the Post-Central Gyrus of the Monkey: a Correlation of Findings Obtained in a Single Unit Analysis with Cytoarchitecture," *Johns Hopkins Hospital Bull.* **105**, 133–162 (1959).

W. A. Rosenblith, "Sensory Performance of Organisms," *Rev. Mod. Phys.* **31**, 485–491 (1959).

W. A. Rosenblith, "Some Quantifiable Aspects of the Electrical Activity of the Nervous System (with emphasis upon responses to sensory stimuli)," *ibid.* **31**, 532–545 (1959).

W. A. Rosenblith and E. B. Vidale, "A Quantitative View of Neuroelectric Events in Relation to Sensory Communication," *Psychology: A Study of a Science*, Vol. 4, S. Koch, Ed., McGraw-Hill Book Co., New York, 1962, pp. 334–379.

R. W. Rodieck, N. Y-s. Kiang, and G. L. Gerstein, "Some Quantitative Methods for the Study of Spontaneous Activity of Single Neurons," *Biophys. J.* **2**, 351–368 (1962).

S. S. Stevens, "The Psychophysics of Sensory Function," *Sensory Communication*, W. A. Rosenblith, Ed., M.I.T. Press, Cambridge, Mass., 1961, pp. 1–33.

T. F. Weiss, "A Model for Firing Patterns of Auditory Nerve Fibers," *Tech. Rep. 418, Research Laboratory of Electronics*, Mass. Inst. Tech., Cambridge, Mass., March, 1964, pp. 1–93.

SUBJECT INDEX

Absorption, optical, 64, 72, 76, 134, 138, 150
Ablation, 177
Acceptors, 145, 151
Acoustic amplification (in CdS), 235
Acoustic stimulus, response to, 257–258
Acoustic waves in solids, 190, 233, 236
Action potentials in neurons, 256, 260
Activation energy in solids, 62, 72
Activators, 131, 141, 142
Active elements, matrices of, 220
Adenosine triphosphate (ATP), 101, 208, 209
Admittance multiplier, 219
Afwillite, 31
Albite, 33
Alkali halide crystals, 8, 25, 70–76, 132, 195
Alkali metals, 20, 156, 195
Alloys, structure of, 24, 52, 161
Amino acids, 100, 199
Amoeba, 208
Amplifiers, feedback impedance, 215
   matrices of, 222
   traveling-wave, 234
   ultrasonic, 233
Amplitude modulation, 228
Anisotropic g factor, 44, 147
Anisotropy, magnetic, 47, 111, 229
Anorthite, 37
Antibonding, 9, 52
Antiferromagnetic interaction, 50–56
Antiferromagnetic state in b.c.c. lattice, 55
a priori theory, 8
Argon, electron mobility in, 189
   Ramsauer effect in, 189, 244
Atmospheres, electronic, ionic, dipolar, 190
Atom, antihydrogen, 151
   electronegative, 24
   electropositive, 24
   misfit, 39
ATP, 101, 208, 209
Avalanche breakdown, 4, 168–172, 183–196
Auditory cortex, 260
Auditory system of cat, 256
Auger effect, 131
Augmented plane-wave method, 18, 156
Axon, 105–108
Axoplasm, 108

Band theory, standard, 15, 156
Bardeen, Cooper, and Schrieffer (BCS) theory, 164–165

Barium titanate, 25, 34–36
Behenic acid, 205
Bimolecular leaflet, 202, 210
Binary crystals, lattice defects in, 122–123
Binary halides and oxides, melting points of, 26
Biological interfaces, 210
Birefringent spindle fibers, 211–214
Bloch wave functions, 16, 51, 160
Body-centered cubic structure, 23
Bohr's circular orbits, 15
Boiling points of elements, 16–23
Bonding bands, 52
Bonds, fixed, 20
   molecular, 9–11
   resonating, 20
Boron emission, 140
Bragg condition, 13
Bragg reflection, 153
Brain function, 107
Brain tissue, organization of, 256
Bravais lattice, 56, 59
Breakdown, in air, 188
   in alkali halides, 194
   in CO_2, 188
   by field emission, 185–187
   in glass, 194
   impurity, 194
   in liquefied gases, 188–189
   in N_2, 169, 170
   single- and multiavalanche, 168–172, 183–191
   thermal, 195, 196
Breakdown strength, temperature dependence of, 194
Brillouin zone, 13, 149
   of b.c.c. lattice, 156
   of f.c.c. lattice, 13, 157
Burgers vector, 79, 86

Cadmium sulfide, 6, 234
Calcium fluoride structure, 25
Calcium ions in packing structures, 30
Calorelectric effect, 177
Canted spin model, 59
Capacitance, cold and hot, 178
Capacitors, ceramic feed-through, 217
   vacuum vs. solid-state, 178
Capture cross sections for electrons, 141, 189
Carbon in iron, 119, 120

Cations, cavity vs. framework, 31
Cavity modes of high-dielectric-constant ceramic, 217
Cell, Danielli-Harvey model of, 202
   definition of, 197
   dimensions of, 198
   division of, 108, 211–217
   hereditary materials of, 200
   molecular control of, 107
Cell dynamics, 202
Cell membranes, 103–106
Cell nucleus and cytoplasm, 199–202
Cell surface, interaction in, 205–207
Cells, dynamic activities of, 201
   mesodermal, 106
   molecular constituents of, 198
   muscle, 106
   Schwann, 105, 106
   structural components of, 197
   on substrates, 203–207
Celsian, 40
Center structures of photoconducting phosphors, 146
Centrioles, 200
Ceramic feed-through capacitors, 217, 218
Ceramic filters, 218
   suppression of resonances in, 217
Childs-Langmuir law, 5
Chloroplasts, 200
Chromium tribromide, 230
Chromosomes, 199–200, 211–212
   movement of, 214
Circulators, 229
Close-packed structures, built from tetrahedral and octahedral modular units, 25
Cloud chamber, 168, 170
Cochlea, 257–261
Coding, genetic, 255
   neural, 255
Coercive fields for fine particles, 116
Coherent emission, 152
Coherent light modulation, 233
Cohesion, 159
Collagen, 100, 209
Collisions of electrons, inelastic, 174
Colloids, 64, 201
Color centers, 71–73
Combustion of coal, 240
Communication processes and sensory coding, 255–262
Computers coupled to brains, 256

Conduction, in condensed systems, 179–180, 189–190
   in gases, 169–170
   in ionic crystals, 3, 72, 122, 195
   in metals, 153–167
   in phosphors, 139–141
   in semiconductors, 149–150
Conductivity of CoO, NiO, and NiO-CoO mixed crystals, 71
Conductivity and crystal defects, 73, 122
Conductors and insulators, 26
Configuration-coordinate diagrams, 132–135
Constant-energy surfaces, 150, 158–160
Contractile fibers, 214
Converters of impedance, 220
Coordination number, 29
Copper alloys, electron/atom ratios for, 155
Copper-colloid formation, 195
Cordierite, 63
Core energy of dislocations, 87, 88
Corrosion, dry, 126
Corundum structure, 26
Coulomb interaction between hole and electron, 151
Countercharges, atmospheres of, 190
Counterdiffusion, 127
Coupling, antiferromagnetic, 50
   magnetomechanical, 236
Crack formation, 3, 94–96
Cristobalite, 31
Crystal, one-electron model for, 15, 26, 153
Crystals, defects in, 69–74, 77–83, 122–124
   domain formation in, 38–39, 113–115
   electro-optically active, 230
   plasticity of, 86
Cubic close-packed structure, 23
Cuprous chloride, 232
Curie point, 36
Curie-Weiss law, 231
Cyclotron mass, 159
Cyclotron resonance, 150, 158–159
Cytology, 197
Cytoplasm, 104, 200, 208
   liquid-crystal model of, 201
Cytoplasmic flow, 208
   matrix, 201, 208

d-electron interaction with nuclear magnetic moment, 44
d-electron states, splitting in octahedral and tetrahedral sites, 45
d electrons, 22, 42–51
Damped precession, 116
de Broglie equation, 13
Debye frequency, 85
De-excitation, radiationless, 131
Defect structures, complex, 79
Defects, in ionic crystals, 69–74
   in metal crystals, 77–83
Defects due to nonstoichiometry, 123
Deformation, elastic, 84
   inelastic, 85

Deformation, Orowan model of viscous, 85
   plastic, 2, 86
Degenerate vibrational modes for octahedral complex, 47
Degradation factor, 242
de Haas–van Alphen effect, 158, 159
Demagnetizing field, 112
Dendrites, 107, 195
Dense-packed structures, 23–25
Density measurements, detection of impurities by, 75
De-oxyribonucleic acid (DNA), 199
Diamond, types of, 75
Dielectric cavities, 216
Dielectric liquids, 190–194
Differential efficiency coefficient in energy conversion, 247
Diffraction phenomena, 13, 161
Diffusion damping in magnets, 119–121
Diffusion in ternary compounds, 127
Diode systems, semiconducting, 152
Dipole field and resonance transfer, 136
Dipole moment reversal in ferroelectric $BaTiO_3$, 35
Discharge, electrodeless, 176
   glow, 185
   plasma, 170, 185–187
Dislocations, density of, 87
   edge, 86–87
   free energy of, 87
   in platinum crystal, 81
   screw, 86–87
   in silicon crystal, 74
   structure of, 77
Disorder, anti-Frenkel, 123
   anti-Schottky, 123
   Frenkel, 123
   interchange, 40
   Schottky, 123
   substitutional, 125
Dispersion law in band theory, 156
Displacive processes, 37
Dissociation by electron collisions, 176
Distillation of sea water, 239
Distortions, induced by electrons, 51
DNA molecule, 199
DNA-RNA-protein system, 107
Dog's bone, belly, and neck of energy surfaces, 158, 159
Domain structures, 115
Domain theory, 112
Domains, antiphase, 38–40
   formation of, 2, 23, 39
   twin, 38–39
   in uniaxial crystal, 113–114
Domain-wall damping, 119
Domain-wall energy, 114
Domain-wall permeability, 120
Domain walls, 39, 113–114
Donors, 145
Doping effect on conductivity, 123, 124
Double injection, 182, 196
Drift mobility, 181

Drift velocity, 169, 173
Dynamic excitation of spins, 116
Dynamic squareness, 121

$\varepsilon(k)$ curves for metallic TiO, 14
Ectoplasmic matrix, 208
Effective mass, 150
Einstein diffusion equation, 43
Elasticity of rubber, 84
Elasticity of solids, 84
Electric strength and external electron supply, 183
Electrode conditioning, 185
Electrode problems in high-temperature range, 252
Electrodialysis, 239
Electrofax (electrostatic photography), 181
Electrolytes, solid, 122–129, 241, 251
Electron-atom collisions, 174
Electron bunching in piezoelectric semiconductor, 233
Electron emission, 191, 246
Electron excitation states, barrier of, 185, 190
Electron injection in rutile, 195
Electron mass, effective, 15, 150, 159
Electron-molecule collisions, 174
Electron motion, influence of long-range order on, 190
Electron-phonon interaction, 164, 190
Electron states in crystals, 13, 154
Electron theory, collective vs. localized, 43
Electron waves, diffraction of, 13, 161
   group velocity of, 15, 154
Electronic heat engines, 245
Electro-optical coefficients, 231
Electro-optical modulator, 230
Electrons, relaxation times of, 159
   thermal excitation of, 27
Electrons and holes in semiconductors, 149–152
Electrostatic valence, 24, 29
Emission, coherent, 152
   fluorescent, for zinc sulfide, 145
   stimulated, 152
Endoplasmic reticulum, 104
Energy, anisotropy, 111
   chemical, 240
   condensation, 20
   of domain-wall trapping, 117
   as function of wave vector $\varepsilon(k)$, 13, 150, 156
   radiant, 241
Energy bands for copper, 157
Energy barriers, 62, 133, 190
Energy conversions, 238, 248
Energy levels in ZnS, 146
Energy of mixing sea water with fresh water, 237–238
Energy transfer in luminescence, 136–138
Engel rule, 23
Entropy, method to minimize, 239
Enzymes, adsorbed to cell surfaces, 205

Enzymes, digestive, 200
    proteolytic, 205
Equilibrium-domain dimension, 114
Etch pits, 73
Eucryptite, 63
Evaporation, controlled field, 78
Exchange coupling, via collective electrons, 52
    indirect, 42
Exchange diffusion, 125
Exchange fields, 162
Exchange interaction, 54, 160
    between $3d$ and $5s$ electrons, 163
Exchange splitting, 44
Excitation, dynamic, of spins, 116
Excitation coefficient, 175
Excitation cross sections, 174
Excitation intensity, 143
Excitation yield in H_2, 176
Excited species in discharges, 173–177
Excitons, 137, 151
Extrinsic luminescence, 140

F band, 70–72
F center, 73, 132
Faraday constant, 229
Faraday rotation of microwave radiation, 229–237
Faraday rotation modulator, 230
Fatty acids, 204
Feedback, series-parallel, 219
Feldspars, displacive changes in, 36
    as framework structures, 31–33
Fermi-Dirac distribution function, 28, 154
Fermi energy, 155–156
Fermi level, 43, 155
Fermi surface, 28, 155, 159
Ferrimagnetic yttrium iron garnet (YIG), 118, 120, 229, 236
Ferrimagnetism, 56
Ferrites, microwave devices, 223
Ferroelectrics, 35, 230
Ferromagnetic spirals, 50–58
Ferromagnetism, induced by parasitic canting, 44
Ferromagnetism in Pd with Co addition, 162
Field distortion, dynamic, 185, 187
    by ionic conduction, 195
Field emission, 192
Field-emission currents, 185
Field-enhanced dissociation, 192
Field evaporation, 78
Field ionization, 77
Field splitting, cubic, 44
Fibers, mechanochemistry of, 209
    metastable spindle, 211–214
    nerve, 105
Fibrils, 199, 201
Filters, active, 223
    cascaded, 218
    frequency effects in, 226
    low-pass, idealized, 223
    simulated-skin-effect, 218

Fine particles, coercive fields for, 116
Fluorescence, infrared quenching of, 144
    infrared stimulation of, 144
    X-ray, 131, 132
Formal ionic charges, 24
Fotoceram, 64
Fotoform glass, 64, 65
Fourpole, active, 225
Fourpole networks, cascading, 219
Fowler-Nordheim equation, 189
Framework structures, 30
    aristotypes, 40
    heterotypes, 40
Franck-Condon principle, 133, 140, 175

g factor, anisotropic, 44, 147
Gain-bandwidth product, 181
Gallium-substituted YIG, 230
Gap parameter of superconductors, 165
Garnet, yttrium iron (YIG), 118, 229, 239
Generators, fuel-cell, 129, 240, 251
    magnetohydrodynamic (MHD), 244, 252
    MHD, nuclear-fired, 253
    Nernst-Ettingshausen, 247
    thermionic, 245, 250
    thermoelectric, 247–249
Genetic code, 263
Germanium, atomic weight of, 75
Glass, chemically machineable, 64
    expansion coefficients of, 63
    inorganic as supercooled liquid, 85
    obsidian, 65
    photochromic, 66 67
    quartz, 195
    shattering by field emission, 194
    silver halide, 66, 67
Glass-ceramic, photosensitive, 63–64
Glial cells, 106
Glow curves, 73
Glow discharge, 185
    hollow-cathode, 177
Gold chloride, 63
Goldschmidt's rules, 23, 29
Golgi apparatus, 104, 200
Group symmetry, 11
Group velocity of electron-wave packets, 15, 154
Guinier-Preston zones, 81
Gyrators, 219, 229
    conditions for parametric transmutation of, 225
Gyromagnetic resonance distribution, 118
Gyroscopic precession, 116

Hair-cell-neural junction of cat, 257
Hamiltonian, exchange, 49
Hamiltonian for isolated ion, 43
Harmonic oscillator and light emission, 134
Hartree-Fock approximation, 18, 156

Heat engines, electronic, 245
    thermoelectric, 245
Heat of fusion, 19
Heisenberg exchange parameter, 44
Heisenberg Hamiltonian, 54
Heisenberg magnets, 54–59
Hexagonal close-packed structure, 23
Hexane, current variation by temperature or addition agent, 192
    current-voltage characteristics in, 191
    gap-width effect in, 192
    polarization effects in, 193
    prebreakdown currents in, 190
High-dielectric-constant ceramics, 215–216
Histone, 199
Holes, in semiconductors, 149–152
Hot-electron devices, 216, 245
Hume-Rothery rule, 24, 155
Hund's rule, 43, 45, 54
Hydrophobic segments, 203
Hyperfine structure in phosphorescence, 147
Hysteresis loop for uniaxial crystal, 113
Hysteresis loops and dynamic squareness, 121
Hysteresis, thermal, 35–36

Impact ionization, 191
Impedance multiplier, 219
Impedance transformation, 223
Imperfections and optical absorption, 76
Impurities, attraction to grain boundaries, 81
    carbon, in iron, 119
    donor and acceptor, 151, 152
Impurities and excitons in a magnetic field, 151
Impurity atoms in metal crystals, 79
Impurity breakdown of oil, 194
Impurity levels in phosphors, 141
Impurity states and stimulated emission, 152
Infinite crystal, magnetization of, 112
Injected carrier, transit time of, 181
Inner-shell transition, 132
Inorganic glass, elasticity of, 86
Insulators, magnetic, 229
    semiconductors, and metals, 27
Insulators and conductors, 26, 183, 196
Interatomic distances, 29
Interdiffusion exchange, 126
Interface, changes in chemical potential of, 209
Interfacial dynamics, 209
Interfacial pressure, 207
Interference-elimination filters, 216
Interstices, tetrahedral and octahedral, 25
Intrinsic luminescence, 140
Inverters of impedance, 220
Ion microscope, 77
Ion mobility and drift velocity, 173
Ionization by collision, 173, 188, 190
Ionization probability, 174, 184

Ionization in stages, 174
Isolators, 224, 225
  conditions for parametric transmutations of, 225
Isotope effect in superconductors, 165

Jahn-Teller effect, 46, 47, 135

KDP, electro-optical properties of, 231
Killers of phosphorescence, 142
Knight shift, 162, 163
Kröger-Vink equations, 142

Landau levels, 150, 158
Landau-Lifshitz equation, 116
Landau-Lifshitz thermodynamic theory, 60
Landé factor, 111
Langmuir trough, 203
Laser modulation, 228
Latency and reaction-time functions, 260
Lattice defects, 69–73, 77–79, 122–124
Lecithin, 202
Leptons, 201
Lichtenberg figures, 185, 187, 189
Lifetime, radiative, 131
Ligand atoms, 10
Ligand-field splitting, 42, 48
Ligand-field theory, 28, 44
Light beams, as information carriers, 227
Light modulation at microwave frequencies, 231
Light modulator, magneto-optical, 229
Lightning discharge, 172, 187
Lipids, 105, 106, 199
Liquefied gases, breakdown strength of, 189
Lithia-alumina-silica system, 63
Lithium metasilicate, 63
Longitudinal spin waves, 57
Lorentz force, 158
Luminescence, blue and red in ZnS, 144, 147
  extrinsic or intrinsic, 140
  impurity-sensitized, 138
  polarized, 135
Luminescence activators and poison, 131–136
Luminescence efficiency, 131
Luminescence of rare-earth ions, 132
Luminescence in solids, 130
Luminescent center, vibrational modes of, 132
Luminescent centers in zinc sulfide, 144, 147
Luttinger-Tisza method to minimize energy, 60
Lysosomes, 200

Machining, chemical, 64
Magnetic coupling, 46
Magnetic discontinuity, 161
Magnetic insulators, 229

Magnetic losses, in ferrites, 118
  in garnets, 118
Magnetic moment, 43
Magnetic ordering, 54
Magnetic pole density, 112
Magnetic resonance, 119
Magnetic sieves, 243, 244
Magnetic state of finite crystal, 114
Magnetic susceptibility tensor, 117
Magnetization, 110
  fundamental mesoscopic equation of, 110
  orbital, 162
  spin, 162
Magnetization reversal modes, 115
Magnetoelastic devices, 235, 236
Magnetohydrodynamic (MHD) generator, 244, 252, 253
Magnetomechanical ratio, 111
Magneto-optical effect, 228
Magneto-optical light modulator, 229
Magneto-optical phenomena, 152
Magnetoresistance of metals, 157, 158
Magnetostriction, 236
Magnets, superconducting, 253
Many-generation (Townsend) mechanism, 170–172
Materials, simulation of, by networks, 215
Matrix algebra, 219
Mechanochemistry, 210
Meissner effect, 166
Melting points, of binary halides and oxides, 26
  of elements completing d shells, 23
  of elements completing s^2p^6 shells, 23
  of rare-earth elements filling the $4f$ shell, 23
Melting temperatures, 16–17
Membranes, biological, 105, 106, 199
Membranous elements, 197
Membranous leaflet, 210
Memory encoding, 107
Mesodermal cells, 106
Metal crystals, defects in, 77
Metal-vacuum interfaces, 245
Metals, cohesion in, 160
  conduction electrons in, 160, 161
  definition of, 154
  equilibrium distances in, 160
  grouping of, in periodic table, 18
  oxidation rate of, 126, 127
  paramagnetism of, 155
  reaction of, with S, Se, Cl, 127
  strength of, 77
Metals and energy bands, 153
Metastables, 174, 175
Metastable states, 53, 61, 62
  triplet, 175
Methyl alcohol, one-generation breakdown in, 171
MHD generators, nuclear-fired, 253
Microexudate, 204
Microfibrils, 200
Microfilaments, 200

Micromagnetics, 112
Microminiaturization, 216
Microscope, field-ion, 77
  polarization, 212
Microspikes, 209
Microtubules, 200–201, 208
Microwave ferrite devices, 229
Microwave radiation, Faraday rotation of, 236
Mind, as a system, 109
Mitochondrion, 104, 200
Mitosis, 214
Mitotic spindle, 211
Modulation of light beams, 228
Modulator, electro-optical, 230
  Faraday rotation, 230
  zigzag, 232, 237
Molecular hydrogen, electronic states of, 175
Molecule-cell control, 107
Moment, magnetic, 43
Mössbauer effect, 4, 162
Multiavalanche breakdown, 4, 168–172, 183–191
Muscle cells, 106
Myelin sheath, 105, 106
Myofibrils, 201

$8\text{-}N$ rule, 18–20
Néel state, 55
Negator, 223
Nernst-Einstein equation, 125
Nernst-Ettingshausen generator, 247
Nerve axon, 108
Networks, negation of, 223
  simulating materials, 215–226
Neural populations, 262
Neural responses to stimuli, 259
Neural spikes or action potentials, 260
Neuroanatomy of auditory system, 256
Neuroelectric activity, patterns of, 255, 258
Neurofilaments, 108
Neurology, molecular, 107
Neurons, 107, 255–257
Nonisothermal transport, 128
Nuclear magnetic resonance, 162
Nuclear membrane, 200
Nuclear reactor, 248
Nucleoli, 199
Nucleoplasm, 104

Obsidian glass, 65
Occupation function in superconducting ground state, 165
Ohmic flow, 181
One-generation (streamer) mechanism, 170–172
Onsager's theory of weak electrolytes, 192
Optical absorption, 67, 72, 134, 177
Orbitals, atomic, 1, 9, 11, 43
  molecular, for octahedral TiO_6 configuration, 12

Subject Index

p-n diode, red luminescence of GaP, 144
p-n junction, Si in forward bias, 140
Packing structures, 30
Pair-bond formation, 18
Paschen curves, 183–187
Paschen's law, failure of, 185–189
Pauli exclusion principle, 54, 111
Pauli magnetism, 155
Pauling's electroneutrality rule, 24
Pauling's electrostatic valence rule, 29
Penning effect, 174
Perovskites, as framework structures, 31
    properties of, 35
    structure of, 25, 27
    structure changes of, 35
Phase discriminator, 228
Phase transitions, 29–36, 63
Phase-velocity matching, 231
Phosphatide, 206
Phospholipids, 199, 202–203, 206, 210
Phospholipid-sterol complex, 210
Phosphors, photoconducting, 139–145
    zinc sulfide type, 139–141
Photoconducting phosphors, effects of excitation intensity in, 143
    paramagnetic studies on, 147
    vacancies in, 146
Photoconductors, 181
    gain process in, 182
Photodiodes, 228
Photoeffects, superlinear, 143
Photoelectrons, 69, 186
Photoexcitation, 3, 4, 174
Photomultipliers, 228
Photon avalanches, 169
Photosynthesis, 138
Photosynthetic mechanism, 104
Piezoelectric semiconductor, acoustically driven, 234
Piezoelectric semiconductor devices, 6, 233–237
Piezo- and ferroelectrics as modulators, 230–231
Planar transformer, 221
Plane wave method, augmented, 18, 156
Plasma discharge, 170, 185–186
Plasma formation, 185–186
Plasma oscillations, 163
Plasma streamer, 170
Plateau-Gibbs-Marangoni effects, 209
Point defects in crystals, 69, 78, 122
Poisson's ratio, 84
Polar molecules, saturated, 27
Polarization rectifier for microscope, 212
Polarons, 190
Polyhedra, regular, 19
Polypeptide chains, 211
Polysaccharides, 199
Polystyrene, 90–91
Positronium, 151
Poststimulus-time (PST) histogram, 260
Potassium dihydrogen phosphate, electro-optical properties of, 231
Prebreakdown currents in hexane, 190

Protein, 207
    neurofilament, 108
    synthesis of, 199
Proteins as elements of cytoplasm, 201
Protoplasmic streaming, 207
Psychophysics, 225, 256
Pyroelectrics, 35

Quantasome, 104
Quantum numbers in k space, 15–17
Quartz, 31
Quartz glass, 195
Quartz-transducer response to echoes, 236
Quencher of luminescence, 136
Quenching of excited species by collisions, 177

Radiative lifetime, 131
Radii, atomic and ionic, 23
Ramsauer effect, 189, 244
Rare-earth ions, luminescence of, 132
Rare-earth metals, 54, 59, 161
Reactor incorporating thermionic generators, 250
Recombination center, nonradiative, 145
Recombination of charge carriers, 140
Regeneration probability, 184
Relaxation, spin-lattice and spin-spin, 118–121
Relaxation spectrum for domain-wall susceptibility, 117
Relaxation time in Ohm's law range, 181
Resonance transfer, 136–138
Resonances, diffusing of, 218
Response to acoustic stimulus, 257–258
Response probability, 260
Reticulum, endoplasmic, 200
Ribonucleic acid (RNA), 199–200
Ribosomes, 199, 200
Richardson equation, 245
RNA, 199–200
Robertson's unit-membrane theory, 104
Rock salt structure, 25
Rubber molecule, 85
Rutile, electron injection in, 195

Sanidine, monoclinic, 32
Schrödinger equation, 13
Schwann cells, 105, 106
Seebeck coefficients, 247, 249
Self-diffusion, coefficient of, 125–127
Semiconductors, compensated, 145
    electric conduction in, 122, 123
    electrons and holes in, 149
    noncentrosymmetric, 233
    piezoelectric, 233
    space-charge-limited currents in, 181
    transport properties of, 150
Semimetals, 247
Sensitizer, 136
Sensory coding, 255–265
Sensory system, information-flow diagram of, 257

Separation distance, critical, for d electrons, 43, 51
Separator, derived from negator, 226
Shattering of glass by field emission, 194
Shells, octet, 18
Silica, forms of, 31
Silicon/aluminum order and disorder, 37
Silver image formation, 65
Silver nucleus and particle growth, 65–68
Site preference, 46
Skin effect, simulated, 218
Smakula equation, 72
Sodium energy bands in k space, 156
Sodium ions, 30
Solar cells, 241–243
    reverse, as semiconductor lasers, 243
Sol-gel transformation, 207
Solid-state sieve (fuel cell), 129, 240–244
Solids, fracture of, 84
Space charge, front of, 181
Space-charge-limited currents, 5, 178, 182, 192, 216, 250
Spark chamber, 171
Spark channel, 186
Specific heat of metals, 157
Spermatocyte, 212
Spikes, in neurons, 256
Spin-lattice relaxation, 118
Spin-orbit coupling, 43–44
Spin-orbit distortion, 46
Spin quenching, 46–47
Spin-spin relaxation, 118, 121
Spin waves, 21, 118, 119
Spindle fibers, 5
    birefringent, 214
    in cell of Easter lily, 214
    formation of, 214
    and mitosis, 211
Spinel, 27, 30, 58, 127
Spiral spin configurations, 56–59
Splitting, ligand-field, 46
    spin-orbit, 45
Splitting of d-electron states in octahedral and tetrahedral sites, 45
Spodumene, 63
Squid, 108
Stainless steel, field-emission properties of, 189
Standing waves in crystal lattice, 154
Starting electrons in gases, 183
States of order in crystals, 37
Static distortion, cooperative, 48
Sticking coefficients, 246
Stimulus coding, 258
Stimulus patterns, complex, 263
Stokes's law, 133, 137
Strain, shear, 85
Streamer mechanism, 170–172, 186
Stress, shear, 85
Strontium titanate, 33, 232
Structure families, aristotypes and heterotypes, 33–34
Structures, competitive, 29
    framework, 1, 30–32

Structures, packing, 11, 30–32
    perovskite, idealized, 33
Superconductors, density of excited states, 165
    (Type II) hard, 163, 167
    (Type I) soft, 163–164
Superconducting films, 166
Superconducting state, transition temperature of, 165
Superconductivity, 163
    microscopic theory of, 164
Superexchange, 50
Surface-layer formation, 189
Susceptibility, domain-wall, 117
Symmetry, reduction of, 34
    translational, 51
Synaptic junctions, 105, 107

Temperatures, boiling and melting, 16–17
Ternary compounds, diffusion and ionic conduction in, 125–127
Thermal breakdown, 195
Thermal efficiency, 250
Thermal ionization, 247, 253
Thermionic converters, 6, 245, 250
Thermocouples, 248
Thermionic measurements by ionic conductors, 128
Thermionic voltage of fuel cells, 251
Thermoelectric converters and generators, 246–254
Thin-film distillation of sea water, 239
Thin-film microelectronics, 216
Three-electrode active elements, 219, 220
Time lag, statistical and formative, 184
Townsend mechanism, 170
Townsend's critical avalanche height, 184, 186
Townsend's law, invalidation of, 188
Transducers, electromechanical, 233
Transfer, probability of, 138
Transfer impedance, 226
    resonance of, 217
Transfer processes, resonant, 138

Transformer, ideal, 218, 221
    planar, 221
Transient currents, 180
Transistor, common base, 220
    common collector, 220
    field-effect, 220
    ideal, 219
Transit time of injected carriers, 181
Transition metals, 26, 42–53, 160
Transition rates of electrons in phosphors, 142
Transitions, displacive, 34–40
    electron-ordering, 43
    first-order, 35
    forbidden, of activator, 138
    from gas to liquid, 190
    high-spin state $\rightleftharpoons$ low-spin state, 43
    of inner-shell electrons, 132
    optical, in Ge, 151
    radiationless, 133
    reconstructive, 34
    semiconductor $\rightleftharpoons$ metal, 52
    from single-bonding to metallic structure (Se $\rightarrow$ Po), 21
Transmission line, nonuniform, 217
Transmutations, parametric (gyrator and isolator), 225
Transport, nonisothermal, 128
Trap densities, 179–181
Traveling waves, 13, 153
Tridymite, 31
Triodes, field-effect, 219
    field-emission, 192
    solid-state, 181
Twinning, 35
Tubules, 199, 201
Tungsten, defects of, 79
    trioxide, 27
Tunneling experiments, 166

Ultrasonic amplifier, 233

Vacancies in crystals, 69, 77, 123
Vacancy, in photoconducting phosphors, 146

Vacancy, in platinum crystal, 78
Valence band, 149–152
Valency, 24, 29
van der Waals–London attraction, 18
Vapor arcs, cold, 177
Vaporization, heat of, 20, 246
Verters, generalized, 220–224
Verters and isolators, matrices of, 223
VHF Cavities, 6, 217
Vibrational levels, 175
Vibrational modes, 49
Vibrations, excitation barrier of, 190
Vidicon (television tube), 181
Virtual bond levels, 161, 162

Wall displacements, 120
Wave functions, Bloch, 18, 51, 160
    hydrogenic, 43, 44
    one-particle, 156
Waveguide, crystal, 14
    cutoff, 14
Wave packages, of electron pairs, 164
Weiss internal field, 55, 59
Werner-Lyman bands, 175, 176
Wheat cell, division of, 212
Wigner-Seitz approximation, 156
Work function, 179, 246
Wüstite, 126

Xerography (electrostatic photography), 181
X-ray fluorescence, 131, 132

Yafet-Kittel state, 56–59
Yttrium iron garnet (YIG), 118–121
    ferrimagnetic, 229, 230
    gallium-substituted, 230

Zeeman effect, 152, 229
Zener's model of d-s electron exchange interaction, 160
Zigzag modulator, 232
Zinc oxide, electron mobility in, 233
Zinc sulfide, luminescent centers in, 144–146

INDEX OF CHEMICALS

A, 174, 244
Ag, 20, 155, 246
AgBr, 66, 122, 123, 124, 125, 127
AgBr-CdBr$_2$, 124
AgCl, 66, 122
AgCl-CdCl$_2$, 126
AgCl-CuCl, 126
AgCl-PbCl$_2$, 126
AgI, 66
AgSbS$_2$, 127, 128
Ag$_2$HgI$_4$, 127
Ag$_2$S, 123, 144
Ag$_3$SbS$_3$, 127, 128
Al, 23, 31, 246
AlMg$_2$O$_4$, 30
Al$_2$O$_3$, 124
Al$_2$S$_3$, 147
Au, 20, 246
AuCl, 63

B, 23, 246
B$_2$O$_3$, 126
Ba, 20, 31, 246
BaAl$_2$Si$_2$O$_8$, 37
BaS, 139
BaSO$_4$, 139
BaTiO$_3$, 33, 34, 215, 216, 235
Be, 20, 246
Be$_2$Te$_3$, 249
Bi, 246

C, 23, 246
CH$_3$(CH$_2$)$_{20}$COOH, 204, 205
CH$_3$OH, 171
CO$_2$, 185
C$_2$H$_2$, 177
C$_6$H$_4$:(CH)$_2$:C$_6$H$_4$, 180, 181
Ca, 20, 24, 30, 31, 246
CaAl$_2$Si$_2$O$_8$, 37
CaCO$_3$, 136
CaF$_2$, 25, 123
CaS, 139
CaTiO$_3$, 32, 35
Ca$_3$Si$_2$(O$_3$OH)$_2$2H$_2$O, 30
Cd, 20, 145, 246
CdBr$_2$, 123
CdS, 131, 139, 180, 233, 234, 235
CdSe, 131, 139, 144, 180
Ce, 246
Cl, 24
Co, 46, 160, 162, 246
CoAl$_2$O$_4$, 126, 127

CoCr$_2$O$_4$, 125, 126, 127
CoFe$_2$O$_4$, 128
CoMnO$_4$, 48
CoO, 49, 71, 125
CoSb$_2$, 52
CoV$_2$O$_4$, 51
Co$_2$TiO$_4$, 127
Co$_3$O$_4$, 46, 48
Cr, 46, 147, 160, 246
CrBr$_3$, 230
CrO$_2$, 43
Cr$_2$O$_3$, 124, 125
Cs, 20, 244, 246
CsCl, 30
Cu, 20, 92, 94, 95, 155, 156, 160, 161, 195, 246
CuCl, 144, 232, 233
CuO, 122
Cu$_2$O, 123, 124, 126
Cu$_2$S, 144

Fe, 24, 46, 47, 147, 162, 246
FeCrO$_4$, 49
FeO, 49
FeP$_2$, 52
FeV$_2$O$_4$, 51
Fe$_3$O$_4$, 126

Ga, 23, 145, 146, 240
GaAs, 145, 152, 233
GaAs-GaP, 152
GaP, 143
GaS$_2$, 140
Ga$_2$O$_3$, 124
Ga$_2$S$_3$, 146
Gd, 22
Ge, 23, 75, 145, 149, 151, 242, 246

He, 174
Hg, 20, 165, 174, 246

I, 181
In, 23, 145
InAs, 152
InAs-GaAs, 152
InAs-InP, 152
InP, 152
InSb, 152
Ir, 246

K, 20, 31, 246
KAlSi$_3$O$_8$, 37

KBr, 72, 73, 74
KCl, 72, 74, 122, 123
KCl-KBr, 70, 72, 73, 74
KCl-RbCl, 70, 73
KH$_2$PO$_4$, 230
KNaNbO$_3$, 36
KNbO$_3$, 33, 34, 35
K$_2$SrCl$_4$, 127
Kr, 174

La, 246
LaCoO$_3$, 46
LaMnO$_3$, 50
Li, 20, 246
LiF, 73, 90
Li$_2$O, 124
Li$_2$O·2SiO$_2$, 66
Li$_2$SiO$_3$, 66
Li$_2$Si$_2$O$_5$, 66

Mg, 20, 246
MgAl$_2$O$_4$, 125
MgO, 30, 95
MgV$_2$O$_4$, 51
Mn, 46, 47, 147, 160, 161, 246
MnAs, 48
MnBi, 48
MnO, 43
MnP, 52
MnSb, 48
Mn$_2$O$_7$, 26
Mo, 78, 81, 165, 246

NO, 177
N$_2$, 176
Na, 20, 24, 30, 31, 156, 158, 160, 246
NaAlSi$_3$O$_8$, 37
NaCl, 25, 70, 194
NaCl-CdCl$_2$, 126
NaNbO$_3$, 32, 33, 34, 35, 36
Nb, 246
Nb$_3$Sn, 52
Nd, 246
Ne, 174, 176
Ni, 47, 49, 81, 143, 160, 162, 246
NiAl$_2$O$_4$, 127
NiAs, 48
NiCrFeO$_4$, 49
NiCr$_2$O$_4$, 126
NiO, 43, 71
NiO·CoO, 71
NiV$_2$O$_4$, 51

O, 24
O_2, 176
O_3, 177
Os, 165, 246

P, 21
Pb, 165, 246
$PbCl_2$, 123
PbTe, 249
Pd, 162, 246
Po, 21
Pr, 246
Pt, 78, 81, 83, 246
Pt-Co, 83

Rb, 20, 246
Re, 246
ReO_3, 48
Rh, 246
Ru, 165, 246

S, 21
SF_6, 26
Sb, 246
Sb_2S_3, 128
Se, 21, 180, 246
Si, 23, 31, 34, 94, 140, 149, 151, 243, 246
SiF_4, 26
SiO_4, 30, 31
Sm, 246
Sn, 23, 165, 246
Sr, 20, 246
$SrTiO_3$, 33, 75, 76, 232

Ta, 246
Te, 21, 246
ThO_2, 129
Ti, 26, 246
TiO, 14, 26
TiO_2, 26, 127
Ti_2O_3, 26
Tl, 165, 246
TlCl, 125

$TlCl_6$, 135
Tl:KCl, 134, 135

U, 246

V, 246
VO_2, 52

W, 78, 81, 90, 96, 246
WO_3, 25, 27, 32

Y_2O_3, 129

Zn, 20, 95, 145, 246
$ZnCl_2$, 147
ZnO, 123, 124, 233
ZnS, 131, 139, 144, 145, 146
ZnS-Mn, 147
ZnSe, 131
ZnV_2O_4, 51
Zr, 246
ZrO_2, 241

MOND H. FOGLER LIBRARY
DATE DUE